Laboratory Manual

Inquiry into Life

Twelfth Edition

Sylvia S. Mader

With significant contributions by

Terry Damron
Citrus College

Eric Rabitoy
Citrus College

Boston Burr Ridge, IL Dubuque, IA New York San Francisco St. Louis
Bangkok Bogotá Caracas Kuala Lumpur Lisbon London Madrid Mexico City
Milan Montreal New Delhi Santiago Seoul Singapore Sydney Taipei Toronto

INQUIRY INTO LIFE LABORATORY MANUAL,
TWELFTH EDITION

Published by McGraw-Hill, a business unit of The McGraw-Hill Companies, Inc., 1221 Avenue of the Americas, New York, NY 10020.

Some of the laboratory experiments included in this text may be hazardous if materials are handled improperly or if procedures are conducted incorrectly. Safety precautions are necessary when you are working with chemicals, glass test tubes, hot water baths, sharp instruments, and the like, or for any procedures that generally require caution. Your school may have set regulations regarding safety procedures that your instructor will explain to you. Should you have any problems with materials or procedures, please ask your instructor for help.

Some ancillaries, including electronic and print components, may not be available to customers outside the United States.

 This book is printed on recycled, acid-free paper containing 10% postconsumer waste.

1 2 3 4 5 6 7 8 9 0 QPD/QPD 0 9 8 7

ISBN 978-0-07-298682-2
MHID 0-07-298682-4

Publisher: *Janice Roerig-Blong*
Sponsoring Editor: *Thomas C. Lyon*
Developmental Editor: *Rose M. Koos*
Marketing Manager: *Tamara Maury*
Project Manager: *Joyce Watters*
Lead Production Supervisor: *Sandy Ludovissy*
Senior Media Project Manager: *Jodi K. Banowetz*
Senior Media Producer: *Eric A. Weber*
Designer: *Laurie B. Janssen*
Cover Designer: *Christopher Reese*
(USE) Cover Image: *© Getty Images/Jim Arbogast*
Senior Photo Research Coordinator: *Lori Hancock*
Photo Research: *Evelyn Jo Hebert*
Compositor: *Aptara, Inc.*
Typeface: *11/13 Slimbach*
Printer: *Quebecor World Dubuque, IA*

The credits section for this book begins on page 485 and is considered an extension of the copyright page.

www.mhhe.com

Contents

***Special Requirements Key**

F Fresh material—local purchase as needed
G Plant growth required
L Living material—order in advance
T Trip to field site—prior planning required

Preface

To the Instructor

The laboratory exercises in this manual are coordinated with *Inquiry into Life,* a general biology text that covers the entire field of biology. The text emphasizes how we can apply biological knowledge to our own lives and to our relationships with other organisms.

Although each laboratory is referenced to the appropriate chapter in *Inquiry into Life,* this manual may be used in coordination with other general biology texts. This laboratory manual can be adapted to a variety of course orientations and designs. There are a sufficient number of laboratories to permit a choice of activities over the length of the course. Many activities may be performed as demonstrations rather than as student activities, thereby shortening the time required to cover a particular concept.

The Exercises

All exercises have been tested for student interest, preparation time, estimated time of completion, and feasibility. The following features are particularly appreciated by adopters.

Integrated Opening. Each laboratory begins with a list of learning outcomes that are organized according to the major sections of the laboratory. The major sections of the laboratory are numbered on the opening page, in the laboratory text material, and in the review. This organization will help students better understand the goals of each laboratory session.

Self-Contained Content. Each laboratory contains all the background information necessary to understand the concepts being studied and to answer the questions asked. This feature will reduce student frustration and increase learning.

Scientific Method. All laboratories stress the scientific method, and many opportunities are given for students to gain an appreciation of the scientific process. The first laboratory of this edition explicitly explains the steps of the scientific method and gives students an opportunity to use them. I particularly recommend this laboratory because it utilizes the pillbug, a living subject.

Student Activities. A color bar is used to designate each student activity. Some student activities are observations and some are experimental procedures. An icon appears whenever a procedure requires a period of time before results can be viewed. Icons are also employed when experiments require advance preparation in order to complete the exercise in a timely manner. Sequentially numbered steps guide students as they perform an activity.

Live Materials. Although students work with living material during some part of almost all laboratories, the exercises are designed to be completed within one laboratory session. This facilitates the use of the manual in multiple-section courses.

Laboratory Safety. Special attention has been given to increasing laboratory safety, and the listing on page viii will assist instructors in their efforts to make the laboratory experience a safe one.

New Artwork. The art in the manual has undergone significant revision. The images are clearer and illustrate the content more effectively for the student.

Revised Laboratories

All laboratories have been revised. New introductory material, student activities, and illustrations appear throughout the manual. The laboratories listed below have been significantly revised. The revisions have been made in such a way as to improve the efficiency of student time on task.

Laboratory 3 Chemical Composition of Cells: Since this is the first experimental laboratory, the experimental procedures have been enhanced to improve student time efficiency and clarify the general relationships of the biochemical molecules under consideration.

Laboratory 8 Photosynthesis: Based upon reviewer comments, the overview of photosynthesis is more elaborate and additional emphasis has been afforded the role of light in the light-dependent reactions.

Laboratory 13 Basic Mammalian Anatomy 1: In response to a number of reviewer concerns, additional emphasis has been directed at the spatial orientation of dissection specimens and additional cautions have been added for greater student safety.

Laboratory 20 Development: The introduction of this exercise was revised to provide more detail about the different patterns of animal development, to emphasize the different patterns of development in terrestrial and non-terrestrial vertebrates, and to elaborate on the early development in humans.

Laboratory 24 Evidence of Evolution: This laboratory has been revised to provide a greater sense of the continuity of evolutionary change from the level of the molecule to that of the population.

Laboratory 32 Effects of Pollution on Ecosystems: This lab exercise has been updated with information about contemporary environmental issues.

Customized Editions

The 32 laboratories in this manual are now available as individual "lab separates," so instructors can custom-tailor the manual to their particular course needs.

Laboratory Resource Guide

The *Laboratory Resource Guide* is essential for instructors and laboratory assistants and is available free to adopters of the *Laboratory Manual* at www.mhhe.com/maderinquiry12.

To the Student

Special care has been taken in preparing the *Laboratory Manual for Inquiry into Life* so that you will ***enjoy*** the laboratory experience as you ***learn*** from it. The instructions and discussion are written clearly so you can understand the material while working through it. Student aids are designed to help you focus on important aspects of each exercise. Where time is a factor in successfully completing an exercise, icons are used to alert you, the student, to effective time management strategies.

Student Learning Aids

Student learning aids are carefully integrated throughout this manual. The *Learning Outcomes* set the goals of each laboratory session and help you review the material for a laboratory practical or any other kind of exam. In this edition, the major topics are numbered, and the learning objectives are grouped according to these topics. The section numbering is used in the text material and in the laboratory review questions. This system allows students to study the chapter in terms of the objectives presented.

The *Introduction* establishes the rationale for coming work and reviews much of the necessary background information required for comprehension of upcoming experiments. *Color bars* bring attention to exercises that require your active participation by highlighting Observations and Experimental Procedures, and an icon indicates a timed experiment. Throughout, *space* is provided for recording answers to questions and the results of investigations and experiments. Each laboratory ends with a set of review questions covering the day's work.

Appendices at the end of the book provide useful information on preparing a laboratory report, the metric system and classification of organisms. Practical examination answer sheets are also provided.

Laboratory Preparation

Read each exercise before coming to the laboratory. *Study* the introductory material and the experimental procedures. If necessary, to obtain a better understanding, *read* the corresponding chapter in your text. If your text is *Inquiry into Life,* by Sylvia S. Mader, see the text *chapter reference* in the table of contents at the beginning of the *Laboratory Manual.*

Explanations and Conclusions

Throughout the laboratory, you are often asked to formulate explanations or conclusions. To do so, you will need to synthesize information from a variety of sources, including the following:

1. Your experimental results and/or the results of other groups in the class. If your data are different from other groups in your class, do not erase your answer; add the other answers in parentheses.
2. Your knowledge of underlying principles. Obtain this information from the laboratory introduction or the appropriate section of the laboratory and the corresponding chapter of your text.
3. Your understanding of how the experiment was conducted and/or the materials that were used. *Note:* Ingredients can be contaminated or procedures incorrectly followed, resulting in reactions that seem inappropriate. If this occurs, consult with other students and your instructor to see if you should repeat the experiment.

In the end, be sure you are truly writing an explanation or conclusion and not just giving a restatement of the observations made.

Color Bars, Icon, and Safety Boxes

Observation—An activity in which you observe models or slides and make identifications or draw conclusions. Observations are designated with a tan color bar.

Experimental Procedure—An activity in which a series of laboratory steps is followed to achieve a learning objective. Experimental procedures are identified with a blue color bar.

Time—An icon is used to designate when time is needed for an experimental procedure. Allow the designated amount of time for this activity. Start these activities at the beginning of the laboratory, proceed to other activities, and return to these when the designated time is up.

Safety—Safety related precautions are highlighted in text boxes. Your health and safety are of utmost concern, and you should follow the safety recommendations.

Laboratory Review

Each laboratory ends with approximately 10-20 short-answer questions that will help you determine if you have accomplished the objectives for the laboratory. The answers to these questions are found in the *Resource Guide.*

Student Feedback

If you have any suggestions for how this laboratory manual could be improved, you can send your comments to

The McGraw-Hill Companies
Introductory Biology
2460 Kerper Blvd.
Dubuque, Iowa 52001

Acknowledgments

We gratefully acknowledge the following reviewers for their assistance in the development of this lab manual:

Gail F. Baker
LaGuardia Community College

Neil R. Baker, Ph.D.
The Ohio State University

Tamatha R. Barbeau
Francis Marion University

Robert P. Benard
American International College

Donna H. Bivans
Pitt Community College

Dr. Carol A. Burkart
Mountain Empire Community College

John R. Capeheart
University of Houston—Downtown

Kelly S. Cartwright
College of Lake County

Ann M. Chiesa
Castleton State College

Carol M. Cleveland
Northwest Mississippi Community College

Susan J. Cook
Indiana University South Bend

Timothy V. Horger
Illinois Valley Community College

Ali M. Jafri
Malcolm X College

Judith Kelly
Henry Ford Community College

Philip C. Lyons
University of Houston—Downtown

Dr. Yaser A. Maksoud
University of Illinois at Chicago/Olive-Harvey College

Jill Nugent
University of North Texas

Emily C. Oaks
State University of New York at Oswego

Sandy Pace
Rappahannock Community College

Kayla Rihani
Northeastern Illinois University

Lewis C. Robertson
Sandhills Community College, Pinehurst, NC

Dr. Rebecca S. Roush
Sandhills Community College

Soma Sanyal
Raymond Walters College, University of Cincinnati

Laboratory Safety

The following is a list of practices required for safety purposes in the biology laboratory and in outdoor activities. Following rules of lab safety and using common sense throughout the course will enhance your learning experience by increasing your confidence in your ability to safely use chemicals and equipment. Pay particular attention to oral and written safety instructions given by the instructor. If you do not understand a procedure, ask the instructor, rather than a fellow student, for clarification. Be aware of your school's policy regarding accident liability and any medical care needed as a result of a laboratory or outdoor accident.

The following rules of laboratory safety should become a habit:

1. Wear safety glasses or goggles during exercises in which glassware and chemical reagents are handled, or when dangerous fumes may be present, creating possible hazards to eyes or contact lenses.
2. Assume that all reagents are poisonous and act accordingly. Read the labels on chemical bottles for safety precautions and know the nature of the chemical you are using. If chemicals come into contact with skin, wash immediately with water.
3. **DO NOT**
   - **a.** ingest any reagents.
   - **b.** eat, drink, or smoke in the laboratory. Toxic material may be present, and some chemicals are flammable.
   - **c.** carry reagent bottles around the room.
   - **d.** pipette anything by mouth.
   - **e.** put chemicals in the sink or trash unless instructed to do so.
   - **f.** pour chemicals back into containers unless instructed to do so.
   - **g.** operate any equipment until you are instructed in its use.
   - **h.** dispose of biological or chemical wastes in regular classroom trash receptacles.
4. **DO**
   - **a.** note the location of emergency equipment such as a first aid kit, eyewash bottle, fire extinguisher, switch for ceiling showers, fire blanket(s), sand bucket, and telephone (911).
   - **b.** be familiar with the experiments you will be doing before coming to the laboratory. This will increase your understanding, enjoyment, and safety during exercises. Confusion is dangerous. Completely follow the procedure set forth by the instructor.
   - **c.** keep your work area neat, clean, and organized. Before beginning, remove everything from your work area except the lab manual, pen, and equipment used for the experiment. Wash hands and desk area, including desk top and edge, before and after each experiment. Use clean glassware at the beginning of each exercise, and wash glassware at the end of each exercise or before leaving the laboratory.
   - **d.** wear clothing that, if damaged, would not be a serious loss, or use aprons or laboratory coats, since chemicals may damage fabrics.
   - **e.** wear shoes as protection against broken glass or spillage that may not have been adequately cleaned up.
   - **f.** handle hot glassware with a test tube clamp or tongs. Use caution when using heat, especially when heating chemicals. Do not leave a flame unattended; do not light a Bunsen burner near a gas tank or cylinder; do not move a lit Bunsen burner; do keep long hair and loose clothing well away from the flame; do make certain gas jets are off when the Bunsen burner is not in use. Use proper ventilation and hoods when instructed.
   - **g.** read chemical bottle labels; be aware of the hazards of all chemicals used. Know the safety precautions for each.
   - **h.** stopper all reagent bottles when not in use. Immediately wash reagents off yourself and your clothing if they spill on you, and immediately inform the instructor. If you accidentally get any reagent in your mouth, rinse the mouth thoroughly, and immediately inform your instructor.
   - **i.** use extra care and wear disposable gloves when working with glass tubing and when using dissection equipment (scalpels, knives, or razor blades), whether cutting or assisting.
   - **j.** administer first aid immediately to clean, sterilize, and cover any scrapes, cuts, and burns where the skin is broken and/or where there may be bleeding. Wear bandages over open skin wounds.
   - **k.** report all accidents to the instructor immediately, and ask your instructor for assistance in cleaning up broken glassware and spills.
   - **l.** report to the instructor any condition that appears unsafe or hazardous.
   - **m.** use caution during any outdoor activities. Watch for snakes, poisonous insects or spiders, stinging insects, poison oak, poison ivy, etc. Be careful near water.
   - **n.** wash your hands thoroughly after handling any preserved biological specimens.

I understand the safety rules as presented above. I agree to follow them and all other instructions given by the instructor.

Name: ______________________________ Date: ____________________

Laboratory Class and Time: ______________________________

1

Scientific Method

Learning Outcomes

1.1 Using the Scientific Method
- State the steps of the scientific method.
- Explain how observations, hypotheses, conclusions, and scientific theories contribute to the scientific method.

1.2 Observing the Pillbug
- Identify how the external anatomy of the pillbug, *Armadillidium vulgare,* contributes to its observed forms of behavior.
- Describe how a pillbug moves.

1.3 Formulating Hypotheses
- Formulate a hypothesis when supplied with appropriate observations.

1.4 Performing an Experiment
- Design an experiment that can be repeated by others.
- Develop a conclusion based upon observation and experimentation.

Introduction

This laboratory will provide you with an opportunity to use the scientific method in the same manner that scientists do. Scientists often begin by making observations about the subject of interest. Today our subject is the pillbug, *Armadillidium vulgare,* a type of crustacean that lives on land (Fig. 1.1).

Pillbugs have overlapping "armored" plates that form what is known as an exoskeleton. The exoskeleton makes pillbugs look like little armadillos. A pillbug can roll up into such a tight ball that its legs and head are no longer visible, earning it the nickname "roly-poly." They are commonly found in damp leaf litter, under rocks, and in basements or crawl spaces under houses. Pillbugs breathe by gills located on the underside of their bodies. The gills must be kept slightly moist, and that is why they are usually found in damp places.

In winter, pillbugs are inactive, but when spring arrives they become active and mate. Females have a pouch on the underside of their body, where they can carry from 7 to 200 eggs. The eggs hatch several weeks after mating, and the young look like miniature adults. The young stay in the

Figure 1.1 Pillbugs (roly-poly bugs). The pillbug, *A. vulgare,* lives on land. Pillbugs are wingless, and coloration varies from brown to slate grey. They breathe by gills, and they have seven pairs of jointed legs for locomotion on land. Pillbugs live in damp places, mostly under rocks, wood, and decaying leaves. They feed at night.

pouch another six weeks, and then they leave and begin to feed. They eat decaying plants and animals and some living plants. They can live up to three years.

Pillbugs molt (shed their exoskeleton) four or five times. They have three body parts: head, thorax, and abdomen. Among crustaceans, pillbugs are classified as isopods because they are dorsoventrally (front to back) flattened, lack a carapace, have compound eyes, and have two pairs of antennae.

Isopods are the only crustaceans that include forms adapted to living their entire life on land, although moisture is required. Currently, it is believed that they do not transmit diseases, nor do they bite or sting. Because they eat dead organic matter, such as leaves, they are easy to find and keep in a moist terrarium with leaf litter, rocks, and wood chips. You are encouraged to collect some for your experiment. Since they live in the same locations as snakes, be careful when collecting them.

First, become acquainted with your subject and how it normally moves. Then you will use your knowledge of the pillbug to hypothesize whether it will be attracted to, repelled by, or indifferent to various substances of your choice. After you have tested your hypotheses, you will conclude whether they are supported or not. Finally, your conclusions may lead to other hypotheses, and if time permits, you may go ahead and test those also.

1.1 Using the Scientific Method

Scientists use the scientific method (Fig. 1.2) to come to a conclusion about the natural world. When a scientist begins a study, he or she uses preliminary observations and previous data to formulate a hypothesis. A **hypothesis** is a tentative explanation of observed phenomena.

After a hypothesis is formulated, it must be tested by doing new experiments and/or making new observations. Experiments are done and observations made in such a way that others can repeat them. Only repeatable observations and experiments are accepted as valid contributions to the field of science.

Data are any factual information that can be observed either independently of, or as a result of, experimentation. On the basis of the data, a scientist comes to a **conclusion**—whether the observations support the hypothesis or prove it false. Scientists are always aware that further observations and experiments could lead to a change in prior conclusions. Therefore, it is never said that the data prove a hypothesis true. The arrow in Figure 1.2 indicates that research often enters a cycle of hypothesis—experiments and observations—conclusion—hypothesis. Explain this cycle. ____________

____

After many years of testing and study, the scientific community may develop a **scientific theory,** which is a concept that ties together many varied conclusions into a generalized statement. For example, after testing the cause of many individual diseases, the germ theory of disease was formulated. It states that infectious diseases are caused by pathogens (e.g., bacteria and viruses) that can be passed from one person to another. How is a scientific theory different from a conclusion? ____________

____

Figure 1.2 Flow diagram for the scientific method.
On the basis of observations, a scientist formulates a hypothesis. The hypothesis is tested by further experiments and/or observations. The scientist then concludes whether the data support or do not support the hypothesis. The return arrow shows that scientists often choose to retest the same hypothesis or test a related hypothesis.

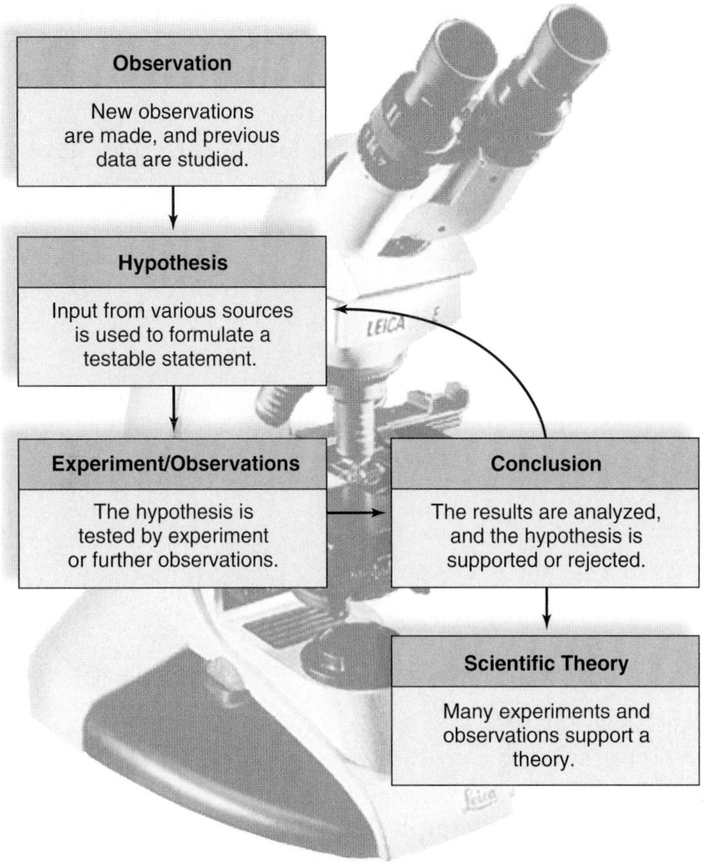

Observation
New observations are made, and previous data are studied.
Hypothesis
Input from various sources is used to formulate a testable statement.
Experiment/Observations
The hypothesis is tested by experiment or further observations.
Conclusion
The results are analyzed, and the hypothesis is supported or rejected.
Scientific Theory
Many experiments and observations support a theory.

1.2 Observing the Pillbug

Wash your hands before and after handling pillbugs. Please handle them carefully so they are not crushed. When touched, pillbugs roll up into a ball or "pill" shape as a defense mechanism. They will soon recover if left alone.

Observation: Pillbug's External Anatomy

1. Obtain a pillbug that has been numbered with white correction fluid or tape tags. First examine the shell and body with the unaided eye and then with a magnifying lens or dissecting microscope. Put the pillbug in a small glass or plastic dish to keep it contained.
2. Examine the shell shape, color, and texture. Note the number of legs and antennae and whether there are any posterior appendages, such as uropods (paired appendages at end of abdomen) or brood pouches. (Females have leaflike growths at the base of some legs where developing eggs and embryos are held in pouches.) Locate the eyes. Count the number of overlapping plates.

3. In the following space, draw a large outline of your pillbug (at least 10–12 cm across). Label the head, thorax, abdomen, antennae, eyes, uropods, and one of the seven pairs of legs.

4. Draw the pillbug rolled into a ball.

Observation: Pillbug's Motion

1. Watch a pillbug's underside as the pillbug moves up a transparent surface, such as the side of a graduated cylinder or beaker. Describe the action of the feet and any other motion you see.

   ____

2. As you watch the pillbug, identify behaviors that might
   - **a.** protect it from predators ______________________________
   - **b.** help it acquire food ______________________________
   - **c.** protect it from the elements ______________________________
   - **d.** allow interaction with the environment ______________________________
3. Allow a pillbug to crawl on your hand. Describe how it feels and how it acts.

   ____

4. Place the pillbug on a graduated cylinder. Experiment with the angle of the cylinder and the position of the pillbug to determine the pillbug's preferred direction of motion. For example, place the cylinder on end, and position the pillbug so that it can move up or down. Try other arrangements also. Repeat this procedure with three other pillbugs. Record the preferred direction of motion and other observations for each pillbug in Table 1.1.

Table 1.1 Preferred Direction of Motion

Pillbug	Direction Moved	Comments
1		
2		
3		
4		

5. Measure the speed of the pillbug. Use what you learned about each pillbug's preferred direction of motion (see Table 1.1) to get maximum cooperation from each of the four pillbugs you worked with in step 4. Place each pillbug on a metric ruler, and use a stopwatch to measure the time it takes for the pillbug to move a certain number of millimeters. Record your results for each pillbug in Table 1.2. Calculate each pillbug's average speed in millimeters (mm) per second here, and record your data in Table 1.2.

Table 1.2 Pillbug Speed

Pillbug	Millimeters Traveled	Time (sec)	Average Speed (mm/sec)
1			
2			
3			
4			

1.3 Formulating Hypotheses

Hypotheses are often stated as "if-then" statements. For example, *if* the pillbug is exposed to ______, *then* it will ______. You will be testing whether pillbugs are attracted to, repelled by, or unresponsive to particular substances. Pillbugs move away from a substance when they are repelled by it, and they move toward and eat a substance they are attracted to. If a pillbug simply rolls into a ball, nothing can be concluded, and you may wish to choose another pillbug or wait a minute or two to check for further response.

1. Choose
   a. two or three powders, such as flour, cornstarch, coffee creamer, baking soda, fine sand.
   b. two or three liquids, such as milk, orange juice, ketchup, applesauce, a carbonated beverage, water.
2. Hypothesize in Table 1.3 how you expect the pillbug to respond, and offer an explanation for your reasoning.

Table 1.3 Hypotheses About Pillbug's Reaction to Common Powders and Liquids

Substance Tested	Hypothesis About How Pillbug Will Respond to Substance	Reasoning for Hypothesis
1		
2		
3		
4		
5		
6		

1.4 Performing an Experiment

Design an experiment to test the pillbug's reaction to the chosen substances. The pillbug must be treated humanely. No substance must be put directly on the pillbug, nor can the pillbug be placed directly onto the substance. Since pillbugs tend to walk around the edge of a petri dish, you could put the wet or dry substance around the edge of the dish. Or for wet substances, you could put liquid-soaked cotton in the pillbug's path.

A good experimental design contains a control. A **control** group goes through all the steps of an experiment but lacks, or is not exposed to, the factor being tested. If you are testing the pillbug's reaction to a liquid, water can be the control substance substituted for the test liquid. If you are testing the pillbug's reaction to a powder, substitute fine sand for the test powder.

Experimental Procedure: Pillbug's Reaction to Common Substances

1. What substances are you testing? Include in your list any controls, and complete the first column in Table 1.4.
2. Obtain a small beaker and fill it with water. Rinse your pillbug between procedures by spritzing with distilled water from a spray bottle. Then put it on a paper towel to dry off.
3. Test the pillbug's reaction using the method described above.

4. Watch the pillbug's reaction to each substance, and record it in Table 1.4.
5. Do your results support your hypotheses? Answer *yes* or *no* in the last column.

Table 1.4 Pillbug's Reaction to Common Substances

Substance Tested	Pillbug's Reaction	Hypothesis Supported?
1		
2		
3		
4		
5		
6		

6. Compare your results with those of other students who tested the same substances. Complete Table 1.5.

Table 1.5 Class Results

Group No.	Exp. 1 Direction	Exp. 2 Speed (mm/sec)	Exp. 3 Dry Control	*Substances*			Exp. 4 Wet Control	*Substances*		
1										
2										
3										
4										
5										
6										
7										
8										

Continuing the Experiment

7. Study your results and those of other students, and decide what factors may have caused the pillbug to be attracted to or repelled by a substance. ____

____

On the basis of your decision, what is your new hypothesis? ______________________________

____

8. Test your hypothesis and describe your results here. If possible, make up a table to display your results.

9. Based on your new data, what is your conclusion?

____

Laboratory Review 1

________ 1. Which is more comprehensive, a conclusion or a theory?

________ 2. What is a tentative explanation of observed phenomena?

________ 3. What do you call the information scientists collect when doing experiments and making observations?

________ 4. What step in the scientific method follows experiments and observations?

________ 5. What do you call a sample that goes through all the steps of an experiment and does not contain the factor being tested?

________ 6. Can data prove a hypothesis true? (Yes or No)

Indicate whether statements 7 and 8 are hypotheses, conclusions, or scientific theories.

________ 7. The data show that vaccines protect people from disease.

________ 8. All living things are made of cells.

________ 9. How many body divisions does a pillbug have?

________ 10. If a pillbug travels 5 mm in 60 seconds, what is its rate of speed?

________ 11. What can be concluded if a pillbug curls into a ball?

________ 12. Pillbugs that back away from a substance are (attracted to/repelled by) the substance.

Thought Questions

15. What is a scientific theory?

16. Why is it important to use one substance at a time when testing a pillbug's reaction?

17. Can the scientific method be used to explain all observations? If not, provide an example of an observation or question that cannot be tested by the scientific method.

LABORATORY

2

Metric Measurement and Microscopy

Learning Outcomes

2.1 The Metric System
- Perform measurements for length, weight, and volume, using metric units of measurement.
- Demonstrate how temperatures can be converted between the Celsius and Fahrenheit temperature scales.

2.2 Microscopy
- Describe three differences between the compound light microscope and the electron microscope.

2.3 Binocular Dissecting Microscope (Stereomicroscope)
- Identify the parts and tell how to focus the binocular dissecting microscope.

2.4 Use of the Compound Light Microscope
- Name and give the function of the basic parts of the compound light microscope.
- Discuss the proper sequence of events used to bring an object into focus with the compound light microscope.
- Describe how a slide of any object provides information on the inversion of the image in the compound light microscope.
- Calculate the diameter of field of view and the total magnification for both low- and high-power lens systems.
- Explain how a slide of colored threads provides information on the depth of field.

2.5 Microscopic Observations
- Identify the three types of cells studied in this exercise.
- State two differences between onion epidermal cells and human epithelial cells.

Introduction

This laboratory introduces you to the metric system, which biologists use to indicate the sizes of cells and cell structures. This laboratory also examines the features, functions, and use of the compound light microscope and the binocular dissecting microscope. Transmission and scanning electron microscopes are explained, and their micrographs produced using these throughout these exercises. The binocular dissecting microscope and the scanning electron microscope view the surface and/or the three-dimensional structure of an object. The compound light microscope and the transmission electron microscope can view only extremely thin sections of a specimen. If a subject was sectioned lengthwise for viewing, the interior of the projections at the top of the cell, called cilia, would appear in the micrograph. A lengthwise cut through any type of specimen is called a **longitudinal section (l.s.).** On the other hand, if the subject in Figure 2.1 was sectioned crosswise below the area of the cilia, you would see other portions of the interior of the subject. A crosswise cut through any type of specimen is called a **cross section (c.s.).**

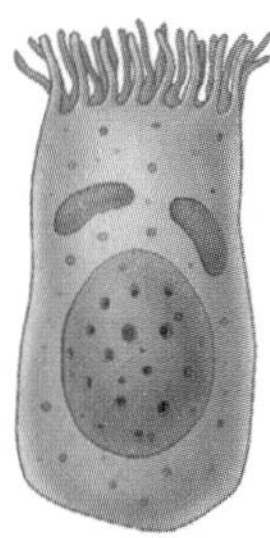

a. Cell

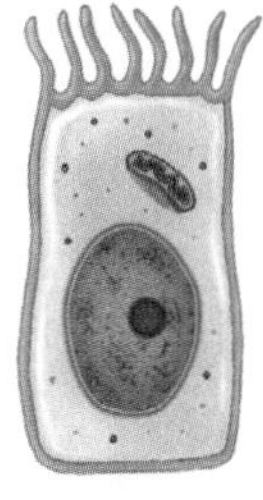

b. Longitudinal section

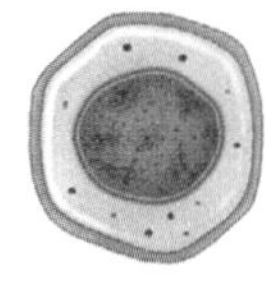

c. Cross section

Figure 2.1 Longitudinal and cross sections. **a.** Transparent view of a cell. **b.** A longitudinal section would show the cilia at the top of the cell. **c.** A cross section shows only the interior where the cut is made.

2.1 The Metric System

The **metric system** is the standard system of measurement in the sciences, including biology, chemistry, and physics. It has tremendous advantages because all conversions, whether for volume, mass (weight), or length, can be in units of ten. For comparative purposes, the table in Appendix B (page 479) lists some of the more common conversion values used to change from metric to English units of measure.

Length

Metric units of length measurement studied in this laboratory include the **nanometer (nm)**, **micrometer (μm)**, **millimeter (mm)**, **centimeter (cm)**, and **meter (m)** (Table 2.1). The prefixes nano- (10^{-9}), micro- (10^{-6}), and milli- (10^{-3}) are used with length, weight, and volume.

Table 2.1 Metric Units of Length Measurement

Unit	Meters	Millimeters	Centimeters	Relative Size
Nanometer (nm)	0.000000001 (10^{-9}) m	0.000001 (10^{-6}) mm	0.0000001 (10^{-7}) cm	Smallest
Micrometer (μm)	0.000001 (10^{-6}) m	0.001 (10^{-3}) mm	0.0001 (10^{-4}) cm	
Millimeter (mm)	0.001 (10^{-3}) m	1.0 mm	0.1 cm	
Centimeter (cm)	0.01 (10^{-2}) m	10 mm	1 cm	
Meter (m)	1 m	1,000 mm	100 cm	Largest

Experimental Procedure: Length

1. Obtain a small ruler marked in centimeters and millimeters. How many centimeters are represented? ________ One centimeter equals how many millimeters? ________ To express the size of small objects, such as cell contents, biologists use even smaller units of the metric system than those on the ruler. These units are the micrometer (μm) and the nanometer (nm). According to Table 2.1, 1 μm = ________ mm, and 1 nm = ________ mm. Therefore, 1 mm = ________ μm = ________ nm.
2. Measure the diameter of the circle shown below to the nearest millimeter. This circle is ________ mm = ________ μm = ________ nm.

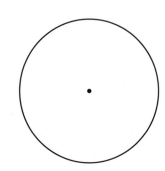

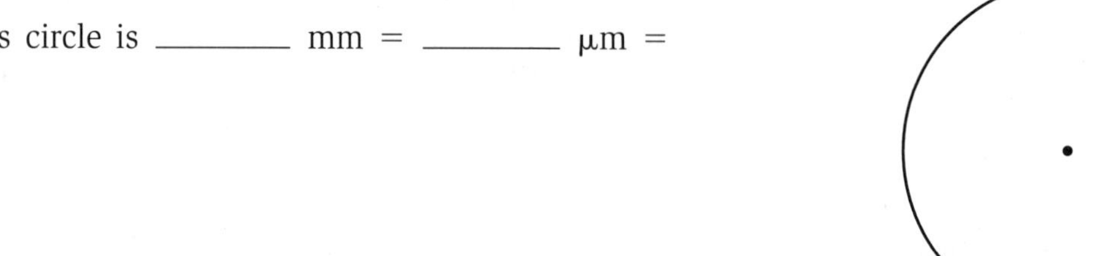

3. Obtain a meterstick. On one side, find the numbers 1 through 39, which denote inches. One meter equals 39.37 inches; therefore, 1 meter is roughly equivalent to 1 yard. Turn the meterstick over, and observe the metric subdivisions. How many centimeters are in a meter? ________ How many millimeters are in a meter? ________ The prefix *milli* means ________.
4. Use the meterstick and the method shown in Figure 2.2 to measure the length of two long bones from a disarticulated human skeleton. Lay the meterstick flat on the lab table. Place a long bone next to the meterstick between two pieces of cardboard (each about 10 cm × 30 cm), which are held upright at right angles to the stick. The narrow end of each piece of cardboard should touch the meterstick. The length between the cards is the length of the bone in centimeters. For example, if the bone measures from the 22 cm mark to the 50 cm mark, the length of the bone is ________ cm. If the bone measures from the 22 cm mark to midway between the 50 cm and 51 cm marks, its length is ________ mm, or ________ cm.
5. Record the length of two bones. First bone: ________ cm = ________ mm. Second bone: ________ cm = ________ mm.

Figure 2.2 Measurement of a long bone.
How to measure a long bone using a meterstick.

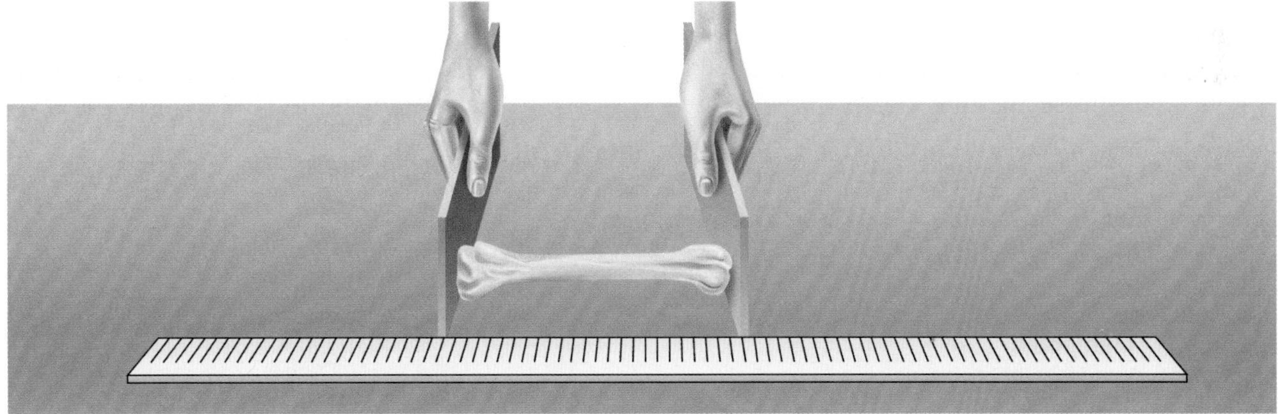

Weight

Two metric units of weight are the **gram (g)** and the **milligram (mg)**. A paper clip weighs about 1 g, which equals 1,000 mg. 2 g = ________ mg; 0.2 g = ________ mg; and 2 mg = ________ g.

Experimental Procedure: Weight

1. Use a balance scale to measure the weight of a wooden block small enough to hold in the palm of your hand.
2. Measure the weight of the block to the tenth of a gram. The weight of the wooden block is ________ g = ________ mg.
3. Measure the weight of an item that is small enough to fit inside the opening of a 50 ml graduated cylinder. The item, a(n) ________, is ________ g = ________ mg.

Volume

Two metric units of volume are the **liter (l)** and the **milliliter (ml)**. One liter = 1,000 ml.

Experimental Procedure: Volume

1. Volume measurements can be related to those of length. For example, use a millimeter ruler to measure the wooden block used in the previous Experimental Procedure to get its length, width, and depth.

   length = ________ cm; width = ________ cm; depth = ________ cm
   The volume, or space, occupied by the wooden block can be expressed in cubic centimeters (cc or cm^3) by multiplying: length × width × depth = ________ cm^3. For purposes of this Experimental Procedure, 1 cubic centimeter equals 1 milliliter; therefore, the wooden block has a volume of ________ ml.
2. In the biology laboratory, liquid volume is usually measured directly in liters or milliliters with appropriate measuring devices. For example, use a 50 ml graduated cylinder to add 20 ml of water to a test tube. First, fill the graduated cylinder to the 20 ml mark. To do this properly, you have to make sure that the lowest margin of the water level, or the **meniscus** (Fig. 2.3), is at the 20 ml mark. Place your eye directly parallel to the level of the meniscus, and add water until the meniscus is at the 20 ml mark. (Having a dropper bottle filled with water on hand can help you do this.) A large, blank, white index card held behind the cylinder can also help you see the scale more clearly. Now, pour the 20 ml of water into the test tube.
3. Hypothesize how you could find the total volume of the test tube. ________________________

   ____

   Now perform the operation you just suggested. What is the test tube's total volume? ________

Figure 2.3 Meniscus.
The proper way to view the meniscus.

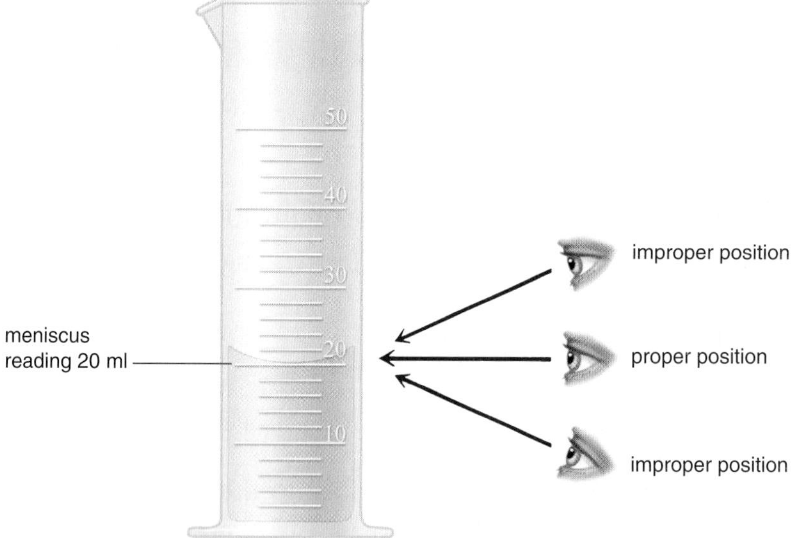

4. Fill a 50 ml graduated cylinder with water to about the 20 ml mark. How could you use this setup to calculate the volume of the item you weighed previously? ______________

____

Now perform the operation you suggested. The item, ________, has a volume of ________ ml.

5. How could you determine how many drops from the pipette of the dropper bottle equal 1 ml?

____

Now perform the operation you suggested. How many drops from the pipette of the dropper bottle equal 1 ml? __________ Some pipettes are graduated and can be filled to a certain level as a way to measure volume directly. Your instructor will demonstrate this. Are pipettes customarily used to measure large or small volumes? ______________

Temperature

There are two temperature scales: the **Fahrenheit (F)** and **Celsius (centigrade, C)** scales (Fig. 2.4). Scientists use the Celsius scale.

Experimental Procedure: Temperature

1. Study the two scales in Figure 2.4, and complete the following information:
   a. Water freezes at either ________ °F or ________ °C.
   b. Water boils at either ________ °F or ________ °C.
2. To convert between the Fahrenheit and Celsius scales, use the following equations:

$$°C = (°F - 32)/1.8$$

or

$$°F = (1.8°C) + 32$$

Human body temperature of 98°F is what temperature on the Celsius scale? ________

3. Record any two of the following temperatures in your lab environment. In each case, allow the end bulb of the Celsius thermometer to remain in or on the sample for one minute.

   Room temperature = ________ °C

   Surface of your skin = ________ °C

   Cold tap water in a 50 ml beaker = ________ °C

   Hot tap water in a 50 ml beaker = ________ °C

   Ice water = ________ °C

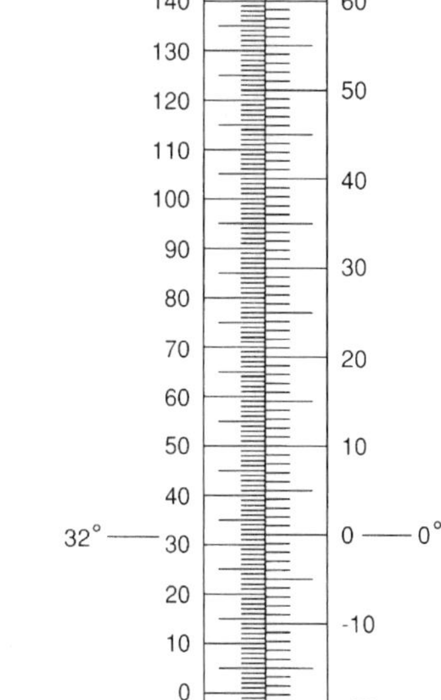

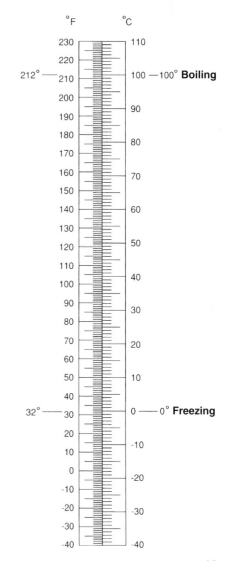

Figure 2.4 Temperature scales.
The Fahrenheit (°F) scale is on the left, and the Celsius (°C) scale is on the right.

2.2 Microscopy

Because biological objects can be very small, we often use a microscope to view them. Many kinds of instruments, ranging from the hand lens to the electron microscope, are effective magnifying devices. A short description of two kinds of light microscopes and two kinds of electron microscopes follows.

Light Microscopes

Light microscopes use visible light rays that are magnified and focused by means of lenses. The **binocular dissecting microscope** (stereomicroscope) is designed to study entire objects in three dimensions at low magnification. The **compound light microscope** is used for examining small or thinly sliced sections of objects under higher magnification than that of the binocular dissecting microscope. The term **compound** refers to the use of two sets of lenses: the ocular lenses located near the eyes and the objective lenses located near the object. Illumination is from below, and visible light passes through clear portions but does not pass through opaque portions. To improve contrast, the microscopist uses stains or dyes that bind to cellular structures and absorb light. Figure 2.5*a* is a **photomicrograph,** a photograph of an image produced by a compound light microscope.

Figure 2.5 Comparative micrographs.
Micrographs of a lymphocyte, a type of white blood cell. **a.** A photomicrograph (light micrograph) shows less detail than a **(b)** transmission electron micrograph (TEM). **c.** A scanning electron micrograph (SEM) shows the cell surface in three dimensions.

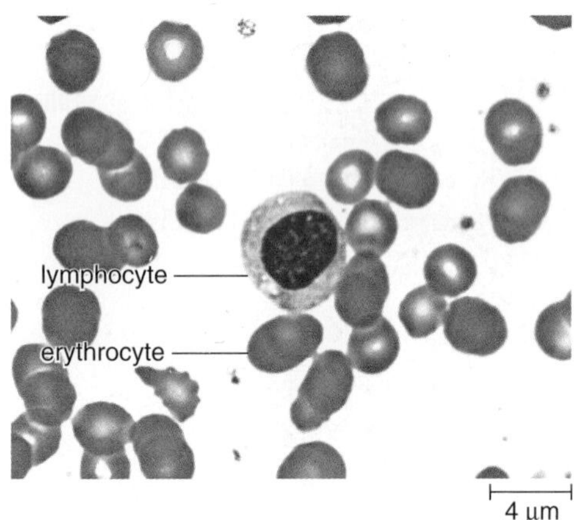

a. Photomicrograph or light micrograph (LM)

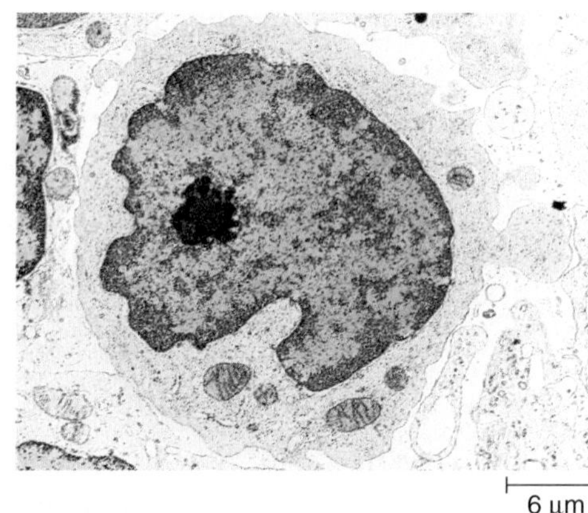

b. Transmission electron micrograph (TEM)

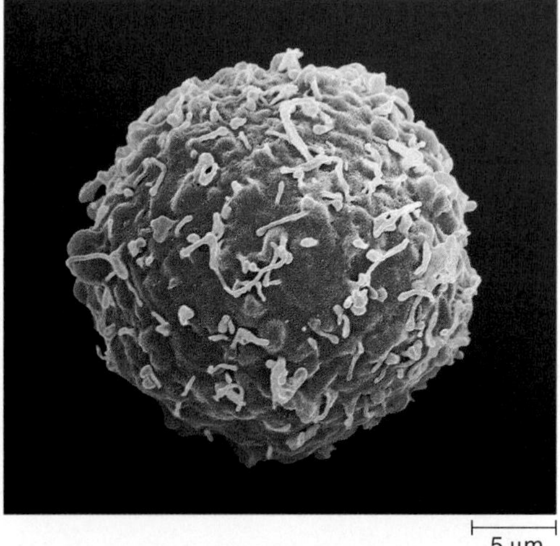

c. Scanning electron micrograph (SEM)

Rules for Microscope Use

Observe the following rules for using a microscope:

1. The lowest power objective (scanning or low) should be in position both at the beginning and at the end of microscope use.
2. Use only lens paper for cleaning lenses.
3. Do not tilt the microscope as the eyepieces could fall out, or wet mounts could be ruined.
4. Keep the stage clean and dry to prevent rust and corrosion.
5. Do not remove parts of the microscope.
6. Keep the microscope dust-free by covering it after use.
7. Report any malfunctions.
8. Do not use coarse focus when viewing a specimen with the high-power objective.

Electron Microscopes

Electron microscopes use a beam of electrons that is magnified and focused on a photographic plate by means of electromagnets. The **transmission electron microscope** is analogous to the compound light microscope. The object is ultra-thinly sliced and treated with heavy metal salts to provide contrast. Figure 2.5*b* is a micrograph produced by this type of microscope. The **scanning electron microscope** is analogous to the dissecting light microscope. It gives an image of the surface and dimensions of an object, as is apparent from the scanning electron micrograph in Figure 2.5*c*.

The micrographs in Figure 2.5 demonstrate that much smaller objects can be viewed with electron microscopes than with compound light microscopes. The difference between these two types of microscopes, however, is not simply a matter of magnification; it is also the electron microscope's ability to show detail. The electron microscope has greater resolving power. **Resolution** is the minimum distance between two objects at which they can still be seen, or resolved, as two separate objects. The use of high-energy electrons rather than light gives electron microscopes a much greater resolving power since two objects that are much closer together can still be distinguished as separate points. Table 2.2 lists several other differences between the compound light microscope and the transmission electron microscope.

Table 2.2 Comparison of the Compound Light Microscope and the Transmission Electron Microscope

Compound Light Microscope	Transmission Electron Microscope
1. Glass lenses	1. Electromagnetic lenses
2. Illumination by visible light	2. Illumination due to a beam of electrons
3. Resolution ≅ 200 nm	3. Resolution ≅ 0.1 nm
4. Magnifies to 2,000×	4. Magnifies to 1,000,000×
5. Costs up to tens of thousands of dollars	5. Costs up to hundreds of thousands of dollars

Conclusions

- Which two types of microscopes view the surface of an object? ____________________
- Which two types of microscopes view objects that have been sliced and treated to improve contrast? ____________________
- Of the microscopes just mentioned, which one resolves the greater amount of detail?

  ____

2.3 Binocular Dissecting Microscope (Stereomicroscope)

The **binocular dissecting microscope** allows you to view objects in three dimensions at low magnifications. It is used to study entire small organisms, any object requiring lower magnification, and opaque objects that can be viewed only by reflected light. It is also called a stereomicroscope because it produces a three-dimensional image.

Identifying the Parts

After your instructor has explained how to carry a microscope, obtain a binocular dissecting microscope and a separate illuminator, if necessary, from the storage area. Place it securely on the table. Plug in the power cord, and turn on the illuminator. There is a wide variety of binocular dissecting microscope styles, and your instructor will discuss the specific style(s) available to you. Regardless of style, the following features should be present:

Figure 2.6 Binocular dissecting microscope (stereomicroscope).
Label this microscope with the help of the text material.

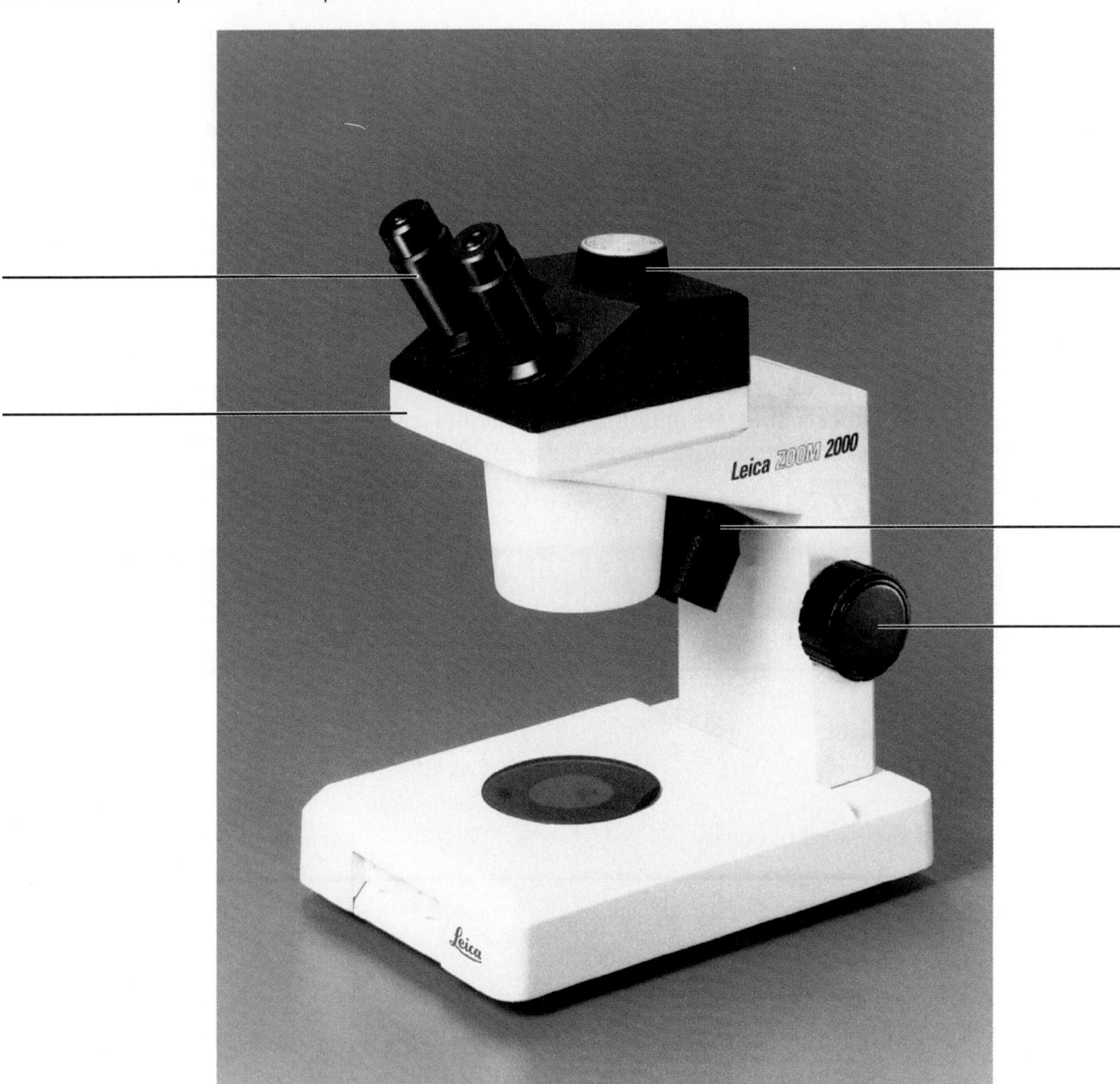

1. **Binocular head:** Holds two eyepiece lenses that move to accommodate for the various distances between different individuals' eyes.
2. **Eyepiece lenses:** The two lenses located on the binocular head. What is the magnification of your eyepieces? ________ Some models have one **independent focusing eyepiece** with a knurled knob to allow independent adjustment of each eye. The nonadjustable eyepiece is called the **fixed eyepiece.**
3. **Focusing knob:** A large black or gray knob located on the arm; used for changing the focus of both eyepieces together.
4. **Magnification changing knob:** A knob often built into the binocular head that is used to change magnification in both eyepieces simultaneously. This may be a **zoom** mechanism or a **rotating lens** mechanism of different powers that clicks into place.
5. **Illuminator:** Used to illuminate an object from above; may be built into the microscope or separate.

Locate each of these parts on your binocular dissecting microscope, and label them on Figure 2.6.

Focusing the Binocular Dissecting Microscope

1. Place a plastomount that contains small organisms in the center of the stage.
2. Adjust the distance between the eyepieces on the binocular head so that they comfortably fit the distance between your eyes. You should be able to see the object with both eyes as one three-dimensional image.
3. Use the focusing knob to bring the object into focus.
4. Does your microscope have an independent focusing eyepiece? ________ If so, use the focusing knob to bring the image in the fixed eyepiece into focus, while keeping the eye at the independent focusing eyepiece closed. Then adjust the independent focusing eyepiece so that the image is clear, while keeping the other eye closed. Is the image inverted? ________________
5. Turn the magnification changing knob, and determine the kind of mechanism on your microscope. A zoom mechanism allows continuous viewing while changing the magnification. A rotating lens mechanism blocks the view of the object as the new lenses are rotated. Be sure to click each lens firmly into place. If you do not, the field will be only partially visible. What kind of mechanism is on your microscope? ________________________________
6. Set the magnification changing knob on the lowest magnification. Sketch the object in the following circle as though this represents your entire field of view:

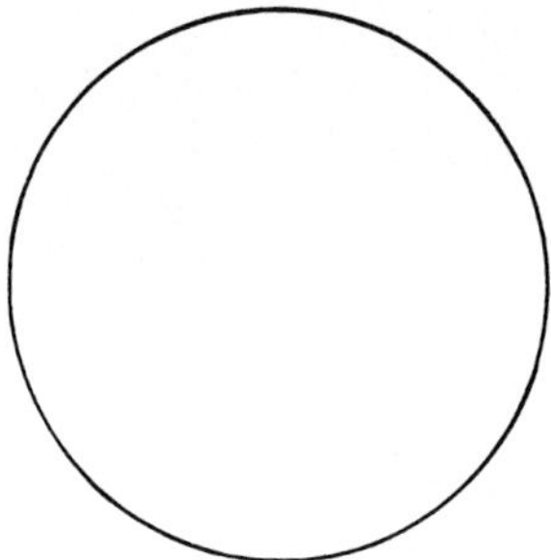

7. Rotate the magnification changing knob to the highest magnification. Draw another circle within the one provided to indicate the reduction of the field of view.
8. Experiment with various objects at various magnifications until you are comfortable with using the binocular dissecting microscope.
9. When you are finished, return your binocular dissecting microscope and illuminator to their correct storage areas.

2.4 Use of the Compound Light Microscope

As mentioned, the name **compound light microscope** indicates that it uses two sets of lenses and light to view an object. The two sets of lenses are the ocular lenses located near the eyes and the objective lenses located near the object. Illumination is from below, and the light passes through clear portions but does not pass through opaque portions. This microscope is used to examine small or thinly sliced sections of objects under higher magnification than would be possible with the binocular dissecting microscope.

Identifying the Parts

Obtain a compound light microscope from the storage area and place it securely on the table. *Identify the following parts on your microscope, and label them in Figure 2.7.*

Figure 2.7 Compound light microscope.
Compound light microscope with binocular head and mechanical stage. Label this microscope with the help of the text material.

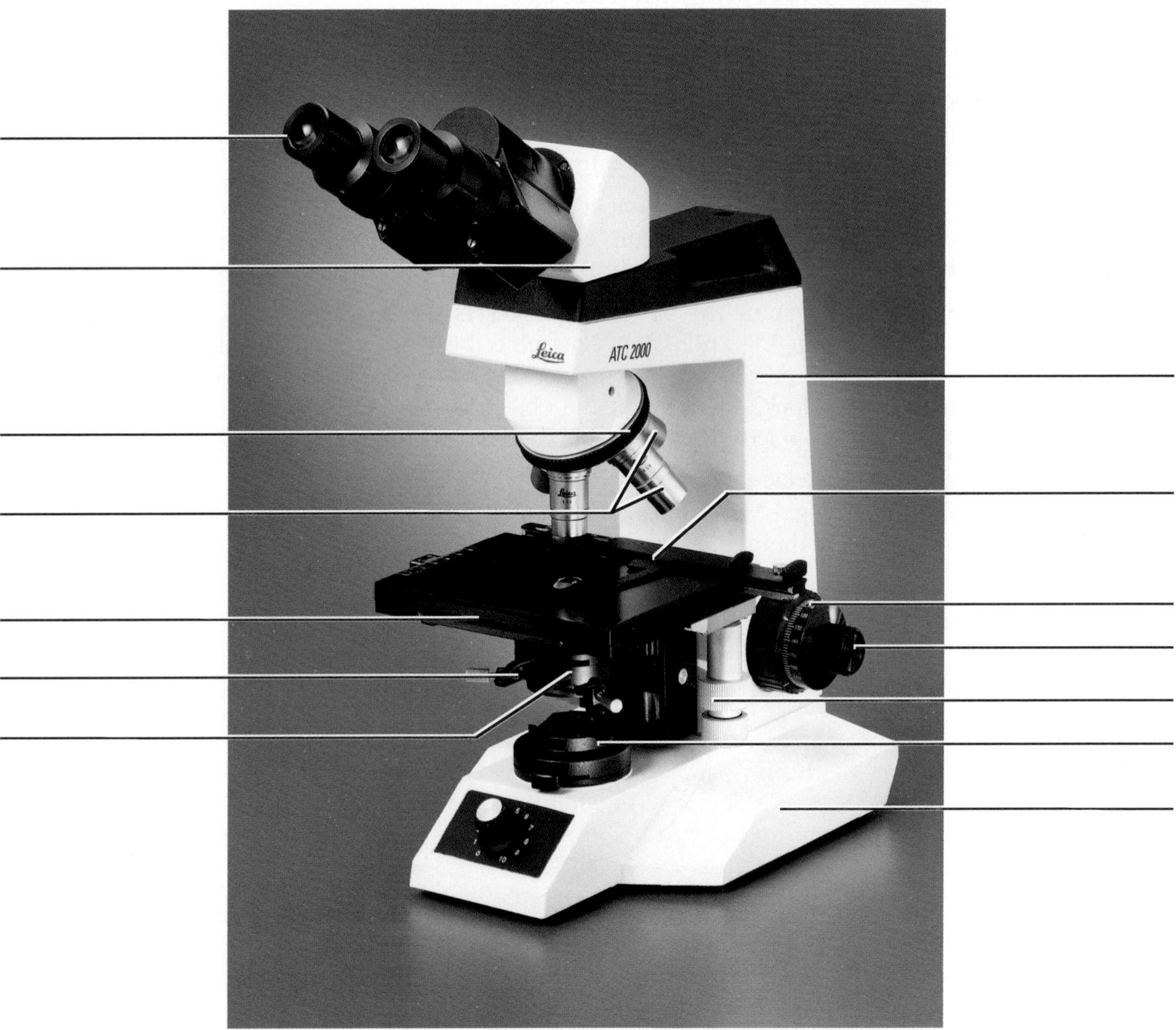

1. **Eyepieces (ocular lenses):** What is the magnifying power of the ocular lenses on your microscope? ____________
2. **Body tube:** Holds nosepiece at one end and eyepiece at the other end; conducts light rays.
3. **Arm:** Supports upper parts and provides carrying handle.
4. **Nosepiece:** Revolving device that holds objectives.
5. **Objectives** (objective lenses):

   a. **Scanning power objective:** This is the shortest of the objective lenses and is used to scan the whole slide. The magnification is stamped on the housing of the lens. It is a number followed by an ×. What is the magnifying power of the scanning lens on your microscope? ________

   b. **Low-power objective:** This lens is longer than the scanning lens and is used to view objects in greater detail. What is the magnifying power of the low-power objective lens on your microscope? ________

   c. **High-power objective:** If your microscope has three objective lenses, this lens will be the longest. It is used to view an object in even greater detail. What is the magnifying power of the high-power objective lens on your microscope? ________

   d. **Oil immersion objective** (on microscopes with four objective lenses): Holds a 95× (to 100×) lens and is used in conjunction with immersion oil to view objects with the greatest magnification. Does your microscope have an oil immersion objective? ________ If this lens is available, your instructor will discuss its use when the lens is needed.

6. **Coarse-adjustment knob:** Knob used to bring object into approximate focus; used only with low-power objective.
7. **Fine-adjustment knob:** Knob used to bring object into final focus.
8. **Condenser:** Lens system below the stage used to focus the beam of light on the object being viewed.
9. **Diaphragm** or **diaphragm control lever:** Controls amount of illumination used to view the object.
10. **Light source:** An attached lamp that directs a beam of light up through the object.
11. **Base:** The flat surface of the microscope that rests on the table.
12. **Stage:** Holds and supports microscope slides.
13. **Stage clips:** Hold slides in place on the stage.
14. **Mechanical stage** (optional): A movable stage that aids in the accurate positioning of the slide. Does your microscope have a mechanical stage? ____________
15. **Mechanical stage control knobs** (optional): Two knobs that are usually located below the stage. One knob controls forward/reverse movement, and the other controls right/left movement.

Focusing the Microscope—Lowest Power

1. Turn the nosepiece so that the *lowest*-power lens is in straight alignment over the stage.
2. Always begin focusing with the *lowest*-power objective lens (4× [scanning] or 10× [low power]).
3. With the coarse-adjustment knob, lower the stage (or raise the objectives) until it stops.
4. Place a slide of the letter *e* on the stage, and stabilize it with the clips. (If your microscope has a mechanical stage, pinch the spring of the slide arms on the stage, and insert the slide.) Center the *e* as best you can on the stage or use the two control knobs located below the stage (if your microscope has a mechanical stage) to center the *e*.

5. Again, be sure that the lowest-power objective is in place. Then, as you look from the side, decrease the distance between the stage and the tip of the objective lens until the lens comes to an automatic stop or is no closer than 3 mm above the slide.
6. While looking into the eyepiece, rotate the diaphragm (or diaphragm control lever) to give the maximum amount of light.
7. Using the coarse-adjustment knob, slowly increase the distance between the stage and the objective lens until the object—in this case, the letter *e*—comes into view, or focus.
8. Once the object is seen, you may need to adjust the amount of light. To increase or decrease the contrast, rotate the diaphragm slightly.
9. Use the fine-adjustment knob to sharpen the focus if necessary.
10. Practice having both eyes open when looking through the eyepiece, as this greatly reduces eyestrain.

Inversion

Inversion refers to the fact that a microscopic image is upside down and reversed.

Observation: Inversion

1. In space 1 provided here, draw the letter *e* as it appears on the slide (with the unaided eye, not looking through the eyepiece).

1.	2.

2. In space 2, draw the letter *e* as it appears when you look through the eyepiece.
3. What differences do you notice? ____________________
4. Move the slide to the right. Which way does the image appear to move? ____________

   Explain. ____________________

Focusing the Microscope—Higher Powers

Compound light microscopes are **parfocal**—that is, once the object is in focus with the lowest power, it should also be almost in focus with the higher power (Fig. 2.8 and Fig. 2.9).

1. Bring the object into focus under the lowest power by following the instructions in the previous section.
2. Make sure that the letter *e* is centered in the field of the lowest objective.
3. Move to the next higher objective (low power [10×] or high power [40×]) by turning the nosepiece until you hear it click into place. Do not change the focus; parfocal microscope objectives will not hit normal slides when changing the focus if the lowest objective is initially in focus. (If you are on low power [10×], proceed to high power [40×] before going on to step 4.)
4. If any adjustment is needed, use only the *fine*-adjustment knob. (*Note:* Always use only the fine-adjustment knob with high power.) On your drawing of the letter *e*, draw a circle around the portion of the letter that you are now seeing with high-power magnification.
5. When you have finished your observations of this slide (or any slide), rotate the nosepiece until the lowest-power objective clicks into place, lower the stage, and then remove the slide.

Figure 2.8 Comparison of slides using reflected light and transmitted light.

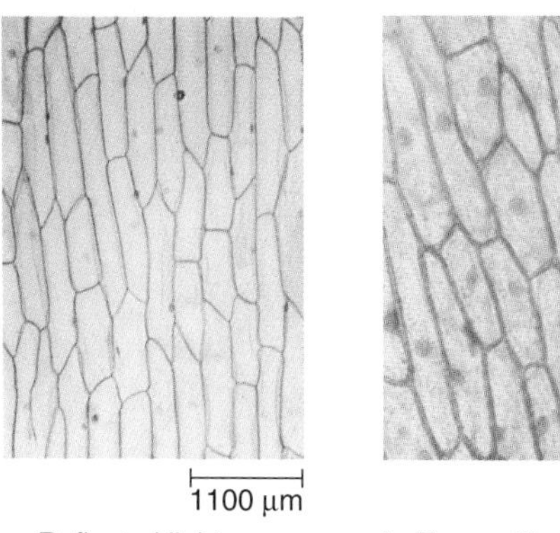

a. Reflected light b. Transmitted light

Figure 2.9 Comparison of unstained cells and stained cells.

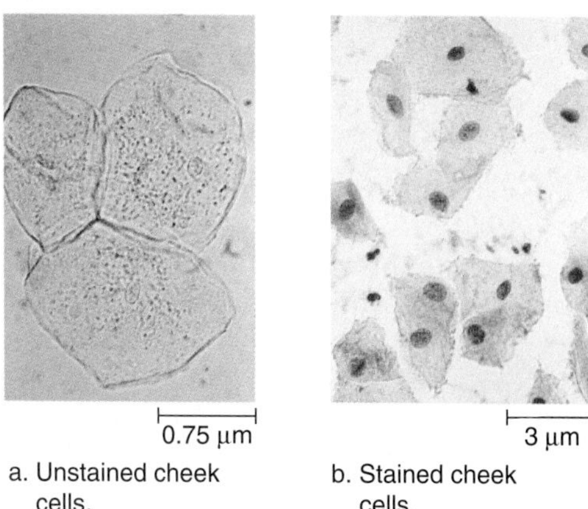

a. Unstained cheek cells. b. Stained cheek cells.

Total Magnification

Total magnification is calculated by multiplying the magnification of the ocular lens (eyepiece) by the magnification of the objective lens.

Observation: Total Magnification

Calculate total magnification figures for your microscope, and record your findings in Table 2.3.

Table 2.3 Total Magnification

Objective	Ocular Lens	Objective Lens	Total Magnification
Scanning power (if present)			
Low power			
High power			
Oil immersion (if present)			

Field of View

A microscope's **field of view** is the circle visible through the lens. The **diameter of field** is the length of the field from one edge to the other.

Observation: Field of View

Low-Power (10×) Diameter of Field

1. Place a clear, plastic ruler across the stage so that the edge of the ruler is visible as a horizontal line along the diameter of the low-power (not scanning) field. Be sure that you are looking at the millimeter side of the ruler.
2. Estimate the number of millimeters, to tenths, that you see along the field: ________ mm. (*Hint:* Start with one of the millimeter markers at the edge of the field.) Convert the figure to micrometers: ________ μm. This is the **low-power diameter of field (LPD)** for your microscope in micrometers.

High-Power (40×) Diameter of Field

1. To compute the **high-power diameter of field (HPD)**, substitute these data into the formula given:
   - **a.** LPD = low-power diameter of field (in micrometers) = ____________
   - **b.** LPM = low-power total magnification (from Table 2.3) = ____________
   - **c.** HPM = high-power total magnification (from Table 2.3) = ____________

Example: If the diameter of field is about 2 μm, then the LPD is 2,000 mm. Using the LPM and HPM values from Table 2.3, the HPD would be 500 μm.

$$\text{HPD} = \text{LPD} \times \frac{\text{LPM}}{\text{HPM}}$$

$$\text{HPD} = (\quad\quad) \times \frac{(\quad\quad)}{(\quad\quad)} = \underline{\hspace{3cm}}$$

Conclusions

- Does low power or high power have a larger field of view (one that allows you to see more of the object)? ____________
- Which has a smaller field but magnifies to a greater extent? ____________
- To locate small objects on a slide, first find them under low power; then place them in the center of the field before rotating to high power.

Depth of Field

When viewing an object on a slide under high power, the **depth of field** (Fig. 2.10) is the area—from top to bottom—that comes into focus while slowly focusing up and down with the microscope's fine-adjustment knob.

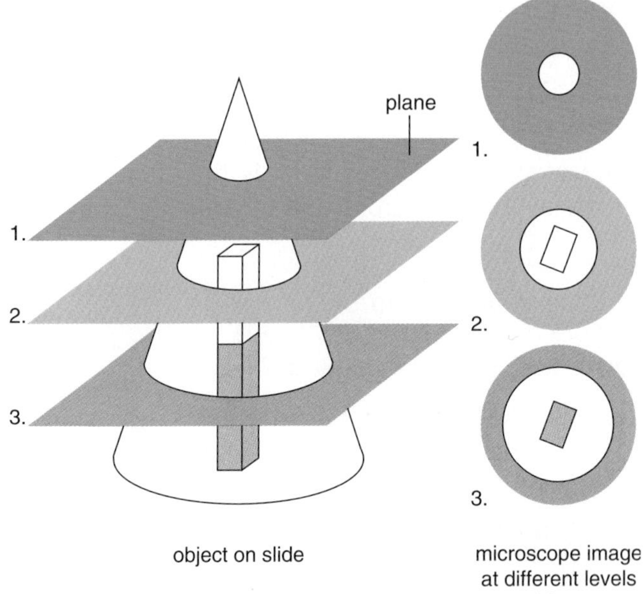

Figure 2.10 Depth of field.
A demonstration of how focusing at depths 1, 2, and 3 would produce three different images (views) that could be used to reconstruct the original three-dimensional structure of the object.

Observation: Depth of Field

1. Obtain a prepared slide with three or four colored threads mounted together, or prepare a wet-mount slide with three or four crossing threads or hairs of different colors. (Directions for preparing a wet mount are given in Section 2.5.)
2. With low power, find a point where the threads or hairs cross. Slowly focus up and down. Notice that when one thread or hair is in focus, the others seem blurred. *Determine the order of the threads or hairs, and complete Table 2.4.* Remember, as the stage moves upward (or the objectives move downward), objects on top come into focus first.

Table 2.4 Order of Threads (or Hairs)

Depth	Thread (or Hair) Color
Top	
Middle	
Bottom	

3. Switch to high power, and notice that the depth of field is more shallow with high power than with low power. Constant use of the fine-adjustment knob when viewing a slide with high power will give you an idea of the specimen's three-dimensional form. For example, viewing a number of sections allows reconstruction of the three-dimensional structure, as demonstrated in Figure 2.10.

2.5 Microscopic Observations

When a specimen is prepared for observation, the object should always be viewed as a **wet mount.** A wet mount is prepared by placing a drop of liquid on a slide or, if the material is dry, by placing it directly on the slide and adding a drop of water or stain. The mount is then covered with a coverslip, as illustrated in Figure 2.11. Dry the bottom of your slide before placing it on the stage.

Figure 2.11 Preparation of a wet mount.

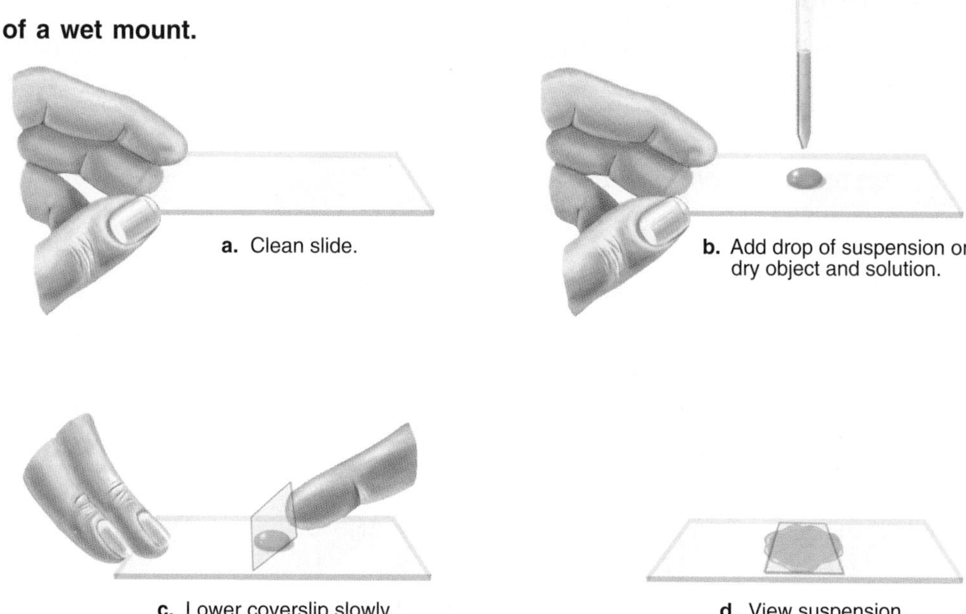
a. Clean slide.
b. Add drop of suspension or dry object and solution.
c. Lower coverslip slowly.
d. View suspension.

Human Epithelial Cells

Epithelial cells cover the body's surface and line its cavities.

> **Caution:** **Methylene blue** Avoid ingestion, inhalation, and contact with skin, eyes, and mucous membranes. Exercise care in using this chemical. If any should spill on your skin, wash the area with mild soap and water. Methylene blue will also stain clothing. Follow your instructor's directions for disposal of this chemical.

Observation: Human Epithelial Cells

1. Obtain a prepared slide, or make your own as follows:
   a. Obtain a prepackaged flat toothpick (or sanitize one with alcohol or alcohol swabs).
   b. Gently scrape the inside of your cheek with the toothpick, and place the scrapings on a clean, dry slide. Discard used toothpicks in the biohazard waste container provided.
   c. Add a drop of very weak *methylene blue* or *iodine solution*, and cover with a coverslip.

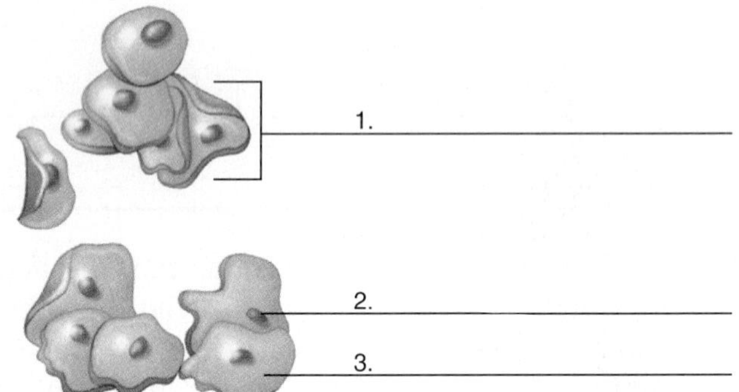

Figure 2.12 Cheek epithelial cells.
Label the nucleus, the cytoplasm, and the plasma membrane.

2. Observe under the microscope.
3. Locate the nucleus (the central, round body), the cytoplasm, and the plasma membrane (outer cell boundary). *Label Figure 2.12.*
4. Because your epithelial slides are biohazardous, they must be disposed of as indicated by your instructor.

Onion Epidermal Cells

Epidermal cells cover the surfaces of plant organs, such as leaves. The bulb of an onion is made up of fleshy leaves.

Observation: Onion Epidermal Cells

> **Caution:** Exercise care when using the scalpel.

1. With a scalpel, strip a small, thin, transparent layer of cells from the inside of a fresh onion leaf.
2. Place it gently on a clean, dry slide, and add a drop of *iodine solution* (or *methylene blue*). Cover with a coverslip.

3. Observe under the microscope.
4. Locate the cell wall and the nucleus. *Label Figure 2.13.*
5. Count the number of onion cells that line up end to end in a single line across the diameter of the high-power (40×) field. ________ Based on what you learned in Section 2.4 about measuring diameter of field, what is your high-power diameter of field (HPD) in micrometers? ________ μm Calculate the length of each onion cell (HPD ÷ number of cells): ________ μm
6. Note some obvious differences between the human cheek cells and the onion cells, and list them in Table 2.5.

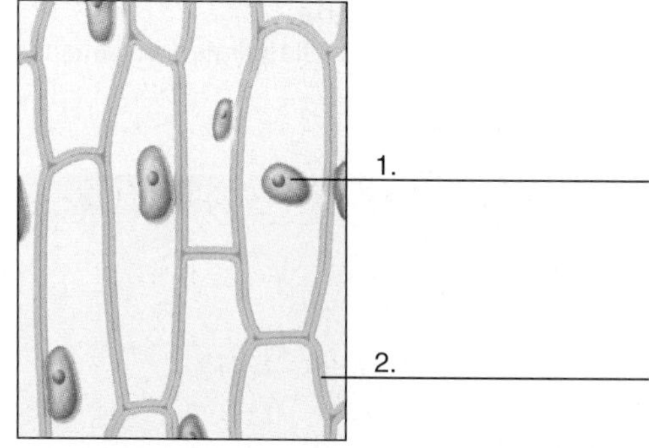

Figure 2.13 Onion epidermal cells.
Label the cell wall and the nucleus.

Table 2.5 Differences Between Human Epithelial and Onion Epidermal Cells

Differences	Human Epithelial Cells (Cheek)	Onion Epidermal Cells
Shape (sketch)		

Euglena

Examination of *Euglena* (a unicellular organism with a flagellum to facilitate movement) will test your ability to observe objects with the microscope, to utilize depth of field, and to control illumination to heighten contrast.

Observation: Euglena

1. Make a wet mount of *Euglena* by using a drop of a *Euglena* culture and adding a drop of Protoslo® (methyl cellulose solution) onto a slide. The Protoslo slows the organism's swimming.
2. Mix thoroughly with a toothpick, and add a coverslip.
3. Scan the slide for *Euglena:* Start at the upper left-hand corner, and move the slide forward and back as you work across the slide from left to right. The *Euglena* may be at the edge of the slide because they show an aversion to Protoslo. Use Figure 2.14 to help identify the structural details of *Euglena.*

Figure 2.14 ***Euglena.***
Euglena is a unicellular, flagellated organism.

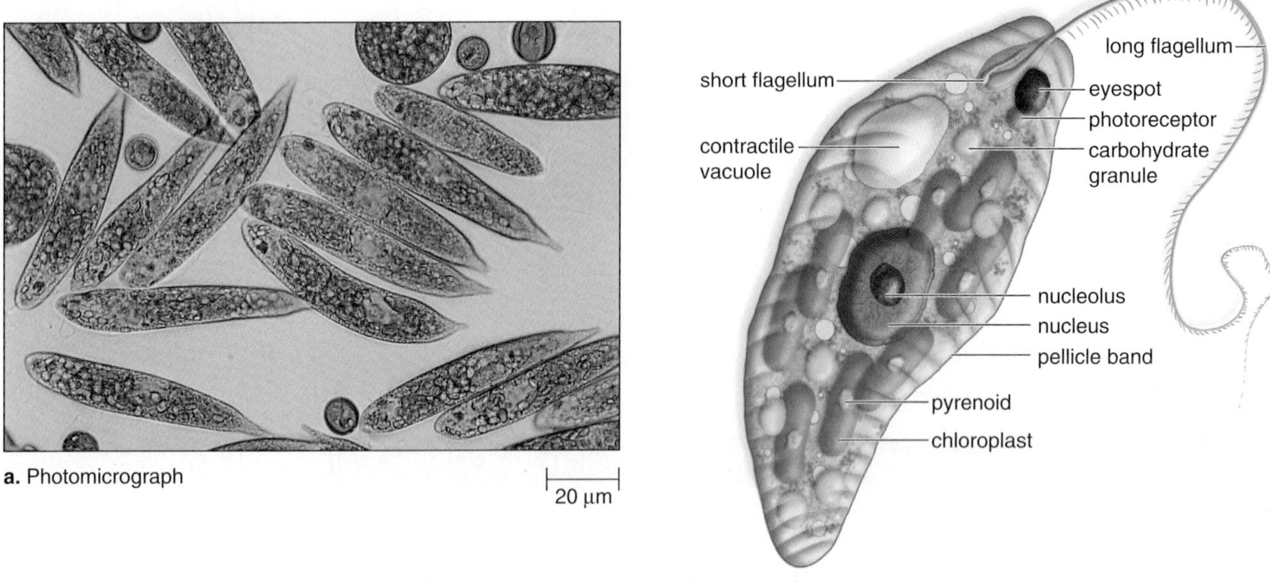

a. Photomicrograph

b. Drawing

4. Experiment by using scanning, low-power, and high-power objective lenses, by focusing up and down with the fine-adjustment knob, and by adjusting the light so that it is not too bright.
5. Compare your *Euglena* specimens with Figure 2.14. List the labeled features that you can actually see: ____

____

Pond Water

Examination of pond water will also test your ability to observe objects with the microscope, to utilize depth of field, and to control illumination to heighten contrast.

Observation: Pond Water

1. Make a wet mount of pond water by taking a drop from the bottom of a container of pond water.
2. Scan the slide for organisms: Start at the upper left-hand corner, and move the slide forward and back as you work across the slide from left to right.
3. Use the pictorial guides provided by your instructor to help identify the organisms present.
4. Experiment by using all available objective lenses, by focusing up and down with the fine-adjustment knob, and by adjusting the light so that it is not too bright.
5. Identify the organisms you see by consulting Figure 2.15.

Figure 2.15 Microorganisms found in pond water. Drawings are not the actual sizes of the organisms.

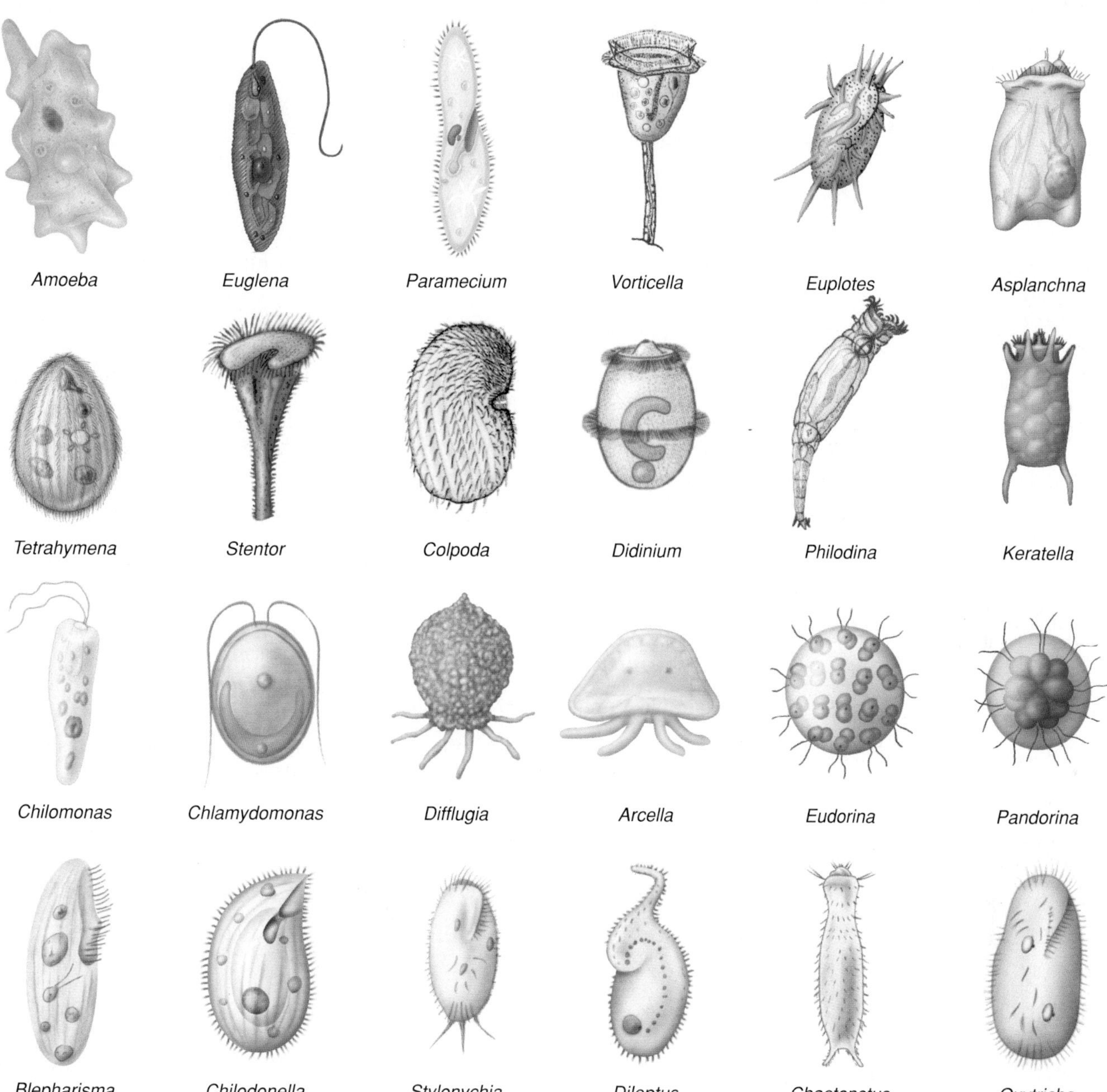

Laboratory Review 2

______ 1. 11 mm equals how many cm?

______ 2. 950 mm equals how many m?

______ 3. 2.1 liters equals how many ml?

______ 4. 122°F equals how many degrees Celsius?

______ 5. 4,100 mg equals how many grams?

______ 6. Which type of microscope would you use to view a *Euglena* swimming in pond water?

______ 7. What are the ocular lenses?

______ 8. Which objective always should be in place both when beginning to use the microscope and also when putting it away?

______ 9. A total magnification of 100× requires the use of the 10× ocular lens with which objective?

______ 10. Which parts of a microscope regulate the amount of light?

______ 11. What word is used to indicate that if the object is in focus at low power it will also be in focus at high power?

______ 12. If the thread layers are red, brown, green, from top to bottom, which layer will come into focus first if you are using the microscope properly?

______ 13. What adjustment knob is used with high power?

______ 14. If a *Euglena* is swimming to the left, which way should you move your slide to keep it in view?

______ 15. What is the final item placed on a wet mount before viewing with a light microscope?

______ 16. What type of object do you study with a binocular dissecting microscope?

______ 17. Why is a binocular dissecting microscope also called a stereomicroscope?

Thought Questions

18. A virus is 50 nm in size. Which type of microscope should be used to view it? Why?

19. Why is locating an object more difficult if you start with the high-power objective rather than the low-power objective?

20. What advantages does the metric system provide over English units of measure?

21. Which objective (scanning, low power, or high power) allows you to observe the greatest number of cells within the field of view?

LABORATORY

3

Chemical Composition of Cells

Learning Outcomes

3.1 Proteins

- Demonstrate an understanding of the relationship of amino acids to the composition of proteins (polypeptides) and explain the role that the peptide bond plays in their assembly.
- Identify the test for the presence of protein and state the basic reason for its ability to detect the presence of larger proteins or polypeptides.
- Cite examples of typical structural and functional proteins and describe their respective functions.

3.2 Carbohydrates

- Demonstrate how monosaccharides, such as glucose, are related to disaccharides and polysaccharides.
- Identify the test that will detect the presence of starch. Distinguish tests that identify starch from those that identify sugars.
- Explain why the Benedict's test produces varied results.

3.3 Lipids

- Identify the structural composition of a common lipid such as fat.
- Describe a simple test for the detection of fat.
- Using the basic structure of a fat as a reference, describe how an emulsifier works.

3.4 Testing the Chemical Composition of Everyday Materials and an Unknown

- Explain how an unknown substance can be tested to determine its composition.

Planning Ahead

Your laboratory instructor may advise you to set up parts of the lab procedure in advance to ensure your ability to complete the exercise. A clock icon will alert you to sections where prior setup may be warranted.

In this laboratory exercise there are two procedures in particular that, with advanced planning, may result in significant time savings: heating the water bath (page 36) and preparing the starch experiment in advance (page 37).

Introduction

Matter is the substance in our universe that is composed of **atoms.** When atoms are linked together by chemical bonds they form **molecules.** In nature, molecules vary from simple to complex. In the case of a molecule of oxygen (O_2), two atoms of oxygen are bound together by a **covalent bond,** creating a single structure. Other molecules can form when two or more different types of atoms combine, as with carbon dioxide (CO_2). Molecules of this type are called **compounds.** Compounds can be simple, containing only two or three atoms per molecule, or they can be much more complex, containing many atoms.

Molecules are routinely subdivided into two categories. Molecules that contain carbon and hydrogen atoms are referred to as **organic molecules.** Molecules that lack carbon and hydrogen atoms in their molecular structure are referred to as **inorganic molecules.** In this laboratory, you will be studying three types of organic molecules: **proteins, carbohydrates** (monosaccharides, disaccharides, polysaccharides), and **lipids.**

Large organic molecules form during the dehydration reaction when smaller molecules bond as water is given off. During hydrolysis, bonds are broken as water is added. (The dehydration reaction and the hydrolysis reaction are illustrated in Figure 3.1.) Dehydration and hydrolysis reactions specific to proteins, carbohydrates, and lipids, respectively, are demonstrated in Figures 3.2, 3.3, and 3.5. A fat contains one glycerol and three fatty acids. Proteins and some carbohydrates (called polysaccharides) are **macromolecules** or *polymers* because they are made up of smaller molecules called **subunits** or *monomers*. Proteins contain a large number of amino acids (the monomer) joined together like the links in a chain. A polysaccharide, such as starch, contains a large number of glucose molecules (the monomer) joined together.

H_2O **dehydration reaction**

A + B ⇌ C

H_2O **hydrolysis reaction**

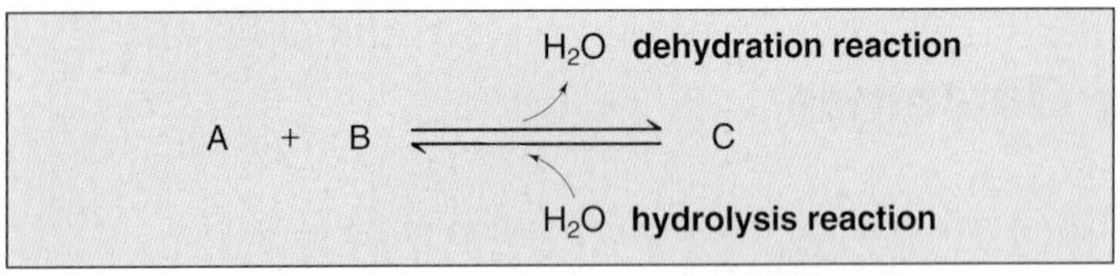

Figure 3.1
Dehydration reactions link molecules together by removing water. During hydrolysis reactions, bonds are broken as water is added.

3.1 Proteins

Proteins have numerous functions in cells; some are present for functional reasons, while others provide structural integrity. Antibodies are **functional proteins** that combine with disease-causing pathogens as part of the body's immune response. Transport proteins combine with and move substances from place to place. Hemoglobin transports oxygen throughout the body. Albumin is another transport protein in our blood that performs several important roles including fatty acid transport. Regulatory proteins control cellular metabolism in some way. For example, the hormone insulin regulates the amount of glucose in blood so that cells have a ready supply of energy. **Structural proteins** include keratin, which is found in hair, and myosin, which is found in muscle. **Enzymes** are functional proteins that speed chemical reactions. A reaction that could take days or weeks to complete can happen within an instant if the correct enzyme is present. Amylase is an enzyme that speeds the breakdown of starch in the mouth and small intestine.

Proteins are polymers of **amino acids** joined together. About 20 different common amino acids are found in cells. All amino acids have an acidic group (—COOH) and an amino group (H_2N—), each linked to a central carbon by a separate covalent bond. They differ by the **R group** (remainder group) attached to the central carbon atom, as shown in Figure 3.2. The R groups have varying sizes, shapes, and chemical activities.

A chain of two or more amino acids is called a **peptide,** and the bond between the two amino acids is called a **peptide bond.** A **polypeptide** is a very long chain of amino acids. A protein can be comprised of one or more polypeptide chains. Insulin consists of two polypeptide chains that function together as a single unit. Hemoglobin contains four polypeptides that also function as one unit. As in all biological molecules, the shape of a protein determines its ability to function properly.

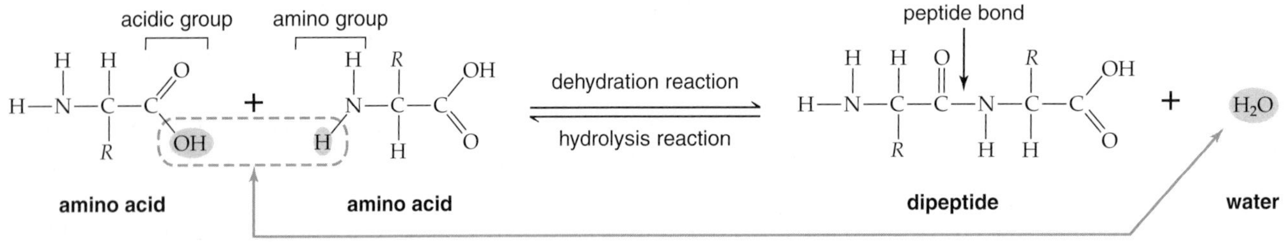

Figure 3.2 Peptide bond.
Peptide bond formation between amino acids creates a dipeptide. This dehydration reaction involves the removal of one water molecule. During a hydrolysis reaction, water is added, and the peptide bond is broken. In a polypeptide, many amino acids are held together by multiple peptide bonds.

Test for Proteins

Biuret reagent (blue color) contains a strong solution of sodium or potassium hydroxide (NaOH or KOH) and a small amount of dilute copper sulfate ($CuSO_4$) solution. The reagent changes color in the presence of proteins or peptides because the peptide bonds of the protein or peptide chemically combine with the copper ions in biuret reagent.

In this reaction, the range of positive results described in Table 3.1 occurs when copper ions react with different amino groups, deforming the adjacent polypeptide chains and causing a resultant change in color.

Table 3.1 Test for Protein

	Protein	Peptides
Biuret reagent (blue)	Purple	Pinkish-purple

Experimental Procedure: Test for Proteins

> **Caution:** **Biuret reagent** Biuret reagent is highly corrosive. Exercise care in using this chemical. If any should spill on your skin, wash the area with mild soap and water. Follow your instructor's directions for disposal of this chemical.

With a millimeter ruler and a wax pencil, label and mark four clean test tubes at the 1 cm level. After filling a tube, cover it with Parafilm, and swirl well to mix. (Do not turn upside down.) The reaction is almost immediate.

*Tube 1
1. Fill to the mark with *distilled water* and add about 5 drops of *biuret reagent.*
2. Record the final color in Table 3.2.

Tube 2
1. Fill to the mark with *albumin solution* and add about 5 drops of *biuret reagent.*
2. Record the final color in Table 3.2.

Tube 3
1. Fill to the mark with *pepsin solution* and add about 5 drops of *biuret reagent.*
2. Record the final color in Table 3.2.

Tube 4
1. Fill to the mark with *starch solution* and add about 5 drops of *biuret reagent.*
2. Record the final color in Table 3.2.

Table 3.2 Biuret Test

Tube	Contents	Final Color	Conclusions
1	Distilled water		
2	Albumin		
3	Pepsin		
4	Starch		

*To test an unknown for a protein, use this procedure. Instead of water, use the unknown. If protein is present, a pinkish-purple color appears.

Conclusions

- From your test results, did you conclude that protein is present in any of the three test solutions? Enter your conclusions in Table 3.2.
- If your results are not as expected, offer an explanation. ____________________ Then inform your instructor, who will advise you how to proceed.
- Which of the four tubes is the negative control sample? ________ Why? ________

  ____

  Why do experimental procedures include control samples? ____________________

  ____
- Pepsin is an enzyme. Enzymes are composed of what type of organic molecule? ________
- Did any of the three test solutions produce a negative result? Offer an explanation for any negative results that may have been obtained from the three test solutions. ____________

  ____
- If two tubes produced positive Biuret's reactions, how would you interpret the results to determine if one tube contained more protein than the other? ____________________

  ____

3.2 Carbohydrates

Carbohydrates include sugars and molecules that are chains (polymers) of sugars. **Glucose,** which has only one sugar unit, is a monosaccharide; **maltose,** which forms when dehydration synthesis joins two glucose units, is a disaccharide (Fig. 3.3). Glycogen, starch, and cellulose are polysaccharides, which are polymers made up of chains of glucose units (Fig. 3.4).

Glucose is used by all organisms as an energy source. Energy is released when glucose is broken down to carbon dioxide and water. This energy is used by the organism to do work. Animals store glucose as glycogen and plants store glucose as starch. Plant cell walls are composed of cellulose.

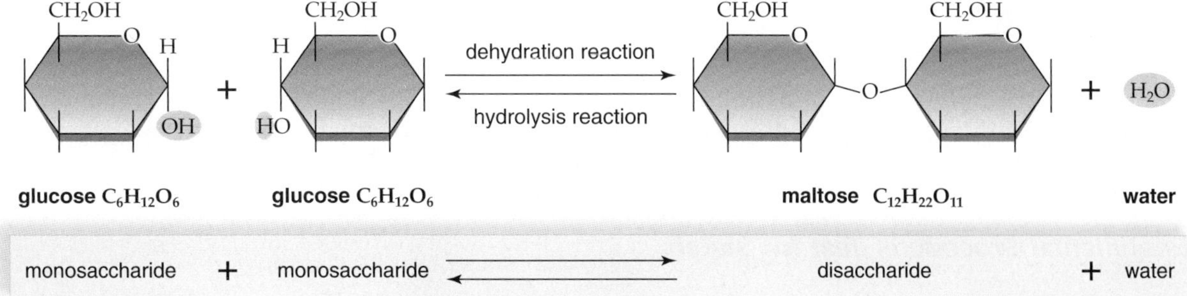

Figure 3.3 Formation of maltose, a disaccharide.
During a dehydration reaction, a bond forms between the two glucose molecules, the components of water are removed, and maltose results. During a hydrolysis reaction, the components of water are added, and the bond is broken.

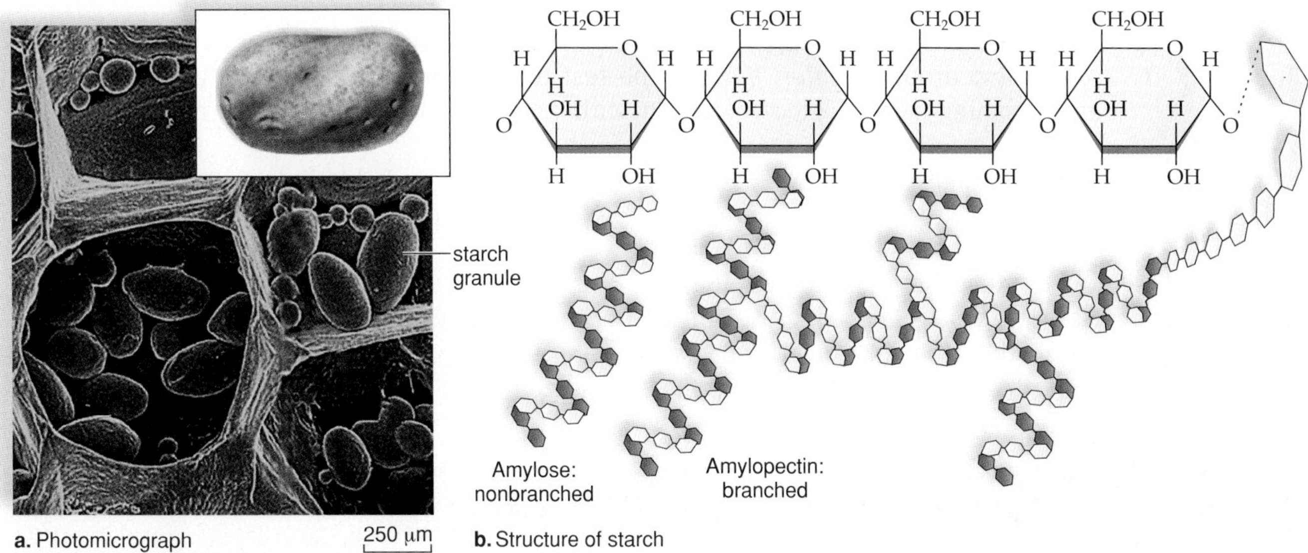

Figure 3.4 Starch.
Starch is a polysaccharide composed of many glucose units. **a.** Photomicrograph of starch granules in plant cells. **b.** Structure of starch. Starch consists of amylose which is nonbranched and amylopectin which is branched.

Test for Starch

In solution, the chain of glucose molecules that make up the structure of starch take on a shape similar to that of a spring or slinky. Iodine molecules lodge in this coiled structure and the result is a change in the way the iodine molecules are able to reflect light. The color shifts from brown to a deep purple to black.

Experimental Procedure: Test for Starch

With a wax pencil, label and mark five clean test tubes at the 1 cm level.

*Tube 1
1. Fill to the 1 cm mark with *water,* and add five drops of *iodine solution.*
2. Note the final color change, and record your results in Table 3.3.

Tube 2
1. It is very important to shake the *starch suspension* well before taking your sample. After shaking, fill this tube to the 1 cm mark with the 1% *starch suspension.* Add five drops of *iodine solution.*
2. Note the final color change, and record your results in Table 3.3.

Tube 3
1. Add a few drops of *onion juice* to the test tube. (Obtain the juice by adding water and crushing a small piece of onion with a mortar and pestle. Clean mortar and pestle after using.) Add five drops of *iodine solution.*
2. Note the final color change, and record your results in Table 3.3.

Tube 4
1. Add a few drops of *potato juice* to the test tube. (Obtain the juice by adding water and crushing a small piece of potato with a mortar and pestle. Clean mortar and pestle after using.) Add five drops of *iodine solution.*
2. Note the final color change, and record your results in Table 3.3.

Tube 5
1. Fill to the 1 cm mark with *glucose solution,* and add five drops of *iodine solution.*
2. Note the final color change, and record your results in Table 3.3.

Table 3.3 Iodine (IKI) Tests for Starch

Tube	Contents	Color	Conclusions
1	Water		
2	Starch suspension		
3	Onion juice		
4	Potato juice		
5	Glucose solution		

*To test an unknown for starch, use this procedure. Instead of water, use the unknown. If starch is present, a blue-black color appears.

Conclusions

- From your test results, draw conclusions about what organic compound is present in each tube. Write these conclusions in Table 3.3.
- Do the potato and onion plants store carbohydrates in different forms? If so, offer a hypothesis for why this would be the case.
- If your results are not as expected, offer an explanation. Then inform your instructor, who will advise you how to proceed.

Experimental Procedure: Microscopic Study

Potato

1. With a sharp razor blade, carefully slice a very thin piece of potato. Place it on a microscope slide, add a drop of *water* and a coverslip, and observe under low power with your compound light microscope. Compare your slide with the photomicrograph of starch granules (see Fig. 3.4a). Find the cell wall (large, geometric compartments) and the starch grains (numerous clear, oval-shaped objects).
2. Without removing the coverslip, place two drops of *iodine solution* onto the microscope slide so that the iodine touches the coverslip. Draw the iodine under the coverslip by placing a small piece of paper towel in contact with the water on the **opposite** side of the coverslip.
3. Microscopically examine the potato again on the side closest to where the iodine solution was applied.

   Is the appearance of your slide similar to that of the image in Figure 3.4*b*? ________________

   What can you conclude from the appearance of the potato cells on your slide? ________________

Onion

Peel a single layer of onion from the bulb. On the inside surface, you will find a thin, transparent layer of onion skin. Peel off a small section of this layer for use on your slide.

1. Place a piece of the onion skin on a microscope slide.
2. Add a large drop of iodine solution.
3. Does onion contain starch?
4. Would you conclude that your results for this experiment are consistent with the results that you obtained and recorded on Table 3.3?
5. What conclusions can you construct when evaluating your data associated with this experiment?

Test for Sugars

Monosaccharides and some disaccharides will react with **Benedict's reagent** (qualitative) after being heated in a boiling water bath. In this reaction, copper ion (Cu^{2+}) in the Benedict's reagent reacts with part of the sugar molecule and is reduced, causing a distinctive color change. The color change can range from green to red, and increasing concentrations of sugar will give a continuum of colored products (Table 3.4). This experiment tests for the presence (or absence) of varying amounts of sugars in a variety of materials and chemicals.

Table 3.4 Benedict's Reagent (Some Typical Reactions)

Chemical	Chemical Category	Benedict's Reagent (After Heating)
Water	Inorganic	Blue (no change)
Glucose	Monosaccharide (carbohydrate)	Varies with concentration: very low—green low—yellow moderate—yellow-orange high—orange very high—orange-red
Maltose	Disaccharide (carbohydrate)	Varies with concentration—see "Glucose"
Starch	Polysaccharide (carbohydrate)	Blue (no change)

> **Caution:** **Benedict's reagent** Benedict's reagent is highly corrosive. Exercise care in using this chemical. If any should spill on your skin, wash the area with mild soap and water. Follow your instructor's directions for disposal of this chemical. Use protective eyewear when performing this experiment.

Experimental Procedure: Test for Sugars

Heating the water bath in advance may maximize your efficient use of time.

With a wax pencil, label and mark five clean test tubes at the 1 cm level. Save your tubes for comparison with Section 3.4.

*Tube 1
1. Fill to the 1 cm mark with *water,* and then add about 5 drops of *Benedict's reagent.*
2. Heat in a boiling water bath for 5 to 10 minutes, note any color change, and record in Table 3.5.

Tube 2
1. Fill to the 1 cm mark with *glucose solution,* and then add about 5 drops of *Benedict's reagent.*
2. Heat in a boiling water bath for 5 to 10 minutes, note any color change, and record in Table 3.5.

Tube 3
1. Place a few drops of *onion juice* in the test tube. (Obtain the juice by adding water and crushing a small piece of onion with a mortar and pestle. Clean mortar and pestle after using.)
2. Fill to the 1 cm mark with *water,* and then add about 5 drops of *Benedict's reagent.*
3. Heat in a boiling water bath for 5 to 10 minutes, note any color change, and record in Table 3.5.

Tube 4
1. Place a few drops of *potato juice* in the test tube. (Obtain the juice by adding water and crushing a small piece of potato with a mortar and pestle.)
2. Fill to the 1 cm mark with *water,* and then add about 5 drops of *Benedict's reagent.*
3. Heat in a boiling water bath for 5 to 10 minutes, note any color change, and record in Table 3.5.

Tube 5
1. Fill to the 1 cm mark with *starch suspension,* and then add about 5 drops of *Benedict's reagent.*
2. Heat in a boiling water bath for 5 to 10 minutes, note any color change, and record in Table 3.5.

Table 3.5 Benedict's Reagent Test

Tube	Contents	Color (After Heating)	Conclusions
1	Water		
2	Glucose solution		
3	Onion juice		
4	Potato juice		
5	Starch suspension		

*To test an unknown for sugars, use this procedure. Instead of water, use the unknown. If sugar is present, a green to orange-red color appears.

Conclusions

- Why were tubes 1, 2, and 5 included within this experiment?
- Based upon the results in Table 3.5, what type of chemical was present in tubes 3 and 4?
- Compare Table 3.3 with Table 3.5. Sugars are an immediate energy source in cells. In plant cells, glucose (a primary energy molecule) is often stored in the form of starch. Is glucose stored as starch in the potato? ______ Is glucose stored as starch in the onion? ______ Does this explain your results in Table 3.5? Why? ______
- What would you conclude if an unknown sample tested positive for both the iodine and the Benedict's test?

Starch Composition

Recall that polysaccharides are chains of monosaccharides joined together. While the bonds holding the monomers to each other release water during dehydration synthesis, the addition of water in the presence of an enzyme can break the bonds, releasing the monomers. Amylase is a starch-digesting enzyme that uses hydrolysis in a process of controlled destruction to break the covalent bonds in a starch polymer, releasing smaller maltose molecules.

$$\text{starch} + \text{water} \xrightarrow{\text{amylase}} \text{maltose}$$

Experimental Procedure: Starch Composition

30 minutes are required for reaction time. You may wish to set this up in advance.

With a wax pencil, label and mark two clean test tubes at the 2 cm, 4 cm, and 6 cm levels. Save your tubes for comparison with Section 3.4.

Tube 1
1. Fill to the 2 cm mark with *water* and to the 4 cm mark with 1% *amylase.*
2. Shake and then wait for 30 minutes.
3. Add *Benedict's reagent* to the 6 cm level and heat in a boiling water bath.
4. Note any color change, and record your results in Table 3.6.

Tube 2
1. Fill to the 2 cm mark with *starch* and to the 4 cm mark with *amylase.*
2. Shake and then wait for 30 minutes.
3. Add *Benedict's reagent* and heat as before.
4. Note any color change, and record your results in Table 3.6.

Save your positive test results for use in Section 3.4.

Table 3.6 Starch Composition

Tube	Contents	Color Change	Conclusions
1	Water Amylase		
2	Starch Amylase		

Conclusions

- From your test results, you may conclude that starch is composed of what kind of chemical? ______
- How do you know? ______

  Enter your conclusions for the previous two questions in Table 3.6.
- What is the function of tube 1? Why is it included within this experiment? ______
- What would you conclude if you performed the iodine test on the sample solution in tube 3 and you found that the solution also tested positive for starch? ______

3.3 Lipids

Lipids are compounds that are insoluble in water and soluble in solvents, such as alcohol and ether. Lipids include fats, oils, phospholipids, steroids, and cholesterol. Typically, fats and oils are composed of three molecules of fatty acids bonded to one molecule of glycerol (Fig. 3.5). Lipids carry out a number of important roles. Fats function as energy storage devices, as insulators in animals, and as raw materials for the synthesis of types of lipids, such as the steroids cholesterol and estrogen.

Fats are common lipids in nature. Two terms are often associated with fats: saturated and unsaturated. If you examine the image at the right-hand side of Figure 3.5, you see a representation of a triglyceride, or fat, molecule. Note the first "tail" that extends to the right. With the exception of the first carbon in the chain (at the left side of the rectangular box), all of the carbons in the chain are bonded to **four** other atoms. Since carbon can form a maximum of four bonds with other atoms, we say that it is **saturated,** or "full." It is the difference in the shape of the two types of fatty acids that allows saturated fats to form solids, whereas **unsaturated** fats, with less than four bonds to four separate atoms, have a different shape that causes them to form liquids or oils.

Figure 3.5 Formation of a fat.
A fat molecule forms when glycerol joins with three fatty acids as three water molecules are removed during a dehydration reaction. During a hydrolysis reaction, water is added, and the bonds are broken.

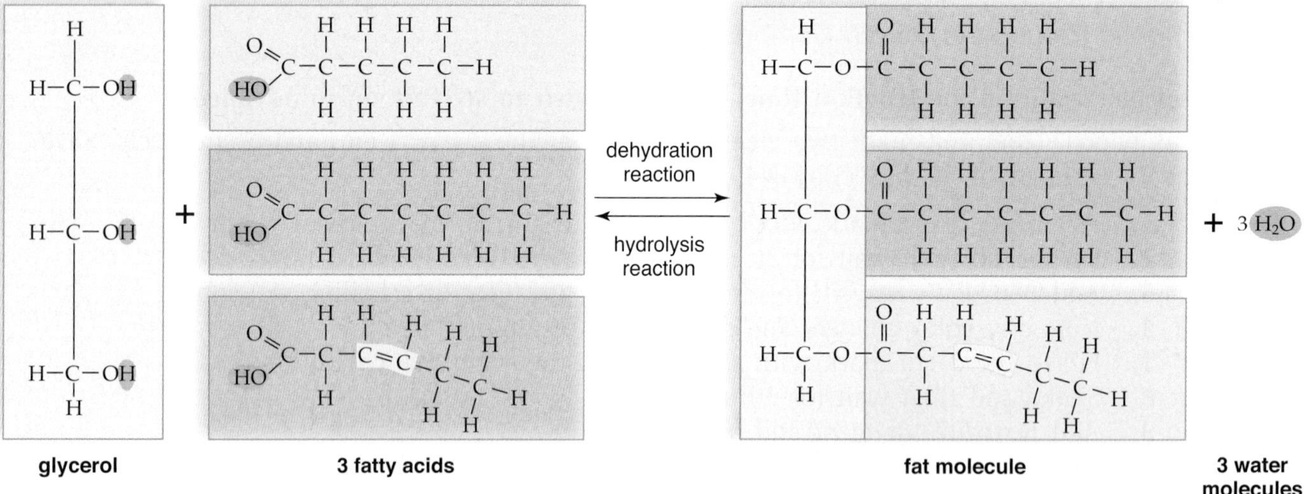

Test for Lipids

Fats do not evaporate from brown paper; instead, they leave an oily spot.

Experimental Procedure: Test for Lipids

*1. Place a small drop of *water* on a square of brown paper. Describe the immediate effect. ________

____

2. Place a small drop of *vegetable oil* on a square of brown paper. Describe the immediate effect.

____

3. Wait at least 15 minutes for the paper to dry. Evaluate which substance penetrates the paper and which is subject to evaporation. Record your observations and conclusions in Table 3.7. Save the paper for comparison with Section 3.4.

Table 3.7 Test for Lipids

Sample	Observations	Conclusions
Water spot		
Oil spot		

Emulsification of Lipids

Some molecules are **polar,** meaning that they have charged groups or atoms, and some are **nonpolar,** meaning that they have no charged groups or atoms. A water molecule is polar, and therefore water is a good solvent for other polar molecules, dissolving them when they come in contact with one another. When the charged ends of water molecules interact with the charged groups of polar molecules, these polar molecules disperse in water.

Water is not a good solvent for nonpolar molecules, such as fats. A fat has no polar groups to interact with water molecules. An **emulsifier,** however, can cause a fat to disperse in water. An emulsifier contains molecules with both polar and nonpolar ends. When the nonpolar ends interact with the fat and the polar ends interact with the water molecules, the fat disperses in water, and an *emulsion* results (Fig. 3.6).

Bile salts (emulsifiers found in bile produced by the liver) are used in the digestive tract. Detergents act as emulsifiers, attracting the nonpolar grease and oils and allowing water to carry them away when clothes or dishes are washed. Tween®, a commercial wetting agent, is also an emulsifier.

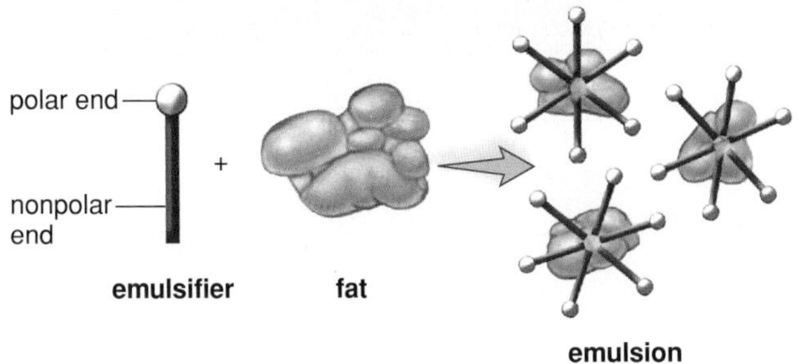

Figure 3.6 Emulsification.
An emulsifier contains molecules with both a polar and a nonpolar end. The nonpolar ends are attracted to the nonpolar fat, and the polar ends are attracted to the water. This causes droplets of fat molecules to disperse.

*To test an unknown for lipids, use this procedure. Instead of water, use the unknown. If lipids are present, an oily spot appears.

Experimental Procedure: Emulsification of Lipids

With a wax pencil, label two clean test tubes 1 and 2. Mark tube 1 at the 3 cm and 4 cm levels. Mark tube 2 at the 2 cm, 3 cm, and 4 cm levels.

Tube 1

1. Fill to the 3 cm mark with *water* and to the 4 cm mark with *vegetable oil.* Shake.
2. Observe for the initial dispersal of oil, followed by rapid separation into two layers.

   Is vegetable oil soluble in water? ______________________________
3. Let the tube settle for 5 minutes. Label a microscope slide as 1.
4. Use a dropper to remove a sample of the solution that is just below the layer of oil. Place the drop on the slide, add a coverslip, and examine with the low power of your compound light microscope.
5. Record your observations in Table 3.8.

Tube 2

1. Fill to the 2 cm mark with *water,* to the 3 cm mark with *vegetable oil,* and to the 4 cm mark with the available emulsifier (*Tween* or *bile salts*). Shake.
2. Describe how the distribution of oil in tube 2 compares with the distribution in tube 1.

   ____

   ____
3. Let the tube settle for 5 minutes. Label a microscope slide as 2.
4. Use a different dropper to remove a sample of the solution that is just below the layer of oil. Place the drop on the slide, add a coverslip, and examine with the low power of your compound light microscope.
5. Record your observations in Table 3.8.

Table 3.8 Emulsification

Tube	Contents	Observations	Conclusions
1	Oil Water		
2	Oil Water Emulsifier		

Conclusions

- From your observations, conclude why the contents of tube 1 and tube 2 appear as they do under the microscope. Record your conclusions in Table 3.8.
- Explain the correlation between your macroscopic observations (how the tubes look to your unaided eye) and your microscopic observations. ______________________________

  ____

Adipose Tissue

Adipose tissue stores droplets of fat. Adipose tissue is found beneath the skin, where it helps insulate and keep the body warm. It also forms a protective cushion around various internal organs.

Observation: Adipose Tissue

1. Obtain a slide of adipose tissue, and view it under the microscope at high power. Refer to Figure 3.7 for help in identifying the structures.
2. Notice how the fat droplets push the cytoplasm to the edges of the cells.

Figure 3.7 Adipose tissue. The cells are so full of fat that the nucleus is pushed to one side.

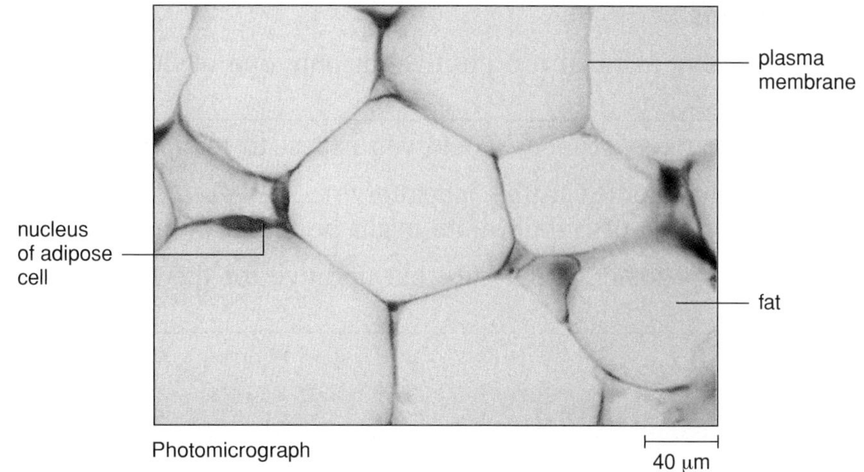

3.4 Testing the Chemical Composition of Everyday Materials and an Unknown

It is common for us to associate the term *organic* with the foods we eat. While we may recognize foods as being organic, often we are not aware of what specific types of compounds are found in what we eat. In the following Experimental Procedure, you will use the same tests you used previously to evaluate the composition of everyday materials (unknowns).

Experimental Procedure: Chemical Composition

1. Your instructor will provide you with several everyday materials. Use the tests in this laboratory for proteins, p. 31; carbohydrates (starch, p. 34; sugar, p. 35), and lipids, p. 38, to determine the macromolecules present in these materials.
2. Write the *name* of each known material, and assign a *letter* to any unknowns provided (Unknown A, Unknown B, etc.).
3. Record your results as positive or negative in Table 3.9.

Table 3.9 Everyday Materials and Unknowns

Sample Name	Sugar (Benedict's)	Starch (Iodine)	Protein (Biuret)	Lipid (Brown Paper)
Unknown A				
Unknown B				

Conclusions

- Did any material test positive for only one of the organic compounds? ______
  Explain. ______
- What types of foods would you expect to test positive for more than one of the organic compounds studied in this laboratory? ______
- What type of carbohydrate might be found in an unknown food source that would test positive for the iodine test but negative for the Benedict's test? ______

______

Laboratory Review 3

______ 1. What type of bond joins amino acids to make peptides?

______ 2. What type of protein speeds chemical reactions?

______ 3. What group is different between types of amino acids?

______ 4. If iodine solution turns blue-black, what substance is present?

______ 5. If Benedict's reagent turns red, what substance is present?

______ 6. What is the function of starch in plant cells?

______ 7. Is starch a monosaccharide or a polysaccharide?

______ 8. What is the function of fat in animal cells?

______ 9. What molecules are released when fat undergoes a hydrolysis reaction?

______ 10. What type of organic molecule requires the action of emulsifiers to be successfully digested?

______ 11. Are fats polar or nonpolar?

______ 12. If biuret reagent turns purple, what substance is present?

______ 13. A student adds iodine solution to egg white and waits for a color change. How long will the student have to wait?

______ 14. To test whether a sample contains glucose, what reagent should be used?

Thought Questions

15. Why is it necessary to shake a bottle of salad dressing before adding it to a salad?

16. An unknown sample is tested with both biuret reagent and Benedict's reagent. Both tests result in a blue color. What has been learned? Why are these called negative results?

17. Starch and water are mixed together as ingredients for making gravy. Why doesn't starch react with water to produce monosaccharides?

LABORATORY

4

Cell Structure and Function

Learning Outcomes

4.1 Prokaryotic Versus Eukaryotic Cells
- Explain the criteria you would use to determine if a cell is prokaryotic or eukaryotic.

4.2 Animal Cell and Plant Cell Structure
- Using a photomicrograph of an animal cell, identify the major organelles and relate their respective functions.
- Using a photomicrograph of a plant cell, identify the major organelles and relate their respective functions.
- Comparing a photomicrograph of a plant cell with a photomicrograph of an animal cell, contrast the differences in structure and function between the two types of cells.

4.3 Diffusion
- Describe how the process of diffusion occurs and why it might occur differently in a living organism than it would in the physical environment.
- Predict which substances will or will not diffuse across a plasma membrane.

4.4 Osmosis
- Define isotonic, hypertonic, and hypotonic solutions, and give examples in terms of NaCl concentrations.
- Predict the effect of different tonicities on animal (e.g., red blood) cells and on plant (e.g., *Elodea*) cells.

4.5 pH and Cells
- Predict the pH before and after the addition of an acid to nonbuffered and buffered solutions.

Planning Ahead

The tonicity experiment using potato strips (see page 53) requires that the experiment run for one hour. Your instructor may advise you to set this experiment up in advance so that you will have adequate time to complete the experiment.

Introduction

The molecules we studied in the previous laboratory are not alive—the basic units of life are cells. The **cell theory** states that all living things are composed of cells and that cells come only from other cells. Cells vary from simple to extremely complex. Some organisms, such as *Euglena,* which you observed in Laboratory 2, are unicellular—their entire existence is housed in a single cell. On the other hand, multicellular organisms are composed of many cells found in different tissues and organs. While we are accustomed to considering the heart, the liver, and the intestines as enabling the human body to function, it is actually cells that do the work of these organs.

4.1 Prokaryotic Versus Eukaryotic Cells

All living cells are classified as either prokaryotic or eukaryotic. One of the basic differences between the two types is that prokaryotic cells do not contain nuclei (*pro* means "before"; *karyote* means "nucleus"), while eukaryotic cells do contain nuclei (*eu* means "true"; *karyote* means "nucleus"). Only bacteria (including cyanobacteria) and archaea are prokaryotes; all other organisms are eukaryotes.

Organelles are small, membranous bodies, each with a specific structure and function. As another distinction, prokaryotes lack the organelles found in eukaryotic cells (Fig. 4.1). Prokaryotes do have **cytoplasm,** which is the material bounded by a **plasma membrane** and **cell wall.** The cytoplasm contains **ribosomes,** small granules that coordinate the synthesis of proteins; **thylakoids** (only in cyanobacteria), which participate in photosynthesis; and innumerable enzymes. Prokaryotes also have a nucleoid region, which contains a chromosome composed largely of DNA with little protein.

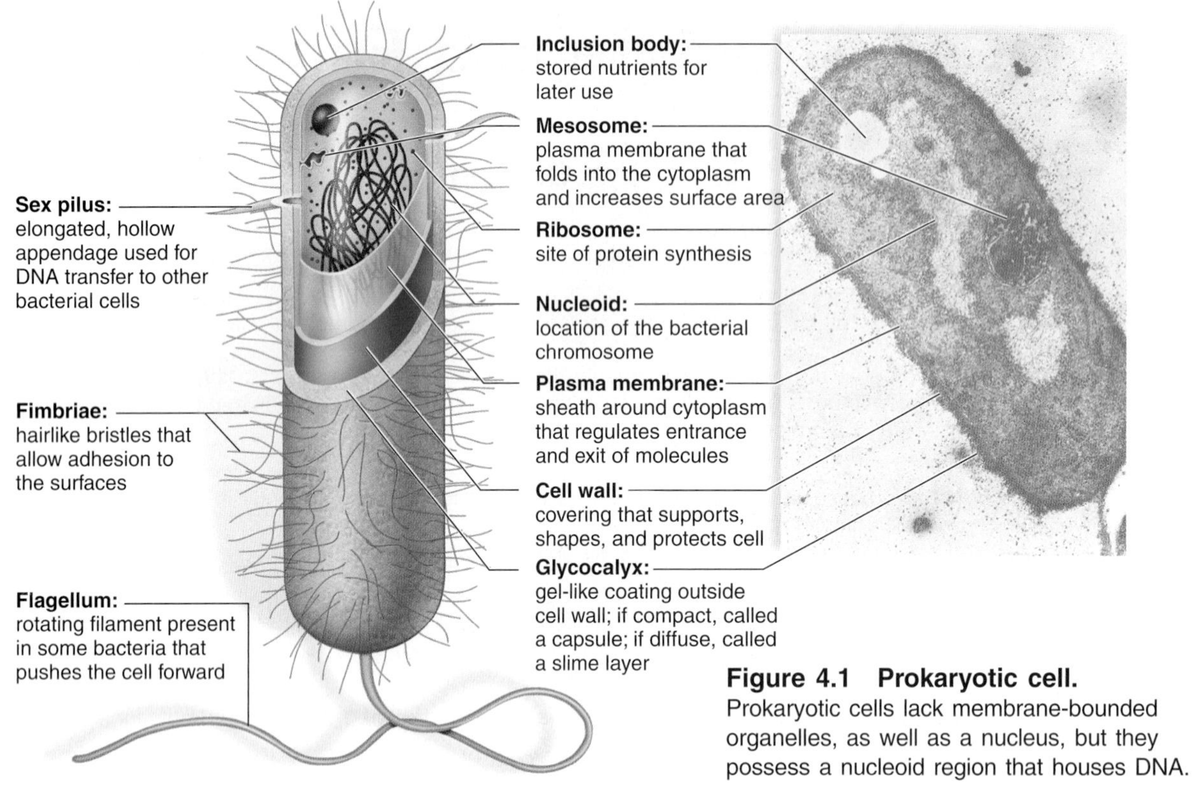
Inclusion body:
stored nutrients for later use
Mesosome:
plasma membrane that folds into the cytoplasm and increases surface area
Ribosome:
site of protein synthesis
Nucleoid:
location of the bacterial chromosome
Plasma membrane:
sheath around cytoplasm that regulates entrance and exit of molecules
Cell wall:
covering that supports, shapes, and protects cell
Glycocalyx:
gel-like coating outside cell wall; if compact, called a capsule; if diffuse, called a slime layer
Sex pilus:
elongated, hollow appendage used for DNA transfer to other bacterial cells
Fimbriae:
hairlike bristles that allow adhesion to the surfaces
Flagellum:
rotating filament present in some bacteria that pushes the cell forward

Figure 4.1 Prokaryotic cell. Prokaryotic cells lack membrane-bounded organelles, as well as a nucleus, but they possess a nucleoid region that houses DNA.

Observation: Prokaryotic/Eukaryotic Cells

Two microscope slides will show you the main difference between prokaryotic and eukaryotic cells.

1. Examine a prepared slide of a prokaryote. Notice that there are no nuclei in these cells. Sketch this cell.
2. Examine a prepared slide of cuboidal cells from a human kidney (see Fig. 11.3). Notice that you can make out a nucleus. Sketch this cell next to the prokaryotic cell.

4.2 Animal Cell and Plant Cell Structure

Figure 4.2 shows a human cheek epithelial cell as viewed by an ordinary compound light microscope available in general biology laboratories. The presence of the nucleus near the center of the cell clearly distinguishes the cheek cell as being eukaryotic. This figure also shows that the content of a cell, also called cytoplasm in eukaryotic cells, is bounded by a plasma membrane similar to those found in prokaryotic cells. Besides serving as the outer boundary of an animal cell, the plasma membrane regulates the movement of molecules into and out of the cytoplasm. In this lab, we will study how the passage of water into or out of a cell depends on the difference in concentration of solutes (particles) between the cytoplasm and the surrounding

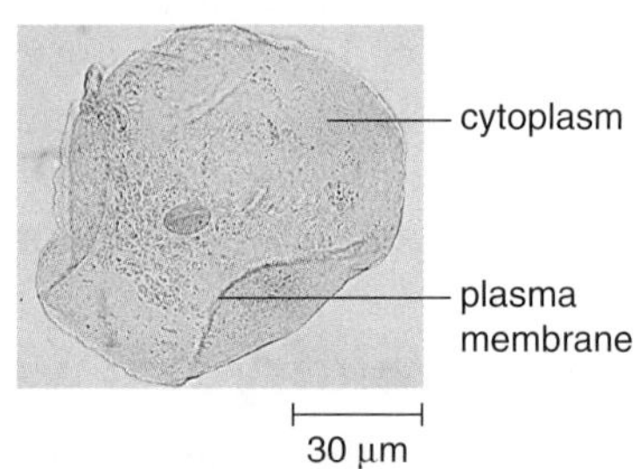
cytoplasm
plasma membrane
30 μm

Figure 4.2 Photomicrograph of an epithelial cell. (Magnification ×250)

Table 4.1 Eukaryotic Structures in Animal Cells and Plant Cells

Name	Composition	Function
Cell wall*	Contains cellulose fibrils	Support and protection
Plasma membrane	Phospholipid bilayer with embedded proteins	Definition of cell boundary; regulation of molecule passage into and out of cell
Nucleus	Nuclear envelope surrounding nucleoplasm, chromatin, and nucleoli	Storage of genetic information; synthesis of DNA and RNA
Nucleolus	Concentrated area of chromatin, RNA, and proteins	Ribosomal formation
Ribosome	Protein and RNA in two subunits	Protein synthesis
Endoplasmic reticulum (ER)	Membranous, flattened channels and tubular canals	Synthesis and/or modification of proteins and other substances; transport by vesicle formation
Rough ER	Studded with ribosomes	Protein synthesis
Smooth ER	Has no ribosomes	Various functions; lipid synthesis in some cells
Golgi apparatus	Stack of membranous saccules	Processing, packaging, and distribution of proteins and lipids
Vacuole and vesicle	Membranous sacs	Storage of substances
Lysosome	Membranous vesicle containing digestive enzymes	Intracellular digestion
Peroxisome	Membranous vesicle containing specific enzymes	Various metabolic tasks
Mitochondrion	Membranous cristae bounded by an outer membrane	Cellular respiration
Chloroplast*	Membranous cristae bounded by two membranes	Photosynthesis
Cytoskeleton	Microtubules, intermediate filaments, and actin filaments	Shape of cell and movement of its parts
Cilia and flagella	9 + 2 pattern of microtubules	Movement of cell
Centriole**	9 + 0 pattern of microtubules	Formation of basal bodies

*Plant cells only

**Animal cells only

medium or solution. The well-being of cells also depends upon the pH of the solution surrounding them. We will see how a buffer can maintain the pH within a narrow range and how buffers within cells can protect them against damaging pH changes.

Because a photomicrograph shows only a minimal amount of detail, it is necessary to turn to the electron microscope to study the contents of a cell in greater depth. The models of plant and animal cells available in the laboratory today are based on electron micrographs.

Study Table 4.1 to determine structures unique to plant cells and unique to animal cells and write them below the examples given.

	Plant Cells	**Animal Cells**
Unique structures:	1. Large central vacuole	Small vacuoles
	2. ____________	____________
	3. ____________	____________

Animal Cell Structure

With the help of Table 4.1, give a function for each of these structures, and label Figure 4.3.

Structure	Function
Plasma membrane	
Nucleus	
Nucleolus	
Ribosomes	
Endoplasmic reticulum	
Rough ER	
Smooth ER	
Golgi apparatus	
Vesicles	
Lysosome	
Mitochondrion	
Centriole	

Figure 4.3 Animal cell structure.

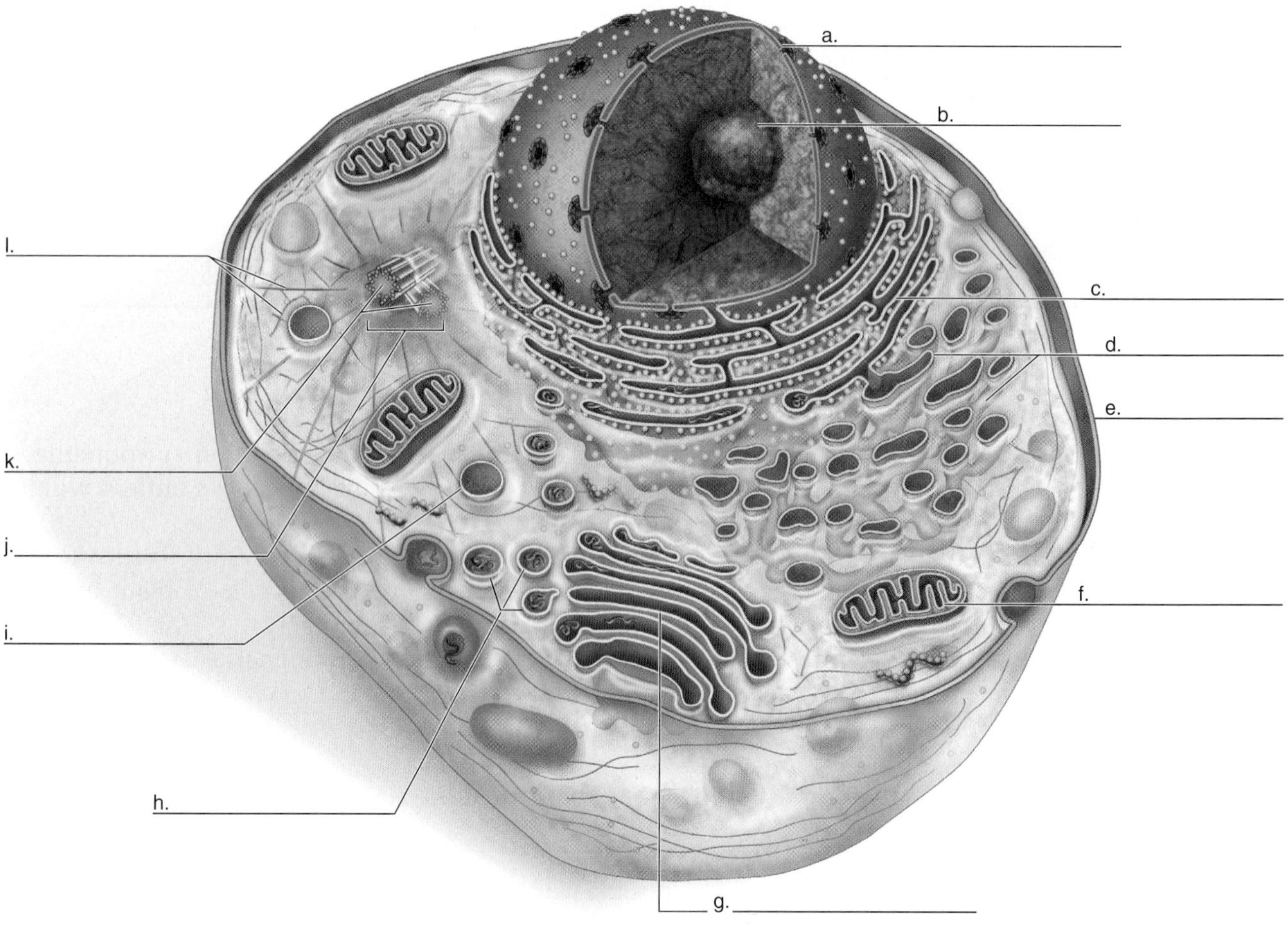

Plant Cell Structure

With the help of Table 4.1, give a function for these structures that are unique to plant cells, and label Figure 4.4.

Structure	Function
Cell wall	
Central vacuole, large	
Chloroplasts	

Figure 4.4 Plant cell structure.

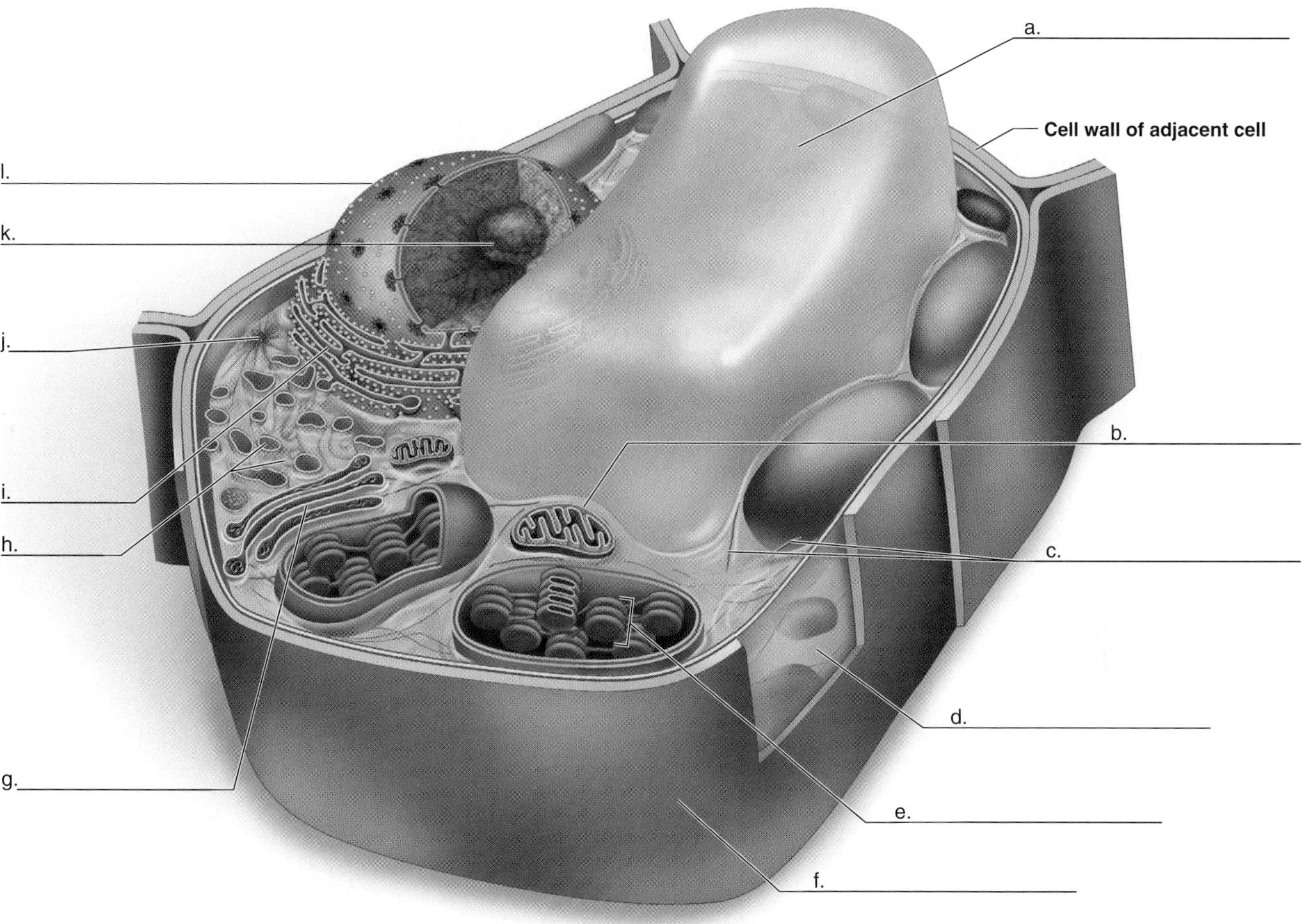

Observation: Plant Cell Structure

1. Prepare a wet mount of a small piece of young *Elodea* leaf in fresh water. *Elodea* is a multicellular, eukaryotic plant found in freshwater ponds and lakes.
2. Have the drop of *water* ready on your slide so that the leaf does not dry out, even for a few seconds. Take care that the leaf is mounted with its top side up.
3. Examine the slide using low power, focusing sharply on the leaf surface.
4. Select a cell with numerous chloroplasts for further study, and switch to high power.
5. Carefully focus on the side and end walls of the cell. The chloroplasts appear to be only along the sides of the cell because the large, fluid-filled, membrane-bounded central vacuole pushes the cytoplasm against the cell walls (Fig. 4.5*a*). Then focus on the surface and notice an even distribution of chloroplasts (Fig. 4.5*b*).
6. Can you locate the cell nucleus? ___________ It may be hidden by the chloroplasts, but when visible, it appears as a faint, grey lump on one side of the cell.
7. Can you detect movement of chloroplasts in this cell or any other cell? ___________ The chloroplasts are not moving under their own power but are being carried by a streaming of the nearly invisible cytoplasm.
8. Save your slide for use later in this laboratory.

Figure 4.5 ***Elodea*** **cell structure.**

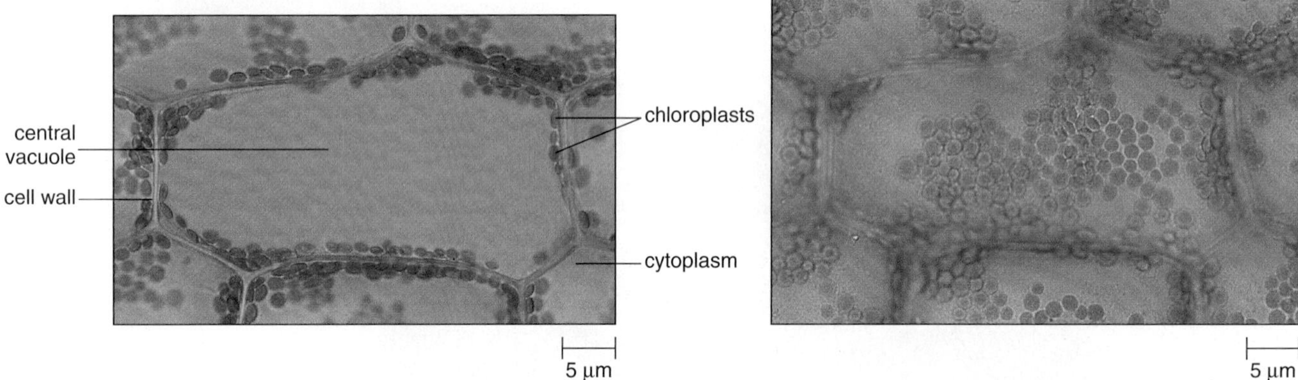

a. Middle of the cell, with chloroplasts visible around the perimeter and not in the center, which is occupied by a membrane-bounded, fluid-filled, central vacuole.

b. Upper surface of cells, showing chloroplasts in the middle, as well as around the perimeter.

4.3 Diffusion

Diffusion is the movement of molecules from a higher to a lower concentration until equilibrium is achieved and the molecules are distributed equally (Fig. 4.6). At this point, molecules may still be moving back and forth, but there is no net movement in any one direction.

Diffusion is a general phenomenon in the environment. The speed of diffusion is dependent on such factors as the temperature, the size of the molecule, and the type of medium.

Figure 4.6 Process of diffusion.
Diffusion is apparent when dye molecules have equally dispersed.

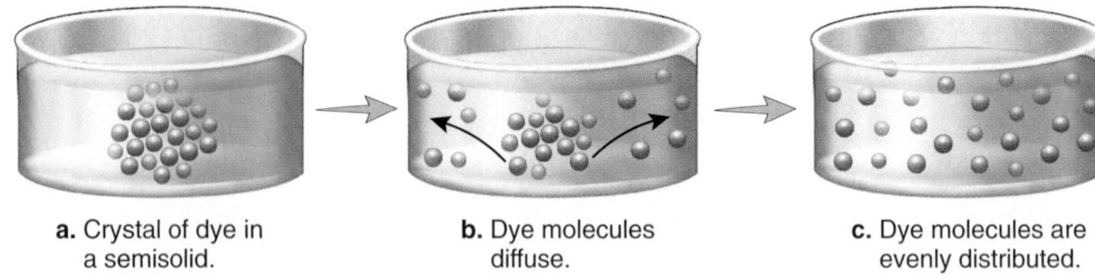

a. Crystal of dye in a semisolid.

b. Dye molecules diffuse.

c. Dye molecules are evenly distributed.

Experimental Procedure: Diffusion

> **Caution:** **Methylene blue** Avoid ingestion, inhalation, and contact with skin, eyes, and mucous membranes. Exercise care in using this chemical. If any should spill on your skin, wash the area with mild soap and water. Methylene blue will also stain clothing. Follow your instructor's directions for disposal of this chemical.
>
> **Potassium permanganate ($KMnO_4$)** $KMnO_4$ is highly poisonous and is a strong oxidizer. Avoid contact with skin and eyes, and with combustible materials. If spillage occurs, wash all surfaces thoroughly. $KMnO_4$ will also stain clothing.

Diffusion Through a Semisolid

1. Observe a petri dish containing 1.5% gelatin (or agar) to which potassium permanganate ($KMnO_4$) was added in the center depression at the beginning of the lab.
2. Obtain *time zero* from your instructor, and record *time zero* and the *final time* (now) in Table 4.2. Calculate the length of time in hours and minutes. Convert the time to hours: ________ hr.
3. Using a ruler placed over the petri dish, measure (in mm) the movement of color from the center of the depression outward in one direction: ________ mm.
4. Calculate the speed of diffusion: ________ mm/hr.
5. Record all data in Table 4.2.

Diffusion Through a Liquid

1. Add enough water to cover the bottom of a glass petri dish.
2. Place the petri dish over a thin, flat ruler.
3. With tweezers, add a crystal of potassium permanganate ($KMnO_4$) directly over a millimeter measurement line. Note the *time zero* in Table 4.2.
4. After 10 minutes, note the distance the color has moved. Record the *final time, length of time,* and *distance moved* in Table 4.2.
5. Multiply the length of time and the distance moved by 6 in order to calculate the *speed of diffusion:* ________ mm/hr. Record in Table 4.2.

Diffusion Through Air

1. Measure the distance from a spot designated by your instructor to your laboratory work area today. Record this distance in the fifth column of Table 4.2.
2. Record *time zero* in Table 4.2 when a perfume or similar substance is released into the air.
3. Note the time when you can smell the perfume. Record this as the *final time* in Table 4.2. Calculate the *length of time* since the perfume was released, and record it in Table 4.2.
4. Calculate the speed of diffusion: ________ mm/hr. Record in Table 4.2.

Table 4.2 Speed of Diffusion

Medium	Time Zero	Final Time	Length of Time (hr)	Distance Moved (mm)	Speed of Diffusion (mm/hr)
Semisolid					
Liquid					
Air					

Conclusions

- In which experiment was diffusion the fastest? ______________________________
- What accounts for the difference in speed? ______________________________

Diffusion Across the Plasma Membrane

While simple diffusion can happen in air or water, as we saw in the previous exercise, when the process occurs within living organisms, the plasma membrane can limit or enhance the process. Some molecules can diffuse across a plasma membrane and some cannot. In general, small, non-charged molecules can cross a membrane by simple diffusion, but large molecules cannot diffuse across a membrane. In organisms, many dissolved substances, known as **solutes,** are transported dissolved in water, a **solvent.** If a solute diffuses across a membrane, we refer to this type of diffusion as **dialysis.** If a solvent diffuses across a membrane, the process is referred to as **osmosis.** These processes can happen together or the membrane may favor one process over the other. The dialysis tube membrane in this experimental procedure simulates a plasma membrane.

Experimental Procedure: Diffusion Across the Plasma Membrane

At the start of the experiment,

1. Cut a piece of dialysis tubing approximately 12 cm long. Soak the tubing in water until it is soft and pliable.
2. Close one end of the dialysis tubing with a plastic clamp.
3. Fill the bag halfway with *glucose solution.*
4. Add 4 full droppers of *starch solution* to the bag.
5. Hold the open end while you mix the contents of the dialysis bag. Rinse off the outside of the bag with *distilled water.*
6. Fill a beaker 2/3 full with *distilled water.*
7. Add droppers of *iodine solution* (IKI) to the water in the beaker until an amber (tealike) color is apparent.
8. Record the color of the solution in the beaker in Table 4.3.
9. Place the bag in the beaker with the open end hanging over the edge. Secure the open end of the bag to the beaker with a rubber band as shown (Fig. 4.7). Make sure the contents do not spill into the beaker.

After about five minutes, at the end of the experiment,

10. You will note a color change. Record the color of the bag contents in Table 4.3.
11. Mark off a test tube at 1 cm and 3 cm.

12. Draw solution from near the bag and at the bottom of the beaker for testing with Benedict's reagent. Fill the test tube to the first mark with this solution. Add *Benedict's reagent* to the 3 cm mark. Heat in a boiling water bath for 5–10 minutes, observe any color change, and record your results as positive or negative in Table 4.3.
13. Remove the dialysis bag from the beaker. Dispose of it and the used Benedict's reagent solution in the manner directed by your instructor.

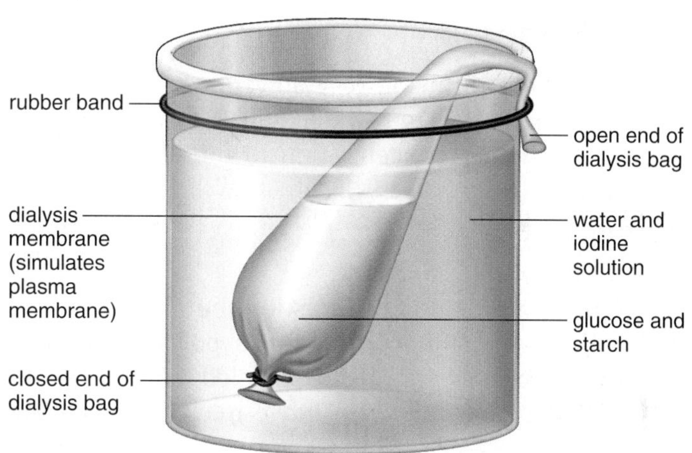

Figure 4.7 Placement of dialysis bag in water containing iodine.

> **Caution:** **Benedict's reagent** Exercise care in using this chemical. It is highly corrosive. If any should spill on your skin, wash the area with mild soap and water. Follow your instructor's directions for disposal of this chemical. Use protective eyewear when heating Benedict's reagent.

Table 4.3 Diffusion Across the Plasma Membrane

	At Start of Experiment		*At End of Experiment*		
	Contents	**Color**	**Color**	**Benedict's Test (+) or (−)**	**Conclusion**
Bag	Glucose Starch	—		—	
Beaker	Water Iodine		—		

Conclusions

- Based on the color change noted in the bag, conclude what solute diffused across the dialysis membrane from the beaker to the bag, and record your conclusion in Table 4.3.
- From the results of the Benedict's test on the beaker contents, conclude what solute diffused across the dialysis membrane from the bag to the beaker, and record your conclusion in Table 4.3.
- Which solute did not diffuse across the dialysis membrane from the bag to the beaker? ______ ______________________ Explain. ____

4.4 Osmosis

Osmosis is the diffusion of a solvent such as water across a differentially permeable membrane. Just like any other molecule, water will follow its concentration gradient and move from the area where water is in higher concentration to the area where it is in lower concentration.

Experimental Procedure: Osmosis

To demonstrate osmosis, a thistle tube is covered with a membrane at its lower opening and partially filled with 50% corn syrup (a polysaccharide) or a similar substance. The whole apparatus is placed in a beaker containing distilled water (Fig. 4.8). The water concentration in the beaker is 100%. Water molecules can move freely between the thistle tube and the beaker.

1. Note the level of liquid in the thistle tube, and measure how far it travels in 10 minutes: ________ mm.
2. Calculate the speed of osmosis under these conditions: ________ mm/hr.

Figure 4.8 Osmosis demonstration.
a. A thistle tube, covered at the broad end by a differentially permeable membrane, contains a corn syrup solution. The beaker contains distilled water. **b.** The solute is unable to pass through the membrane, but the water (arrows) passes through in both directions. There is a net movement of water toward the inside of the thistle tube, where there is a lower percentage of water molecules. **c.** Due to the incoming water molecules, the level of the solution rises in the thistle tube.

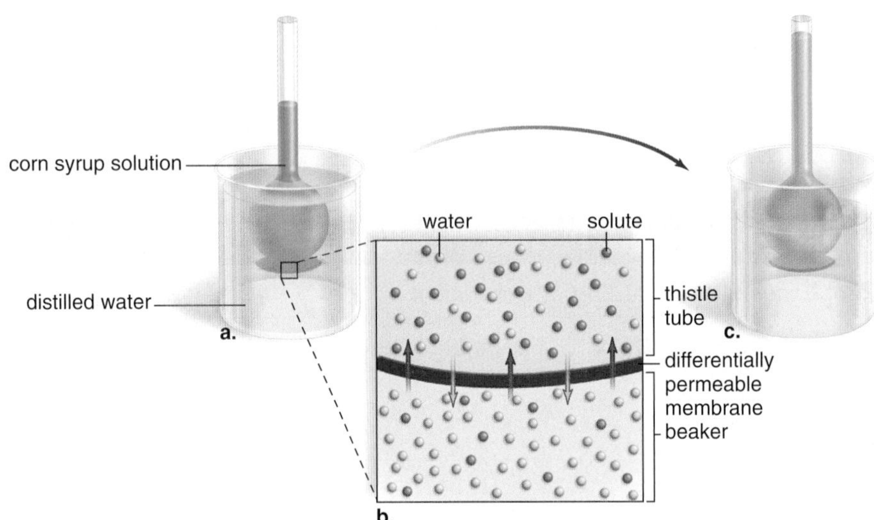

Conclusions

- In which direction was there a net movement of water? ________________________
  Explain what is meant by "net movement" after examining the arrows in Figure 4.8*b*.
  ____
- If the sugar molecules in corn syrup moved from the thistle tube to the beaker, would there have been a net movement of water into the thistle tube? ________ Why wouldn't large sugar molecules be able to move across the membrane from the thistle tube to the beaker?
  ____
- Explain why the water level in the thistle tube rose: In terms of solvent concentration, water moved from the area of ________ water concentration to the area of ________ water concentration across a differentially permeable membrane.

Tonicity

Tonicity is the relative concentration of solute (particles), and therefore also of solvent (water), outside the cell compared with inside the cell.

- An **isotonic solution** has the same concentration of solute (and therefore of water) as the cell. When cells are placed in an isotonic solution, there is no net movement of water.
- A **hypertonic solution** has a higher solute (therefore, lower water) concentration than the cell. When cells are placed in a hypertonic solution, water moves out of the cell into the solution.
- A **hypotonic solution** has a lower solute (therefore, higher water) concentration than the cell. When cells are placed in a hypotonic solution, water moves from the solution into the cell.

Experimental Procedure: Potato Strips

This procedure runs for one hour and prior setup can maximize your time efficiency.

1. Cut two strips of potato, each about 7 cm long and 1.5 cm wide.
2. Label two test tubes 1 and 2. Place one *potato strip* in each tube.
3. Fill tube 1 with *water* to cover the potato strip.
4. Fill tube 2 with 10% *sodium chloride* (NaCl) to cover the potato strip.
5. After 1 hour, remove the potato strips from the test tubes and place them on a paper towel. Observe each strip for limpness (water loss) or stiffness (water gain). Which tube has the limp potato strip? ________ Why did water diffuse out of the potato strip in this tube?

   ____

   Which tube has the stiff potato strip? ________ Why did water diffuse into the potato strip in this tube? ____

Red Blood Cells (Animal Cells)

A solution of 0.9% NaCl is isotonic to red blood cells. In such a solution, red blood cells maintain their normal appearance (Fig. 4.9*a*). A solution greater than 0.9% NaCl is hypertonic to red blood cells. In such a solution, the cells shrivel up, a process called **crenation** (Fig. 4.9*b*). A solution of less than 0.9% NaCl is hypotonic to red blood cells. In such a solution, the cells swell to bursting, a process called **hemolysis** (Fig. 4.9*c*).

Figure 4.9 Tonicity and red blood cells.

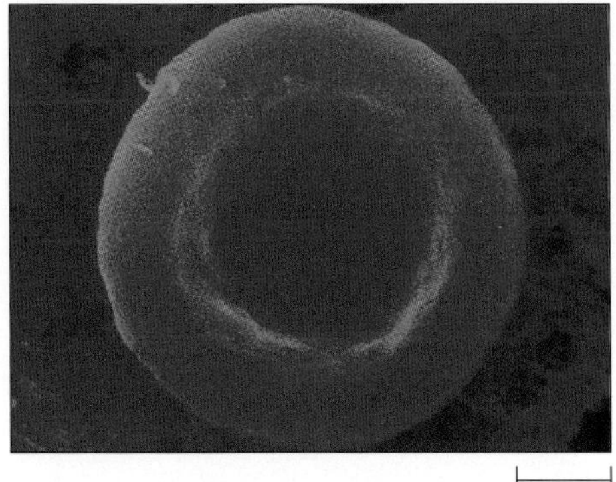

a. Isotonic solution. Red blood cell has normal appearance due to no net gain or loss of water.

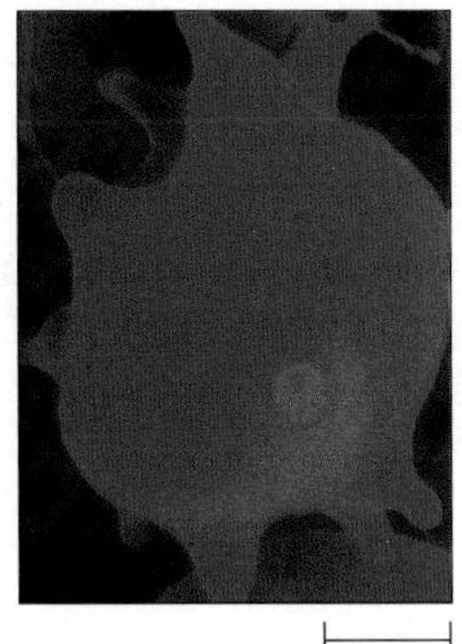

b. Hypertonic solution. Red blood cell shrivels due to loss of water.

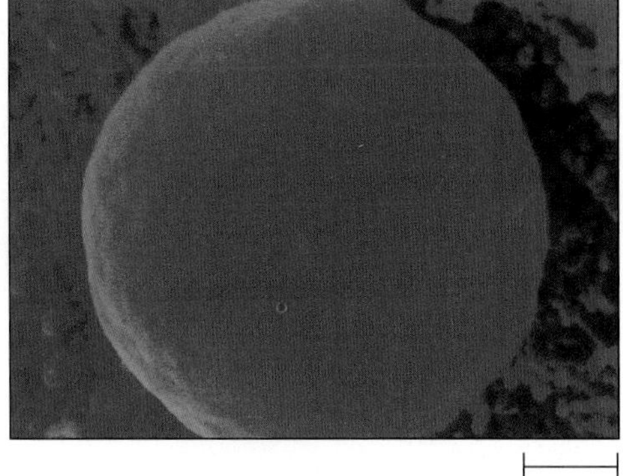

c. Hypotonic solution. Red blood cell fills to bursting due to gain of water.

Experimental Procedure: Demonstration of Tonicity in Red Blood Cells

> **Caution:** Do not remove the stoppers of test tubes during this procedure. Exercise care in handling sheep blood and wash hands thoroughly after preparing tubes.

Three stoppered test tubes on display have the following contents:

Tube 1: 0.9% NaCl plus a few drops of whole sheep blood
Tube 2: 10% NaCl plus a few drops of whole sheep blood
Tube 3: 0.9% NaCl plus distilled water and a few drops of whole sheep blood

1. In the second column of Table 4.4, record the tonicity of each tube.
2. Hold each tube in front of one of the pages of your lab manual. Determine whether you can see the print on the page through the tube. Record your findings in the third column of Table 4.4.

Table 4.4 Tonicity and Print Visibility

Tube	Tonicity	Print Visibility	Explanation
1			
2			
3			

Conclusion

- Explain in the fourth column of Table 4.4 why you can or cannot see the print.

Elodea (Plant Cells)

When plant cells are in a hypotonic solution, the large central vacuole gains water and exerts pressure, called **turgor pressure.** The cytoplasm, including the chloroplasts, is pushed up against the cell wall (Fig. 4.10*a*).

When plant cells are in a hypertonic solution, the central vacuole loses water, and the cytoplasm, including the chloroplasts, pulls away from the cell wall. This is called **plasmolysis** (Fig. 4.10*b*).

Experimental Procedure: Elodea Cells

Hypotonic Solution

1. If possible, use the *Elodea* slide you prepared earlier in this laboratory. If not, prepare a new wet mount of a small *Elodea* leaf using fresh water.
2. After several minutes, focus on the surface of the cells, and compare your slide with Figure 4.10*a*.
3. Complete the portion of Table 4.5 that pertains to a hypotonic solution.

Hypertonic Solution

1. Prepare a new wet mount of a small *Elodea* leaf using a 10% NaCl solution.
2. After several minutes, focus on the surface of the cells, and compare your slide with Figure 4.10*b*.
3. Complete the portion of Table 4.5 that pertains to a hypertonic solution.

Table 4.5 Effect of Tonicity on *Elodea* Cells

Tonicity	Appearance of Cells	Scientific Term
Hypotonic		
Hypertonic		

Conclusions

- In a hypotonic solution, the large central vacuole of plant cells exerts ________ pressure, and the chloroplasts are seen ________ the cell wall.
- In a hypertonic solution, the central vacuole loses water, and the chloroplasts are seen ________ the cell wall.

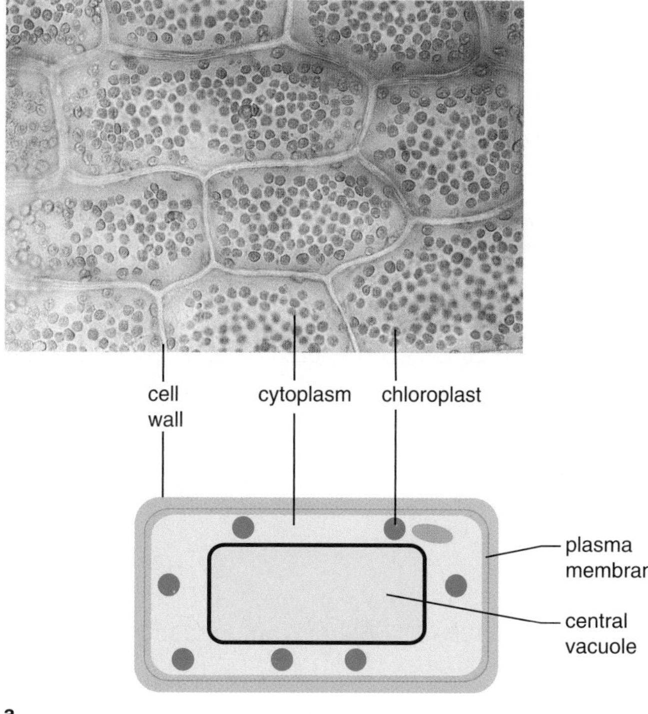

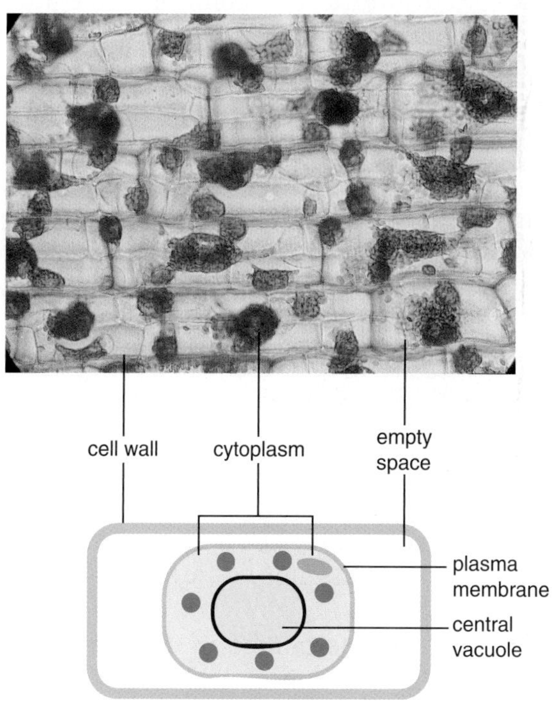

Figure 4.10 *Elodea* cells.
a. Surface view of cells in a hypotonic solution (*above*) and longitudinal section diagram (*below*). The large, central vacuole, filled with water, pushes the cytoplasm, including the chloroplasts, right up against the cell wall. **b.** Surface view of cells in a hypertonic solution (*above*) and longitudinal section diagram (*below*). When the central vacuole loses water, cytoplasm, including the chloroplasts, piles up in the center of the cell because the cytoplasm has pulled away from the cell wall. (*a:* Magnification ×400)

4.5 pH and Cells

The pH of a solution tells its hydrogen ion concentration $[H^+]$. The **pH scale** ranges from 0 to 14. A pH of 7 is neutral. A pH lower than 7 indicates that the solution is acidic (has more hydrogen ions than hydroxide ions), whereas a pH greater than 7 indicates that the solution is basic (has more hydroxide ions than hydrogen ions). A **buffer** is a substance that resists changes in pH. Maintaining stable pH levels in their cells and tissues is a major concern for living organisms and buffers help to accomplish this.

As an example, consider that in humans the pH of the blood must be maintained at about 7.4 or we become ill.

Why are cells and organisms buffered? ______________________________

____

Experimental Procedure: pH and Cells

Caution: Hydrochloric acid (HCl) used to produce an acid pH is a strong, caustic acid. Exercise care in using this chemical. If any HCl spills on your skin, rinse immediately with clear water. Follow your instructor's directions for disposal of tubes that contain HCl. Use protective eyewear when performing this experiment.

1. Label three test tubes, and fill them to the halfway mark as follows: tube 1: *water;* tube 2: *buffer* (inorganic) solution; and tube 3: *simulated cytoplasm* (buffered protein solution).
2. Use pH paper to determine the pH of each tube. Dip the end of a stirring rod into the solution, and then touch the stirring rod to a 5 cm strip of pH paper. Read the current pH by matching the color observed with the color code on the pH paper package. Record your results in the "pH Before Acid" column in Table 4.6.
3. Add one drop of 0.1 N hydrochloric acid (HCl) to each tube. Shake or swirl. Use pH paper as in step 2 to determine the new pH of each solution. (Do as many pH determinations as possible on your pH paper before using a clean strip.) Record your results in the "pH After Acid" column in Table 4.6.

Table 4.6 pH and Cells

Tube	Contents	pH Before Acid	pH After Acid	Explanation
1	Water			
2	Buffer			
3	Cytoplasm			

Conclusion

- Use the information in column two to explain your test results. Write your explanations in the last column of Table 4.6.

Experimental Procedure: Buffer Strength

1. Pour the contents of tube 2 into a clean 50 ml beaker labeled "2." Pour the contents of tube 3 into a clean 50 ml beaker labeled "3."
2. Add hydrochloric acid in 5-drop increments to beakers 2 and 3. Mix and test the pH after every 5 drops, and record these values in Table 4.7.
3. Record the patterns of pH changes on the graphs provided. Use the top graph for the pattern observed with the inorganic buffer and the bottom graph for the pattern observed with simulated cytoplasm.

Table 4.7 Drops of HCl and pH Values

# Drops	*pH* **Beaker 2 (Buffer)**	*pH* **Beaker 3 (Cytoplasm)**
0		
5		
10		
15		
20		
25		
30		
35		
40		
45		
50		

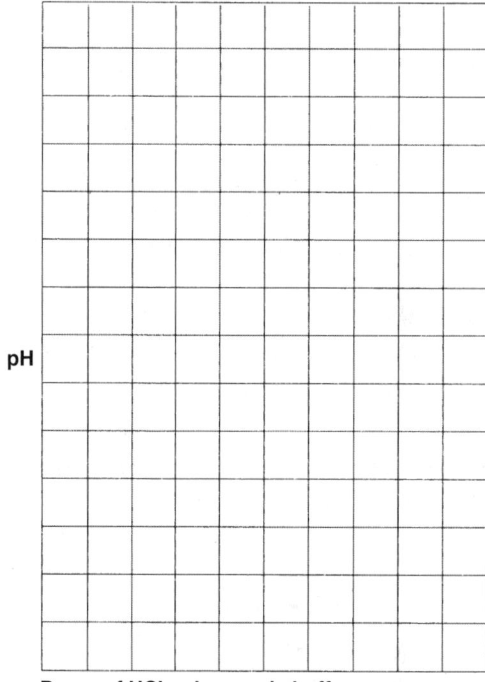

Drops of HCl — Inorganic buffer

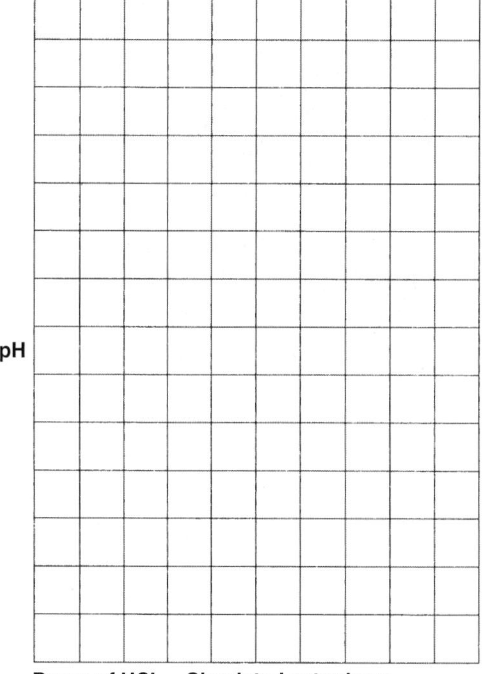

Drops of HCl — Simulated cytoplasm

Conclusion

- Do the two graphs have a similar pattern?

  ____

  Explain.

  ____

Laboratory Review 4

______________ 1. What is the name of the large, often central organelle in eukaryotic cells that contains chromosomes?

______________ 2. Prokaryotes have what structures necessary for protein synthesis?

______________ 3. What is the function of the nucleus?

______________ 4. What is the function of rough endoplasmic reticulum?

______________ 5. Which organelle carries on intracellular digestion?

______________ 6. Name a structure present in an animal cell but not in a plant cell.

______________ 7. Name a structure present in a plant cell but not in an animal cell.

______________ 8. What term describes the movement of molecules from an area of higher concentration to one of lower concentration?

______________ 9. What is the name for the movement of water across the plasma membrane?

______________ 10. In what direction does water move when cells are placed in a hypertonic solution?

______________ 11. Is 10% NaCl isotonic, hypertonic, or hypotonic to red blood cells?

______________ 12. What appearance will red blood cells have when they are placed in 9.0% NaCl?

______________ 13. What scientific term is used to refer to the condition of cells described in question 12?

______________ 14. What type of molecule prevents extensive changes in the pH of living organisms?

______________ 15. If acid is added to water, does the pH increase or decrease?

______________ 16. What is a pH of 7 called?

______________ 17. Name two features or cellular components that all cells have in common.

Thought Questions

18. The police are trying to determine whether material removed from a crime scene is plant or animal matter. What would you suggest they look for?

19. Your grandmother asks you to fertilize her favorite plant. Without reading the directions on the box, you pour some fertilizer into the pot and then water the plant. The next time you see your grandmother, she tells you the plant died. In terms of osmosis, explain what happened to the plant.

20. Explain what happens to both plant and animal cells when they are placed into a solution that is hypotonic to the interior of the cell. If the two cells meet different fates, explain why.

LABORATORY

5

Mitosis and Meiosis

Learning Outcomes

5.1 The Cell Cycle

- Explain how mitosis is related to the cell cycle.
- Discuss the phases of mitosis and describe the characteristic events in each phase.
- Recognize the specific phases of mitosis on microscope slides, models, and illustrations.
- Contrast the events of mitosis in plant cells and animal cells, providing an explanation for any noted differences.

5.2 Meiosis

- Describe the behavior of the chromosomes during each phase of meiosis I.
- Describe the behavior of the chromosomes during each phase of meiosis II.

5.3 Mitosis Versus Meiosis

- Based upon a comparison of the behavior of the chromosomes in the phases of mitosis with the behavior of the chromosomes in each of the two stages of meiosis, provide evidence for which meiotic division has more similarity to mitosis.
- Establish why crossing-over can only occur in meiosis, and not in mitosis.

5.4 Gametogenesis in Animals

- Using models, slides, and illustrations, demonstrate your knowledge of the events that occur in spermatogenesis and oogenesis.

Introduction

Dividing eukaryotic cells experience nuclear division, cytoplasmic division, and a period of time between divisions called interphase. During **interphase,** the nucleus appears normal, and the cell is performing its usual cellular functions. Also, the cell is increasing all of its components, including such organelles as the mitochondria, ribosomes, and centrioles, if present. DNA replication (making an exact copy of the DNA) occurs during interphase. Thereafter, the chromosomes, which contain DNA, are duplicated and contain two chromatids held together at a **centromere.** These chromatids are called **sister chromatids.**

During nuclear division, called **mitosis,** the new nuclei receive the same number of chromosomes as the parental nucleus. When the cytoplasm divides, a process called **cytokinesis,** two daughter cells are produced. In multicellular organisms, mitosis permits growth and repair of tissues. In eukaryotic, unicellular organisms, mitosis is a form of asexual reproduction. Sexually reproducing organisms utilize another form of nuclear division, called **meiosis.** In animals, meiosis is a part of gametogenesis, the production of gametes (sex cells). The gametes are sperm in male animals and eggs in female animals. As a result of meiosis, the daughter cells have half the number of chromosomes as the parental cell. Because crossing-over of genetic material takes place and the chromosomes occur in various combinations in the daughter cells, meiosis contributes to recombination of genetic material and to variation among sexually reproducing organisms.

This laboratory examines both mitotic and meiotic cell division to show their similarities and differences. At the start of both types of divisions, the parental nucleus, surrounded by a double membrane (the nuclear envelope), contains one or more **nucleoli** (concentrated regions of RNA) and **chromatin** (threadlike strands of DNA) suspended in a transparent liquid called **nucleoplasm.** During division, chromatin condenses so that the chromosomes are visible, the nuclear envelope fragments, and a spindle appears. Spindle fibers assist the movement of chromosomes, which occurs at this time.

5.1 The Cell Cycle

As stated in the Introduction, the period of time between cell divisions is known as interphase. Because early investigators noted little visible activity between cell divisions, they dismissed this period of time as a resting state. But when they discovered that DNA replication and chromosome duplication occur during interphase, the **cell cycle** concept was proposed. Investigators have also discovered that cytoplasmic organelle duplication occurs during interphase, as does synthesis of the proteins involved in regulating cell division. Thus, the cell cycle can be broken down into four stages (Fig. 5.1). State the event of each stage on the line provided:

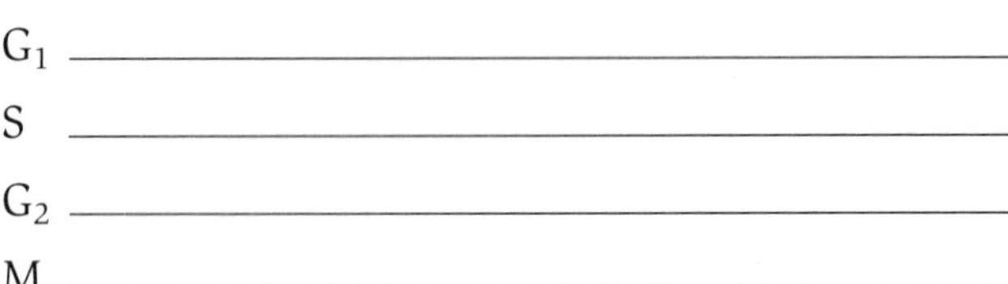

G_1 ______________________________

S ______________________________

G_2 ______________________________

M ______________________________

The length of time required for the entire cell cycle varies according to the organism, but 18–24 hours is typical for animal cells. Mitosis (including cytokinesis, if it occurs) lasts less than an hour to slightly more than 2 hours; for the rest of the time, the cell is in interphase.

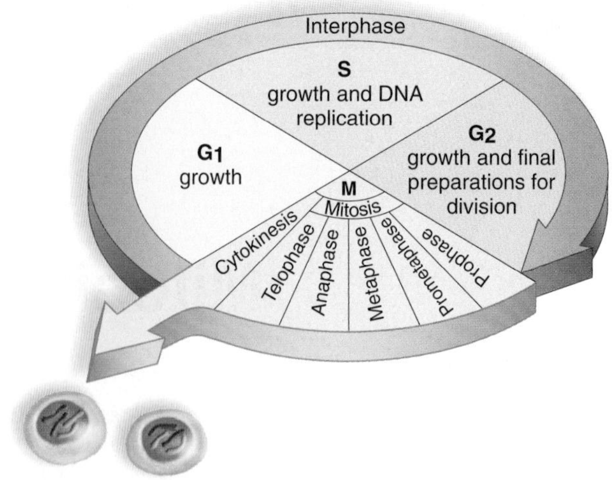

Figure 5.1 The cell cycle.
Immature cells go through a cycle that consists of four stages: G_1, S (for synthesis), G_2, and M (for mitosis). Eventually, some daughter cells "break out" of the cell cycle and become specialized cells.

Table 5.1 Structures Associated with Mitosis

Structure	Description
Nucleus	A large organelle containing the chromosomes and acting as a control center for the cells
Chromosome	Rod-shaped body in the nucleus that is seen during mitosis and meiosis and that contains DNA and therefore the hereditary units, or genes
Nucleolus	An organelle found inside the nucleus; composed largely of RNA for ribosome formation
Spindle	Microtubule structure that brings about chromosome movement during cell division
Chromatids	The two identical parts of a chromosome following DNA replication
Centromere	A constriction where duplicates (sister chromatids) of a chromosome are held together
Centrosome	The central microtubule-organizing center of cells; consists of granular material; in animal cells, contains two centrioles
Centriole*	A short, cylindrical organelle in animal cells that contains microtubules and is associate with the formation of the spindle during cell division
Aster*	Short, radiating fibers produced by the centrioles; important during mitosis and meiosis

*Animal cells only

Mitosis

Mitosis is nuclear division that results in two new nuclei, each having the same number of chromosomes as the original nucleus. The **parental cell** is the cell that divides, and the resulting cells are called **daughter cells.**

When cell division is about to begin, chromatin starts to condense and compact to form visible, rodlike sister chromatids that are held together at the centromere (Fig. 5.2*a*). Label the sister chromatids and the centromere in the drawing of a duplicated chromosome in Figure 5.2*b*. This illustration represents a chromosome as it would appear just before nuclear division occurs.

Figure 5.2 Duplicated chromosomes.
DNA replication results in a duplicated chromosome that consists of two sister chromatids held together at a centromere. **a**. Scanning electron micrograph of a duplicated chromosome. **b**. Drawing of a duplicated chromosome.

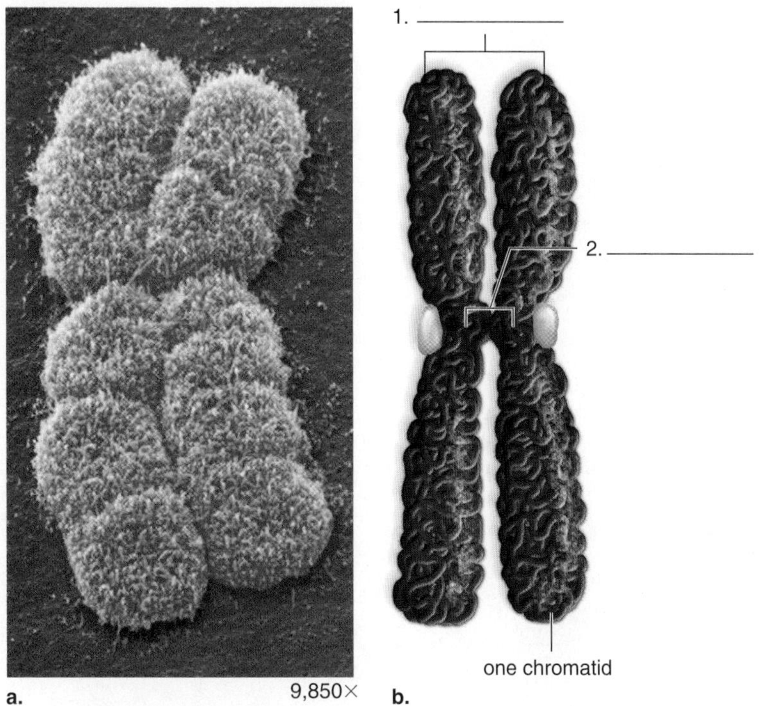

Spindle

Table 5.1 lists the structures that play a role during mitosis. The spindle is a structure that appears and brings about an orderly distribution of chromosomes to the daughter cell nuclei. A spindle has fibers that stretch between two poles (ends). Spindle fibers are bundles of microtubules, which are protein cylinders found in the cytoplasm that can assemble and disassemble. The **centrosome,** which is the main microtubule-organizing center of the cell, divides before mitosis so that each pole of the spindle has a pair of centrosomes. Animal cells contain two barrel-shaped organelles called centrioles in each centrosome and asters, which are arrays of short microtubules radiating from the poles (see Fig. 5.3). The fact that plant cells lack centrioles suggests that centrioles are not required for spindle formation.

Animal Mitosis Phases

The phases of mitosis are **prophase, metaphase, anaphase,** and **telophase**—in that order (Fig. 5.3). Early and transitional stages of prophase are also shown in this figure.

Figure 5.3 Phases of mitosis in animal and plant cells.
The colors signify that the chromosomes were inherited from different parents.

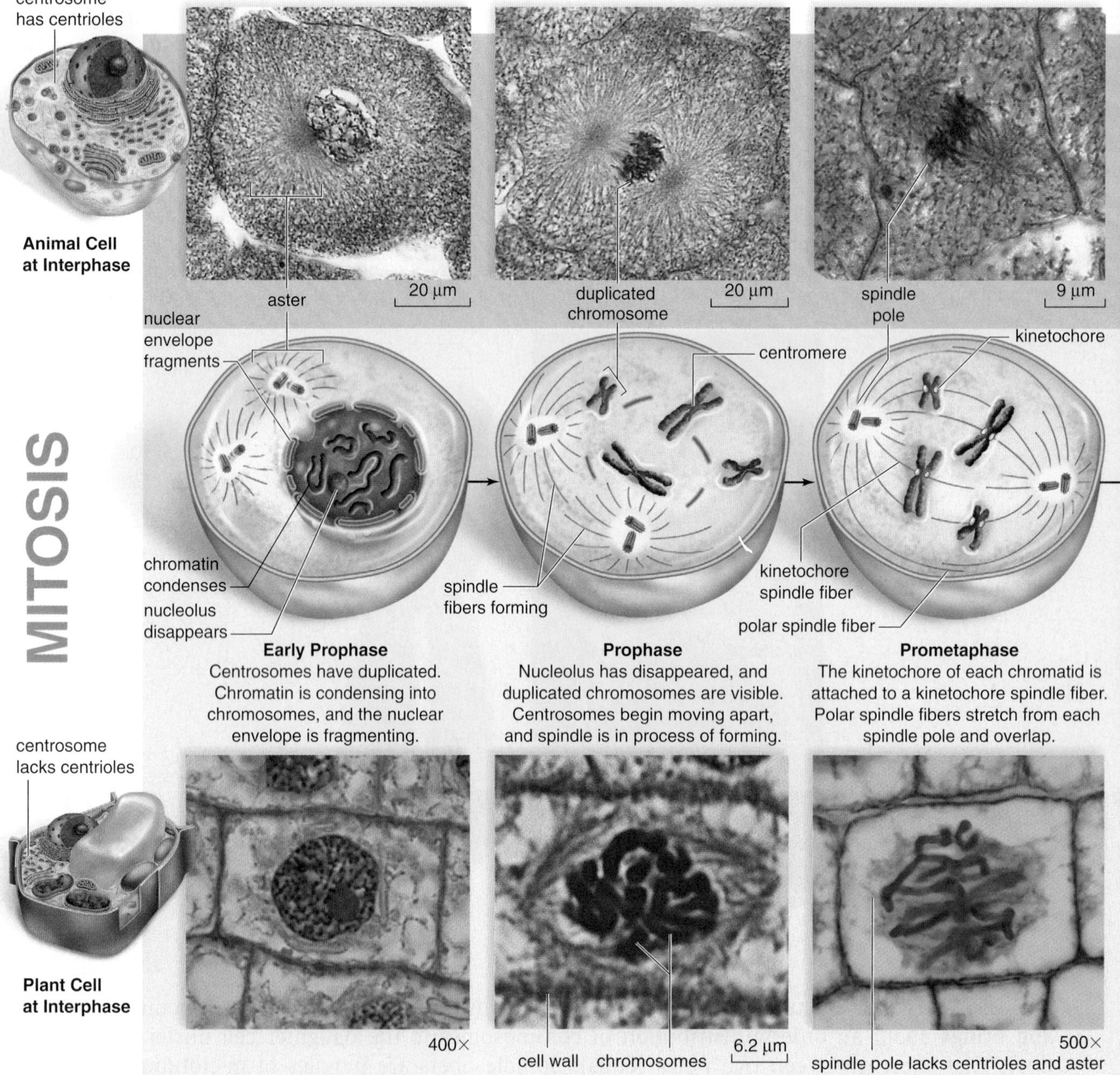

Prophase

During early prophase, the chromosomes continue to condense, the nucleolus disappears, and the nuclear envelope fragments. The spindle begins to assemble as the centrosomes, each containing two centrioles, migrate to the poles.

During prophase, the chromosomes have no apparent orientation within the cell. The already duplicated chromosomes are composed of two sister chromatids held together at a centromere. Counting the number of *centromeres* in diagrammatic drawings gives the number of chromosomes for the cell. *What is the chromosome number for the cells shown as artist's images between the plant and animal photomicrographs in Figure 5.3?* ________

During late prophase, the mitotic spindle occupies the region formerly occupied by the nucleus. Short microtubules radiate out in a starlike aster from the pair of centrioles located in each centrosome.

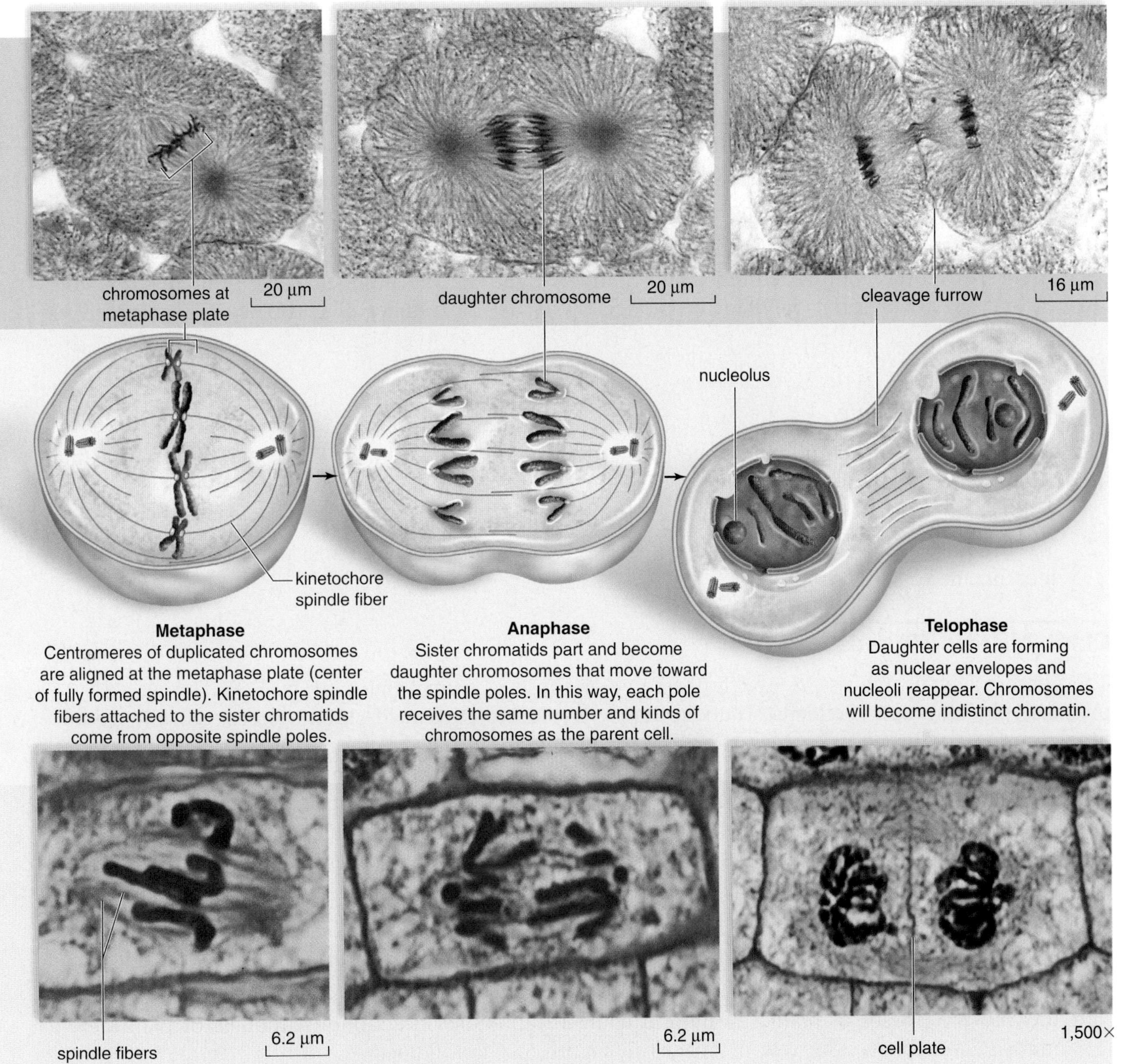

Metaphase
Centromeres of duplicated chromosomes are aligned at the metaphase plate (center of fully formed spindle). Kinetochore spindle fibers attached to the sister chromatids come from opposite spindle poles.

Anaphase
Sister chromatids part and become daughter chromosomes that move toward the spindle poles. In this way, each pole receives the same number and kinds of chromosomes as the parent cell.

Telophase
Daughter cells are forming as nuclear envelopes and nucleoli reappear. Chromosomes will become indistinct chromatin.

The spindle consists of poles, asters, and fibers, which are bundles of parallel microtubules. The chromosomes become attached to spindle fibers coming from opposite poles.

Metaphase

The sister chromatids are now attached to the spindle and the chromosomes are aligned at the equator of the spindle.

Anaphase

At the start of anaphase, the centromeres split, and the sister chromatids of each chromosome separate, giving rise to two daughter chromosomes. The daughter chromosomes begin to move toward opposite poles of the spindle. Each pole receives the diploid number of daughter chromosomes.

Telophase

New nuclear envelopes form around the daughter chromosomes at the poles. Each daughter nucleus contains the same number and kinds of chromosomes as the parental cell. The chromosomes become more diffuse chromatin once again, and a nucleolus appears in each daughter nucleus. Division of the cytoplasm by formation of a **cleavage furrow** is nearly complete.

Observation: Animal Mitosis

Animal Mitosis Models

1. Using the previous descriptions and Figure 5.3 as a guide, identify the phases of animal cell mitosis in models of animal cell mitosis.
2. Each species has its own chromosome number. Counting the number of centromeres tells you the number of chromosomes in the models. What is the number of chromosomes observed in each nucleus of the cells? ___________

Whitefish Blastula Slide

The blastula is an early embryonic stage in the development of animals. The **blastomeres** (blastula cells) that make up the top row in Figure 5.3 are in different phases of mitosis.

1. Examine a prepared slide of whitefish blastula cells undergoing mitotic cell division.
2. Try to find a cell in each phase of mitosis. Have a partner or your instructor check your identification.

Plant Mitosis Phases

The phases of plant mitosis are exactly the same as those of animal mitosis (Fig. 5.3). During early prophase, the chromatin condenses into scattered, previously duplicated chromosomes, and the spindle forms; during late prophase, chromosomes attach to spindle fibers; during metaphase, the chromosomes are at the equator of the spindle; during anaphase, the daughter chromosomes move to the poles of the spindle; and during telophase, division of the cytoplasm by formation of a **cell plate** begins.

Notice that plant cells do not have centrioles or asters. Exactly how plant cells are able to form the mitotic spindle without centrioles is a topic of considerable interest among cellular biologists.

Observation: Plant Mitosis

Plant Mitosis Models

1. Identify the phases of plant cell mitosis using the previous descriptions, models of plant cell mitosis, and the bottom row of images in Figure 5.3 as a guide.
2. What is the number of chromosomes in each of the cells in this model series? ______________

Onion Root Tip Slide

1. In plants, the root tip contains tissue that is continually dividing and producing new cells. Examine a prepared slide of onion root tip cells undergoing mitotic cell division. Try to find the phases that correspond to those shown in Figure 5.3.
2. Using high power, focus up and down on a cell in telophase. You may be able to just make out the cell plate, the region where a plasma membrane is forming between the two prospective daughter cells. Later, cell walls appear in this area.

Cytokinesis

Cytokinesis, division of the cytoplasm, usually accompanies mitosis. During cytokinesis, each daughter cell receives a share of the organelles that duplicated during interphase. Cytokinesis begins in anaphase, continues in telophase, and reaches completion by the start of the next interphase.

Cytokinesis in Animal Cells

In animal cells, a **cleavage furrow,** which is an indentation of the membrane between the daughter nuclei, begins as anaphase draws to a close (Fig. 5.4). The cleavage furrow deepens as a band of actin filaments called the contractile ring slowly constricts the cell, forming two daughter cells.

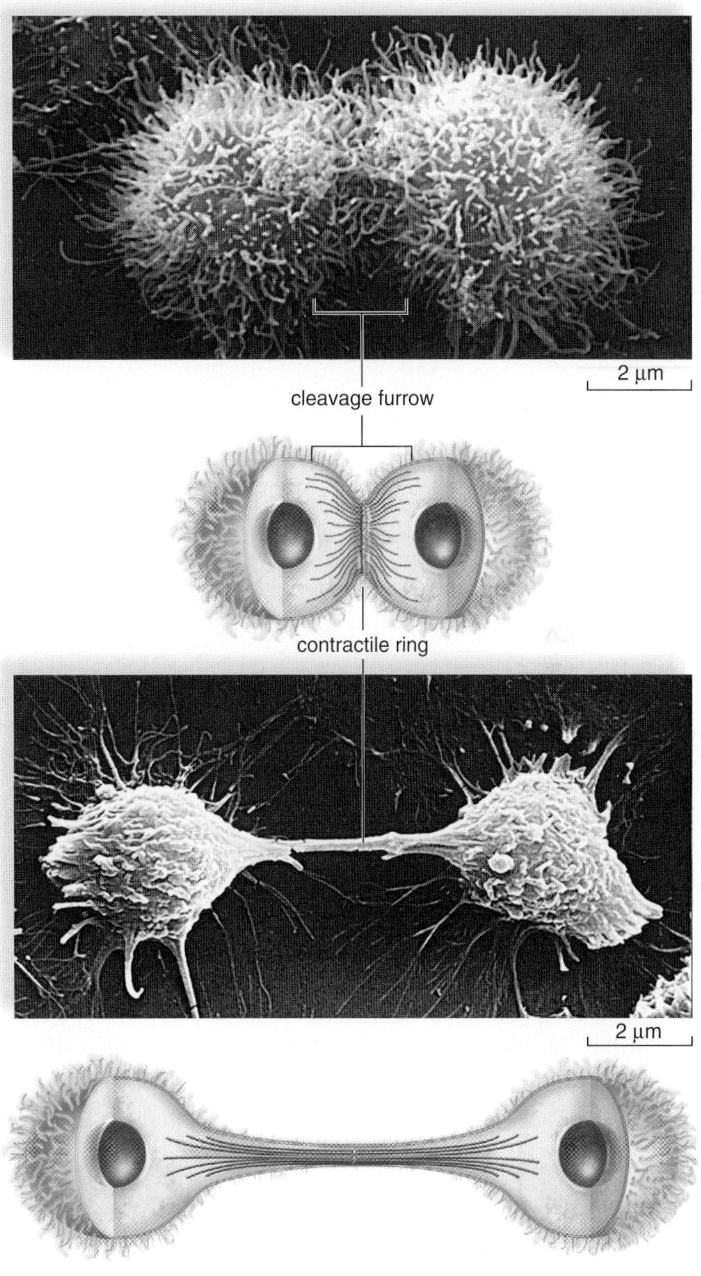

Figure 5.4 Cytokinesis in animal cells. A single cell becomes two cells by a furrowing process. A contractile ring composed of actin filaments gradually gets smaller, and the cleavage furrow pinches the cell into two cells.

Cytokinesis in Plant Cells

After mitosis, the cytoplasm divides by cytokinesis. In plant cells, membrane vesicles derived from the Golgi apparatus migrate to the center of the cell and form a **cell plate** (Fig. 5.5), the location of a new plasma membrane for each daughter cell. Later, individual cell walls appear in this area.

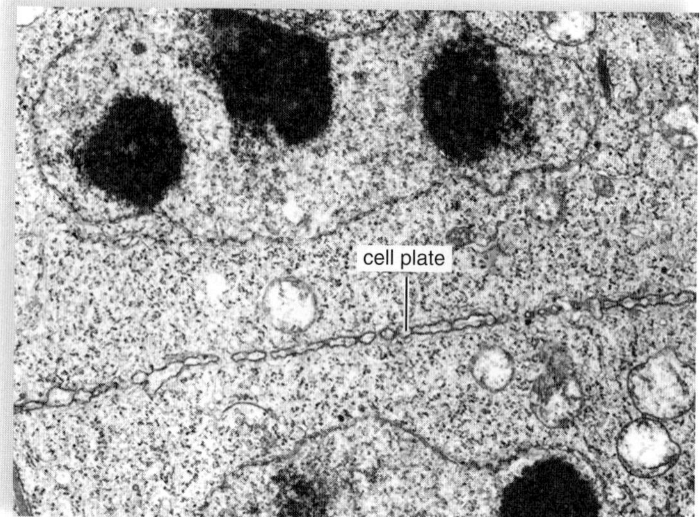

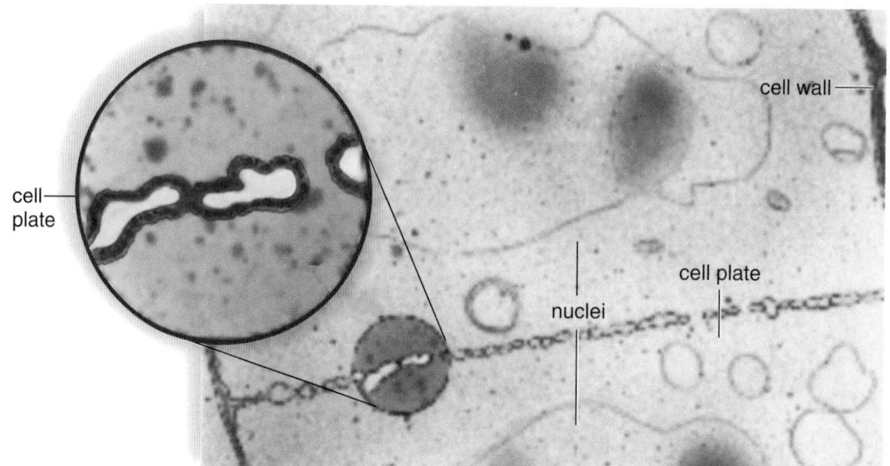

Figure 5.5 Cytokinesis in plant cells.
During cytokinesis in a plant cell, a cell plate forms midway between two daughter nuclei and extends to the plasma membrane. Vesicles containing cell wall components fuse to form cell plate.

Summary of Mitotic Cell Division

1. The nuclei in the daughter cells have the ________ number of chromosomes as the parental cell had.
2. Mitosis is cell division in which the chromosome number ____________________.

Complete Table 5.2 to compare animal and plant cell mitosis.

Table 5.2 Comparison of Mitosis in Plant and Animal Cells

Cell Type	Centrioles (Yes or No)	Aster (Yes or No)	Cytokinesis by Means of What Structure?
Animal			
Plant			

5.2 Meiosis

Meiosis is a form of nuclear division in which the chromosome number is reduced by half. While the nucleus of the parental cell has the diploid number of chromosomes, the daughter nuclei, after meiosis is complete, have the haploid number of chromosomes. In sexually reproducing species, meiosis must occur, or the chromosome number would double with each generation.

A diploid cell nucleus contains **homologues,** also called homologous chromosomes. Homologues look alike and carry the genes for the same traits. Before meiosis begins, the chromosomes are already duplicated—that is, they contain sister chromatids. Meiosis requires two divisions, called **meiosis I** and **meiosis II.**

Experimental Procedure: Meiosis

In this exercise, you will use pop beads to construct chromosomes and move the chromosomes to simulate meiosis.

Building Chromosomes to Simulate Meiosis

1. Obtain the following materials: 48 pop beads of one color (e.g., red) and 48 pop beads of another color (e.g., blue) for a total of 96 beads in all; eight magnetic centromeres; and four centriole groups.
2. Build a homologous pair of duplicated chromosomes using Figure 5.6*a* as a guide. Each chromatid will have 16 beads. Be sure to bring the centromeres of two units of the same color together so that they attract and link to form one duplicated chromosome. (One member of the pair will be red, and the other will be blue.)
3. Build another homologous pair of duplicated chromosomes using Figure 5.6*b* as a guide. Each chromatid will have eight beads. Be sure to bring the centromeres of two units of the same color together so that they attract. (One member of the pair will be red, and the other will be blue.)
4. Note that your chromosomes are the same as those in Figure 5.6. The red chromosomes were inherited from one parent, and the blue chromosomes were inherited from the other parent.

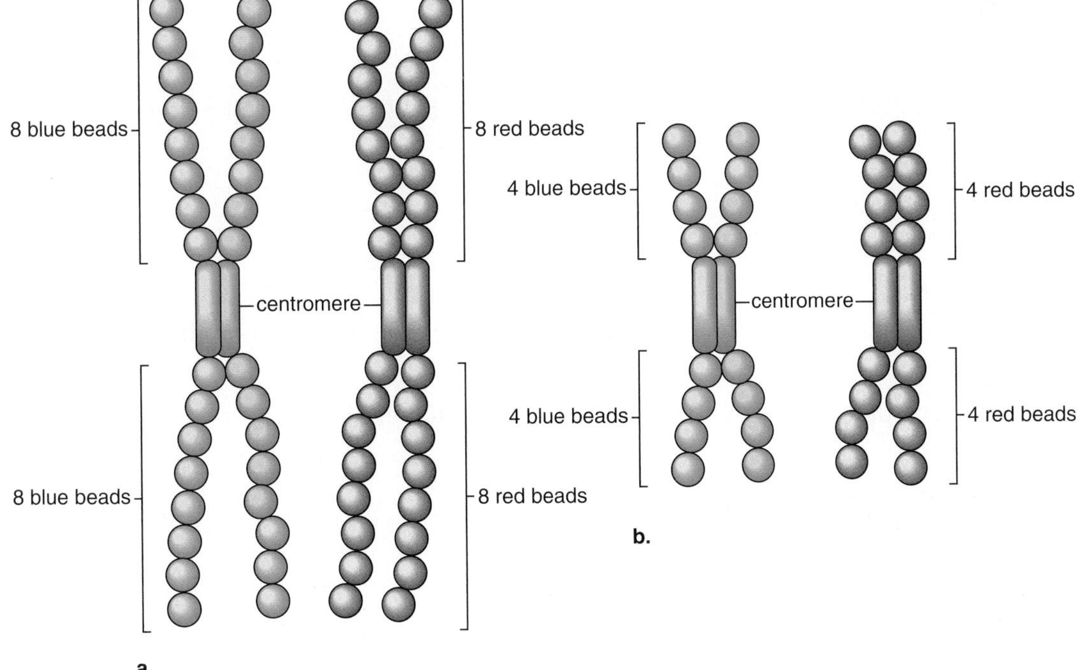

Figure 5.6 Two pairs of homologues.
The chromosomes of these homologous pairs are duplicated.

Meiosis I

During prophase of meiosis I, the spindle appears while the nuclear envelope and nucleolus disappear. Homologues line up next to one another during a process called synapsis. During **crossing-over,** the nonsister chromatids of a homologue pair exchange genetic material. At metaphase I, the homologue pairs line up at the equator of the spindle. During anaphase I, homologues separate and the chromosomes (still composed of two chromatids) move to each pole. In telophase I, the nuclear envelope and the nucleolus reappear as the spindle disappears. Each new nucleus contains one from each pair of chromosomes.

Prophase I

5. Using Figure 5.7 as a guide, put all four of the chromosomes you built in the center of your work area, which represents the nucleus. Place your two pairs of centrioles outside the nucleus.
6. Separate the pairs of centrioles, and move one pair to opposite poles of the nucleus.
7. Synapsis is the pairing of homologues during prophase I. Simulate synapsis by bringing the homologues together.
8. **Crossing-over** is an exchange of genetic material between two homologues. It is a way to achieve genetic recombination during meiosis. Simulate crossing-over by exchanging the exact segments of two nonsister chromatids of a single homologous pair. Why use nonsister chromatids and not sister chromatids? ______________________________

Metaphase I

Position the homologues at the equator in such a way that the homologues are prepared to move apart toward the centrioles.

Anaphase I

Separate the homologues, and move each one toward the opposite pole.

Telophase I

9. During telophase I, the chromosomes are at the poles. What combinations of chromosomes are at the poles? Fill in the following blanks with the words *red-long*, *red-short*, *blue-long*, and *blue-short:*

   Pole A: ______________ and ______________

   Pole B: ______________ and ______________
10. What other combinations would have been possible? (*Hint:* Alternate the colors at metaphase I.)

   Pole A: ______________ and ______________

   Pole B: ______________ and ______________

Conclusions

- Do the chromosomes inherited from the mother or father have to remain together following meiosis I? ________
- Name two ways that meiosis contributes to genetic recombination:

  a. ______________________________

  b. ______________________________

Interkinesis

Interkinesis is the period of time between meiosis I and meiosis II. In some species, daughter cells do not form, and meiosis II follows right after meiosis I. *Does DNA replication occur during interkinesis?* ________ Explain. ____________________

Meiosis II

During prophase of meiosis II, a spindle appears. Each chromosome attaches to the spindle independently. During metaphase II, the chromosomes are lined up at the equator. During anaphase II, the centromeres divide and the chromatids separate, becoming daughter chromosomes that move toward the poles. In telophase II, the spindle disappears as the nuclear envelope reappears. Notice that meiosis II is exactly like mitosis except that the nuclei of the parental cell and the daughter cells are haploid.

Prophase II

1. Using Figure 5.7 as a guide, choose the chromosomes from one pole to represent those in the new parental cell undergoing meiosis II.
2. Place two pairs of centrioles at opposite sides of these chromosomes to form the new spindle.

Metaphase II

Move the duplicated chromosomes to the metaphase II equator. *How many chromosomes are at the metaphase II equator?* ____________________

Anaphase II

Pull the two magnets of each duplicated chromosome apart. *What does this action represent?* ________
____

Telophase II

Put the chromosomes—each having one chromatid—at the poles (the new centrioles).

Conclusions

- You worked with only one daughter cell from meiosis I as the new parental cell. Suppose you had worked with both daughter cells. How many cells would have been present when meiosis II was complete? ________
- How many chromosomes are in the parental cell undergoing meiosis II? ________
- How many chromosomes are in the daughter cell? ______ Explain. ____________________
____

Summary of Meiotic Cell Division

1. The parental cell has the diploid (2n) number of chromosomes, and the daughter cells have the ______ (n) number of chromosomes.
2. Meiosis is cell division in which the chromosome number ____________________. Whereas meiosis reduces the chromosome number, fertilization restores the chromosome number.
3. A zygote contains the same number of chromosomes as the parent, but are these exactly the same chromosomes? ______
4. What is another way that sexual reproduction results in genetic recombination?
____

Figure 5.7 Meiosis I and II in plant cell micrographs and animal cell drawings. Crossing-over occurred during meiosis I.

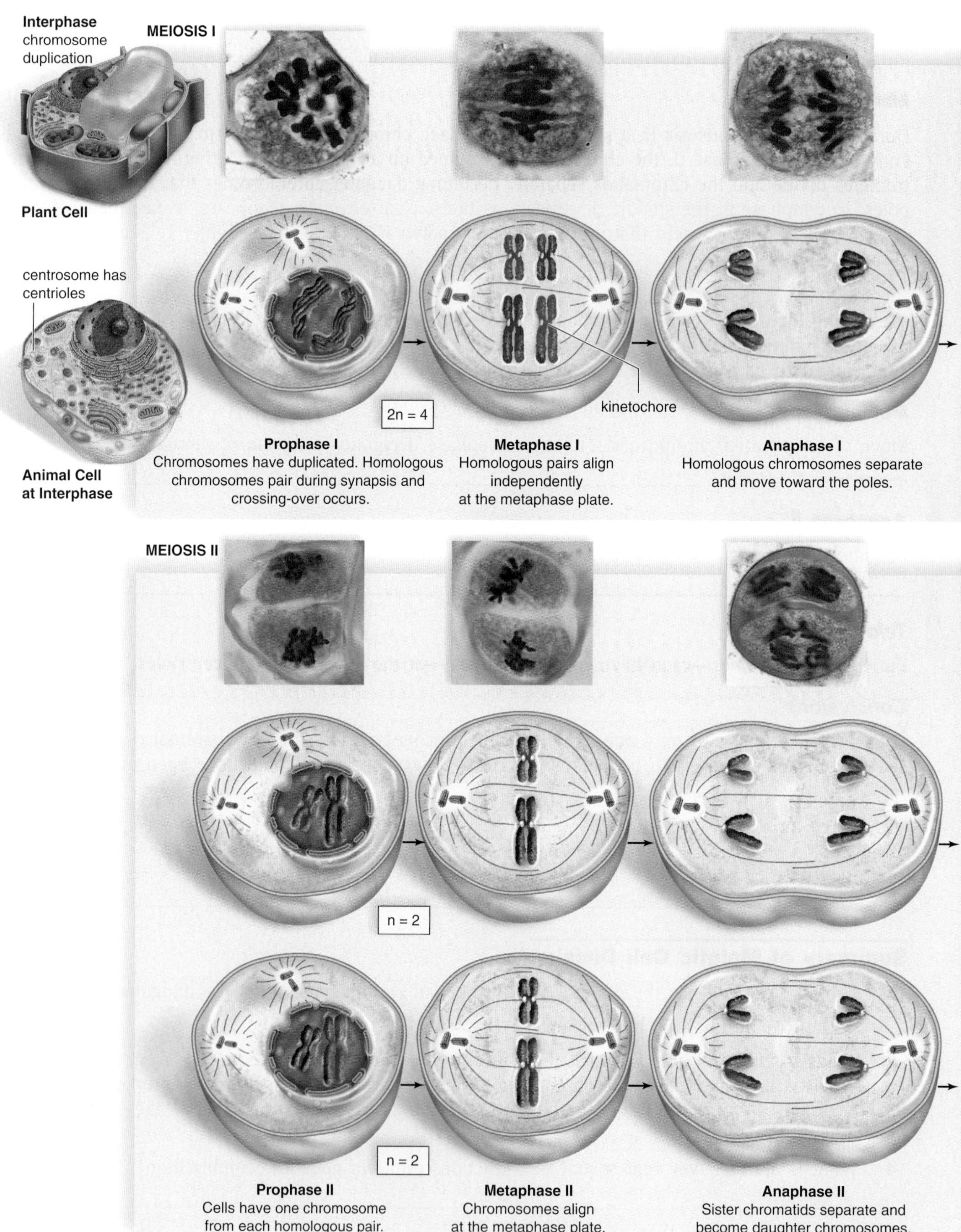

Interphase
chromosome duplication
Plant Cell
centrosome has centrioles
Animal Cell at Interphase
MEIOSIS I
2n = 4
kinetochore
Prophase I
Chromosomes have duplicated. Homologous chromosomes pair during synapsis and crossing-over occurs.
Metaphase I
Homologous pairs align independently at the metaphase plate.
Anaphase I
Homologous chromosomes separate and move toward the poles.
MEIOSIS II
n = 2
n = 2
Prophase II
Cells have one chromosome from each homologous pair.
Metaphase II
Chromosomes align at the metaphase plate.
Anaphase II
Sister chromatids separate and become daughter chromosomes.

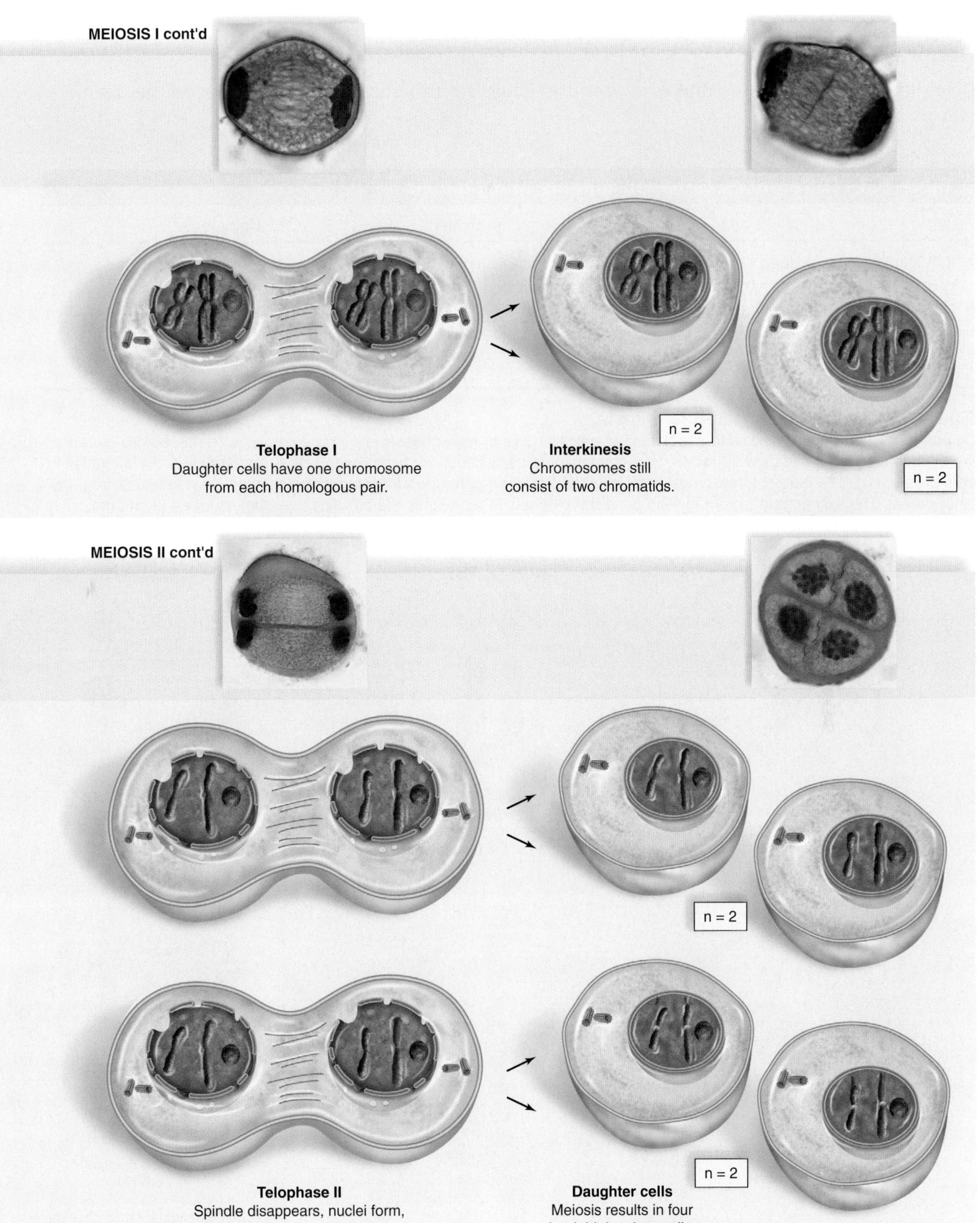

Telophase I
Daughter cells have one chromosome from each homologous pair.

Interkinesis
Chromosomes still consist of two chromatids.

Telophase II
Spindle disappears, nuclei form, and cytokinesis takes place.

Daughter cells
Meiosis results in four haploid daughter cells.

5.3 Mitosis Versus Meiosis

Examine Figure 5.8, and note the differences between mitosis and meiosis.

General Differences

Given that a parental cell is diploid (2n), fill in Table 5.3 to indicate general differences between mitosis and meiosis.

Table 5.3 Differences Between Mitosis and Meiosis

	Mitosis	Meiosis
1. Number of divisions		
2. Chromosome number in daughter cells		
3. Number of daughter cells		

Figure 5.8 Meiosis I compared to mitosis.
Compare metaphase I of meiosis to metaphase of mitosis. Only in metaphase I are the homologous chromosomes paired at the metaphase plate. Members of homologous chromosome pairs separate during anaphase I, and therefore the daughter cells are haploid. The blue chromosomes were inherited from one parent, and the red chromosomes were inherited from the other parent. The exchange of color between nonsister chromatids represents the crossing-over that occurred during meiosis I.

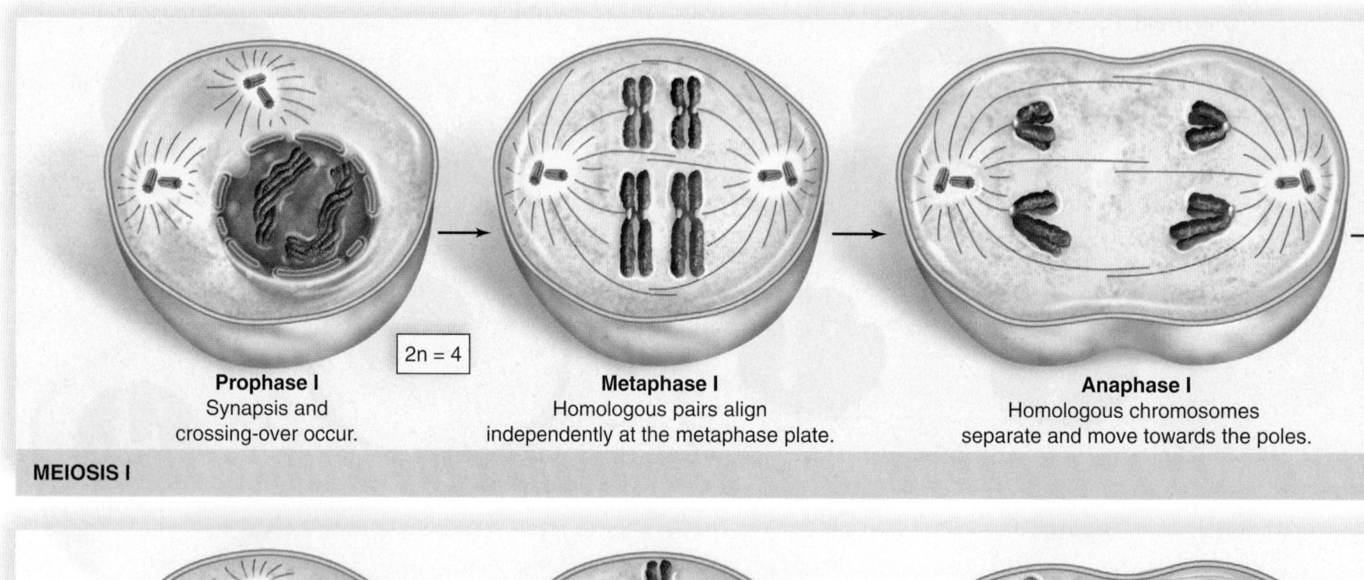
2n = 4
Prophase I
Synapsis and crossing-over occur.
Metaphase I
Homologous pairs align independently at the metaphase plate.
Anaphase I
Homologous chromosomes separate and move towards the poles.
MEIOSIS I

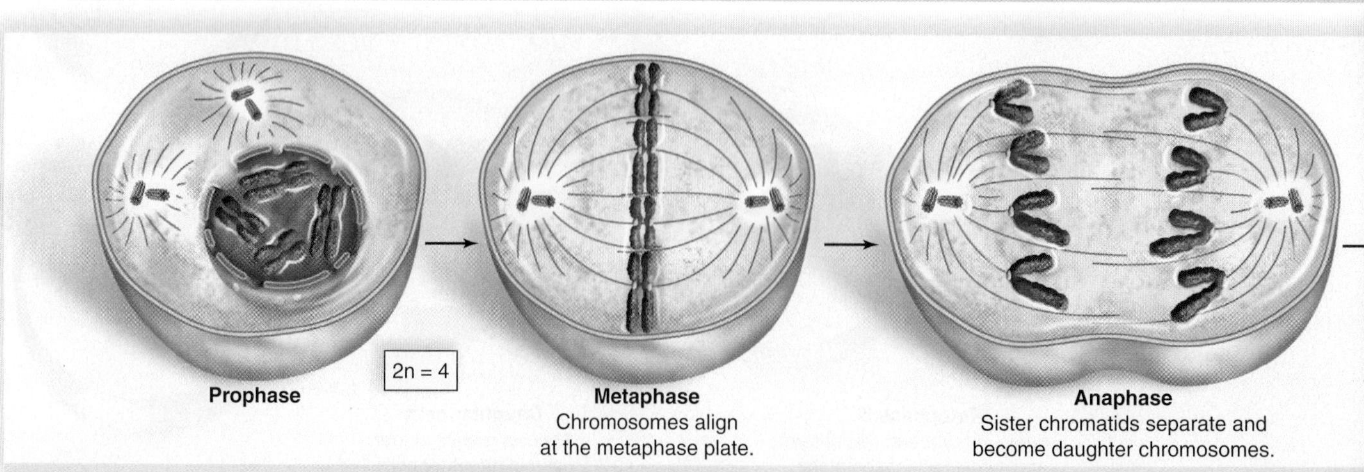
2n = 4
Prophase
Metaphase
Chromosomes align at the metaphase plate.
Anaphase
Sister chromatids separate and become daughter chromosomes.
MITOSIS

Specific Differences

1. Complete Table 5.4 to indicate specific differences between mitosis and meiosis I.

Table 5.4 Mitosis Compared with Meiosis I

Mitosis	Meiosis I
Prophase: no pairing of chromosomes	Prophase I: ______
Metaphase: duplicated chromosomes at equator	Metaphase I: ______
Anaphase: sister chromatids separate	Anaphase I: ______
Telophase: chromosomes have one chromatid	Telophase I: ______

2. Complete Table 5.5 to indicate specific differences between mitosis and meiosis II.

Table 5.5 Mitosis Compared with Meiosis II

Mitosis	Meiosis II
Prophase: no pairing of chromosomes	Prophase II: ______
Metaphase: duplicated chromosomes at equator	Metaphase II: ______
Anaphase: sister chromatids separate	Anaphase II: ______
Telophase: two diploid daughter cells	Telophase II: ______

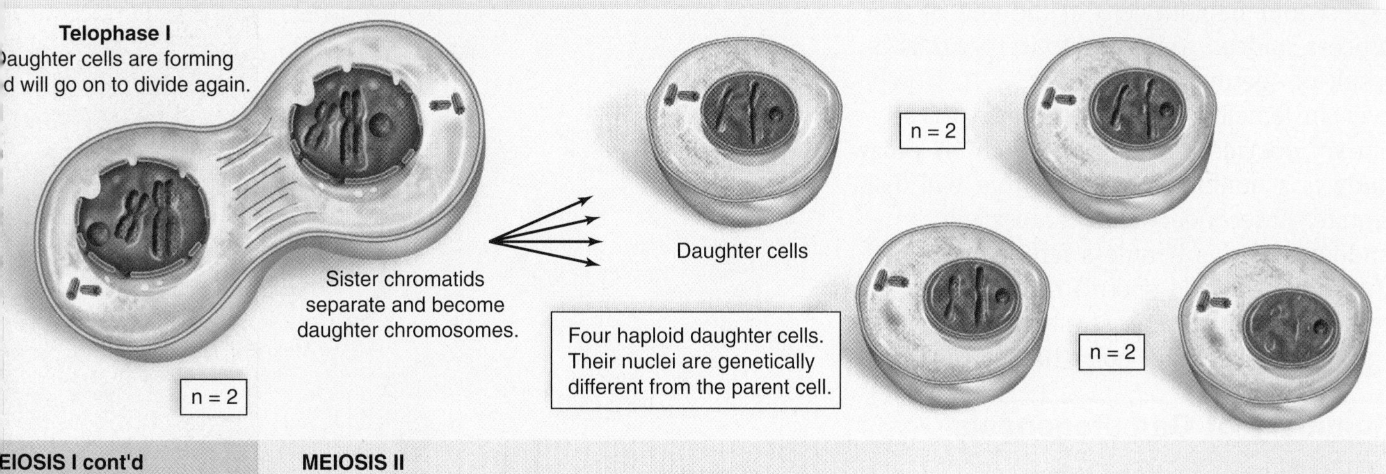
Telophase I
aughter cells are forming
d will go on to divide again.
Sister chromatids separate and become daughter chromosomes.
n = 2
Daughter cells
n = 2
n = 2
Four haploid daughter cells. Their nuclei are genetically different from the parent cell.
EIOSIS I cont'd
MEIOSIS II

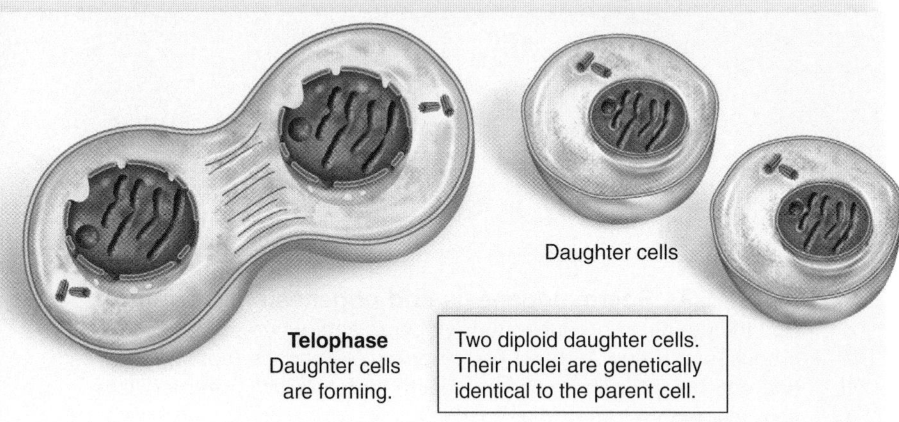
Daughter cells
Telophase
Daughter cells are forming.
Two diploid daughter cells. Their nuclei are genetically identical to the parent cell.
ITOSIS cont'd

5.4 Gametogenesis in Animals

Gametogenesis is the formation of **gametes** (sex cells) in animals. In humans and other mammals, the gametes are sperm and eggs. **Fertilization** occurs when the nucleus of a sperm fuses with the nucleus of an egg.

Gametogenesis in Mammals

Gametogenesis occurs in the testes of males, where **spermatogenesis** produces sperm. Gametogenesis occurs in the ovaries of females, where **oogenesis** produces oocytes (eggs).

A **diploid** (2n) nucleus contains the full number of chromosomes, and a **haploid** (n) nucleus contains half as many. Gametogenesis involves **meiosis,** the process that reduces the chromosome number from 2n to n. In sexually reproducing species, if meiosis did not occur, the chromosome number would double with each generation. Meiosis consists of two divisions: the first meiotic division (meiosis I) and the second meiotic division (meiosis II). Therefore, you would expect four haploid cells at the end of the process. Indeed, there are four sperm as a result of spermatogenesis (Fig. 5.9). However, in females, meiosis I results in a secondary oocyte and one polar body. A **polar body** is a nonfunctioning cell that will disintegrate. A secondary oocyte does not undergo meiosis II unless fertilization (fusion of egg and sperm) occurs. At the completion of oogenesis, there is a single egg and at least two polar bodies (Fig. 5.9).

Summary of Gametogenesis

1. What is gametogenesis? ________
   In general, how many chromosomes are in a gamete? ________
2. What is spermatogenesis? ________
   How many chromosomes does a human sperm have? ________
3. What is oogenesis? ________
   How many chromosomes does a human egg have? ________
4. Following fertilization, how many chromosomes does the zygote, the first cell of the new individual, have? ________

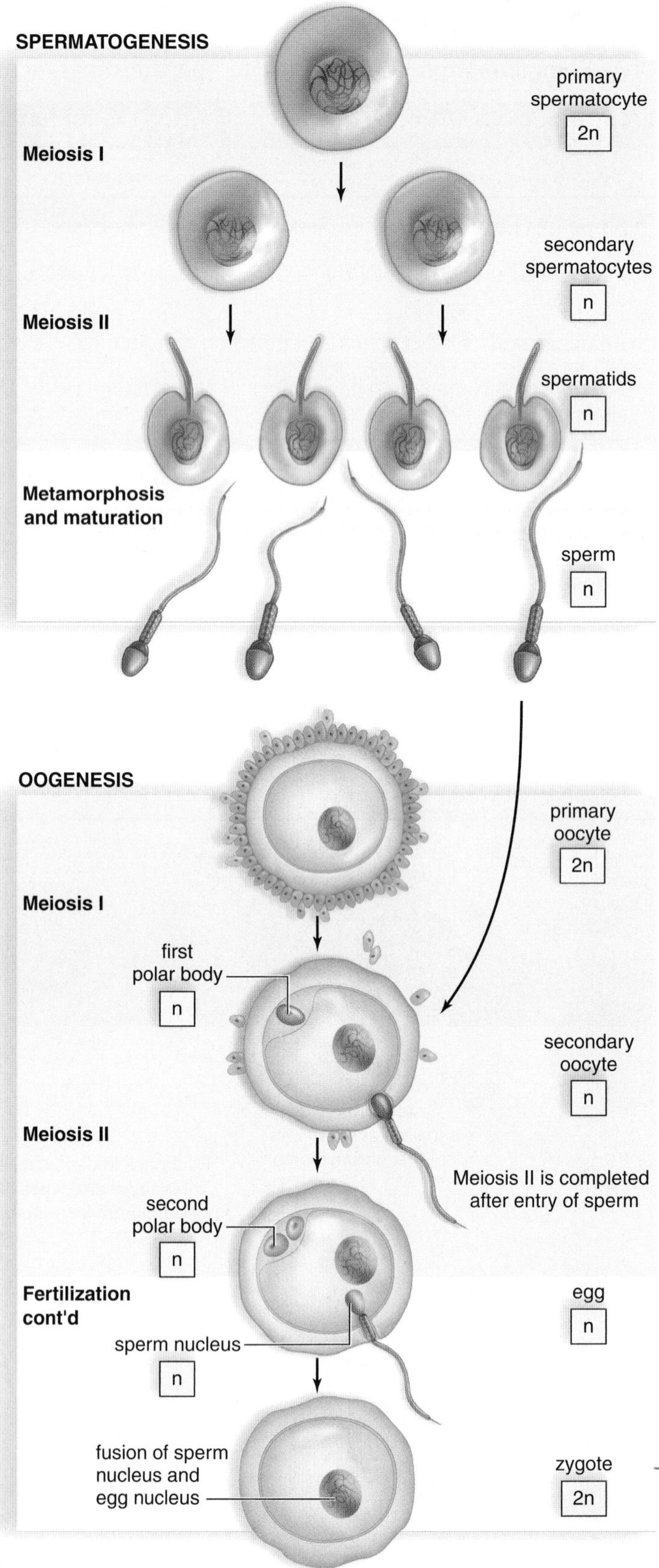

Figure 5.9 Spermatogenesis and oogenesis in mammals. Spermatogenesis produces four viable sperm, whereas oogenesis produces one egg and two polar bodies. In humans, both sperm and egg have 23 chromosomes each; therefore, following fertilization, the zygote has 46 chromosomes.

Gametogenesis Models

Examine any available gametogenesis models, and determine the diploid number of the parental cell and the haploid number of a gamete. Remember that counting the number of centromeres tells you the number of chromosomes.

Slide of Testis

1. With the help of Figure 5.10, examine a prepared slide of a testis. Under low power, note the many circular structures. These are the **seminiferous tubules,** where sperm formation takes place.
2. Switch to high power, and observe one tubule in particular. Find mature sperm (which look like thin, fine, dark lines) in the middle of the tubule. **Interstitial cells,** which produce the male sex hormone testosterone, are between the tubules.

Figure 5.10 Microscopic testis anatomy.
a. A testis contain many seminiferous tubules. **b.** Scanning electron micrograph of a cross section of the seminiferous tubules, where spermatogenesis occurs. Note the location of interstitial cells in clumps among the seminiferous tubules in this light micrograph.

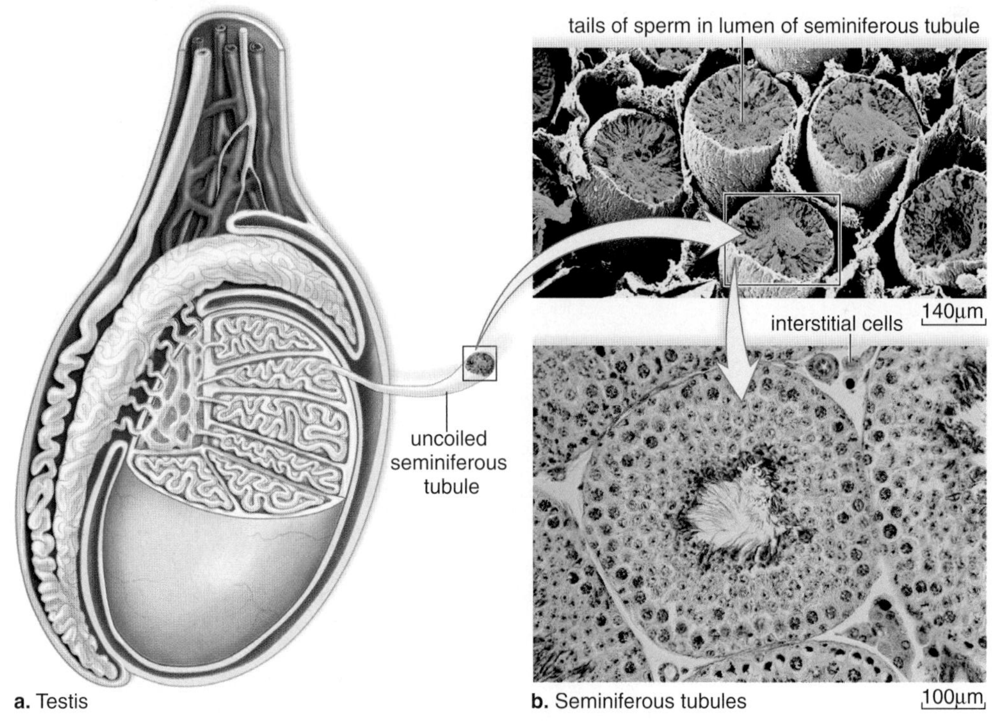

Slide of Ovary

1. With the help of Figure 5.11, examine a prepared slide of an ovary. Under low power, you will see a large number of small, primary follicles near the outer edge. A primary follicle contains a primary oocyte.
2. Find a secondary follicle, and switch to high power. Note the secondary oocyte (egg), surrounded by numerous cells, to one side of the liquid-filled follicle.
3. Also look for a large, fluid-filled vesicular (Graafian) follicle, which contains a mature secondary oocyte to one side. The vesicular follicle will be next to the outer surface of the ovary because this type of follicle releases the egg during ovulation.
4. How many secondary follicles can you find on your slide? _____ How many vesicular follicles can you find? _____ How does this number compare with the number of sperm cells seen in the testis cross section? ______________________________

Figure 5.11 Microscopic ovary anatomy.
The stages of follicle and oocyte (egg) development are shown in sequence. Each follicle goes through all the stages. Following ovulation, a follicle becomes the corpus luteum.

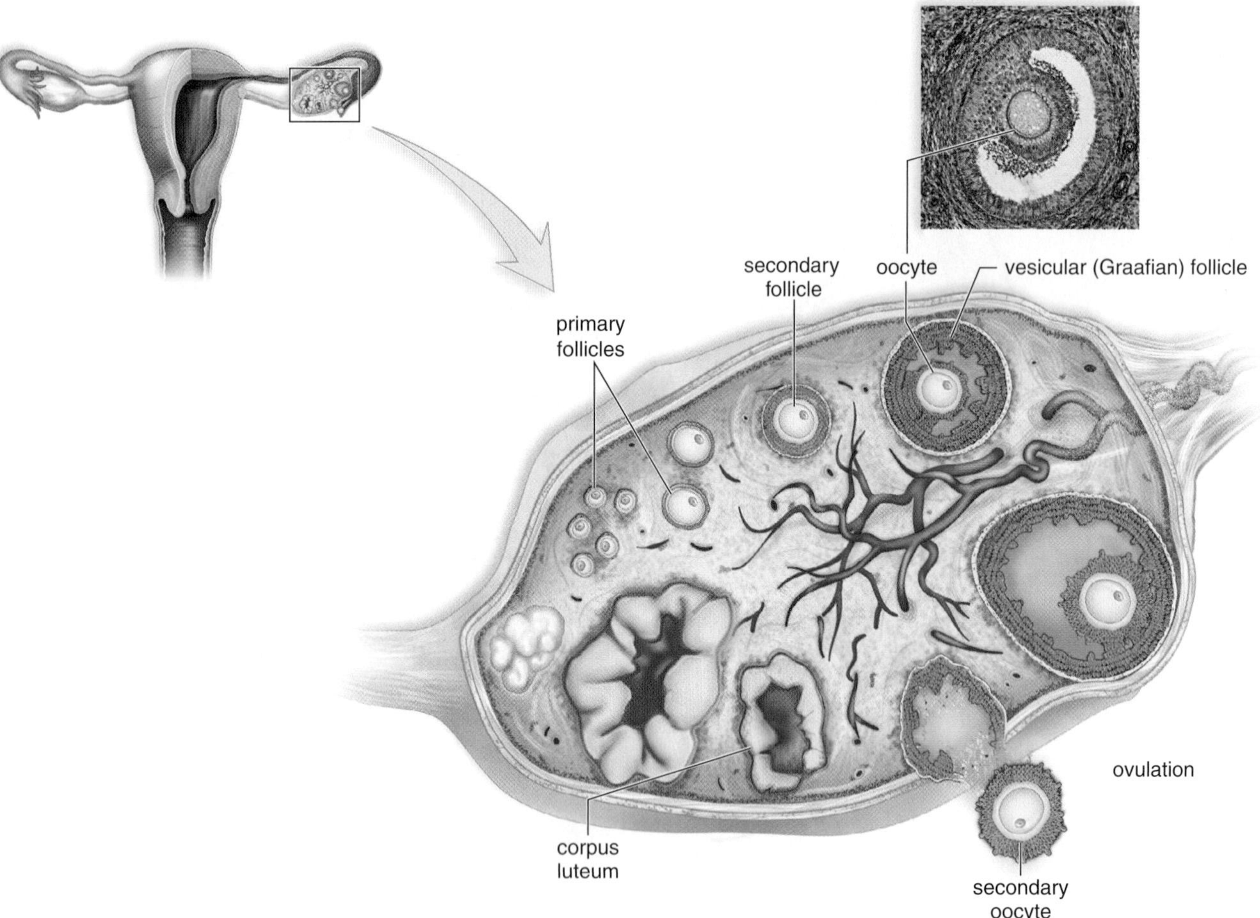

Laboratory Review 5

______________ 1. During what stage of the cell cycle does DNA replication occur?

______________ 2. Name the phase of cell division during which separation of sister chromatids occurs.

______________ 3. By what process does the cytoplasm of a human cell separate?

______________ 4. Name the phase of cell division when duplicated chromosomes first appear.

______________ 5. What structure forms in plant cells during cytokinesis?

______________ 6. Where in humans would you expect to find meiosis taking place?

______________ 7. If there are 13 pairs of homologues in a primary spermatocyte, how many chromosomes are there in a sperm?

______________ 8. What term refers to the production of an egg?

______________ 9. During which type of gametogenesis would you see polar bodies?

______________ 10. What do you call chromosomes that look alike and carry genes for the same traits?

______________ 11. If homologues are separating, what phase is this?

______________ 12. If the parental cell has 24 chromosomes, how many does each daughter cell have at the completion of meiosis II?

______________ 13. Name the type of cell division during which homologues pair.

______________ 14. Name the type of cell division described by $2n \rightarrow 2n$.

______________ 15. Does metaphase of mitosis, meiosis I, or meiosis II have the haploid number of chromosomes at the equator of the spindle?

Thought Questions

16. Meiosis functions to reduce chromosome number. When, during the human life cycle, is the diploid number of chromosomes restored?

17. List four similarities and four differences when comparing mitosis with meiosis.

18. How does the alignment of chromosomes differ between metaphase of mitosis and metaphase of meiosis I?

19. A student is simulating meiosis I with homologues that are red-long and yellow-long. Why would you *not* expect to find both red-long and yellow-long in one resulting daughter cell?

20. With reference to the same homologues, describe the appearance of two nonsister chromatids following crossing-over.

21. What would be the outcome of a situation in which a cell completes mitosis without completing cytokinesis?

LABORATORY

6

Enzymes

Learning Outcomes

6.1 Catalase Activity
- Describe how an enzyme functions.
- Demonstrate why the shape of the active site is critical to enzyme specificity and activity.
- Discuss the relationship of the substrate, the enzyme, and the product in the reaction studied today.

6.2 Effect of Temperature on Enzyme Activity
- Drawing upon the knowledge gained from this laboratory exercise, develop a generalized statement regarding the effect of temperature upon an enzyme-catalyzed reaction.

6.3 Effect of Concentration on Enzyme Activity
- Drawing upon the knowledge gained from this laboratory exercise, develop a generalized statement regarding the effects of enzyme and substrate concentrations upon an enzyme-catalyzed reaction.

6.4 Effect of pH on Enzyme Activity
- Drawing upon the knowledge gained from this laboratory exercise, develop a generalized statement regarding the effect of pH on an enzyme-catalyzed reaction.

Introduction

The cell carries out many chemical reactions. All the chemical reactions that occur in a cell are collectively called **metabolism.** A possible chemical reaction can be indicated like this:

$$\underbrace{A + B}_{\text{reactants}} \longrightarrow \underbrace{C + D}_{\text{products}}$$

In all chemical reactions, the **reactants** are molecules that undergo a change, which results in the **products.** The arrow stands for the change that produced the product(s). The number of reactants and products can vary; in the one you are studying today, a single reactant breaks down to two products.

All the reactions that occur in a cell have an enzyme. **Enzymes** are organic catalysts that speed metabolic reactions. Because enzymes are specific and speed only one type of reaction, they are given names. In today's laboratory, you will be studying the action of the enzyme **catalase.** The reactants in an enzymatic chemical reaction are called **substrate(s).**

Enzymes are specific because they have a shape that accommodates the shape of their substrates as a key fits a lock. Enzymatic reactions can be indicated like this:

$$E + S \longrightarrow ES \longrightarrow E + P$$

In this reaction, E = enzyme, ES = enzyme-substrate complex, and P = product.

Two types of enzymatic reactions common to cells are shown in Figure 6.1. During degradation reactions, the substrate is broken down to the product(s), and during synthesis reactions, the substrates are joined to form a product. Notice in each case how the shape of the enzyme accommodates its substrate. The location where the enzyme and substrate form an enzyme-substrate complex is called the **active site** because the reaction occurs here. Unlike a simple lock and key, in the induced fit model for enzyme reactions, the substrate alters the enzyme, improving the "fit" between enzyme and substrate and overall reaction efficiency.

Figure 6.1 Enzymatic action.
The enzyme and substrate come together, and the reaction occurs on the surface of the enzyme at the active site. The product leaves the enzyme, and then the enzyme can be used again. **a.** During degradation, the substrate is broken down. **b.** During synthesis, substrates combine to produce the product.

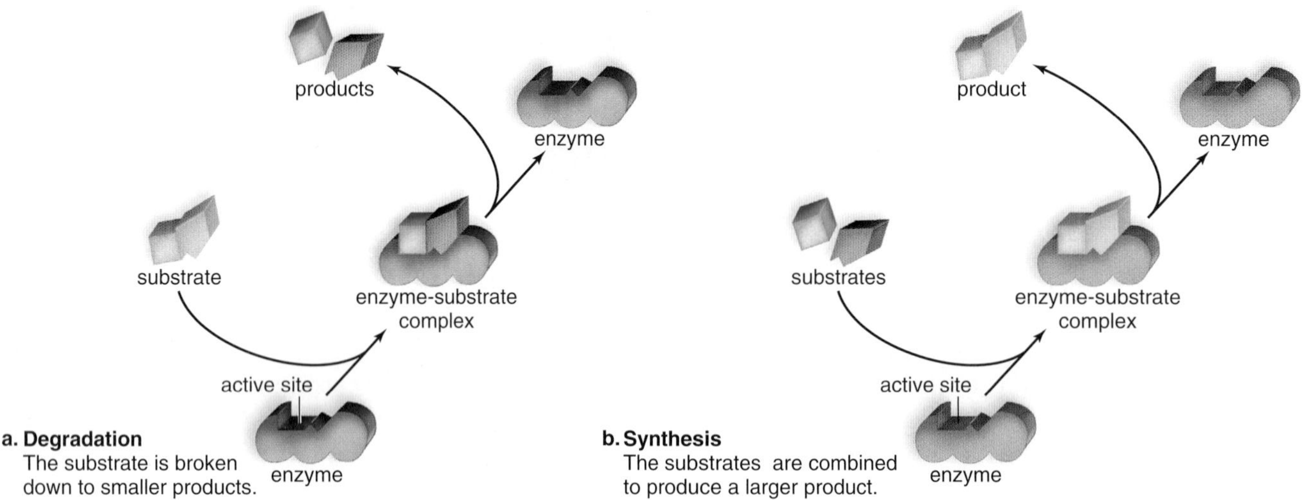

At the end of the reaction, the product is released and the enzyme can then combine with the substrate again. A cell needs only a small amount of an enzyme because enzymes are used over and over again. Some enzymes have turnover rates well in excess of a million product molecules per minute. As enzyme-catalyzed reactions occur, the product molecules may contain less energy than the original substrate molecules. Reactions of this type are referred to as **exergonic reactions** and are used to release energy that may be used by cells. In other reactions, the product molecules may contain more energy than the original substrate, a means of storing energy for future use. Reactions of this latter type are called **endergonic reactions.**

All enzymes are complex proteins that generally act in an organism's closely controlled internal environment, where the temperature and pH remain within a rather narrow range. *Extremes* in pH or temperature may denature the enzyme by permanently altering its chemical structure. Even a small change in the protein's structure will change the enzyme's shape enough to prevent formation of the enzyme-substrate complex and thus keep the reaction from occurring. As more substrate molecules fill active sites, more product results per unit time. Therefore, in general, an increase in enzyme or substrate speeds enzymatic reactions. In this laboratory, you will test the effect of temperature, enzyme concentration, and pH on an enzymatic reaction.

6.1 Catalase Activity

In the Experimental Procedures that follow, you will be working with the enzyme catalase. Catalase is present in cells, where it speeds the breakdown of the toxic chemical hydrogen peroxide to water and oxygen:

$$\underset{\text{hydrogen peroxide}}{2H_2O_2} \xrightarrow{\text{catalase}} \underset{\text{water}}{2H_2O} + \underset{\text{oxygen}}{O_2}$$

What is the reactant in this reaction? ________ What is the substrate for catalase? ____________

What are the products in this reaction? ________ and ________ Bubbling occurs as the reaction proceeds. Why? ____

Experimental Procedure: Catalase Activity

> **Caution:** Use protective eyewear and exercise caution when performing this experiment.

With a wax pencil, label and mark three clean test tubes at the 1 cm and 5 cm levels.

Tube 1
1. Fill to the first mark with *catalase.*
2. Fill to the second mark with *hydrogen peroxide.* Swirl well to mix, and wait at least 20 seconds for bubbling to develop.
3. Measure the height of the bubble column (in millimeters), and record your results in Table 6.1.

Tube 2
1. Fill to the first mark with *water.*
2. Fill to the second mark with *hydrogen peroxide.* Swirl well to mix, and wait at least 20 seconds.
3. Measure the height of the bubble column (in millimeters), and record your results in Table 6.1.

Tube 3
1. Fill to the first mark with *catalase.*
2. Fill to the second mark with *sucrose solution.* Swirl well to mix; wait 20 seconds.
3. Measure the height of the bubble column, and record your results in Table 6.1.

Table 6.1 Catalase Activity

Tube	Contents	Bubble Column Height	Explanation
1	Catalase Hydrogen peroxide		
2	Water Hydrogen peroxide		
3	Catalase Sucrose solution		

Conclusions

- Which tube showed the bubbling you expected? ______ Conclude why this tube showed bubbling, and record your explanation in Table 6.1.
- Which tube was a control? ______ If this tube showed bubbling, what could you conclude about your procedure? ________________ Record your explanation in Table 6.1 for this tube.
- Enzymes are specific; they speed only a reaction that contains their substrate. Which tube exemplifies this characteristic of an enzyme? ___________ Record your explanation in Table 6.1 for this tube.

6.2 Effect of Temperature on Enzyme Activity

In general, cold temperatures slow chemical reactions, and warm temperatures speed chemical reactions. Boiling, however, causes an enzyme to denature in a way that inactivates it.

Experimental Procedure: Effect of Temperature

With a wax pencil, label and mark three clean test tubes at the 1 cm and 5 cm levels.

1. Fill each tube to the first mark with *catalase.*
2. Place tube 1 in a refrigerator or cold water bath, tube 2 in an incubator or warm water bath, and tube 3 in a boiling water bath. Complete the second column in Table 6.2. Wait 15 minutes.
3. As soon as you remove the tubes one at a time from the refrigerator, incubator, and boiling water, fill to the second mark with *hydrogen peroxide.*
4. Swirl well to mix, and wait 20 seconds.
5. Measure the height of the bubble column (in millimeters) in each tube, and record your results in Table 6.2.
6. Plot your results in Figure 6.2. Put temperature (°C) on the X-axis and bubble column height (mm) on the Y-axis.

Table 6.2 Effect of Temperature on Enzyme Activity

Tube	Temperature °C	Bubble Column Height (mm)	Explanation
1 Refrigerator			
2 Incubator			
3 Boiling water			

Conclusions

- The amount of bubbling corresponds to the degree of enzyme activity. Explain in Table 6.2 the degree of enzyme activity per tube.
- What is your conclusion concerning the effect of temperature on enzyme activity?

____

- What effect would a fever have on enzymatic activity in the human body?

____

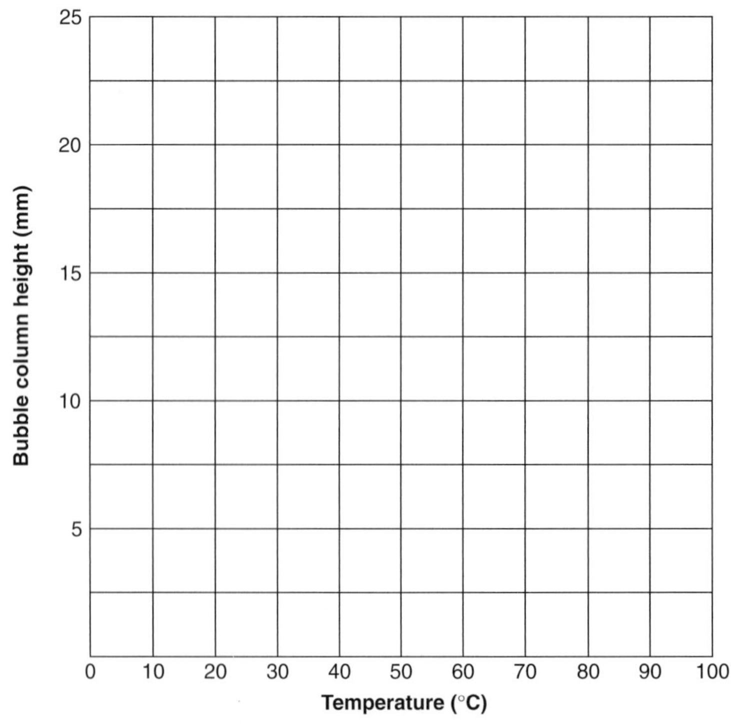

Figure 6.2 Effect of temperature on enzyme activity.

6.3 Effect of Concentration on Enzyme Activity

In general, a higher enzyme/substrate concentration results in faster enzyme activity—that is, the amount of product per unit time for any particular reaction will increase.

Experimental Procedure: Effect of Enzyme Concentration

With a wax pencil, label three clean test tubes.

Tube 1

1. Mark this tube at the 1 cm and 5 cm levels.
2. Fill to the first mark with *catalase* and to the second mark with *hydrogen peroxide.*
3. Swirl well to mix, and wait 20 seconds.
4. Measure the height of the bubble column (in millimeters), and record your results in Table 6.3.

Tube 2

1. Mark this tube at the 2 cm and 6 cm levels.
2. Fill to the first mark with *catalase* and to the second mark with *hydrogen peroxide.*
3. Swirl well to mix, and wait 20 seconds.
4. Measure the height of the bubble column (in millimeters), and record your results in Table 6.3.

Tube 3

1. Mark this tube at the 3 cm and 7 cm levels.
2. Fill to the first mark with *catalase* and to the second mark with *hydrogen peroxide.*
3. Swirl well to mix, and wait 20 seconds.
4. Measure the height of the bubble column (in millimeters), and record your results in Table 6.3.

Table 6.3 Effect of Enzyme Concentration

Tube	Amount of Enzyme	Bubble Column Height (mm)	Explanation
1	1 cm		
2	2 cm		
3	3 cm		

Conclusions

- The amount of bubbling corresponds to the degree of enzyme activity. Explain in Table 6.3 the degree of enzyme activity per tube.
- If unlimited time were allotted, would the results be the same in all tubes? ____________

  Explain why or why not. ____
- Would you expect similar results if the substrate concentration were varied in the same manner as the enzyme concentration? ______ Why or why not? ____________________

6.4 Effect of pH on Enzyme Activity

Each enzyme has a pH at which the speed of the reaction is optimum (occurs best). Any higher or lower pH affects hydrogen bonding and the structure of the enzyme, leading to reduced activity.

Experimental Procedure: Effect of pH

> **Caution:** Hydrochloric acid (HCl) used to produce an acid pH is a strong, caustic acid, and sodium hydroxide (NaOH) used to produce a basic pH is a strong, caustic base. Exercise care in using these chemicals, and follow your instructor's directions for disposal of tubes that contain these chemicals. If any acidic or basic solutions spill on your skin, rinse immediately with water. Use protective eyewear when performing this experiment.

With a wax pencil, label and mark three clean test tubes at the 1 cm, 3 cm, and 7 cm levels. Fill each tube to the 1 cm level with *catalase.*

Tube 1

1. Fill to the second mark with *water* adjusted to pH 3 by the addition of *HCl.*
2. Fill to the third mark with *hydrogen peroxide.* Wait one minute.
3. Carefully swirl to mix, and wait 20 seconds.
4. Measure the height of the bubble column (in millimeters), and record your results in Table 6.4.

Tube 2

1. Fill to the second mark with *water* adjusted to pH 7.
2. Wait one minute. Fill to the third mark with *hydrogen peroxide.*
3. Carefully swirl to mix, and wait 20 seconds.
4. Measure the height of the bubble column (in millimeters), and record your results in Table 6.4.

Tube 3

1. Fill to the second mark with *water* adjusted to pH 11 by the addition of *NaOH.*
2. Fill to the third mark with *hydrogen peroxide.* Wait one minute.
3. Carefully swirl to mix, and wait 20 seconds.
4. Measure the height of the bubble column (in millimeters), and record your results in Table 6.4.
5. Plot your results in Figure 6.3. Put pH on the X-axis and column height (mm) on the Y-axis.

Table 6.4 Effect of pH

Tube	pH	Bubble Column Height (mm)	Explanation
1	3		
2	7		
3	11		

Figure 6.3 Effect of pH on enzyme activity.

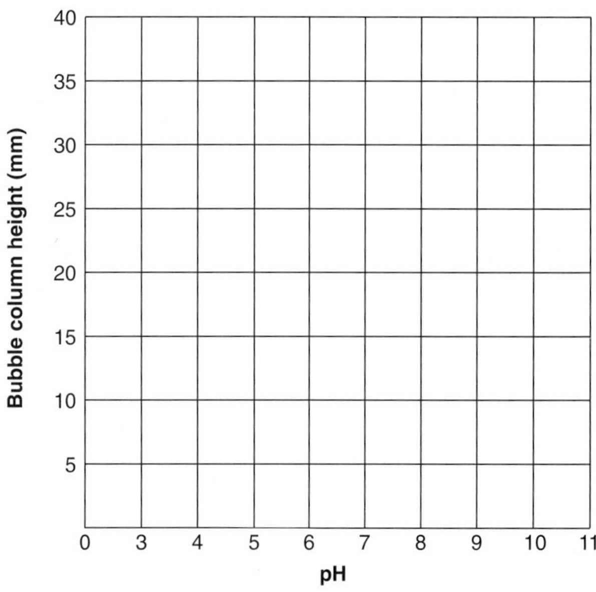

Conclusions

- The amount of bubbling corresponds to the degree of enzyme activity. Explain in Table 6.4 the degree of enzyme activity per tube.
- The results of which tube in Table 6.1 could be used as a control for Table 6.4? ____________

  Why could this tube be considered a control? ____________

Factors That Affect Enzyme Activity

In Table 6.5, summarize what you have learned about factors that affect the speed of an enzymatic reaction. For example, in general, what type of temperature promotes enzyme activity and what type inhibits enzyme activity? Answer similarly for enzyme-substrate concentration and pH.

Table 6.5 Factors That Affect Enzyme Activity

Factors	Promote Enzyme Activity	Inhibit Enzyme Activity
Temperature		
Enzyme/substrate concentration		
pH		

Conclusions

- Why does a warm temperature promote enzyme activity? ____________
- Why does increasing enzyme concentration promote enzyme activity? ____________
- Why does optimum pH promote enzyme activity? ____________

Laboratory Review 6

_______________ 1. In the representation of a chemical reaction, what does the arrow stand for?

_______________ 2. The reactants in an enzymatic reaction are called what?

_______________ 3. Where on an enzyme do substrates come together?

_______________ 4. If an enzyme is boiled, what happens to the enzyme?

_______________ 5. If an enzyme is warmed, what happens to its activity?

_______________ 6. A control for the effect of a warm temperature would lack what substance?

_______________ 7. If more enzyme is used, what happens to the amount of product per unit time?

_______________ 8. What must be held constant when testing the effect of enzyme concentration on enzyme activity?

_______________ 9. What product of the catalase reaction causes bubbling?

_______________ 10. What is varied when testing the effect of pH on enzyme activity?

_______________ 11. If the pH is unfavorable, what happens to enzyme activity?

_______________ 12. What allowed you to measure the amount of enzyme activity?

_______________ 13. What would happen to the overall reaction rate if a substance were added to the test tube that binds to the active site of catalase?

Thought Questions

14. Lipase is a digestive enzyme that digests fat droplets in the small intestine. Lipase requires a slightly basic pH, which the presence of $NaHCO_3$ provides. Indicate which of the following test tubes would show digestion following incubation, and explain why the others would not.

Tube 1 Water, fat droplets _______________

Tube 2 Water, fat droplets, lipase _______________

Tube 3 Water, fat droplets, lipase, $NaHCO_3$ _______________

Tube 4 Water, lipase, $NaHCO_3$ _______________

15. In what way does an enzyme speed its reaction? How does this explain why enzymes are specific?

16. As temperature increases, enzymatic activity increases up to a certain point beyond which enzymatic activity decreases. Explain.

LABORATORY

7

Cellular Respiration

Learning Outcomes

Introduction

- Use equations to demonstrate the differences between yeast cell fermentation and cellular respiration.
- Contrast the role played by oxygen in each process.
- Using ATP production as your point of reference, demonstrate which of the two processes is more desirable for energy production.

7.1 Fermentation

- Explain how the inverted tube in the experimental design relates to the fermentation experiment and the example fermentation equation shown.
- State and explain the effects of food source on fermentation by yeast.

7.2 Cellular Respiration

- Explain how the experimental apparatus and procedure are used to interpret cellular respiration.
- Relate the overall equation for cellular respiration to the cellular respiration experiment.
- Explain why the use of KOH is important for accuracy of the experimental design for this experiment.
- State and explain the effects of germination and nongermination of soybeans on the results of the cellular respiration experiment.

Introduction

In this laboratory, you will study **fermentation** and **cellular respiration,** two processes that cells use to release energy stored in the chemical bonds of the organic molecules found in food. Fermentation is an **anaerobic** process, one in which oxygen is not required. Cellular respiration is **aerobic** or dependent upon the presence of oxygen. During fermentation, glucose is incompletely broken down, and much energy remains in the organic molecule that results. In cellular respiration, glucose is completely broken down to inorganic molecules. During fermentation, a small amount of chemical energy is converted to ATP molecules, and during cellular respiration much more chemical energy is converted to ATP molecules for use by the cell. ATP molecules are absolutely essential to cellular metabolism because they supply energy whenever a cell carries on activities, such as active transport, synthesis, muscle contraction, and nerve conduction.

In yeast cells, fermentation occurs in the following way:

$$\underset{\text{glucose}}{C_6H_{12}O_6} \longrightarrow 2\ \underset{\text{carbon dioxide}}{CO_2} + 2\ \underset{\text{ethanol}}{C_2H_5OH} + 2\ ATP$$

Yeast is used by the wine and beer industries to ferment the carbohydrates in fruits and grains to alcohol. In baking, the carbon dioxide given off from yeast fermentation causes bread to rise. Fermentation in the cells of animals, including humans, and in certain microbes produces lactic acid ($C_3H_6O_3$) rather than ethanol. Lactic acid fermentation produces only lactic acid—there is no release of carbon dioxide as is seen in ethanol fermentation. Lactic acid fermentation helps produce yogurt and many cheeses, as well as such products as sourdough breads, chocolate, and pickled foods.

In most organisms, cellular respiration occurs in the following way:

$$\underset{\text{glucose}}{C_6H_{12}O_6} + 6\ \underset{\text{oxygen}}{O_2} \longrightarrow 6\ \underset{\text{carbon dioxide}}{CO_2} + 6\ \underset{\text{water}}{H_2O} + 36\text{–}38\ ATP$$

Notice that oxygen is consumed during cellular respiration. In eukaryotic cells, fermentation occurs in the cytoplasm of the cell, whereas cellular respiration takes place within the mitochondrion.

7.1 Fermentation

Yeasts (unicellular fungi) can use several types of sugars as an energy source. Glucose and fructose are monosaccharides; sucrose is a disaccharide that contains glucose and fructose. Fructose can be converted to glucose, which is the usual molecule acted on by yeast. In the following Experimental Procedure, you will test several of these food sources for their ability to ferment by measuring the amount of carbon dioxide given off in a respirometer. A **respirometer** is a device for measuring the amount of gas given off and/or consumed.

Experimental Procedure: Fermentation

Respirometer Practice

1. Completely fill a small tube (15 × 125 mm) with water (Fig. 7.1).
2. Invert a large tube (20 × 150 mm) over the small tube, and with your finger or a pencil, push the small tube up into the large tube until the upper lip of the small tube is in contact with the bottom of the large tube.
3. Quickly invert both tubes. Do not permit the small tube to slip away from the bottom of the large tube. A little water will leak out of the small tube and be replaced by an air bubble.
4. Practice this inversion until the bubble in the small tube is as small as you can make it.

Figure 7.1 Respirometer for yeast experiment.
Place a small tube inside a large tube. Hold the small tube in place as you rotate the entire apparatus, and an air bubble will form in the small tube.

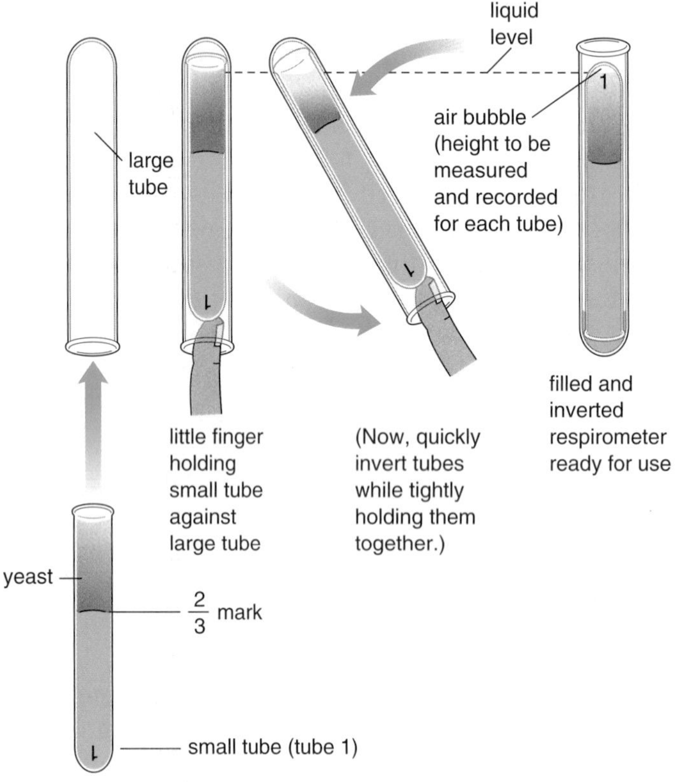

Testing Food Samples

Have ready four large test tubes. With a wax pencil, label and mark off a small test tube at the 2/3-full level. Use this tube to mark off three other small tubes at the same level.

1. Label and fill the small tubes as directed, and record the contents in Table 7.1.

   Tube 1 Fill to the mark with *glucose solution.*
   Tube 2 Fill to the mark with *fructose solution.*
   Tube 3 Fill to the mark with *sucrose solution.*
   Tube 4 Fill to the mark with *distilled water.*

2. Re-suspend a yeast solution each time, and fill all four tubes to the top with yeast suspension (Fig. 7.1).
3. Slide the large tubes over the small tubes, and invert them in the way you practiced. This will mix the yeast and sugar solutions.
4. Place the respirometers in a tube rack, and measure the initial height of the air space in the rounded bottom of the small tube. Record the height in Table 7.1.
5. Place the respirometers in an incubator or in a warm water bath maintained at 37°C. Note the time, and allow the respirometers to incubate about 20 minutes (incubator) or one hour (water bath). However, watch your respirometers and if they appear to be filling with gas quite rapidly, stop the incubation when appropriate. During the incubation period, begin the Experimental Procedure on cellular respiration (see p. 90).
6. At the end of the incubation period, measure the final height of the gas bubble, and record it in Table 7.1. Calculate the net change, and record it in Table 7.1.

Table 7.1 Fermentation by Yeast

Tube	Contents	Initial Gas Height	Final Gas Height	Net Change	Conclusion
1					
2					
3					
4					

Conclusions

- From your results, conclude how the sugars tested compare as a food source for yeast fermentation. Enter your conclusions in Table 7.1.
- How do you know that the yeast cells were respiring anaerobically (in the absence of oxygen) and not aerobically (in the presence of oxygen)?

  ___

- Why did the gas bubbles increase in size? ___
- Speculate on why sucrose is not as good a food source as fructose and glucose. ___

  ___

- Which respirometer was the control? ___
- Why were the respirometers placed in a warm water bath? ___

7.2 Cellular Respiration

As indicated in the cellular respiration equation given in the introduction to this laboratory, oxygen gas is consumed during cellular respiration, and carbon dioxide gas is given off. The uptake of oxygen is the evidence that an organism is carrying on cellular respiration. It is possible to use potassium hydroxide (KOH) to remove carbon dioxide as it is given off. The equation for this reaction is:

$$CO_2 + 2\ KOH \longrightarrow \underset{\text{solid potassium carbonate}}{K_2CO_3} + H_2O$$

Experimental Procedure: Cellular Respiration

1. Obtain a volumeter, an apparatus that measures changes in gas volumes. Remove the three vials from the volumeter. Remove the stoppers from the vials and the vials from the volumeter. Label the vials 1, 2, and 3.

> **Caution:** Potassium hydroxide (KOH) is a strong, caustic base. Exercise care in using this chemical, and follow your instructor's directions for disposal of these tubes. If any potassium hydroxide should spill on your skin, rinse immediately with water. Use protective eyewear when performing this experiment.

2. Using the same amounts, place a small wad of absorbent cotton in the bottom of each vial. Without getting the sides of the vials wet, use a dropper to saturate the cotton with 15% potassium hydroxide (KOH). Place a small wad of dry cotton on top of the KOH-soaked absorbent cotton (Fig. 7.2).
3. Obtain a 100 ml graduated cylinder, and add 50 ml of *water.* Drop in 25 *germinating* soybean or pea seeds, and measure the amount of water displaced (how much water rises above 50 ml).

   Record this number here: ______ ml. Remove the water, and dry the germinating seeds on a paper towel. Place the germinating seeds in vial 1 (Fig. 7.2), and return the vial to the volumeter.
4. Again, add 50 ml of *water* to the 100 ml graduated cylinder. Drop in 25 *nongerminating* seeds. Add glass beads until the same amount of water is displaced, as in step 3. Remove the water, and dry the glass beads and nongerminating seeds on a paper towel. Place both the nongerminating seeds and the glass beads in vial 2, and return the vial to the volumeter.
5. Use the same procedure to measure the number of glass beads only, needed for vial 3. Return vial 3 to the volumeter. This tube is the thermobarometer.

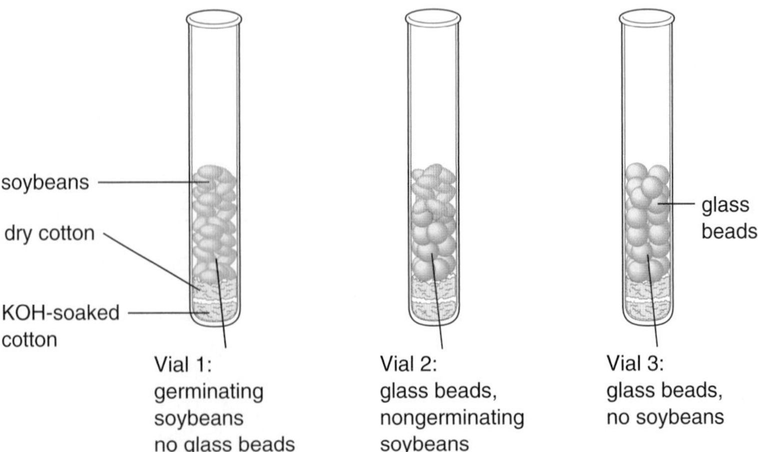

Figure 7.2 Vials. In this experiment, three vials are filled as noted.

6. Each stopper should have a vent (rubber tube) with a clamp and a graduated side arm. Remove the clamps from the vents. Adjust the graduated side arm until only about 5 mm to 1 cm protrudes through the stopper (Fig. 7.3).
7. With a dropper, add a drop of *Brodie manometer fluid* (or water colored with vegetable dye and a small amount of detergent) to each side arm.
8. Firmly place the stoppers in the vials. Adjust the stoppers until the side arms are parallel to the lab bench.
9. Adjust the location of the marker drop in the side arms with the assistance of a dry dropper in the vent. The marker drop of vial 3 (thermobarometer) should be in the middle of the side arm. The marker drop of vials 1 and 2 should be between 0.80 and 0.90 ml. Close the vent with the clamp when the marker drop is at the correct location.
10. Allow the respirometers to equilibrate for 5 minutes, and then record in Table 7.2, to the nearest 0.01 ml, the initial position of the marker drop in each graduated side arm.

Figure 7.3 Volumeter containing three respirometers. In this experiment, the respirometers are vials filled as per Fig. 7.2 with graduated side arms attached. Oxygen uptake is measured by movement of a marker drop in each side arm.

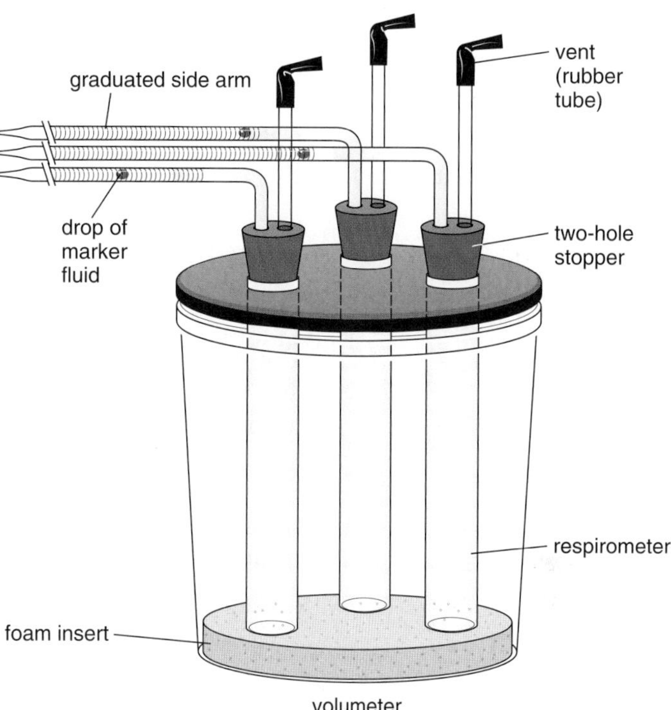

11. Wait 10 minutes, and then record in Table 7.2 any change in the position of the marker drop. Wait 10 more minutes, and then record in Table 7.2 any change in the position of the marker drop. Then record in Table 7.2 the net change for each vial—that is, the initial reading for each vial minus the vial's 20-minute reading.
12. Did the marker drop change in vial 3 (glass beads)? ______ By how much? ______ Enter this number in the "Correction" column of Table 7.2, and use this number to correct the net change you observed in vials 1 and 2. (This is a correction for any change in volume due to atmospheric pressure changes or temperature changes.) This will complete Table 7.2.

Table 7.2 Cellular Respiration

Vial	Contents	Initial Reading	Reading After 10 Minutes	Reading After 20 Minutes	Net Change	Correction	(Corrected) Net Change
1	Germinating seeds						
2	Nongerminating seeds and glass beads						
3	Glass beads						

Conclusions

- In which vial did the water recede? ________ State the vial contents. ________________

  Is this the vial that carried on cellular respiration? ______

- Why did the water recede? ____________________

____

- Why was it necessary to absorb the carbon dioxide? ____________________

____

- In the germinating seed experiment, you were measuring the change in volume of what gas? __________
- Which respirometer in the seed respiration experiment was the control? ____________

Laboratory Review 7

____________ 1. Both cellular respiration and fermentation begin with what molecule?

____________ 2. What reactant needed for cellular respiration is absent from the fermentation reaction?

____________ 3. What gas do organisms give off when they carry out cellular respiration?

____________ 4. Both fermentation and cellular respiration provide what molecule needed by cells?

____________ 5. Do plant cells or animal cells carry on cellular respiration?

____________ 6. Which process, fermentation or cellular respiration, results in an end product that contains C—H bonds?

____________ 7. Name the device that can measure the amount of gas given off by yeast.

____________ 8. Yeast cells carry out fermentation when they are supplied with what type of molecule?

____________ 9. During the fermentation experiment, the gas bubble got larger. What gas was causing this increase?

____________ 10. What role was played by KOH in the soybean experiment?

____________ 11. In the germinating seed experiment, what do you call the tube that contains only glass beads?

____________ 12. What gas is being taken up when the marker in the side arm of a respirometer moves toward a tube that contains germinating seeds?

Thought Questions

13. Why is it reasonable that, of the three sugars (glucose, fructose, and sucrose), glucose would result in the most activity during the fermentation experiment?

14. If you performed the cellular respiration experiment without soaking the cotton with KOH, what results would you predict? Why?

15. What would you expect to happen to the validity of your results if you had performed the cellular respiration experiment without using the glass beads?

LABORATORY

8

Photosynthesis

Learning Outcomes

8.1 Plant Pigments

- From the data produced by the chromatography experiment, evaluate the relative solubilities of the four pigments in the solvents used.

8.2 Solar Energy

- Drawing upon experimental data and background information from the laboratory manual, discuss why white light is able to promote photosynthesis.
- Cite evidence that white light actually contains several different colors of light.
- Offer evidence to support the conclusion that chloroplasts are selective in their use of light during photosynthesis.

8.3 Carbon Dioxide Uptake

- Describe an experiment that shows how carbon dioxide is utilized during photosynthesis.

8.4 Carbon Cycle

- Generalize the relationship between cellular respiration and photosynthesis.

Introduction

The process of photosynthesis allows plants and other organisms to collect the energy of sunlight and convert it into energy that can be stored. Photosynthesis is arguably the most important series of biological reactions on the earth. Without photosynthesis, most living organisms would not have a source of energy to fuel their metabolic needs. In addition, the oxygen released as a by-product of photosynthesis is the primary source of oxygen in our atmosphere.

The overall equation for **photosynthesis** is:

$$CO_2 + H_2O \xrightarrow{\text{solar energy}} (CH_2O)_n + O_2$$

In this equation (CH_2O) represents any general carbohydrate where energy can be stored in the bonds between carbon and hydrogen atoms. Sometimes, this equation is multiplied by 6 so that glucose $(C_6H_{12}O_6)$ appears as an end product of photosynthesis. Figure 8.1 illustrates the overall relationship of energy storage brought about by photosynthesis and its subsequent release through cellular respiration. Upon careful examination of the relationships shown in Figure 8.1, you can see that cellular respiration and photosynthesis have opposite purposes.

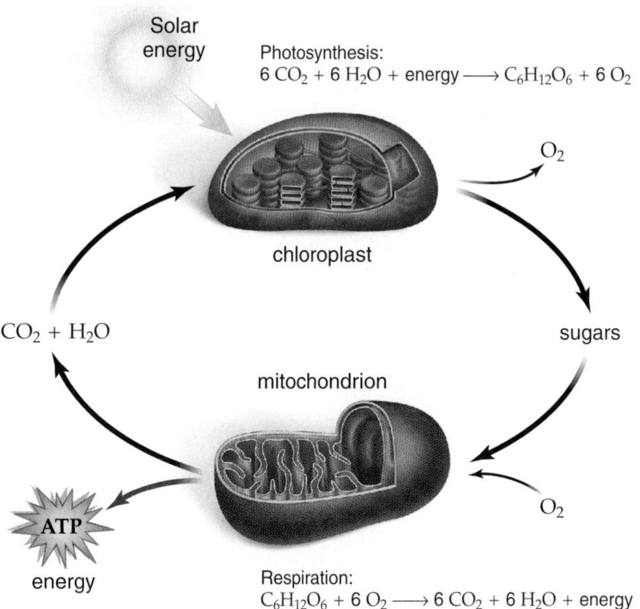

Figure 8.1 Comparison of photosynthesis and respiration.

Photosynthesis stores energy in the chemical bonds of organic molecules like glucose. Cellular respiration, which was evaluated in the previous lab exercise (Exercise 7), begins with organic molecules like glucose and breaks them down to release their stored energy.

Photosynthesis takes place in chloroplasts (Fig. 8.2). Here membranous thylakoids are stacked in grana surrounded by the stroma. The events of photosynthesis can be summarized as two distinct processes. First, light is converted from solar energy into a temporary form of chemical energy during the **light reactions.** The temporary energy storage molecules from the light reactions are then converted into molecules better suited to the long-term storage of energy by a series of chemical reactions known as the **Calvin cycle.** During the light reactions, pigments within the membranes of thylakoids absorb solar energy, water (H_2O) is split, oxygen (O_2) is released, and short-term energy storage molecules are formed. The Calvin cycle reactions occur within the stroma. During these reactions, carbon dioxide (CO_2) is reduced using energy from the energy storage molecules formed in the light reactions and solar energy is now stored in a carbohydrate (CH_2O). It is important to note that light is only necessary for the light reactions. The Calvin cycle relies on chemical energy from the light reactions to power its reactions.

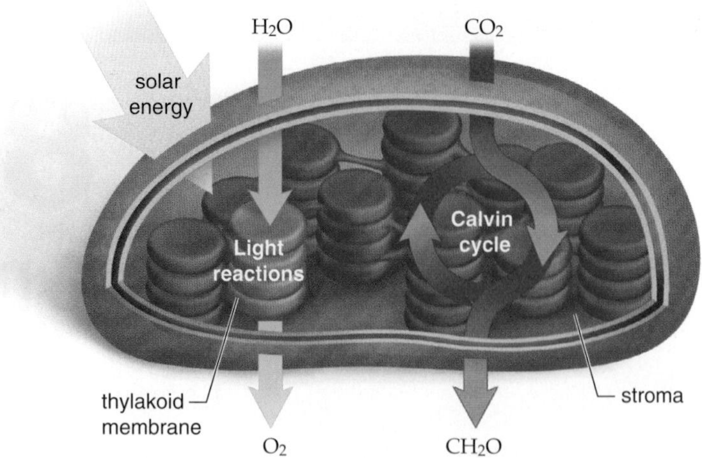

Figure 8.2 Overview of photosynthesis.
Photosynthesis includes the light reactions when energy is collected and O_2 is released and the Calvin cycle reactions when carbohydrate (CH_2O) is formed.

8.1 Plant Pigments

The principal pigment in the thylakoids of plants is **chlorophyll *a*. Chlorophyll *b*, carotenes,** and **xanthophylls** play a secondary role by transferring the energy they absorb to chlorophyll *a* for use in photosynthesis. Extracting and separating the pigments reveals that the green plant extract contains more than just green chlorophyll molecules.

Chromatography is a technique that separates molecules from each other on the basis of their solubility in particular solvents. The solvents used in the following Experimental Procedure are petroleum ether and acetone, which have no charged groups and are, therefore, nonpolar. As a nonpolar solvent moves up the chromatography paper, the pigment moves along with it. The more nonpolar a pigment, the more soluble it is in a nonpolar solvent, and the faster and farther it proceeds up the chromatography paper.

Experimental Procedure: Plant Pigments

Caution: Ether (which is part of the chromatography solution) is toxic and extremely flammable. Do not breathe the fumes, and do not place the chromatography solution near any source of heat. A fume (ventilation) hood is recommended. Use protective eyewear when performing this experiment.

1. Assemble a chromatography apparatus (large, dry test tube and cork with a hook) and a strip of precut chromatography paper (handle by the top only) (Fig. 8.3). Attach the paper strip to the hook, and test for fit. The paper should hang straight and barely touch the bottom of the test tube; trim if necessary. Measure 2 cm from the bottom of the paper, and place a small dot with a pencil (not a pen). With a wax pencil, mark the test tube 1 cm below where the dot is with the stopper in place. Set the apparatus in a test tube rack.

Figure 8.3 Paper chromatography.
The paper must be cut to size and arranged to hang down without touching the sides of a dry tube. Then the pigment (chlorophyll) solution is applied to a designated spot. The chromatogram develops after the spotted paper is suspended in the chromatography solution.

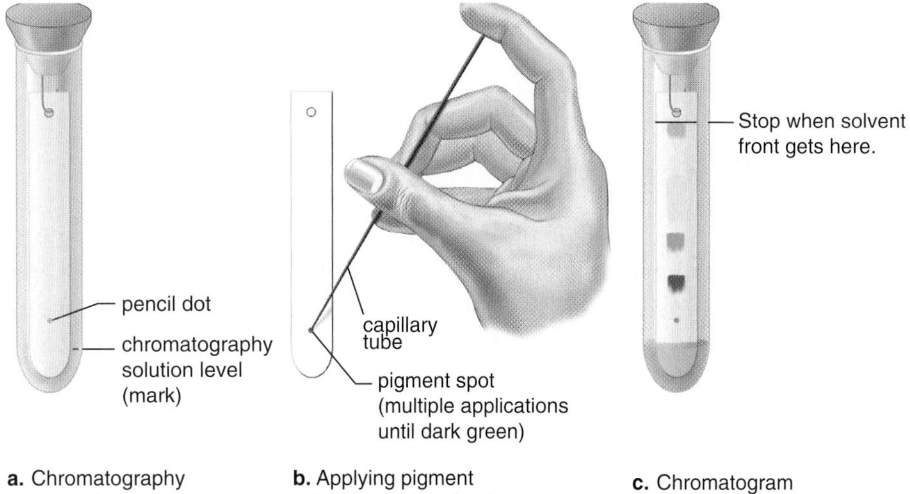

a. Chromatography apparatus **b.** Applying pigment (chlorophyll) extract **c.** Chromatogram

2. Prepare (or obtain) a plant *pigment extract,* as directed by your instructor.
3. Place the premarked chromatography paper strip onto a paper towel.
4. Fill a capillary tube by placing it into the extract. (It will fill by its own capillary action.)
5. Repeatedly apply the *pigment extract* to the pencil dot on the chromatographic strip. Let the spot dry between each application. Try to obtain a small dark green spot. (Placing your index finger over the end of the capillary tube will help keep the dot small.)
6. In a **fume hood, or well-ventilated area away from a source of heat or flame,** add *chromatography solution* to the mark you made earlier. Do not submerge the pigment spot. Set the apparatus in a test-tube rack and close the chromatography apparatus tightly. Do not shake the test tube during the chromatography.
7. Allow approximately 10 minutes for your chromatogram to develop, but check it frequently so that the pigments do not reach the top of the paper.
8. When the solvent front has moved to within 1 cm of the upper edge of the paper (Fig. 8.3*c*), remove your chromatogram. Close the apparatus tightly. With a pencil, lightly mark the location of the solvent front, and allow the chromatogram to dry in the fume hood.
9. Identify the pigment bands. Beta-carotenes are represented by the bright orange-yellow band at the top. Xanthophylls are yellow and may be represented in multiple bands. The blue-green band is chlorophyll *a,* and the lowest, olive-green band is chlorophyll *b.* Which pigment is the most nonpolar (that is, has the greatest affinity for the nonpolar solvent)? ____________
10. Calculate the R_f (ratio-factor) values for each pigment. For these calculations, mark the center of the initial pigment spot. This will be the starting point for all measurements. Also mark the midpoints of each pigment and the solvent front. Measure the distance between points for each pigment in millimeters, and record these values in Table 8.1. Then use the following formula, and enter your R_f values in Table 8.1:

$$R_f = \frac{\text{distance moved by pigment}}{\text{distance moved by solvent}}$$

Table 8.1 R_f (Ratio-Factor) Values for Each Pigment

Pigments	Distance Moved (mm)	R_f Values
Beta-carotenes		
Xanthophylls		
Chlorophyll *a*		
Chlorophyll *b*		
Solvent		

11. Do your results suggest that the chemical characteristics of these pigments might differ? ______
Why? ______

8.2 Solar Energy

During the light reactions of photosynthesis, solar energy is transformed into chemical energy and stored as intermediary molecules, NADPH and ATP. Without solar energy, photosynthesis would be impossible and most life as we know it could not exist. The release of oxygen from a plant indicates that photosynthesis is occurring. *Verify that photosynthesis releases oxygen by writing the equation for photosynthesis here:* ______

Role of White Light

White (sun) light contains different colors of light, as is demonstrated when white light passes through a prism (Fig. 8.4). Plants rely on specific colors of light to conduct photosynthesis. Since white light contains all of the necessary colors of light for photosynthesis, it is the preferred light for photosynthesis experiments. Some of the oxygen released from photosynthesis is taken up by a plant when cellular respiration occurs. This must be taken into account when the rate of photosynthesis is calculated.

Plants require a readily available source of carbon dioxide to conduct photosynthesis. Evaluate the experimental procedure below. Since the experimental tubes will be stoppered, cutting off atmospheric carbon dioxide from the experimental plants, what is the source of carbon dioxide in this experiment?

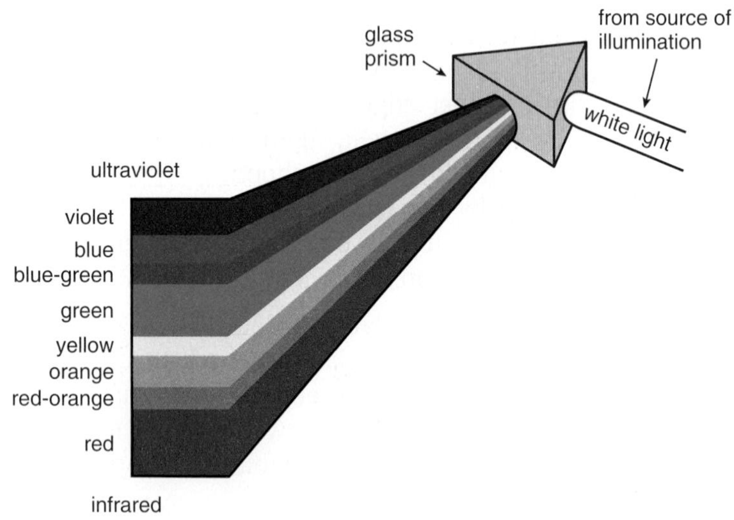

Figure 8.4 White light.
White light is made up of various colors, as can be seen when white light passes through a prism.

Experimental Procedure: White Light

1. Place a generous quantity of *Elodea* with the cut end up (make sure the cuts are fresh) in a test tube with a rubber stopper containing a bent piece of glass tubing, as illustrated in Figure 8.5. When assembled, this is your volumeter for studying the need for light in photosynthesis. (Do not hold the volumeter in your hand, as body heat will also drive the reaction forward). Your instructor will show you how to fix the volumeter in an upright position.
2. Before stoppering the test tube, add sufficient 3% *sodium bicarbonate* ($NaHCO_3$) solution so that, when the rubber stopper is inserted into the tube, the solution comes to rest at about 1/4 the length of the bent glass tubing. Mark this location on the glass tubing with a wax pencil.

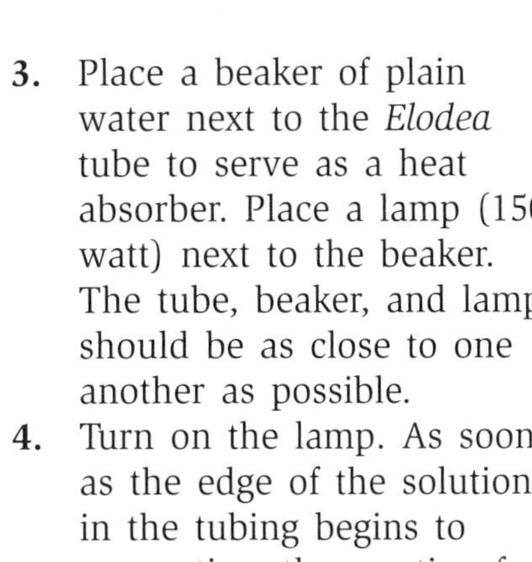

3. Place a beaker of plain water next to the *Elodea* tube to serve as a heat absorber. Place a lamp (150 watt) next to the beaker. The tube, beaker, and lamp should be as close to one another as possible.

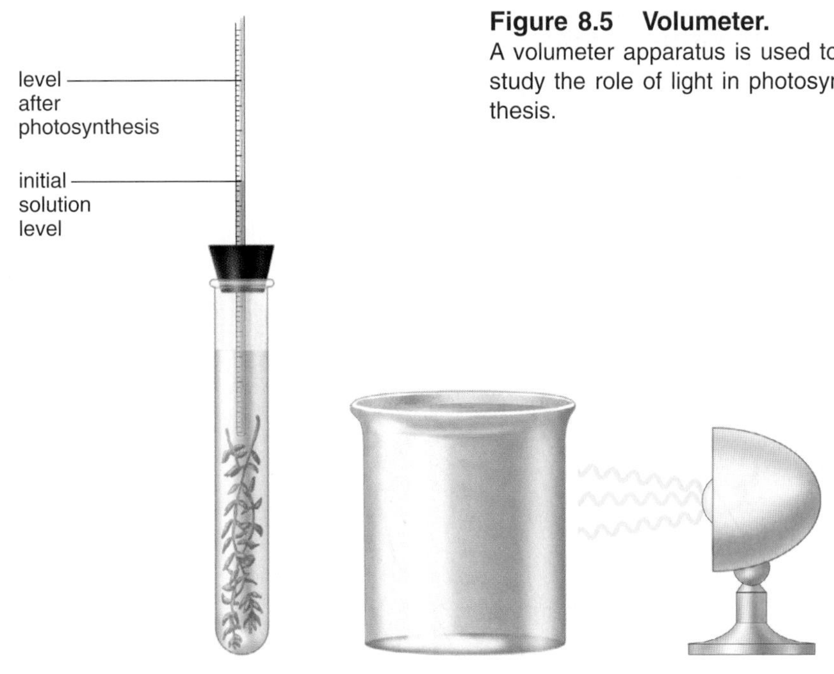

Figure 8.5 Volumeter. A volumeter apparatus is used to study the role of light in photosynthesis.

4. Turn on the lamp. As soon as the edge of the solution in the tubing begins to move, time the reaction for 10 minutes. Be careful not to bump the tubing or to readjust the stopper, or your readings will be altered. After 10 minutes, mark the edge of the solution, and measure in millimeters the distance the edge moved: _______ mm/10 min. This is **net photosynthesis,** a measurement that does not take into account the oxygen that was used up for cellular respiration. Record your results in Table 8.2. Why did the edge move forward? _______________

5. Carefully wrap the tube containing *Elodea* in aluminum foil, and record here the length of time it takes for the edge of the solution in the tubing to recede 1 mm: _______. Convert your measurement to _______ mm/10 min., and record this value for **cellular respiration** in Table 8.2. (Do not use a minus sign, even though the edge receded.) Why does cellular respiration, which occurs in a plant all the time, cause the edge to recede? _______________________________

____

6. If the *Elodea* had not been respiring in step 4, how far would the edge have moved? _______ mm/10 min. This is **gross photosynthesis** (net photosynthesis + cellular respiration). Record this number in Table 8.2.

7. Calculate the **rate of photosynthesis** (mm/hr) by multiplying gross photosynthesis (mm/10 min) by 6 (that is, 10 min × 6 = 60 min = 1 hr): _______ mm/hr. Record this value in Table 8.2.

Table 8.2 Rate of Photosynthesis (White Light)

	Data
Net photosynthesis (white light)	
Cellular respiration (no light)	
Gross photosynthesis (net + cellular respiration)	(mm/10 min)
Rate of photosynthesis	(mm/hr)

Role of Green Light

Green light is only one part of white light (see Fig. 8.4). As can be seen from Figure 8.6, plant pigments absorb certain colors of light better than other colors. According to Figure 8.6, what color light do the chlorophylls absorb best? ________________ Least? ________________

What color light do the carotenoids (carotenes and xanthophylls) absorb best? ________________ Least? ________________

Does photosynthesis use green light? ________________________________

The following Experimental Procedure will test your answer.

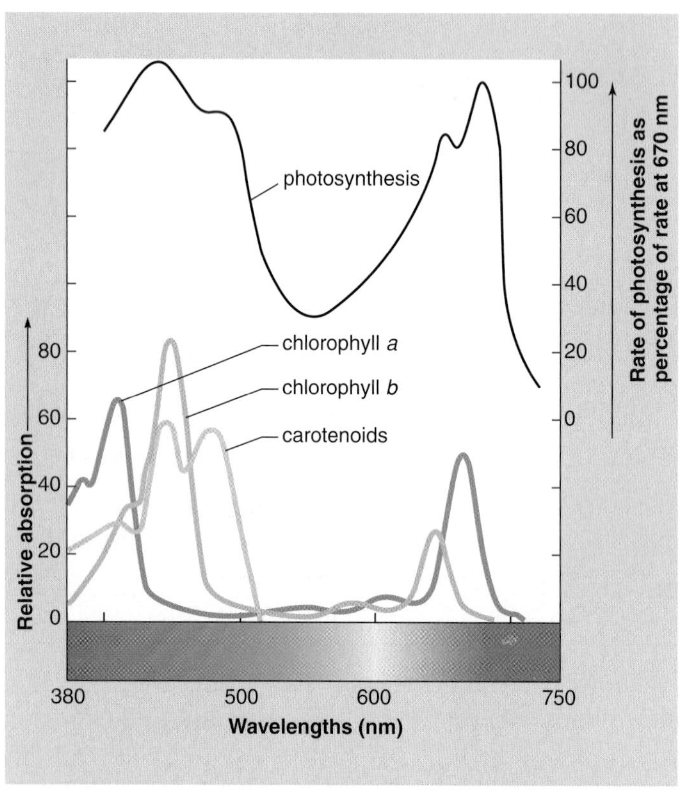

Figure 8.6 Action spectrum for photosynthesis.
The action spectrum for photosynthesis is the sum of the absorption spectrums for the pigments chlorophyll *a*, chlorophyll *b*, and carotenoids.

Experimental Procedure: Green Light

1. Add three drops of green dye (or use a green cellophane wrapper) to the beaker of water used in the previous Experimental Procedure until there is a distinctive green color. Remove all previous wax pencil marks from the glass tubing.
2. Record in Table 8.3 your data for gross photosynthesis (mm/10 min) and for rate of photosynthesis for white light (mm/hr) from Table 8.2.
3. Turn on the lamp. Mark the location of the edge of the solution on the glass tubing. As soon as the edge begins to move, time the reaction for 10 minutes. After 10 minutes, mark the edge of the solution, and measure in millimeters the distance the edge moved. Net photosynthesis for green light = ______ mm/10 min.
4. Carefully wrap the tube containing *Elodea* in aluminum foil, and record here the length of time it takes for the edge of the solution in the tubing to recede 1 mm: ______. Convert your measurement to ______ mm/10 min.

5. Calculate gross photosynthesis for green light (mm/10 min) as you did for white light, and record your data in Table 8.3.

6. Calculate rate of photosynthesis for green light (mm/hr) as you did for white light, and record your data in Table 8.3.

7. Average and record the Table 8.3 class data for both white light and green light, and record these averages in Table 8.3.
8. The following equation shows the rate of photosynthesis (green light) as a percentage of the rate of photosynthesis (white light):

$$\text{percentage} = \frac{\text{rate of photosynthesis (green light)}}{\text{rate of photosynthesis (white light)}} \times 100$$

This percentage, based on your data in Table 8.3 = ________. This percentage, based on class data in Table 8.3 = ________. Record these values in Table 8.3.

Table 8.3 Rate of Photosynthesis (Green Light)

	Your Data	Class Data
Gross Photosynthesis (mm/10 min)		
White (from Table 8.2)		
Green		
Rate of Photosynthesis (mm/hr)		
White (from Table 8.2)		
Green		

Conclusions

- Explain why the rate of photosynthesis with green light is only a portion of the rate of photosynthesis with white light. ____
____
- How does the percentage based on your data differ from that based on class data?
____
____
- Based on Figure 8.6, what colors of light are most important for the overall process of photosynthesis? ____
____

8.3 Carbon Dioxide Uptake

During the **Calvin cycle** of photosynthesis, the plant takes up carbon dioxide (CO_2) and uses the ATP and NADPH from the light reactions to reduce it to a carbohydrate, such as glucose ($C_6H_{12}O_6$). Therefore, the carbon dioxide in the solution surrounding *Elodea* should disappear as photosynthesis takes place.

Experimental Procedure: Carbon Dioxide Uptake

Caution: **Phenol red** Avoid ingestion, inhalation, and contact with skin, eyes, and mucous membranes. If necessary, flush thoroughly with water and wash with mild soap and water. Phenol red will also stain clothing. Exercise care in using this chemical, and follow your instructor's directions for disposal of this chemical. Use protective eyewear when performing this experiment.

1. Temporarily remove the *Elodea* from the test tube. Empty the sodium bicarbonate (NaHCO3) solution from the test tube, rinse the test tube thoroughly, and fill with a phenol red solution diluted to a faint pink. (Add more water if the solution is too dark.) Phenol red is a pH indicator that turns yellow in an acid and red in a base.
2. Blow *lightly* on the surface of the solution. Stop blowing as soon as the surface color changes to yellow. Then shake the test tube until the rest of the solution turns yellow. Blowing onto the solution adds what gas to the test tube? ________ When carbon dioxide combines with water, it forms carbonic acid. What causes the color change?

   ____
3. Thoroughly rinse the *Elodea* with distilled water, return it to the test tube with the phenol red solution, and assemble your volumeter as before.
4. If you used green dye, change the water in the beaker to remove the green solution.
5. Turn on the lamp, and wait until the edge of the solution just begins to move. Note the time. Observe until you note a change in color. Record your results in the appropriate column of Table 8.4.
6. Hypothesize why the solution in the test tube eventually turned red. ____________________

   ____

Use of a Control

Scientists are more confident of their results when an experimental procedure includes a control. Controls undergo all the steps in the experiment except the one being tested.

- Considering the test sample in Table 8.4, suggest a possible control sample for this experiment: ____
- Ask your instructor if you can actually perform this procedure. Both the control and test sample should be done at the same time.
- Record your results in Table 8.4. Why should all experiments have a control? ____________________

  ____

Table 8.4 Carbon Dioxide Uptake	
Tube	**Time for Color Change**
Test sample: *Elodea* + phenol red solution + CO_2	
Control sample:	

8.4 Carbon Cycle

In this laboratory, you have demonstrated a relationship between cellular respiration and photosynthesis. Animals produce the carbon dioxide used by plants to carry out photosynthesis. Plants produce the food and oxygen that they and animals require to carry out cellular respiration. This relationship is again summarized in Figure 8.7 and can be represented by the following equation:

$$C_6H_{12}O_6 + 6\ O_2 \underset{\text{photosynthesis}}{\overset{\text{cellular respiration}}{\rightleftharpoons}} 6\ CO_2 + 6\ H_2O + \text{energy}$$

1. Which organelle in plants carries out the reaction in the equation above in the reverse (right-to-left) direction? ______________________
2. Pertaining to photosynthesis, the energy in the equation is provided by ______________________.
3. Which organelle in plants and animals is involved in carrying out the reaction in this equation in the forward direction? ______________________

Figure 8.7 Photosynthesis and cellular respiration.
Animals are dependent on plants for a supply of oxygen, and plants are dependent on animals for a supply of carbon dioxide.

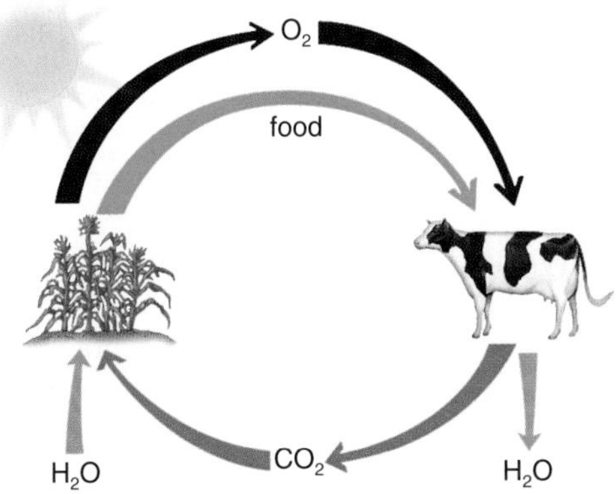

4. Pertaining to cellular respiration, the energy in the equation becomes chemical bond energy in what molecule? ______________________
5. Would it be correct to say that solar energy eventually becomes the chemical bond energy in ATP? __________ Why? ______________________
6. Considering that both plants and animals carry on cellular respiration, revise Figure 8.7 to improve its accuracy.

Laboratory Review 8

1. Where do the light reactions of photosynthesis take place?
2. What procedure did you use to separate plant pigments?
3. What determines the speed with which a pigment moves up the chromatography paper?
4. Where do plants ordinarily get the energy they need to carry on photosynthesis?
5. Green plants do not absorb what color of light?
6. Blue, red, and green light are all present in what color of light?
7. Why do blue and red light, but not green, promote photosynthesis?
8. Does *Elodea* respire in the light or in the dark?
9. If net photosynthesis is 5 mm/10 min and cellular respiration is .5 mm/10 min, how much is gross photosynthesis?
10. Phenol red turns what color when carbon dioxide is added?
11. What happens to carbon dioxide during photosynthesis?
12. What two substances do plants provide for us?
13. Where, within the chloroplast, does the carbon cycle take place?

Thought Questions

14. Some plants are colorless. Do you predict that they carry on photosynthesis? Explain.

15. Suppose there were single-celled protozoans (nonphotosynthetic) in the test tube with *Elodea* when you did the white light experiment. Could you still calculate *Elodea*'s rate of photosynthesis? Explain.

16. You should note that the light reactions of photosynthesis do not directly produce carbohydrates. If this is the case, why are these reactions required for the photosynthetic process?

LABORATORY

9

Organization of Flowering Plants

Learning Outcomes

9.1 Plant Organs

- Distinguish between the shoot system and the root system of a plant.
- Identify the external anatomical features of a flowering plant.
- State the functions of a leaf, a stem, and a root.
- List five differences between monocots and eudicots.

9.2 Organization of Leaves

- Distinguish between a monocot and a eudicot leaf.
- Identify a cross section and the specific tissues of a eudicot leaf.

9.3 Organization of Stems

- Identify a cross section and the specific tissues of herbaceous eudicot and herbaceous monocot stems and a woody stem.
- Distinguish between primary and secondary growth of a stem.
- Explain the occurrence of annual rings and determine the age of a tree from a trunk cross section.

9.4 Organization of Roots

- Name the zones of a eudicot root tip.
- Identify a cross section and the specific tissues of a eudicot root.

9.5 Xylem Transport

- Explain the continuous water column in xylem.
- Hypothesize the rate of transpiration under varied conditions.

Planning Ahead

Your instructor may advise you to set up the transpiration experiment (see page 115) early to make the best use of laboratory time.

Introduction

Despite their great diversity in size and shape, flowering plants all have three vegetative organs that have nothing to do with reproduction: the leaf, the stem, and the root. Leaves carry on photosynthesis and thereby, produce the nutrients that sustain a plant. A stem usually supports the leaves so that they are exposed to sunlight. Roots anchor a plant and absorb water and minerals from the soil.

Each of these organs contains various tissues, which are arranged differently depending on whether a flowering plant is a monocot or a eudicot. The arrangement of tissues is distinctive enough that you should be able to identify the plant as a monocot or eudicot when examining a slide of a leaf, stem, or root. Only eudicots are ever woody.

Plants may also be grouped according to whether they are herbaceous or woody. All flowering trees are woody; their stems contain wood. Many flowering garden plants and all grasses are herbaceous (nonwoody). Herbaceous plants have only primary growth, which increases their height. Woody plants have both primary and secondary growth. Secondary growth increases the girth of a tree. In addition to groupings based on structural similarities, plants may be grouped on the basis of their lifespan. Annual plants live for only a single season. At the other extreme, perennial plants may live for many years. In fact, plants hold the record for the longest-living organisms on earth. Some live for several thousand years.

Xylem is the vascular tissue that transports water up from the roots to the leaves. The cohesion-tension model of xylem transport explains how the continuous column of water in xylem is able to rise to the top of a tall tree. During so-called transpiration, water evaporates at openings in leaves called stomata (sing., stoma). This creates a tension that pulls the water column up only because of the cohesive property of water.

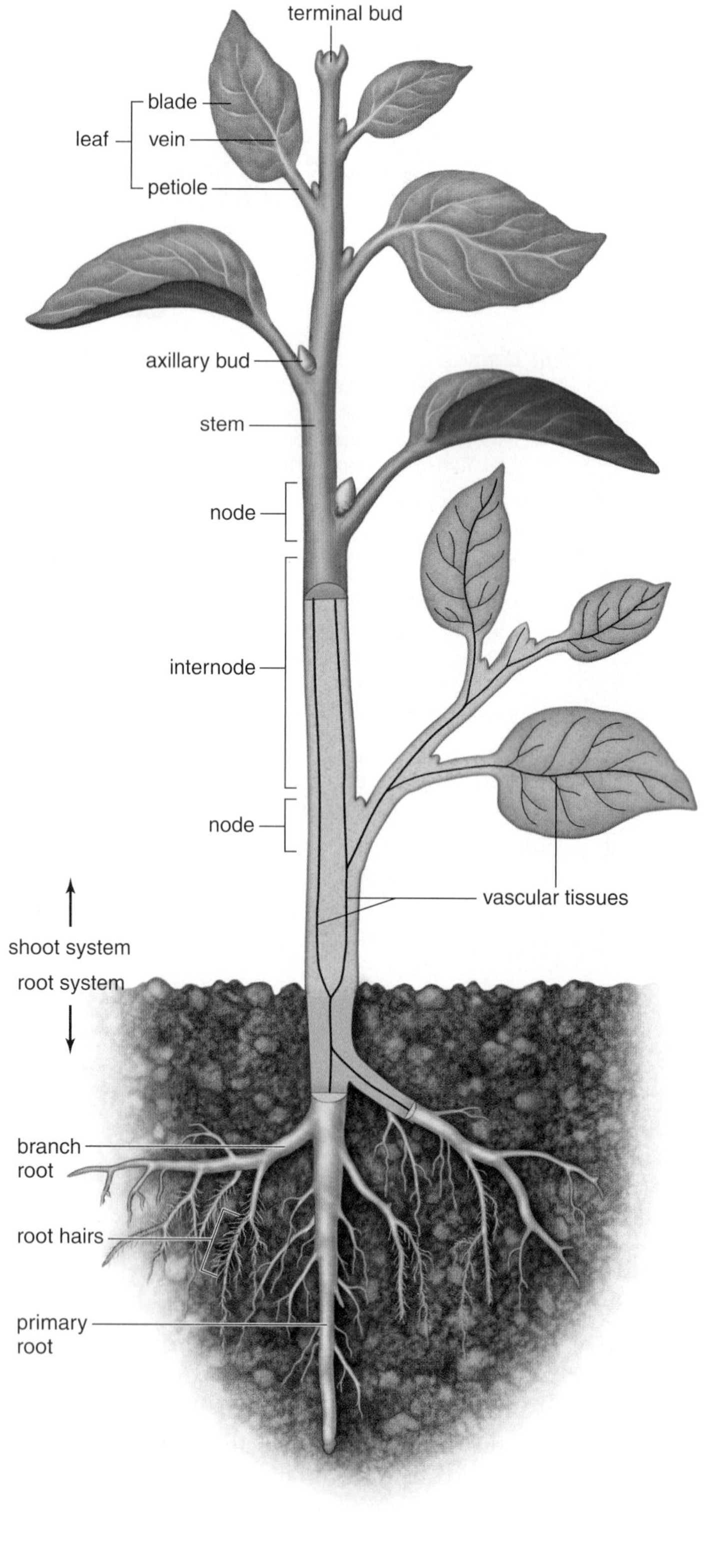

Figure 9.1 Organization of a plant.
Roots, stems, and leaves are vegetative organs. The flower and fruit are reproductive structures.

9.1 Plant Organs

Figure 9.1 shows that a plant has a root system and a shoot system. The **root system** consists of a primary root and all of its lateral (side) roots. The **shoot system** consists of the stem and leaves.

Observation: A Living Plant

Shoot System

What is the primary function of the shoot system? ______________________

The Leaves

1. Describe the **blade.** ______________________
2. Describe the **petiole.** ______________________

The Stem

1. Observe the **stem.** Locate a **node** and an **internode.**
2. Measure the length of the internode in the middle of the stem. Does the internode get larger or smaller toward the apex of the stem? __________ Toward the roots? __________ Based on the fact that a stem elongates as it grows, explain your observation. ______________________

   ____
3. Where is the **terminal bud** of a stem?

   ____

   Where is the **axillary bud?**

   ____

Root System

Observe the root system of a living plant if the root system is exposed. Does the plant have a taproot system—that is, a main root many times larger than the lateral roots? ________ Or, does the plant have a fibrous root system—that is, all the roots approximately the same size? ________ What is the primary function of the root system? ______________________________

What advantages does a taproot provide to a plant?______________________________

Monocots Versus Eudicots

Flowering plants are classified into two groups: **monocots** and **eudicots.** In this laboratory, you will be studying the differences between monocots and eudicots with regard to the leaves, stems, and roots, as noted in Figure 9.2.

Experimental Procedure: Monocot Versus Eudicot

1. Observe the leaves of the plant you are studying. Based on Figure 9.2, is this plant a monocot or a eudicot? __________ Explain. ______________________________
2. Observe any other available types of leaves, and note in Table 9.1 the name of the plant and whether it is a monocot or a eudicot.

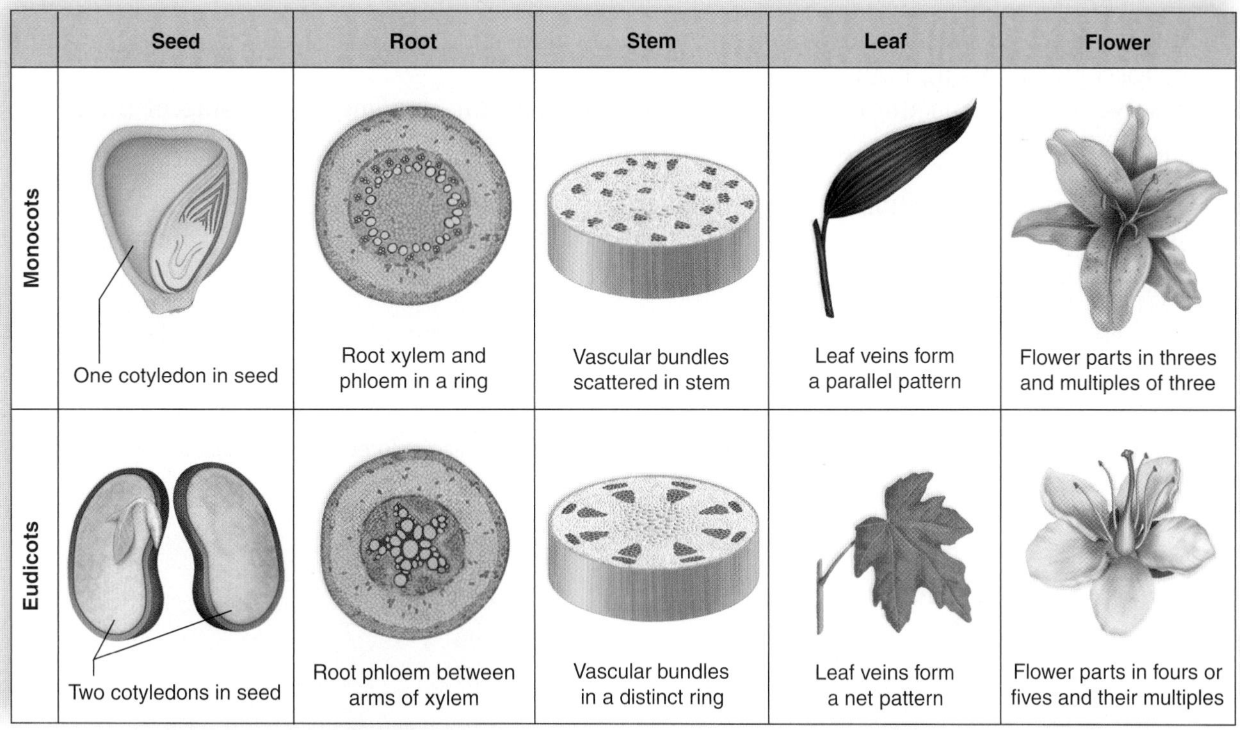

	Seed	Root	Stem	Leaf	Flower
Monocots	One cotyledon in seed	Root xylem and phloem in a ring	Vascular bundles scattered in stem	Leaf veins form a parallel pattern	Flower parts in threes and multiples of three
Eudicots	Two cotyledons in seed	Root phloem between arms of xylem	Vascular bundles in a distinct ring	Leaf veins form a net pattern	Flower parts in fours or fives and their multiples

Figure 9.2 Monocots versus eudicots.
The five features illustrated here are used to distinguish monocots from eudicots.

Table 9.1 Monocots Versus Eudicots		
Name of Plant	**Organization of Leaf Veins**	**Monocot or Eudicot?**
1		
2		
3		
4		

9.2 Organization of Leaves

Leaves are generally broad and quite thin to better capture solar energy. Carbon dioxide enters a leaf at openings called **stomata** (sing., *stoma*), and water enters by way of **leaf veins,** which are extensions of the vascular bundles from the stem.

Observation: Stomata

1. Obtain a *leaf* from a plant designated by your instructor, and put a drop of *distilled water* on a slide.
2. Using the technique shown in Figure 9.3*a*, obtain a strip of outer tissue from a leaf. This tissue is epidermis.
3. Put the tissue in the drop of water on the slide, outer side up. Add a coverslip, and examine it microscopically, using both low power and high power.
4. Observe a stoma and the two guard cells that regulate the opening and closing of the stoma (Figure 9.3*b*). What gas enters a leaf at the stomata? ________ What gas exits the leaf? ________
5. Count the number of stomata in the high-power field of view: ________.
6. Assume that the area of the high-power field is 0.10 mm. Divide the number of stomata by this area to determine the number of stomata in 1 square millimeter: ________.
7. Does your leaf contain a large number of stomata per square millimeter? ________.

Figure 9.3 Stomata.
a. Method of obtaining a strip of epidermis from the underside of leaf. **b.** False-colored scanning electron micrograph of leaf surface.

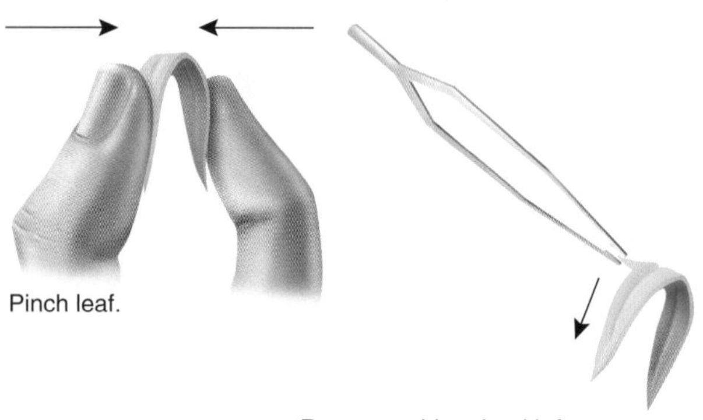
Pinch leaf.
Remove epidermis with forceps.

a.

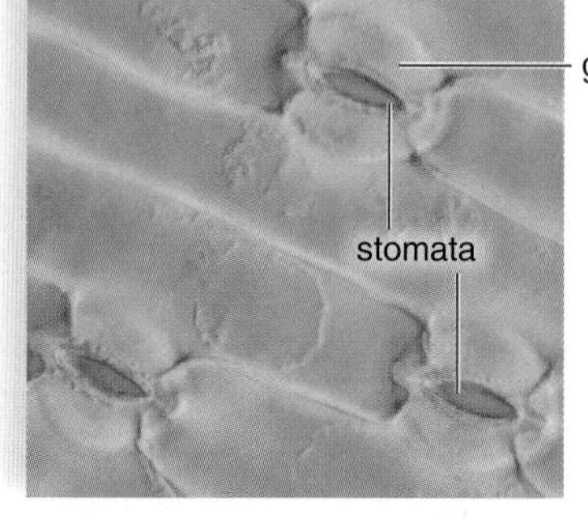
guard cell
stomata

b.

Observation: Leaves

1. Obtain a slide of *Liqustrum* (common privet) leaf or *Syringa* (barley) leaf in cross section.
2. With the help of Figure 9.4, identify:
   a. **Cuticle:** The outermost layer that protects the leaf and prevents water loss.
   b. **Upper and lower epidermis:** A single layer of protective cells at the upper and lower surfaces.
   c. **Leaf vein:** Transports water and organic nutrients. What tissue does a leaf vein contain? ____________ and ____________
   d. **Palisade mesophyll:** Located near the upper epidermis. *Label the palisade mesophyll in Figure 9.4.*
   e. **Spongy mesophyll:** Located near the lower epidermis. *Label the spongy mesophyll in Figure 9.4.*

      Which type of mesophyll has chloroplasts? ____________________

      Which type of mesophyll carries on photosynthesis? ____________________

      Which type of mesophyll has air spaces that facilitate exchange of gases? ____________

   f. **Stomata:** The openings you examined in the previous Observation.

Figure 9.4 Leaf anatomy.
Complete the labeling as directed by the Observation.

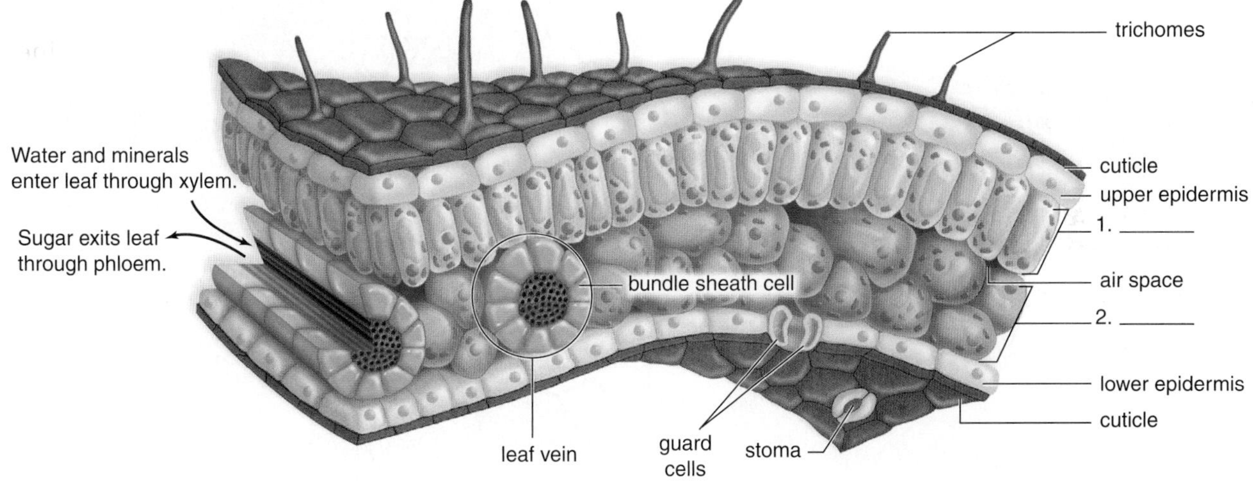

9.3 Organization of Stems

Stems can be nonwoody (herbaceous), or they can be woody. Monocots are usually herbaceous, whereas eudicots can be either herbaceous or woody. Herbaceous plants usually die back during the winter, while woody stems reflect growth over a period of years.

Observation: Herbaceous Stems

Herbaceous Eudicot Stem

1. Obtain a slide of *Helianthus* (sunflower) stem, in cross section.
2. In a herbaceous stem, xylem and phloem occur in a vascular bundle. Use the images of **vascular bundles** in Figure 9.5*a* and *b* to identify a vascular bundle on your slide. With the help of Figure 9.2, describe the arrangement of the vascular bundles. ______________________

   ____

   Which tissue—xylem or phloem—is closer to the stem surface? ______________________
3. Also identify the **cortex** and the **pith** in your slide.

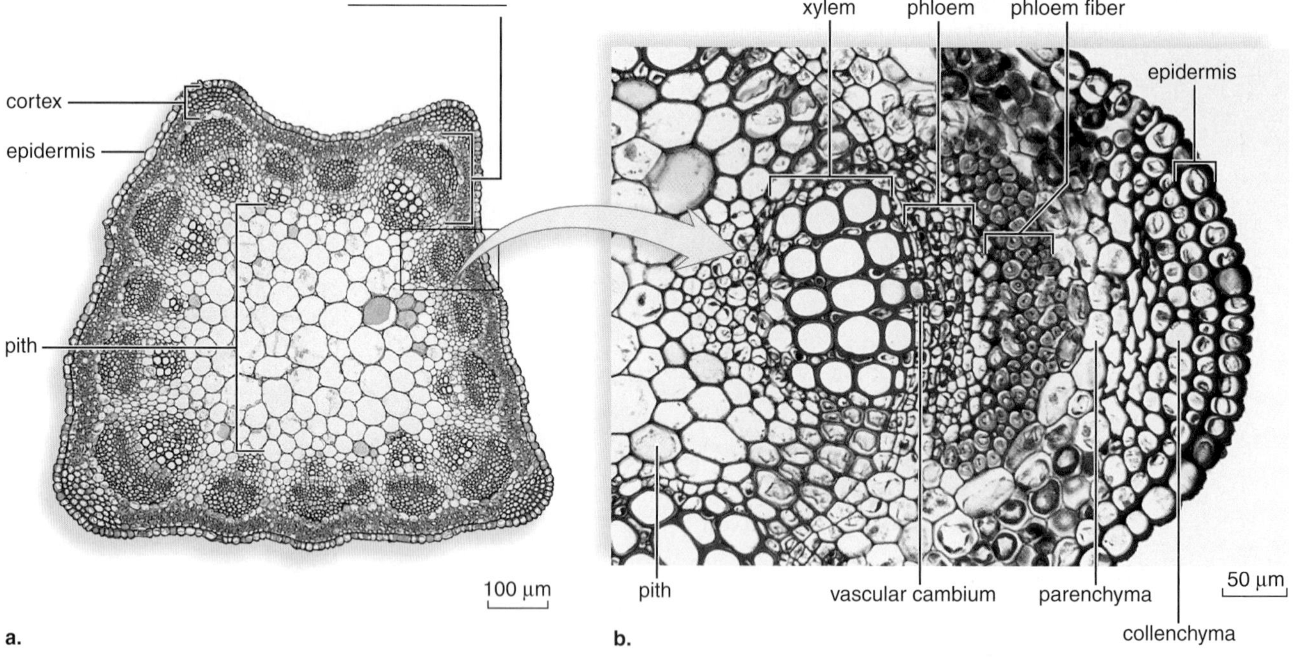

Figure 9.5 Eudicot herbaceous stem.
The vascular bundles are in a definite ring in this photomicrograph of an eudicot herbaceous stem. Complete the labeling as directed by the Observation.

Monocot Stem

1. Obtain a slide of *Zea mays* (corn) stem, in cross section.
2. With the help of Figure 9.6, locate the same four tissues that you found in the herbaceous eudicot stem.
3. Describe the arrangement of the vascular bundles. ______________________

Figure 9.6 Monocot stem.
The vascular bundles, one of which is enlarged, are scattered in this photomicrograph of a monocot herbaceous stem.

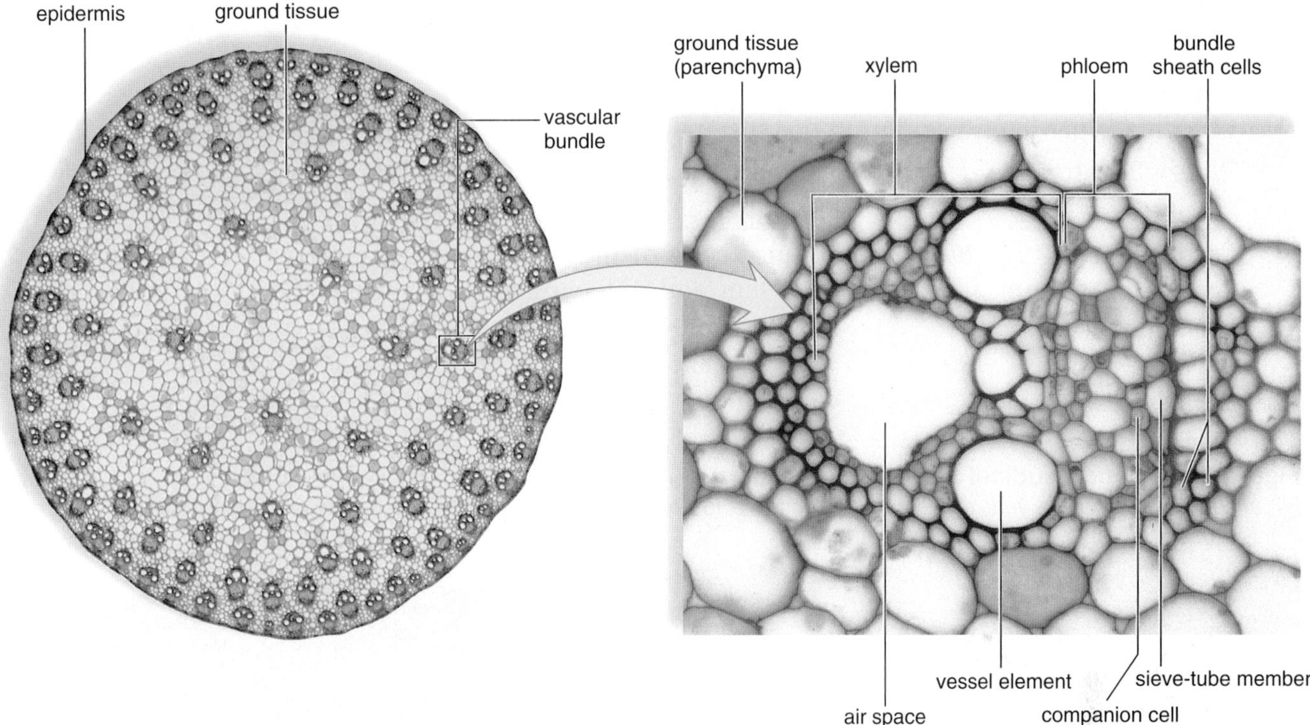

Woody Eudicot Stem

Woody eudicot stems undergo both primary and secondary growth. Primary growth occurs as the stem grows taller. Secondary growth occurs as the girth of a woody stem increases.

Observation: Woody Eudicot Stem

1. Obtain a slide of *Tilia* (tulip tree) stem, in cross section.
2. With the help of Figure 9.7, identify the three main parts of a woody stem: the bark, the wood, and the pith.
3. The **bark** contains

   Cork, a protective outer layer

   Cortex, which stores nutrients

   Phloem, which is in use during the current year

   Why will a tree die if you remove the bark all the way around the trunk? ____________________

   ____
4. Locate *vascular cambium* at the inner edge of the bark, between the bark and the wood. Vascular cambium produces new xylem and phloem every year. Only the xylem builds up year after year.

Figure 9.7 Woody eudicot stem cross section.
Because xylem builds up year after year, it is possible to count the annual rings to determine the age of a tree. This tree is three years old.

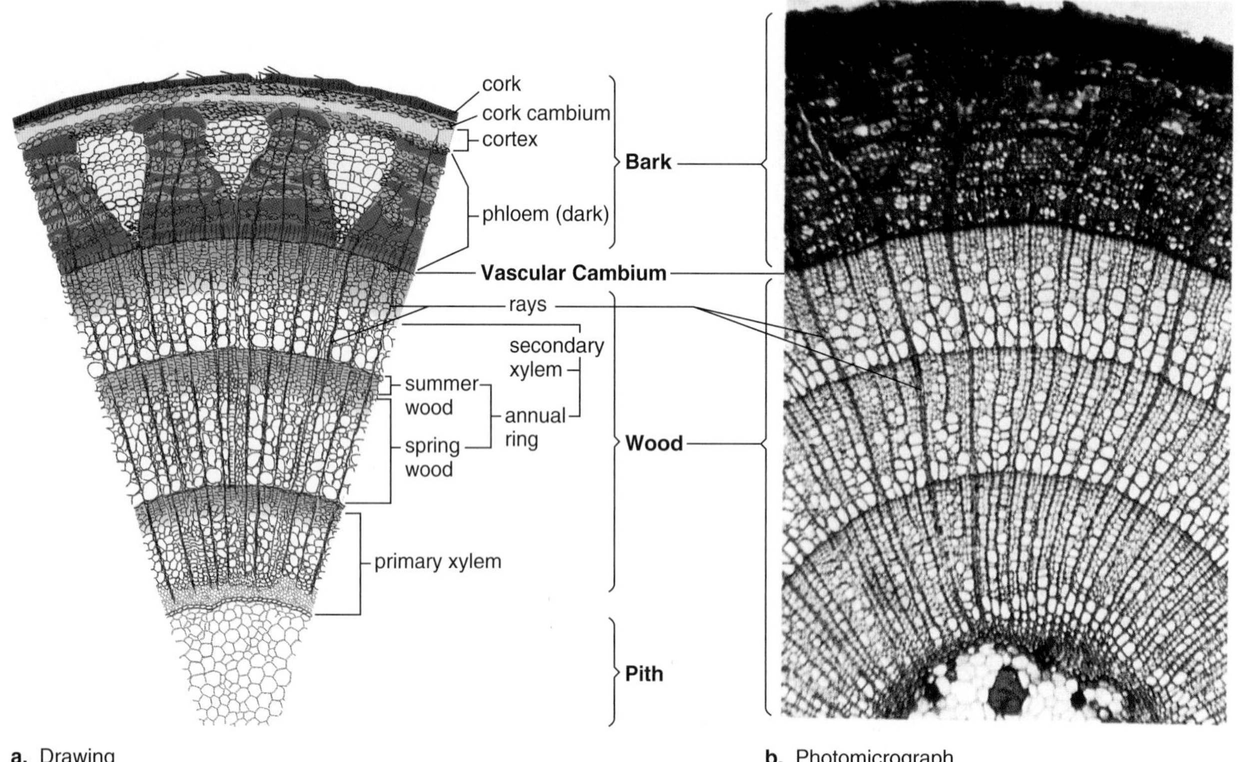

5. The **wood** contains annual rings. An **annual ring** is the amount of xylem added to a tree during one growing season. Rings appear to be present because spring wood has large xylem vessels and looks light in color, while summer wood has much smaller vessels and appears much darker. How old is the stem you are observing? ________

   Are all the rings the same width? ________
6. Identify the **pith,** a tissue that stores organic nutrients and may disappear.
7. Locate **rays,** groups of small, almost cuboidal cells that extend out from the pith laterally.

Observation: Anatomy of a Winter Twig

1. A winter twig typically shows several years' past growth. Examine several examples of winter twigs (Fig. 9.8), and identify the **terminal bud** located at the tip of the twig. This is where new growth will originate.
2. Locate a **terminal bud scar.** These scars encircle the twig and indicate where the terminal bud was located in previous years. The distance between two adjacent terminal bud scars equals one year's growth.
3. Find a **leaf scar.** This is where a leaf was attached to the stem.
4. Note the **vascular bundle scars.** Complete this sentence: Vascular bundle scars appear where the vascular bundles ____________________.
5. Identify a **node.** This is where a leaf was attached.
6. Locate an **axillary bud.** This is where new branch growth can occur.

Figure 9.8 External structure of a winter twig.
Counting the terminal bud scars tells the age of a particular branch.

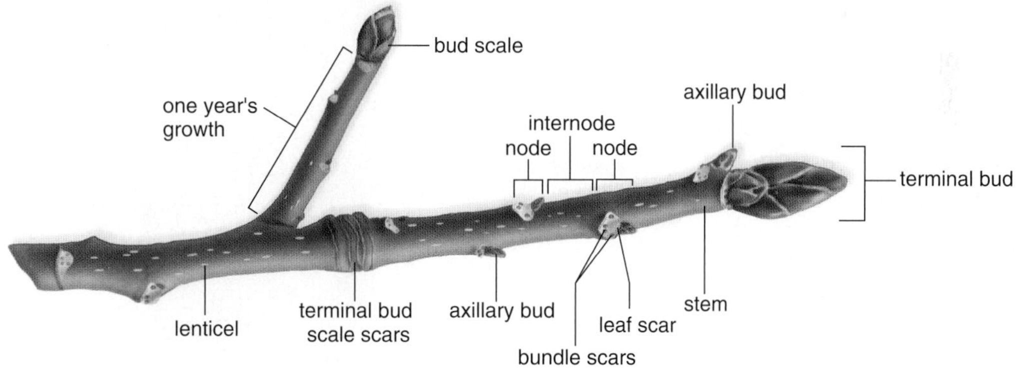

9.4 Organization of Roots

First you will study a eudicot root tip in longitudinal section and then a eudicot root in cross section. You will also examine the root hairs of a living plant.

Observation: Eudicot Root Tip

1. Obtain a model and/or a slide of a eudicot root tip.
2. With the help of Figure 9.9, identify:
   a. **Root cap:** Covers the growing tip. What is the function of the root cap? ______________
   b. **Zone of cell division:** In this zone, new cells are being produced.
   c. **Zone of elongation:** In this zone, rows of newly produced cells elongate. Which two zones are responsible for growth of the root tip?

      ____

      ____
   d. **Zone of maturation:** In this zone, the cells are specialized to carry on a particular function. You can recognize the zone of maturation because of the presence of root hairs. **Root hairs** increase the area for absorption of what by a root? ______________

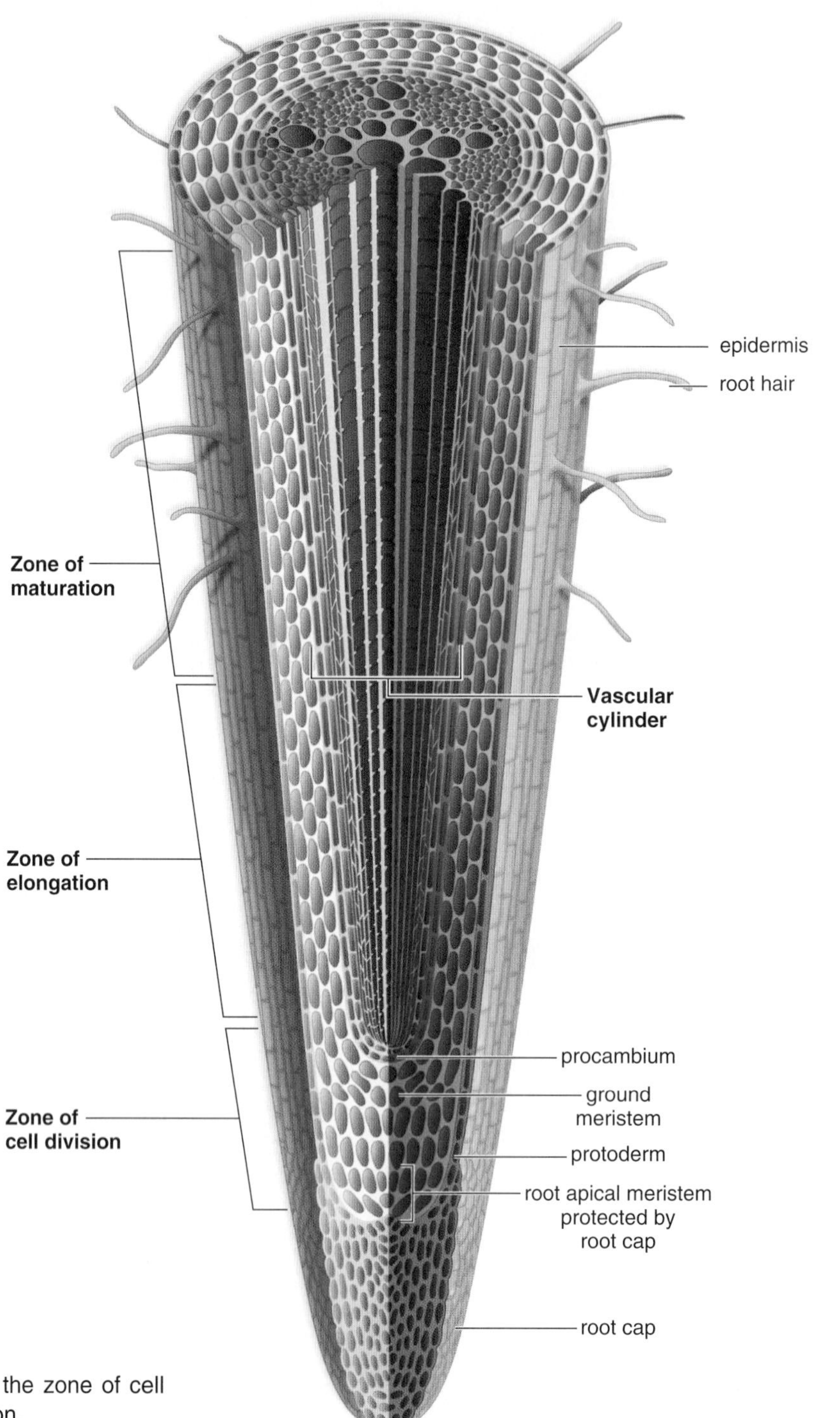

Figure 9.9 Eudicot root tip.
In longitudinal section, the root cap is followed by the zone of cell division, zone of elongation, and zone of maturation.

Figure 9.10 Eudicot root cross section.
The vascular cylinder of a dicot root contains the vascular tissue. Xylem is typically star-shaped, and phloem lies between the points of the star. **a.** Drawing. **b.** Micrograph.

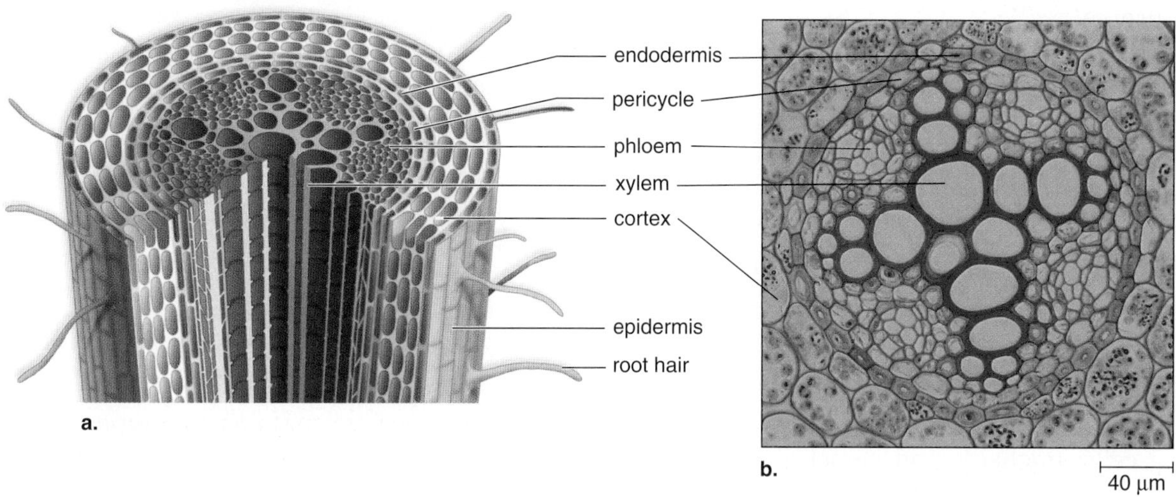

Observation: Eudicot Root Slide

1. Obtain a slide of *Ranunculus* (buttercup) root, in cross section.
2. With the help of Figure 9.10 identify:
   a. **Epidermis:** The outermost layer of small cells that gives rise to **root hairs.** Notice in Figure 9.10 that the epidermis is a single layer of cells.
   b. **Cortex:** Located just inside the epidermis. How many layers of thin-walled cells are present? ________ The cortex stores the products of photosynthesis. Label starch grains in Figure 9.10*b.*
   c. **Endodermis:** A single layer of cells following the cortex. Endodermal cells have a waxy layer, called the Casparian strip, on four sides. This means that water and minerals have to pass through endodermal cells and not around them. For this reason, the endodermis regulates what materials enter a plant through the root.
   d. **Pericycle:** A layer one or two cells thick, just inside the endodermis. Lateral roots originate from this tissue.
   e. **Vascular tissue** (xylem and phloem): **Xylem** has several "arms" that extend like the spokes of a wheel. This tissue conducts water and minerals from the roots to the stem. **Phloem** is located between the arms of xylem. Phloem conducts organic nutrients from the leaves to the roots and other parts of the plant.
3. Trace the path of water as it crosses a root from a root hair to xylem: ____________________

____

____

Monocot Root

You will not examine a monocot root. Refer back to Figure 9.2, and state here how monocot and eudicot roots differ in the organization of vascular tissue: ____________________________

____

Observation: Root Hairs

1. Obtain a young germinated seedling and float it in some water in a petri dish while you observe it.
2. Use the binocular dissecting microscope and locate the root tip. Note the root cap and the region where the root hairs have formed on the root surface. What proportion of the root has root hairs? ______________________
3. Remove the root from the seedling, and make a wet mount of the root, using 0.1% neutral red.
4. Observe your slide under the microscope. Does every epidermal cell have a root hair? ________
   How do root hairs aid absorption? ______________________

9.5 Xylem Transport

Xylem (Fig. 9.11), which transports water from the roots to the leaves, contains two types of conducting cells: tracheids and vessel elements. Both types of conducting cells are hollow and nonliving, the vessel elements are larger, they lack transverse end walls, and they are arranged to form a continuous pipeline for water and mineral transport.

The Water Column

The water column in xylem is continuous because water molecules are cohesive (they cling together) and because water molecules adhere to the sides of xylem cells. Therefore, water evaporation from leaf surfaces creates a negative pressure that pulls water upward.

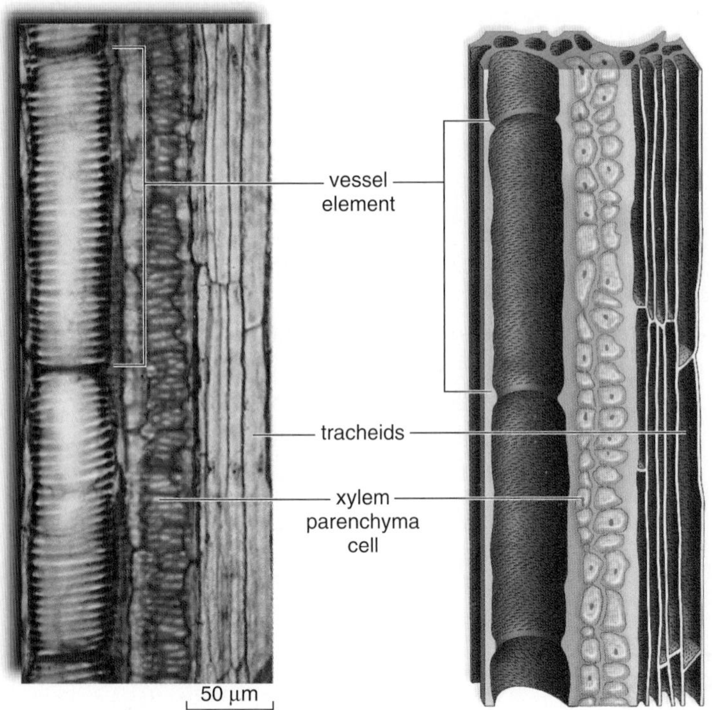

Figure 9.11 Xylem structure. Xylem contains two types of conducting cells: tracheids and vessel elements. Tracheids have pitted walls, but vessel elements are larger and form a continuous pipeline from the roots to the leaves.

Experimental Procedure: The Water Column

1. Place a small amount of *red-colored water* in two beakers. Label one beaker "wet" and the other beaker "dry."
2. Transfer a stalk of *celery* (which was cut and then immediately placed in a container of water) into the "wet" beaker so that the large end is in the colored water.
3. Transfer a stalk of *celery* of approximately the same length and width (but that was kept in the air after being cut) into the "dry" beaker so that the large end is in the colored water.
4. With scissors, cut off the top end of each stalk, leaving about 10 cm total length.
5. Time how long it takes for the red-colored water to reach the top of each stalk, and record these data in Table 9.2.
6. In which celery stalk was the water column broken?

   ____

   ____

   Use this information to write a conclusion in Table 9.2.
7. Using the directions that follow for cutting freehand sections, make a cross-sectional wet mount of the stalk in the "wet" beaker. Observe this slide under the microscope. What type of tissue has been stained by the dye? ______________________

Table 9.2 Celery Stalk Experiment

Stalk	Speed of Dye (Minutes)	Conclusion
Cut end placed in water prior to experiment		
Cut end kept in air prior to experiment		

Transpirational Pull

Evaporation of water from leaves is called **transpiration.** As transpiration occurs, the continuous water column is pulled upward—first within the leaf, then from the stem, and finally from the roots.

Experimental Procedure: Transpiration

Assembling a Transpirometer

1. Tightly fit one end of a 4 cm piece of *rubber tubing* over one end of a 15–20 cm long *glass tube* (or glass pipette) (Fig. 9.12*a*).
2. Immerse the assembled glass and rubber tube in a large tub of *water* so that it will fill. Check that there are no air bubbles in the tube.
3. Cut off a portion of a *geranium plant* (stem with five to seven leaves), and place the cut end of this stem into the tub of water, but keep the leaves dry (Fig. 9.12*a*). With a sharp razor blade held under water, cut off a 1.5 cm piece of the stem at an oblique angle. (If this were done in the air, a vacuum would occur in the vascular system.)
4. Keeping the leaves dry and the stem end submerged, fit the cut stem end into the rubber tubing that has been filled with water. Squeeze out all of the air bubbles. If necessary, tightly wind a rubber band or tie a string around the juncture (while still submerged) for added tightness, but do not crush the stem.

Figure 9.12 Assembling a transpirometer.
The text gives complete directions for **(a)** assembling the transpirometer apparatus and **(b)** shows how the finished assembly should be clamped to a ring stand.

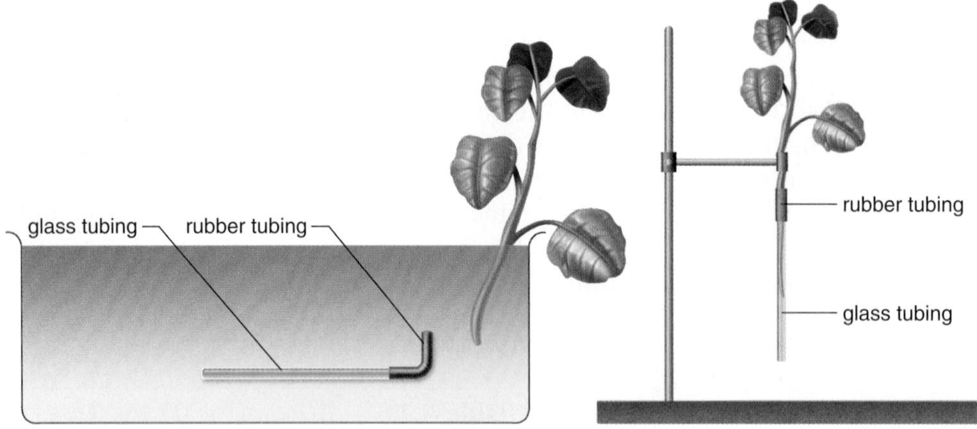

a. Assemble the transpirometer apparatus under water. **b.** Clamp the finished assembly to a ring stand.

5. Remove the assembled apparatus from the water. Hold it so that the geranium stem cutting is upright.
6. Clamp the assembly to a *ring stand* (Fig. 9.12*b*). The water will not run out of the tube due to water's adhesive and cohesive properties.

Determining Transpiration Rate Under Standard Conditions

1. Wait 5 minutes, and then mark with a wax pencil the water level at the lower end of the transpirometer.
2. Then, every 10 minutes for the next 40 minutes, mark the water level, and measure (in millimeters) the distance the water has moved. Record your data in the first two columns of Table 9.3. These figures indicate the millimeters of water transpired.

Determining Transpiration Rates Under Varied Environmental Conditions

1. Your instructor will assign you one of the following conditions under which to repeat the preceding experiment:
   a. Focus a light source on the plant (to simulate heat). The plant should be located at least 25 cm away from the light source. How do you predict an increase in *temperature* will affect the rate of transpiration?

      ____

   b. Spray the plant and the inside of a plastic bag with water (to simulate a rise in humidity). Put the plastic bag over the plant, and use string to draw it closed around the tubing. How do you predict *humidity* will affect the rate of transpiration?

      ____

   c. Use a small fan to gently blow air across the plant (to simulate wind). How do you predict *wind* will affect the rate of transpiration?

      ____

2. Remove your previous marks from the glass tubing. Again wait 5 minutes and then, as before, measure the distance the water level moves every 10 minutes for a total of 40 minutes. Record your measurements in Table 9.3.
3. Plot the results of your two experiments on the graph provided, using one color to show standard conditions and a different color to show the varied environmental condition you tested. The transpiration rate will be the total change in millimeters between readings.

Table 9.3 Effect of [Temperature, Humidity, Wind]* on Transpiration Rate				
	Standard Conditions		*Test Conditions*	
Time	**Reading (mm)**	**Total Change (mm)**	**Reading (mm)**	**Total Change (mm)**
After 10 minutes				
After 20 minutes				
After 30 minutes				
After 40 minutes				

* Circle the condition you tested.

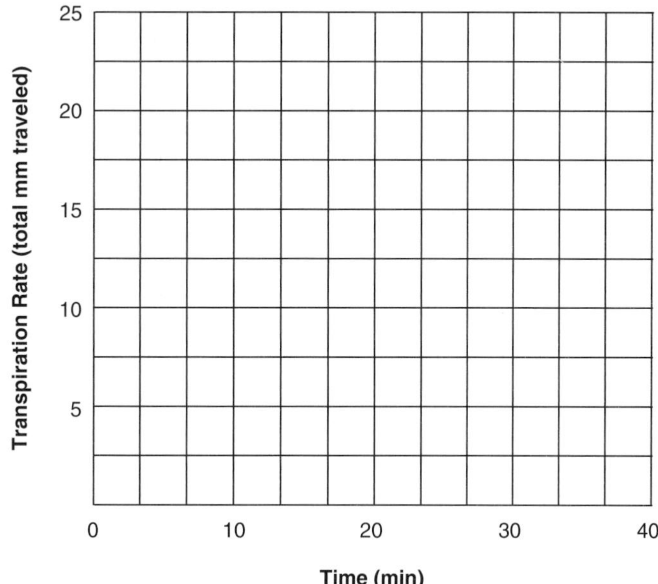

Conclusions

Fill in one of the following according to your own data. Fill in the other two according to data collected by other laboratory groups.

- Effect of temperature on transpirational rate. In general, what effect did an increase in temperature have on transpiration rate? ______________________

  Is this what you predicted? __________ Why or why not? ______________________

  ____
- Effect of humidity on transpirational rate. In general, what effect did an increase in humidity have on transpiration rate? ______________________

  Is this what you predicted? __________ Why or why not? ______________________

  ____
- Effect of wind on transpirational rate. In general, what effect did an increase in wind have on transpiration rate? ______________________

  Is this what you predicted? __________ Why or why not? ______________________

  ____

Laboratory Review 9

_______________ 1. The leaves attach to what portion of a stem?

_______________ 2. What type of venation do monocot leaves have?

_______________ 3. State the function of the structures called stomata that are present in leaf epidermis.

_______________ 4. What are the cells between the upper and lower epidermis of the leaf called?

_______________ 5. On a slide, what structures cause stomata to appear either open or closed?

_______________ 6. What is the pattern for vascular bundle distribution in a monocot stem?

_______________ 7. In woody stems, the bark is divided from the wood by what tissue?

_______________ 8. Identify the slide if there are annual rings present.

_______________ 9. Pith is most likely present in a _______________.

_______________ 10. What zone follows the zone of cell division in a root tip?

_______________ 11. Lateral roots develop from which cell layer?

_______________ 12. Epidermis is modified in a root by the addition of _______________.

_______________ 13. Identify the slide if the center tissue has several "arms" that extend like the spokes of a wheel.

_______________ 14. What type of tissue transports materials in a plant?

_______________ 15. List two similarities and two differences between monocots and eudicots.

Thought Questions

16. If only slides of root and stem were available to you, how could you identify a plant as an herbaceous eudicot?

17. Contrast the manner in which water reaches the inside of a leaf with the manner in which carbon dioxide reaches the inside of a leaf.

18. What advantage does the waxy cuticle provide to the leaf?

LABORATORY

10

Reproduction in Flowering Plants

Learning Outcomes

10.1 Introduction
- State the function of fruits and seeds.

10.2 Flowers
- Identify the parts of a flower.
- Distinguish between monocot and eudicot flowers.
- Describe the life cycle of flowering plants.
- Describe the developmental stages of a eudicot embryo.

10.3 Fruits
- Classify simple fruits as fleshy or dry, and then use a dichotomous key to identify the specific fruit type.

10.4 Seeds
- Identify the parts of a seed and of an embryonic plant, and distinguish between eudicot and monocot seeds.
- Describe the germination of eudicot and monocot seeds.

10.5 Seed Germination
- Predict the effect of acidic conditions on seed germination.

Introduction

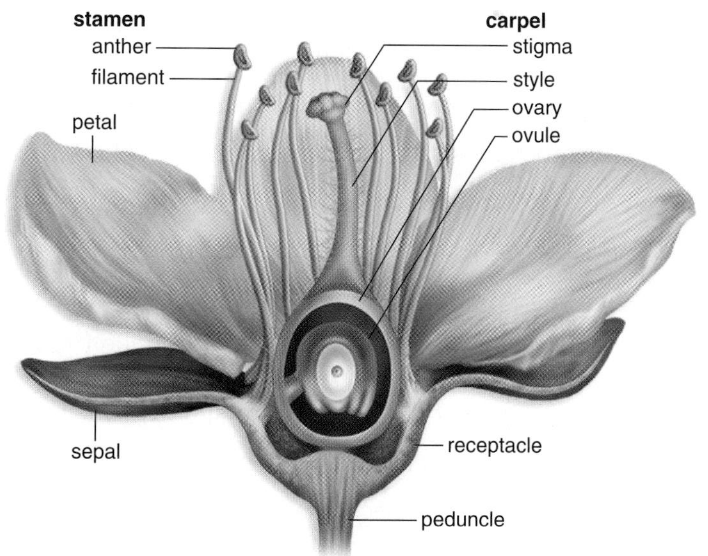

Flowers are the reproductive structure of angiosperms. **Pollination** involves the transport of pollen from an **anther,** where pollen is produced, to the **stigma** portion of the **carpel.** While some flowers are able to pollinate themselves, transferring pollen from their own anthers to their own carpels, most plants rely on cross pollination, wherein the pollen of one plant is transferred to another plant of the same species. Plants with bright, showy flowers generally rely on insects, birds, or other animals to accomplish their pollination. Plants with inconspicuous flowers generally rely on the wind to disperse their pollen. Once transferred to the stigma, pollen germinates and produces a pollen tube. The tube grows through the **style** and into the **ovary,** where there are **ovules,** each containing an egg. One sperm from the pollen tube fertilizes an egg in the ovule. The ovule becomes a seed containing an embryo enclosed within a fruit. The **fruit** develops from the ovary and, at times, from accessory structures. Fruits assist the dispersal of angiosperm seeds. When animals eat fleshy fruits, they may ingest the seeds and defecate them some time later. Coconuts are dispersed by oceanic currents. Some lightweight fruits with wings and some with seed hairs are dispersed by wind. Seeds contain an embryonic plant. When seeds germinate, a new plant begins to develop.

10.1 Flowers

Angiosperms are divided into two groups: monocots and eudicots. **Monocots** have flower parts in threes or multiples thereof. **Eudicots** have flower parts in fours or fives or multiples thereof. Complete flowers contain stamens, carpels, sepals, and petals. Flowers that lack one of these basic parts are called incomplete.

Observation: A Flower

1. Examine a flower model or living flower. Use Figure 10.1 to help you identify:
   a. **Sepals:** The outermost set of modified leaves, collectively termed the **calyx.** Sepals are green in most flowers.
   b. **Petals:** The inner leaves that collectively constitute the **corolla.** Petals often have a design and color that attract specific pollinators, such as bees and butterflies.
   c. **Stamen:** A swollen terminal **anther,** where pollen is produced, and the slender **filament** that supports it.
   d. **Carpel:** This structure consists of a swollen basal ovary; a long, slender style (stalk); and a terminal stigma (sticky knob).
   e. **Ovary:** The enlarged part of the carpel that develops into a fruit.
2. Is the flower you are examining a monocot or eudicot? ____________________

   Explain. ____________________
3. Is the specimen you are examining a complete or an incomplete flower? ____________________

Life Cycle of Flowering Plants

Figure 10.1 describes the life cycle of flowering plants. Use the figure to identify the six major steps in this life cycle.

1. The parts of the flower involved in reproduction are the ______ and the ______.
2. The anther at the top of the stamen has ______, which contain numerous microsporocytes that undergo meiosis to produce ______. The carpel contains an ovary that encloses ______.
3. Within the ovule, a megasporocyte undergoes meiosis to produce four ______. Each microspore becomes a(n) ______.
4. One megaspore develops into a(n) ______. After pollination, the pollen cell of a pollen grain divides to produce two ______. The pollen grain develops a pollen tube that travels down the style of the carpel.
5. During double fertilization, one sperm from the pollen tube fertilizes the egg within the embryo sac, and the other joins with two ______ nuclei.
6. The fertilized egg becomes a(n) ______, and the joining of polar nuclei and sperm becomes the 3n ______. A seed contains the three parts that are labeled in Figure 10.1. In angiosperms, seeds are enclosed by fruits.

Figure 10.1 Flowering plant life cycle.
In flowering plants, meiosis produces microspores that develop into male gametophytes (pollen grains) and megaspores that develop into female gametophytes (embryo sacs).

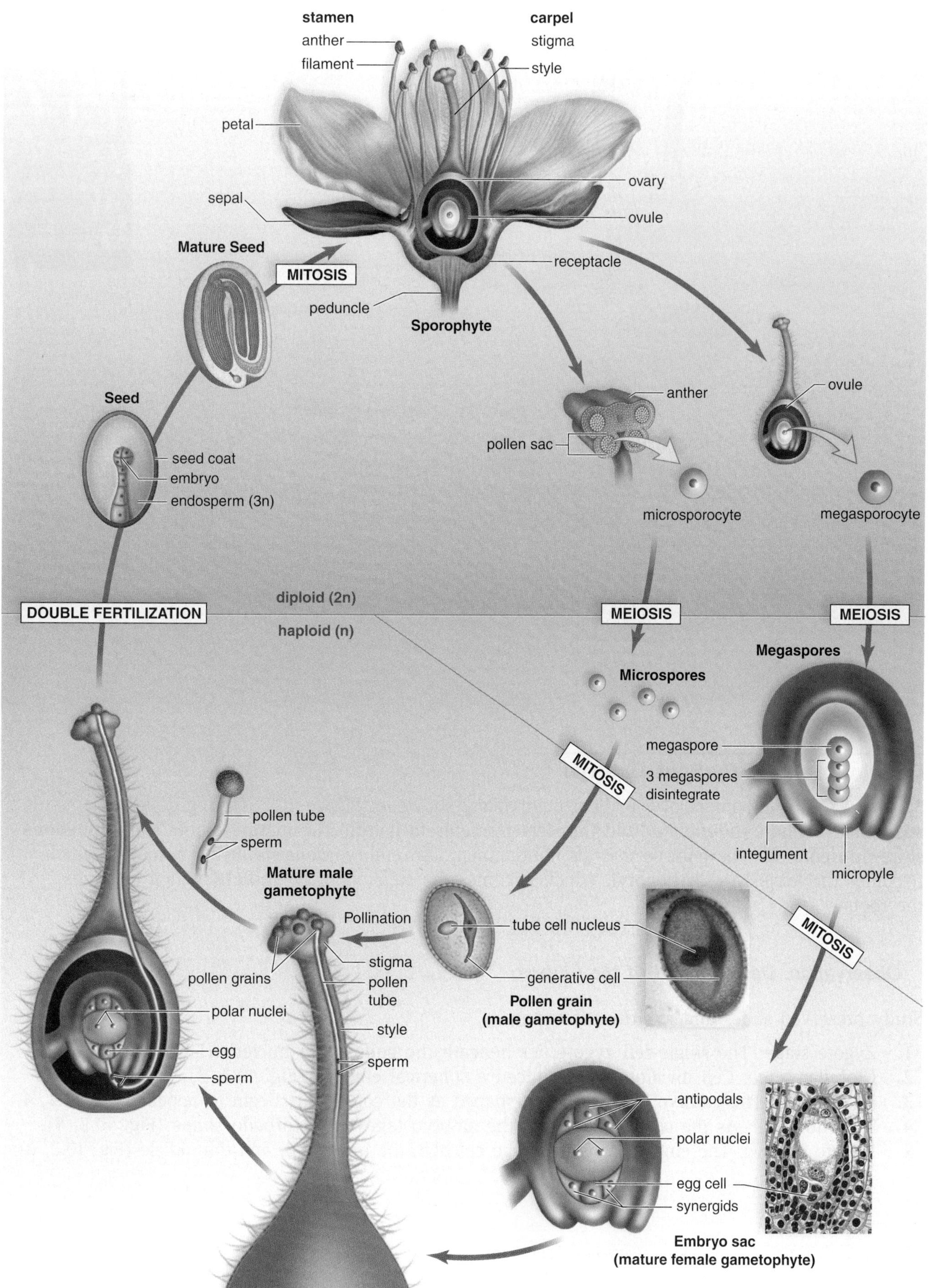

Figure 10.2 Development of a eudicot embryo.
Embryogenesis consists of these stages, described in the text.

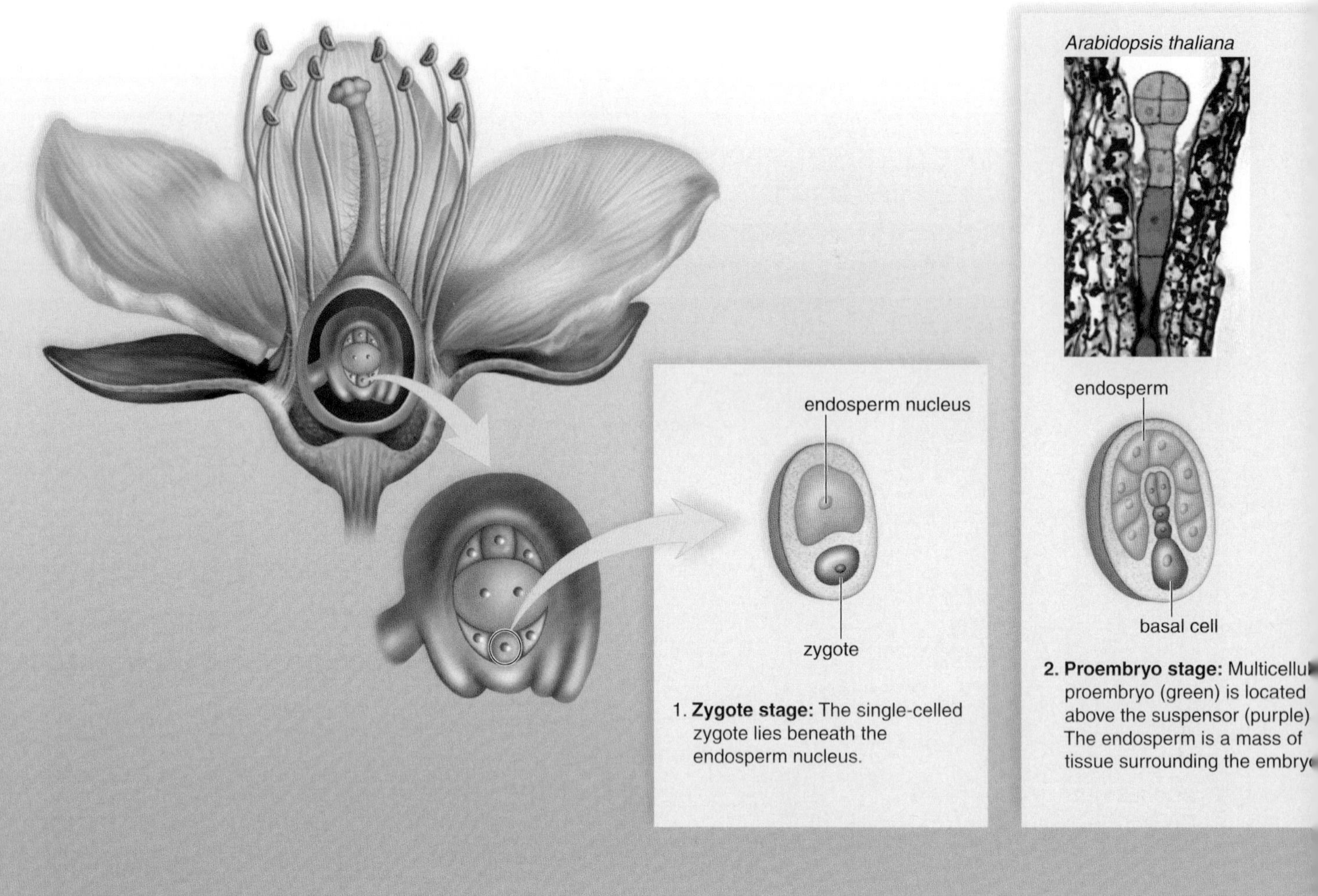

Arabidopsis thaliana
endosperm nucleus
zygote
1. Zygote stage: The single-celled zygote lies beneath the endosperm nucleus.
endosperm
basal cell
2. Proembryo stage: Multicellu
proembryo (green) is located above the suspensor (purple) The endosperm is a mass of tissue surrounding the embry

Development of Eudicot Embryo

Stages in the development of a eudicot embryo are shown in Figure 10.2. During development, the **suspensor** anchors the embryo and transfers nutrients to it from the mature plant. The **cotyledons** store nutrients that the embryo uses as nourishment. An embryo consists of the **epicotyl,** which becomes the leaves; the **hypocotyl,** which becomes the stem; and the **radicle,** which becomes the roots.

Observation: Development of the Embryo

Study preserved slides and identify these stages:

1. Zygote stage: The single-cell zygote lies beneath the endosperm nucleus (Fig. 10.2, 1).
2. Globular stage: Cell division has produced a spherical embryo (Fig. 10.2, 3).
3. Heart stage: The embryo becomes heart-shaped as the cotyledons begin to appear (Fig. 10.2, 4).
4. Torpedo stage: As the cotyledons bend, the embryo takes on a torpedo shape (Fig. 10.2, 5).
5. Mature embryo: The embryo consists of the epicotyl, the hypocotyl, and the radicle (Fig. 10.2, 6).

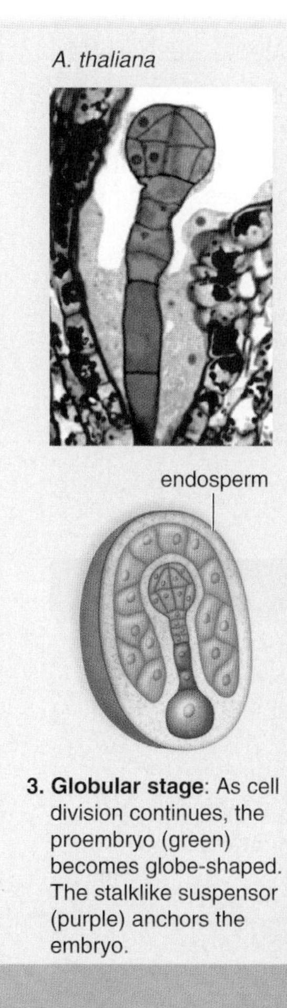

3. Globular stage: As cell division continues, the proembryo (green) becomes globe-shaped. The stalklike suspensor (purple) anchors the embryo.

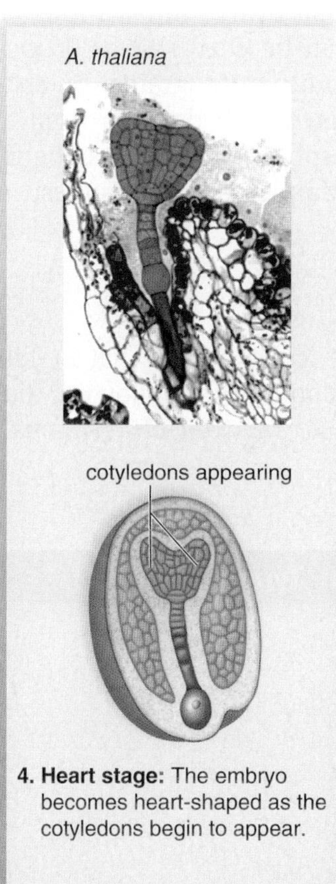

4. Heart stage: The embryo becomes heart-shaped as the cotyledons begin to appear.

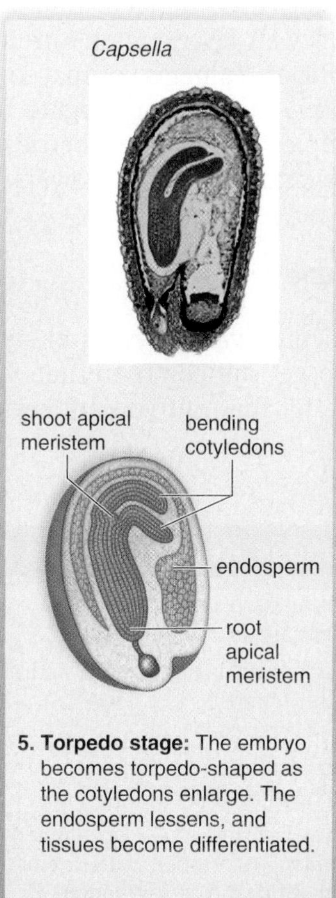

5. Torpedo stage: The embryo becomes torpedo-shaped as the cotyledons enlarge. The endosperm lessens, and tissues become differentiated.

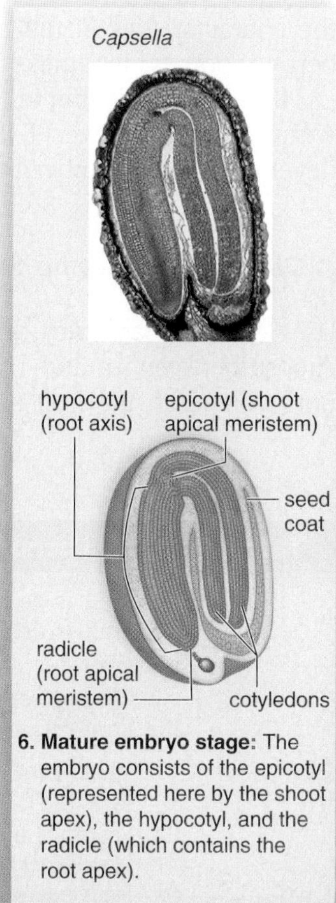

6. Mature embryo stage: The embryo consists of the epicotyl (represented here by the shoot apex), the hypocotyl, and the radicle (which contains the root apex).

10.2 Fruits

A fruit is derived from an ovary or from an ovary and closely associated tissues, whereas "vegetables" are not derived from floral tissues and would never contain seeds. As an ovary develops into a fruit, the ovarian wall thickens and becomes a **pericarp** with three layers (Fig. 10.3).

1. **Exocarp:** The outermost layer of the fruit wall
2. **Mesocarp:** The middle layer of the fruit wall
3. **Endocarp:** The innermost layer of the fruit wall

Figure 10.3 Fleshy fruits versus dry fruits.
Fleshy fruits are represented by **(a)** peach and **(b)** apple. Dry fruits are represented by **(c)** pea which is dehiscent because it splits open; and **(d)** maple which is an indehiscent fruit.

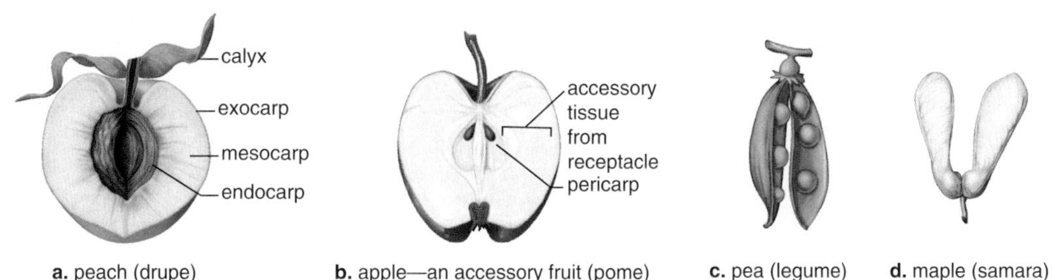

a. peach (drupe) **b.** apple—an accessory fruit (pome) **c.** pea (legume) **d.** maple (samara)

Kinds of Fruits

Biologists have divided **simple fruits** (derived from a single ovary) into various types according to the characteristics of the fruit wall layers. Simple fruits can be classified as drupes, pepos, hesperidia, berries, pomes, legumes, follicles, samaras, drupaceous nuts, nuts, grains, or achenes.

In addition to simple fruits, there are **aggregate fruits,** which develop from a number of ovaries within a single flower. Examples include blackberries, raspberries, and strawberries. **Multiple fruits** develop from a number of ovaries of several flowers. Examples are pineapples, mulberries, and figs.

Observation: Keying Simple Fruits

Use Table 10.1 as a key to identify various types of simple fruits. Notice that in Table 10.1 you must choose between 1*a* and 1*b* to get started. Thereafter, you continue by following the "Go to" instructions until you reach the fruit type for the fruit you are keying. *Record your observations in Table 10.2.*

Table 10.1 A Dichotomous Key for Simple Fruit Types

1*a*.	Fruit is fleshy; pericarp is soft.	Go to 2*a*/2*b*.
1*b*.	Fruit is dry; pericarp is dry.	Go to 6*a*/6*b*.
2*a*.	Fruit has a single seed inside a hard and stony pit. Examples: plum, cherry.	DRUPE
2*b*.	Fruit contains several seeds.	Go to 3*a*/3*b*.
3*a*.	Fruit has a firm rind.	Go to 4*a*/4*b*.
3*b*.	Fruit does not have a hard rind.	Go to 5*a*/5*b*.
4*a*.	Fruit has a firm rind and is not segmented. Examples: squash, cucumber, watermelon.	PEPO
4*b*.	Fruit has a firm rind and is segmented. Examples: lemon, lime, orange.	HESPERIDIUM
5*a*.	Entire wall of fruit is fleshy, and seeds may be eaten. Examples: tomato, grape.	BERRY
5*b*.	Fruit has a papery core. Examples: apple, pear.	POME
6*a*.	Dry fruit is dehiscent (splits open).	Go to 7*a*/7*b*.
6*b*.	Dry fruit is indehiscent (does not split open).	Go to 8*a*/8*b*.
7*a*.	Fruit splits at two seams. Examples: pea, soybean, locust.	LEGUME
7*b*.	Fruit splits at one seam. Examples: milkweed, larkspur.	FOLLICLE
8*a*.	Fruit has one or more wings. Examples: maple, elm, ash.	SAMARA
8*b*.	Fruit does not have wings.	Go to 9*a*/9*b*.
9*a*.	Pericarp has three complete layers. Examples: coconut, hickory.	DRUPACEOUS NUT
9*b*.	Pericarp does not have three complete layers.	Go to 10*a*/10*b*.
10*a*	Fruit is relatively large; pericarp is thick and stony; seed separates from ovarian wall. Examples: walnut, oak.	NUT
10*b*.	Fruit is relatively small; pericarp is thin; seed is at least partially attached to ovarian wall.	Go to 11*a*/11*b*.
11*a*.	Pericarp is completely fused to seed coat. Examples: wheat, corn, oats.	GRAIN
11*b*.	Pericarp attaches to seed coat at only one point. Examples: sunflower, dandelion.	ACHENE

Table 10.2 Identification of Simple Fruits

Common Name	Fleshy or Dry?	Eaten as a Vegetable, Fruit, Other?	Type of Fruit (from Key in Table 10.1)
1			
2			
3			
4			
5			
6			
7			
8			
9			
10			

Laboratory Notes

10.3 Seeds

The seeds of flowering plants develop from ovules. A seed contains an embryonic plant, stored food, and a seed coat. Monocot seeds have one **cotyledon** (seed leaf); eudicots have two cotyledons.

Observation: Eudicot and Monocot Seeds

Bean Seed

1. Obtain a presoaked bean seed (eudicot). Carefully dissect it, using Figure 10.4 to help you identify:
   a. **Seed coat:** The outer covering. Remove the seed coat with your fingernail.
   b. **Cotyledons:** Food storage organs. The endosperm was absorbed by the cotyledons during development. What is the function of these cotyledons? ______________________________

   ____
   c. **Epicotyl:** The small portion of the embryo located above the attachment of the cotyledons. The first true leaves **(plumules)** develop from the epicotyl. (See seedling.)
   d. **Hypocotyl:** The small portion of embryo located below the attachment of the cotyledons. The lower end develops into the embryonic root, or **radicle.** (See seedling.)

Figure 10.4 Eudicot seed structure and germination.
a. A seed contains an embryo, stored food, and a seed coat as exemplified by a bean seed. A eudicot seed has two cotyledons. Following **(b)** germination, a seedling grows to become a mature plant.

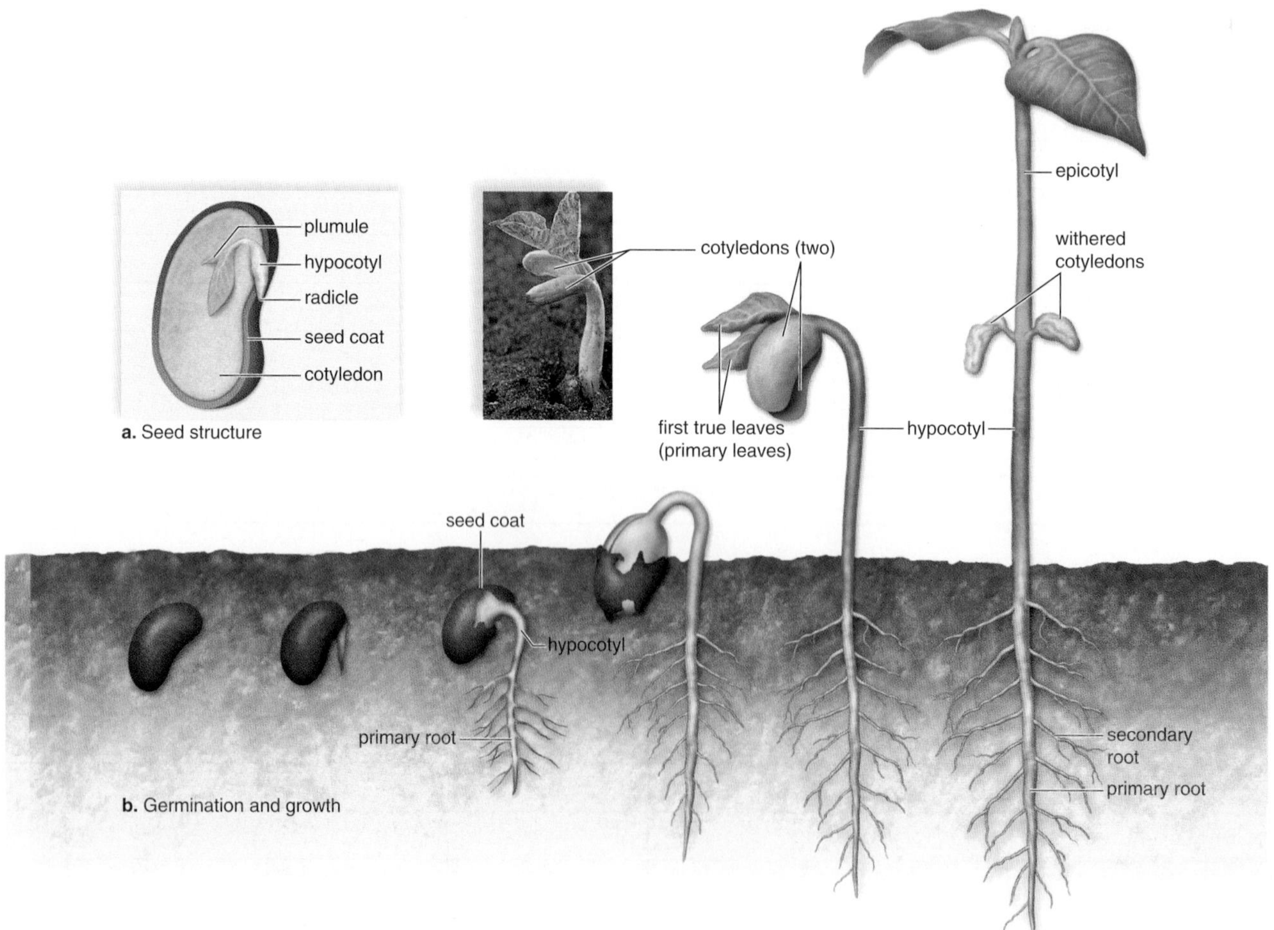

2. Observe seedlings in various stages of development. Which organ emerges first from the seed—the plumule or the radicle? ______

   Of what advantage is this to the plant? ______

   ______

3. The hypocotyl is the first part to emerge from the soil. What is the advantage of the hypocotyl pulling the plumule up out of the ground instead of pushing it up through the ground? ______

   ______

4. Do cotyledons stay beneath the ground? ______

Corn Kernel

1. Obtain a presoaked corn kernel (monocot). Lay the seed flat, and with a razor, carefully slice it in half, as shown in Figure 10.5*a*. A corn kernel is a fruit, and the seed coat is tightly attached to the pericarp.
2. Identify the cotyledon, plumule, and radicle. In addition, identify
   a. **Endosperm:** Stored food for the embryo; passes into the cotyledon as the seedling grows
   b. **Coleoptile:** A sheath that covers the emerging leaves
3. Examine corn seedlings in various stages of development. Does the cotyledon of a corn seed stay beneath the ground? ______

Figure 10.5 Monocot seed structure and germination.
a. A monocot seed has only one cotyledon as exemplified by a corn kernel. A corn kernel is a fruit—the seed is covered by a pericarp. Following **(b)** germination, a corn seedling grows to become a mature plant.

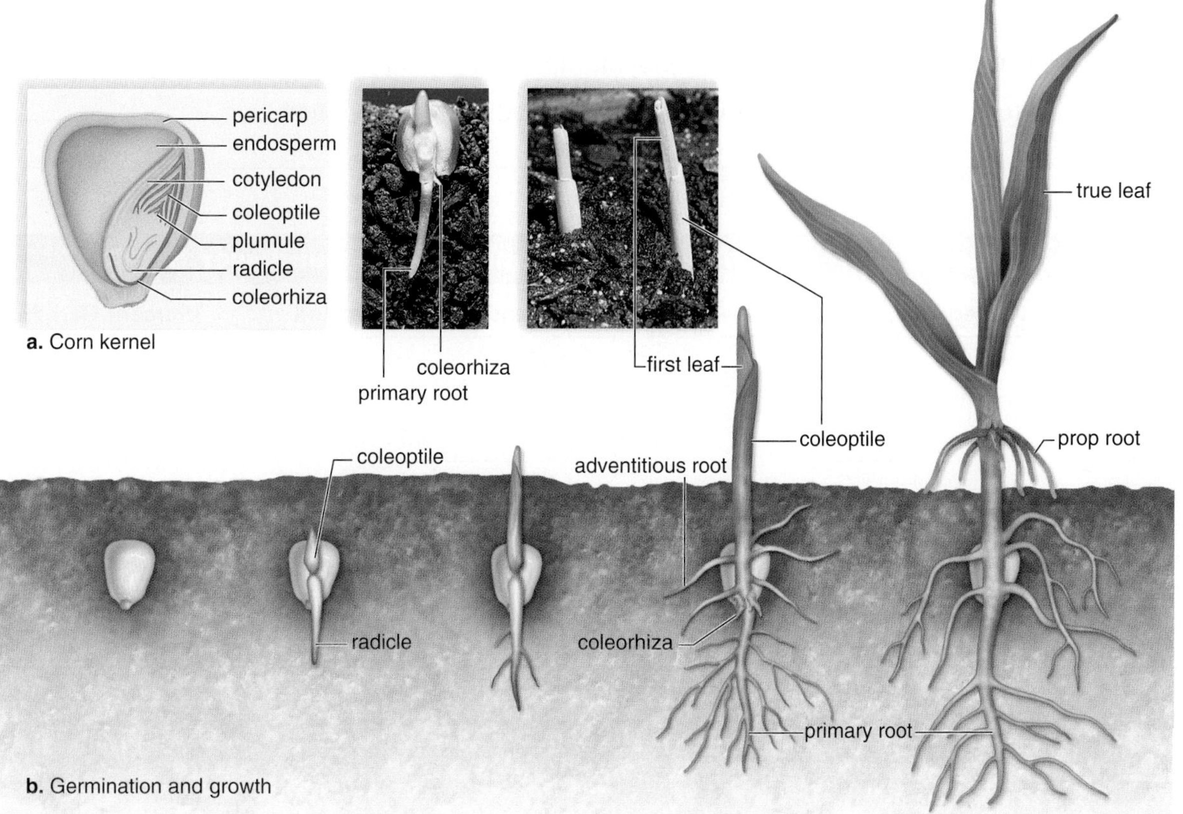

10.4 Seed Germination

Mature seeds contain an embryo that does not resume growth until after germination, which requires the proper environmental conditions. Mature seeds are dry, and for germination to begin, the dry tissues must take up water in a process called **imbibition.** After water has been imbibed, enzymes break down the food source into small molecules that can provide energy or be used as building blocks until the seedling is ready to photosynthesize.

Explain why the cotyledons of a bean seedling shrivel as the seedling grows. ______________________

____

Experimental Procedure: Effect of Acid Rain

In certain parts of the United States, rain has a much lower pH than normal. This is known to have detrimental effects on the sustainability of forests. Would you predict that acid rain also affects germination of seeds? ____

Your instructor has placed 20 sunflower seeds in each of five containers with water of increasing acidity: 0% vinegar (tap water), 1% vinegar, 5% vinegar, 20% vinegar, and 100% vinegar.

1. What hypothesis do you propose regarding the effect of these solutions on the germination of sunflower seeds? ____

   ____
2. Count the number of germinated sunflower seeds in each container, and complete Table 10.3.
3. Record the pH of each container as directed by your instructor.
4. Do the data support or falsify your hypothesis? __________ Explain. ______________________

   ____

   ____
5. You learned in Laboratory 5 that each enzyme has an optimum pH. Explain why acid deposition is expected to inhibit metabolism, and therefore seedling development. ______________________

   ____

Table 10.3 Effect of Increasing Acidity on Germination of Sunflower Seeds

Concentration of Vinegar	pH	Number of Seeds That Germinated	Percent Germination
0%			
1%			
5%			
20%			
100%			

Laboratory Review 10

_______________ 1. What structure transports sperm to the ovule in flowering plants?

_______________ 2. What are the two angiosperm groups?

_______________ 3. What type of cell division produces microspores and megaspores in flowering plants?

_______________ 4. Which angiosperm group has flower parts in threes or multiples of three?

_______________ 5. Name the part of a flower that has a filament topped by the anther.

_______________ 6. What kinds of spores are produced by flowering plants?

_______________ 7. What structure disperses the offspring in flowering plants?

_______________ 8. How many sperm are in the pollen grain of a flowering plant?

_______________ 9. What is the 3n nutritive tissue in the seed of angiosperms called?

_______________ 10. In flowering plants, the ovule becomes the seed, and the ovary becomes what structure?

_______________ 11. Name the three pericarp layers in a fruit. Which layer is the fleshy one we eat?

_______________ 12. Blackberries, raspberries, and strawberries are examples of what kind of fruit?

_______________ 13. Pineapples and figs are examples of what kind of fruit?

_______________ 14. Apples and tomatoes are examples of what kind of fruit?

_______________ 15. Peas and milkweeds are dry fruits that are dehiscent/indehiscent.

Thought Questions

16. What is the difference between pollination and fertilization in angiosperms?

17. Why do you expect acidic conditions to affect the ability of seeds to germinate?

18. Flowering plants can produce both male and female gametes. What mechanisms might these plants employ to prevent self-fertilization?

Laboratory Notes

LABORATORY

11

Animal Organization

Learning Outcomes

11.1 Epithelial Tissue
- Identify slides and models or diagrams of various types of epithelium.
- Tell where a particular type of epithelium is located in the body, and state a function.

11.2 Connective Tissue
- Identify slides and models or diagrams of various types of connective tissue.
- Tell where a particular connective tissue is located in the body, and state a function.

11.3 Muscular Tissue
- Identify slides and models or diagrams of three types of muscular tissue.
- Tell where each type of muscular tissue is located in the body, and state a function.

11.4 Nervous Tissue
- Identify a slide and model or diagram of a neuron.
- Tell where nervous tissue is located in the body, and state a function.

11.5 Tissues Form Organs
- Identify a slide of the intestinal wall and any particular tissue in the wall. State a function for each tissue.
- Identify a slide of skin and any particular tissue or structure in skin. State a function for each tissue or structure.

Introduction

Humans, as well as all other living things, are made up of **cells.** Groups of cells that have the same structural characteristics and perform the same functions are called **tissues.** Figure 11.1 shows the four categories of tissues in the human body. An **organ** is composed of different types of tissues, and various organs form **organ systems.** Humans thus have the following levels of biological organization: cells → tissues → organs → organ systems.

The photomicrographs of tissues in this laboratory were obtained by viewing prepared slides with a light microscope. Preparation required the following sequential steps:

1. **Fixation:** The tissue is immersed in a preservative solution to maintain the tissue's existing structure.
2. **Embedding:** Water is removed with alcohol, and the tissue is impregnated with paraffin wax.
3. **Sectioning:** The tissue is cut into extremely thin slices by an instrument called a microtome. When the section runs the length of the tissue, it is called a longitudinal section (l.s.); when the section runs across the tissue, it is called a cross section (c.s.).
4. **Staining:** The tissue is immersed in dyes that stain different structures. The most common dyes are hematoxylin and eosin stains (H & E). They give a differential blue and red color to the basic and acidic structures within the tissue. Other dyes are available for staining specific structures.

Figure 11.1 The major tissues in the human body.
The many kinds of tissues in the human body are grouped into four types: epithelial tissue, muscular tissue, nervous tissue, and connective tissue.

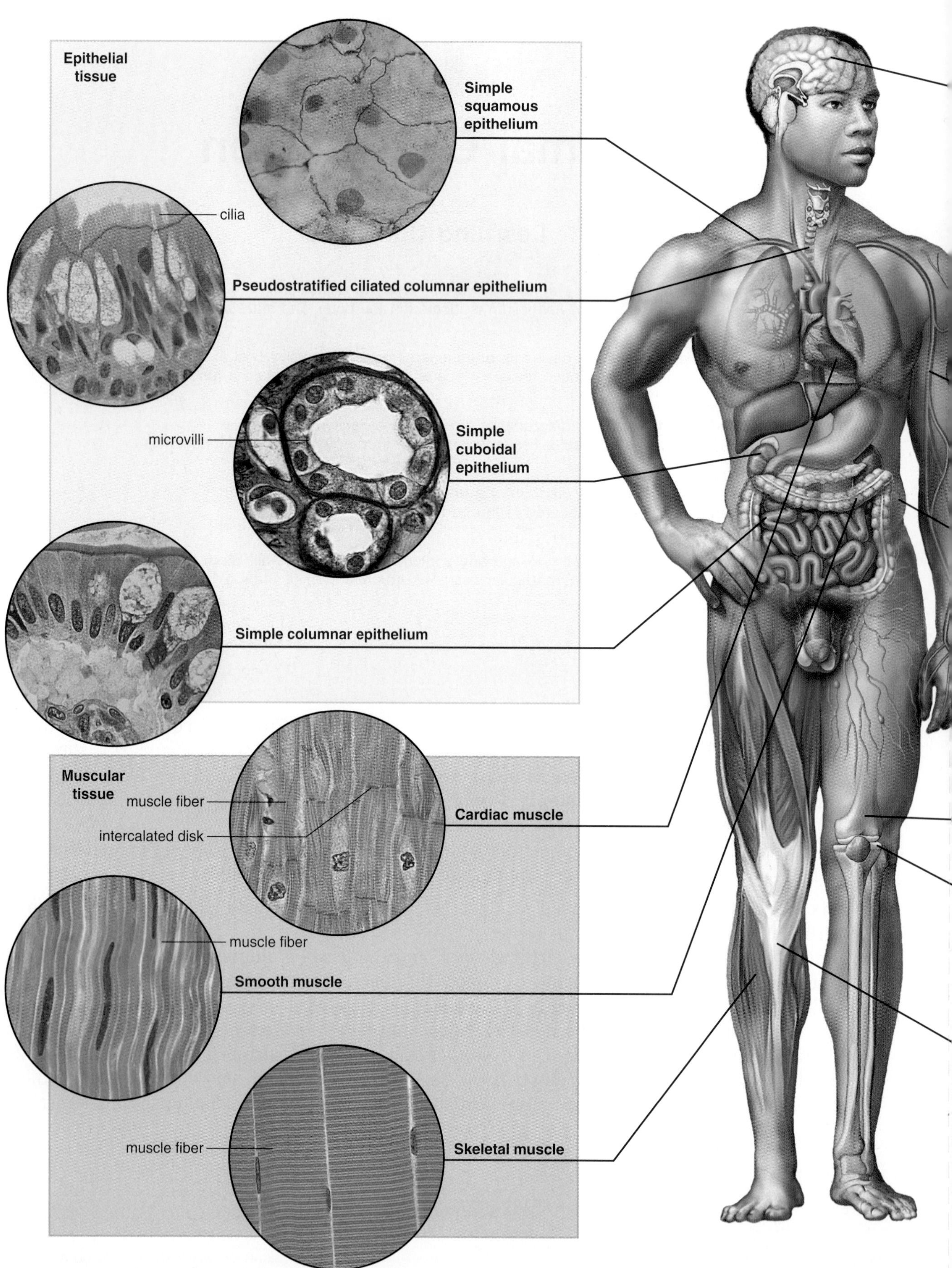

Nervous tissue
neuron
Nervous tissue
neuroglial cell

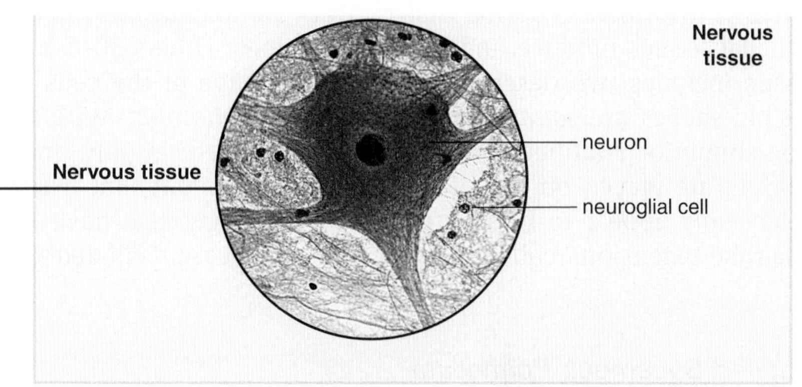

Connective tissue
platelets
Blood
white blood cells
red blood cells
fat
Adipose
osteocytes
Bone
central canal
matrix
Cartilage
fibroblast
Dense fibrous
protein fibers

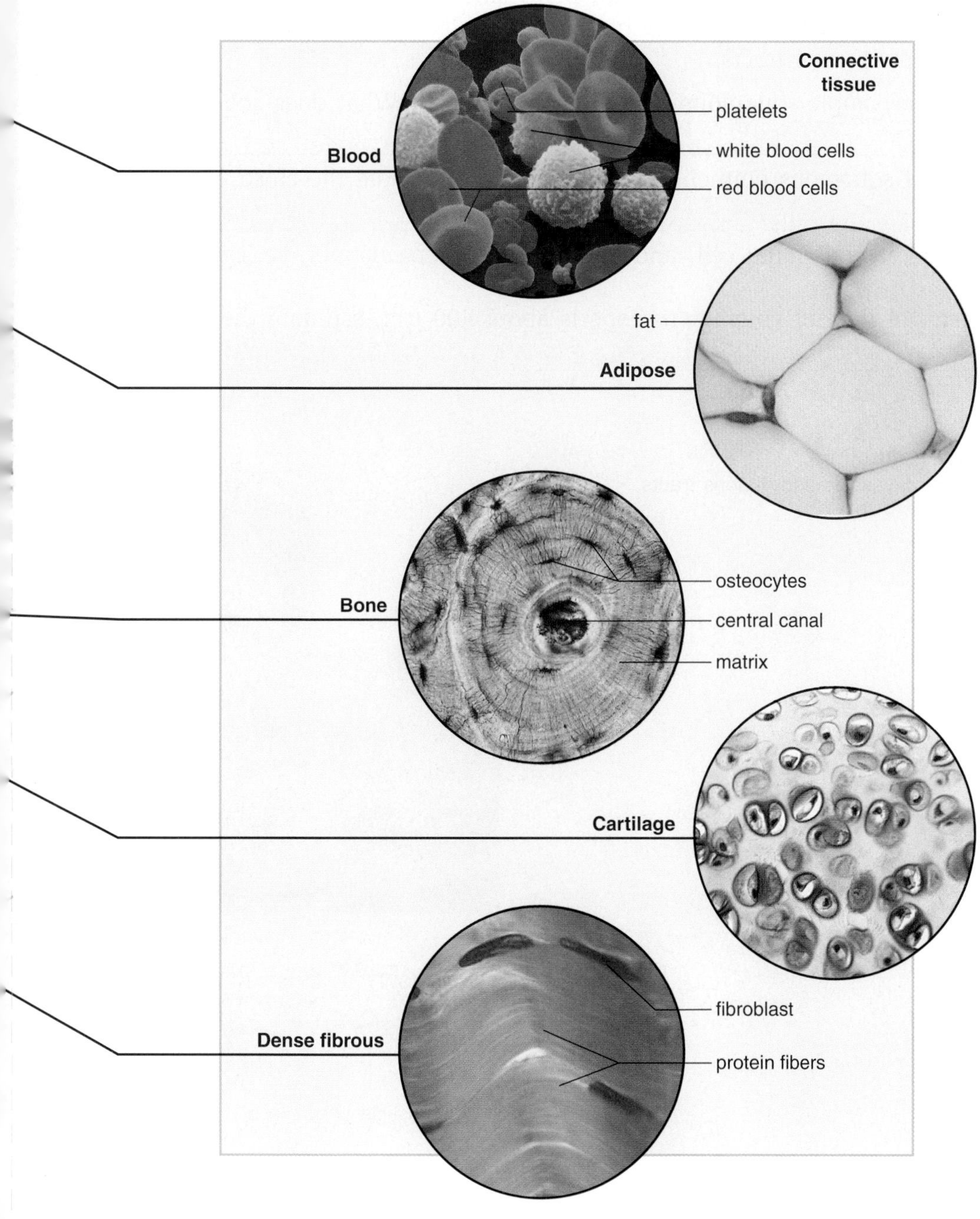

11.1 Epithelial Tissue

Epithelial tissue (epithelium) forms a continuous layer, or sheet, over the entire body surface and most of the body's inner cavities. Externally, it forms a covering that protects the animal from infection, injury, and drying out. Some epithelial tissues produce and release secretions. Others absorb nutrients.

The name of an epithelial tissue includes two descriptive terms: the shape of the cells and the number of layers. The three possible shapes are *squamous, cuboidal,* and *columnar*. With regard to layers, an epithelial tissue may be simple or stratified. **Simple** means that there is only one layer of cells; **stratified** means that cell layers are placed on top of each other. Some epithelial tissues are **pseudostratified,** meaning that they only appear to be layered. Epithelium may also have cellular extensions called **microvilli** or hairlike extensions called **cilia.** In the latter case, "ciliated" may be part of the tissue's name.

Observation: Simple and Stratified Squamous Epithelium

Simple Squamous Epithelium

Simple squamous epithelium is a single layer of thin, flat, many-sided cells, each with a central nucleus. It lines internal cavities, the heart, and all the blood vessels. It also lines parts of the urinary, respiratory, and male reproductive tracts.

1. Study a model or diagram of simple squamous epithelium (Fig. 11.2). What does *squamous* mean? ____________________
2. Examine a prepared slide of squamous epithelium. Under low power, note the close packing of the flat cells. What shapes are the cells? ____________________
3. Under high power, examine an individual cell, and identify the plasma membrane, cytoplasm, and nucleus.
4. Knowing that the diameter of field of your microscope is about 400 μm, estimate the size of an epithelial cell. ____________________

Figure 11.2 Simple squamous epithelium.
Simple squamous epithelium lines blood vessels and various tracts.

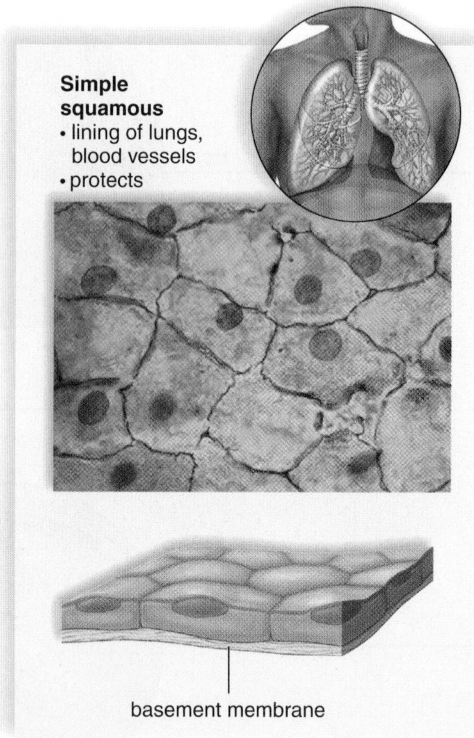

Stratified Squamous Epithelium

As would be expected from its name, stratified squamous epithelium consists of many layers of cells. The innermost layer produces cells that are first cuboidal or columnar in shape, but as the cells push toward the surface, they become flattened.

The outer region of the skin, called the epidermis (see p. 147), is stratified squamous epithelium. As the cells move toward the surface, they flatten, begin to accumulate a protein called **keratin,** and eventually die. Keratin makes the outer layer of epidermis tough, protective, and able to repel water.

The linings of the mouth, throat, anal canal, and vagina are stratified epithelium. The outermost layer of cells surrounding the cavity is simple squamous epithelium. In these organs, this layer of cells remains soft, moist, and alive.

1. Either now or when you are studying skin in Section 11.5, examine a slide of skin and find the portion of the slide that is stratified squamous epithelium.
2. Approximately how many layers of cells make up this portion of skin? ________
3. Which layers of cells best represent squamous epithelium? ________

Observation: Simple Cuboidal Epithelium

Simple cuboidal epithelium is a single layer of cube-shaped cells, each with a central nucleus. It is found in tubules of the kidney and in the ducts of many glands, where it has a protective function. It also occurs in the secretory portions of some glands—that is, where the tissue produces and releases secretions.

1. Study a model or diagram of simple cuboidal epithelium (Fig. 11.3).
2. Examine a prepared slide of simple cuboidal epithelium. Move the slide until you locate cube-shaped cells that line a lumen (cavity). Are these cells ciliated? ________

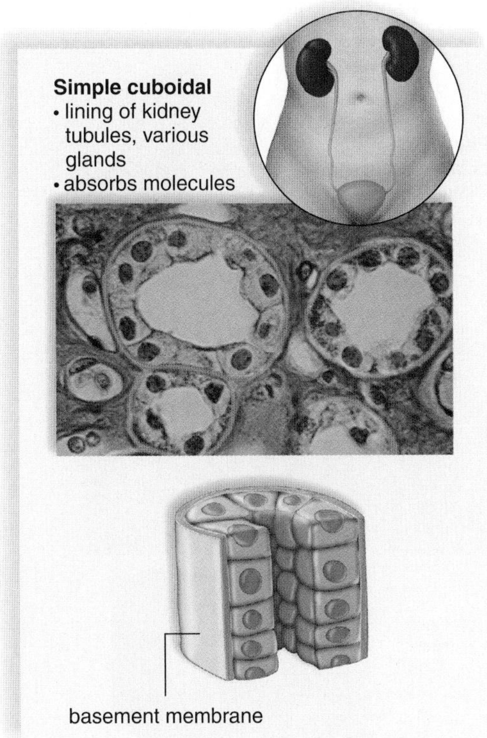

Figure 11.3 Simple cuboidal epithelium.
Simple cuboidal epithelium lines kidney tubules and the ducts of many glands.

Observation: Simple Columnar Epithelium

Simple columnar epithelium is a single layer of tall, cylindrical cells, each with a nucleus near the base. This tissue, which lines the digestive tract from the stomach to the anus, protects, secretes, and allows absorption of nutrients.

1. Study a model or diagram of simple columnar epithelium (Fig. 11.4).
2. Examine a prepared slide of simple columnar epithelium. Find tall and narrow cells that line a lumen. Under high power, focus on an individual cell. Identify the plasma membrane, the cytoplasm, and the nucleus. Epithelial tissues are attached to underlying tissues by a basement membrane composed of extracellular material containing protein fibers.
3. The tissue you are observing contains mucus-secreting cells. Search among the columnar cells until you find a **goblet cell,** so named because of its goblet-shaped, clear interior. This region contains mucus, which may be stained a light blue. In the living animal, the mucus is discharged into the gut cavity and protects the lining from digestive enzymes.

Observation: Pseudostratified Ciliated Columnar Epithelium

Pseudostratified ciliated columnar epithelium appears to be layered, while actually all cells touch the basement membrane. Many cilia are located on the free end of each cell (Fig. 11.5). In the human trachea, the cilia wave back and forth, moving mucus and debris up toward the throat so that it cannot enter the lungs. Smoking destroys these cilia, but they will grow back if smoking is discontinued.

1. Study a model or diagram of pseudostratified ciliated columnar epithelium (Fig. 11.5).
2. Examine a prepared slide of pseudostratified ciliated columnar epithelium. Concentrate on the part of the slide that resembles the model. Identify the cilia.

Figure 11.4 Simple columnar epithelium.
Simple columnar epithelium lines the digestive tract. Goblet cells among the columnar cells secrete mucus.

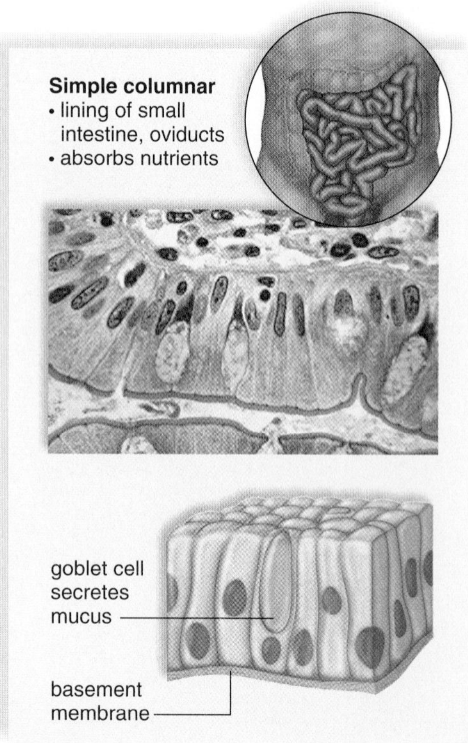

Figure 11.5 Pseudostratified ciliated columnar epithelium.
Pseudostratified ciliated columnar epithelium lines the trachea. The cilia help keep the lungs free of debris.

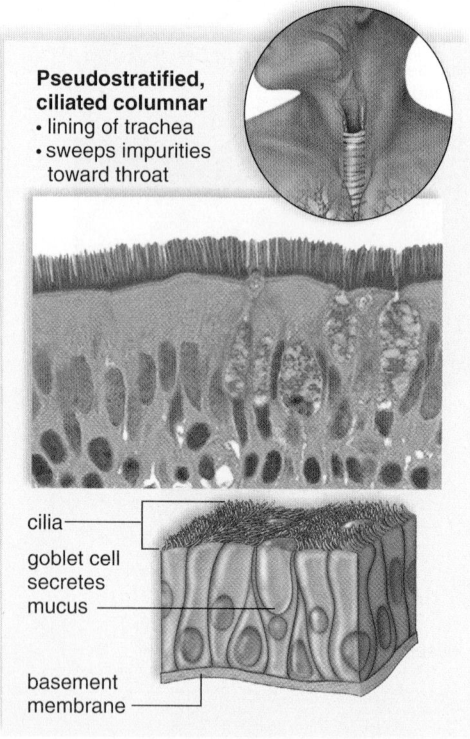

Summary of Epithelial Tissue

Complete Table 11.1 to summarize your study of epithelial tissue.

Table 11.1 Epithelial Tissue

Type	Appearance	Function	Location
Simple squamous			Walls of capillaries, lining of blood vessels, air sacs of lungs, lining of internal cavities
Stratified squamous	Innermost layers are cuboidal or columnar; outermost layers are flattened	Protection, repel water	
Simple cuboidal		Secretion, absorption	
Simple columnar	Columnlike—tall, cylindrical nucleus at base		Lining of uterus, tubes of digestive tract
Pseudostratified ciliated columnar		Protection, secretion, movement of mucus and sex cells	

Laboratory Notes

11.2 Connective Tissue

Connective tissue joins different parts of the body together. There are four general classes of connective tissue: connective tissue proper, cartilage, bone, and blood. All types of connective tissue consist of cells surrounded by a matrix that usually contains fibers. Elastic fibers are composed of a protein called elastin. Collagenous fibers contain the protein collagen.

Observation: Connective Tissue

There are several different types of connective tissue. We will study loose fibrous connective tissue, dense fibrous connective tissue, adipose tissue, bone, cartilage, and blood. **Loose fibrous connective tissue** supports epithelium and many internal organs, such as muscles, blood vessels, and nerves (Fig. 11.6). Its presence allows organs to expand. **Dense fibrous connective tissue** contains many collagenous fibers packed together, as in tendons, which connect muscles to bones, and in ligaments, which connect bones to other bones at joints (Fig. 11.7).

1. Examine a slide of loose fibrous connective tissue, and compare it with Figure 11.6. What is the function of loose fibrous connective tissue? ______________________
2. Examine a slide of dense fibrous connective tissue, and compare it with Figure 11.7. What two kinds of structures in the body contain dense fibrous connective tissue?

   ____

Figure 11.6 Loose fibrous connective tissue. Loose fibrous connective tissue supports epithelium and many internal organs.

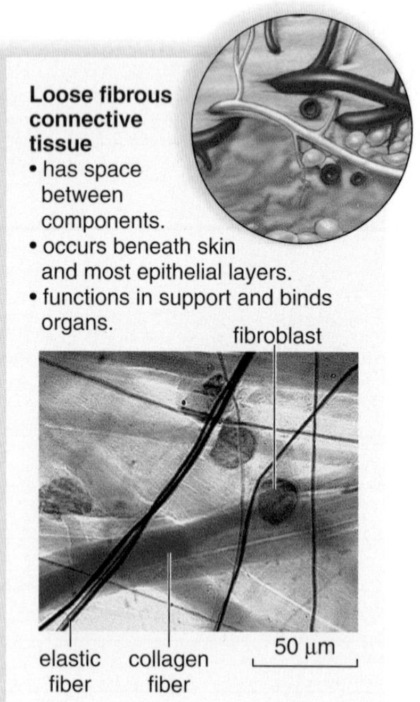

Figure 11.7 Dense fibrous connective tissue. Dense fibrous connective tissue is found in tendons and ligaments.

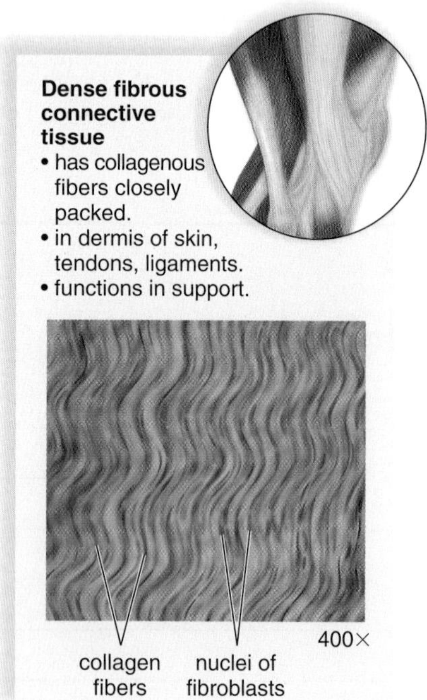

Observation: Adipose Tissue

In **adipose tissue,** the cells have a large, central, fat-filled vacuole that causes the nucleus and cytoplasm to be at the perimeter of the cell (Fig. 11.8). Adipose tissue occurs beneath the skin, where it insulates the body, and around internal organs, such as the kidneys and heart. It cushions and helps protect these organs.

1. Examine a prepared slide of adipose tissue. Why is the nucleus pushed to one side? ______________________________

2. State a location for adipose tissue in the body. ______________________________

   What are two functions of adipose tissue at this location? ______________________________

Observation: Compact Bone

Compact bone is found in the bones that make up the skeleton. It consists of **osteons** (Haversian systems), with a **central canal,** and concentric rings of spaces called **lacunae,** which are connected by tiny crevices called **canaliculi.** The central canal contains a nerve and blood vessels, which service bone. The lacunae contain bone cells called **osteocytes,** whose processes extend into the canaliculi. Separating the lacunae is a matrix that is hard because it contains minerals, notably calcium salts. The matrix also contains collagenous fibers.

1. Study a model or diagram of compact bone (Fig. 11.9). Then look at a prepared slide and identify the central canal, lacunae, and canaliculi.
2. What is the function of the central canal and canaliculi? ______________________________

Figure 11.8 Adipose tissue.
Adipose tissue is composed of cells filled with fat droplets.

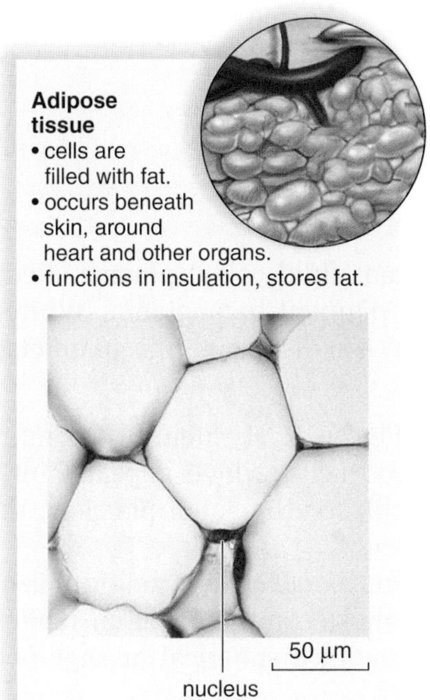
Adipose tissue
• cells are filled with fat.
• occurs beneath skin, around heart and other organs.
• functions in insulation, stores fat.
50 μm
nucleus

Figure 11.9 Compact bone.
Compact bone contains osteons, in which osteocytes within lacunae are arranged in concentric circles.

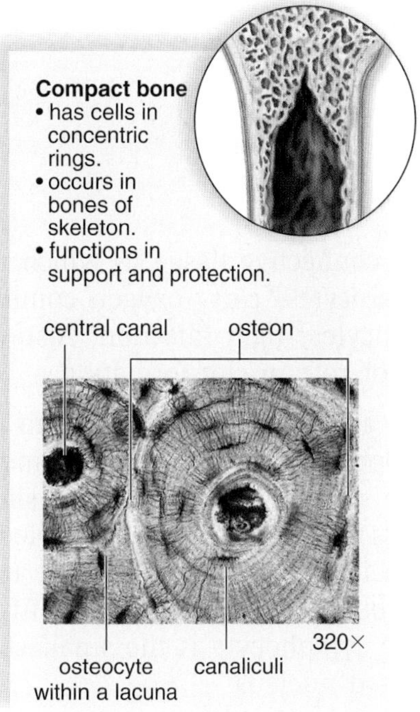
Compact bone
• has cells in concentric rings.
• occurs in bones of skeleton.
• functions in support and protection.
central canal
osteon
320×
osteocyte within a lacuna
canaliculi

Observation: Hyaline Cartilage

In **hyaline cartilage,** cells called **chondrocytes** are found in twos or threes in lacunae. The lacunae are separated by a flexible matrix containing weak collagenous fibers.

1. Study the diagram and photomicrograph of hyaline cartilage in Figure 11.10. Then study a prepared slide of hyaline cartilage, and identify the matrix, lacunae, and chondrocytes.
2. Compare compact bone and hyaline cartilage. Which of these types of connective tissue is more organized?

   ______________________ Why? ______________________________________
3. Which of these two types of connective tissue lends more support to body parts? ___________

   ____

Figure 11.10 Hyaline cartilage.
In cartilage, chondrocytes lie in lacunae, which are separated by a flexible matrix.

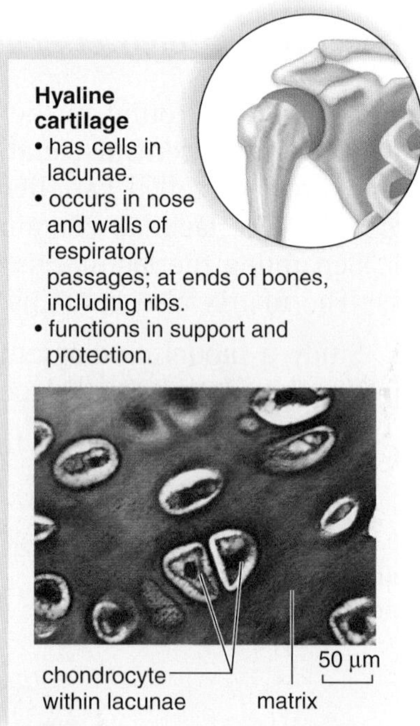
Hyaline cartilage
• has cells in lacunae.
• occurs in nose and walls of respiratory passages; at ends of bones, including ribs.
• functions in support and protection.
50 µm
chondrocyte within lacunae
matrix

Observation: Blood

Blood is a connective tissue in which the matrix is an intercellular fluid called **plasma. Red blood cells** (erythrocytes) carry oxygen combined with the respiratory pigment hemoglobin. **White blood cells** (leukocytes) fight infection. Also present in blood are many small bodies, the **platelets,** which play a major role in clot formation.

1. Study a prepared slide of human blood. With the help of Figure 11.11, identify the numerous red blood cells and the less numerous but larger white blood cells, which appear faint because of the stain. As you scan your slide on high power, also look for the small platelets, the small objects scattered between the blood cells.
2. Try to identify a neutrophil, the most common type of white blood cell. A neutrophil has a multilobed nucleus. Try to identify a lymphocyte, the next most common type of white blood cell. A lymphocyte is the smallest of the white blood cells, with a spherical or slightly indented nucleus.

Figure 11.11 Blood cells.
Red blood cells are more numerous than white blood cells. White blood cells can be separated into five distinct types. If you have blood work done that includes a complete blood count (CBC), the doctor is getting a count of each of these types of WBCs. (*a–e:* Magnification ×1,050)

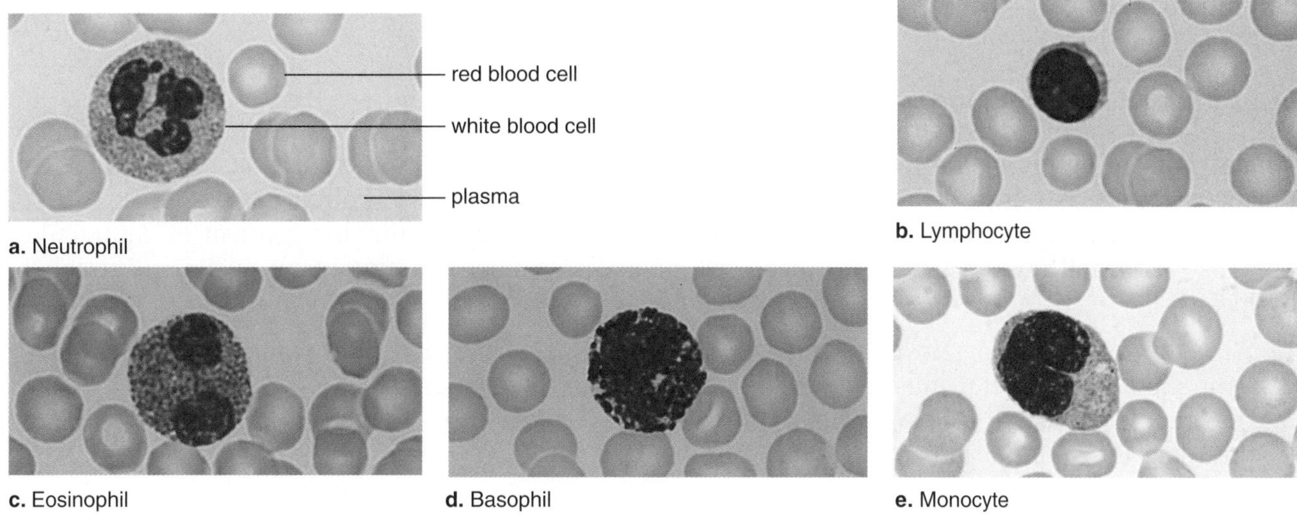

a. Neutrophil **b.** Lymphocyte **c.** Eosinophil **d.** Basophil **e.** Monocyte

Summary of Connective Tissue

Complete Table 11.2 to summarize your study of connective tissue.

Table 11.2 Connective Tissue

Type	Appearance	Function	Location
Loose fibrous connective tissue			Between the muscles; beneath the skin; beneath most epithelial layers
Dense fibrous connective tissue		Binds organs together, binds muscle to bones, binds bone to bone	
Adipose			Beneath the skin; around the kidney and heart; in the breast
Compact bone		Support, protection	
Hyaline cartilage	Cells in lacunae		Nose; ends of bones; rings in walls of respiratory passages; between ribs and sternum
Blood	Red and white cells floating in plasma		Blood vessels

11.3 Muscular Tissue

Muscular (contractile) tissue is composed of cells called muscle fibers. Muscular tissue has the ability to contract, and contraction usually results in movement. The body contains skeletal, cardiac, and smooth muscle.

Observation: Skeletal Muscle

Skeletal muscle occurs in the muscles that are attached to the bones of the skeleton. The contraction of skeletal muscle is said to be **voluntary** because it is under conscious control. Skeletal muscle is striated; it contains light and dark bands. The striations are caused by the arrangement of contractile filaments (actin and myosin filaments) in muscle fibers. Each fiber contains many nuclei, all peripherally located.

1. Study a model or diagram of skeletal muscle (Fig. 11.12), and note that striations are present. You should see several muscle fibers, each marked with striations.
2. Examine a prepared slide of skeletal muscle. The striations may be difficult to make out, but bringing the slide in and out of focus may help.

Figure 11.12 Skeletal muscle.
Skeletal muscle is striated and voluntary. Its cells are tubular and contain many nuclei.

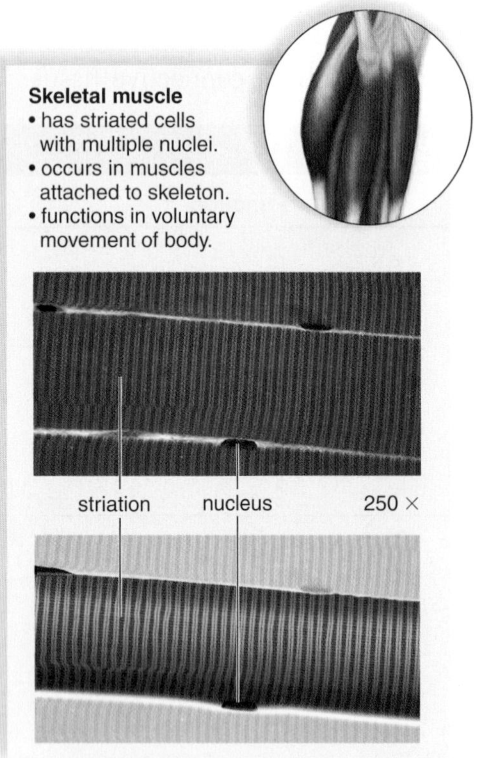
Skeletal muscle
• has striated cells with multiple nuclei.
• occurs in muscles attached to skeleton.
• functions in voluntary movement of body.
striation
nucleus
250 ×

Observation: Cardiac Muscle

Cardiac muscle is found only in the heart. It is called **involuntary** because its contraction does not require conscious effort. Cardiac muscle is striated in the same way as skeletal muscle. However, the fibers are branched and bound together at **intercalated disks,** where their folded plasma membranes touch. This arrangement aids communication between fibers.

1. Study a model or diagram of cardiac muscle (Fig. 11.13), and note that striations are present.
2. Examine a prepared slide of cardiac muscle. Find an intercalated disk. What is the function of cardiac muscle? ______________________________

   ____

Figure 11.13 Cardiac muscle.
Cardiac muscle is striated and involuntary. Its branched cells join at intercalated disks.

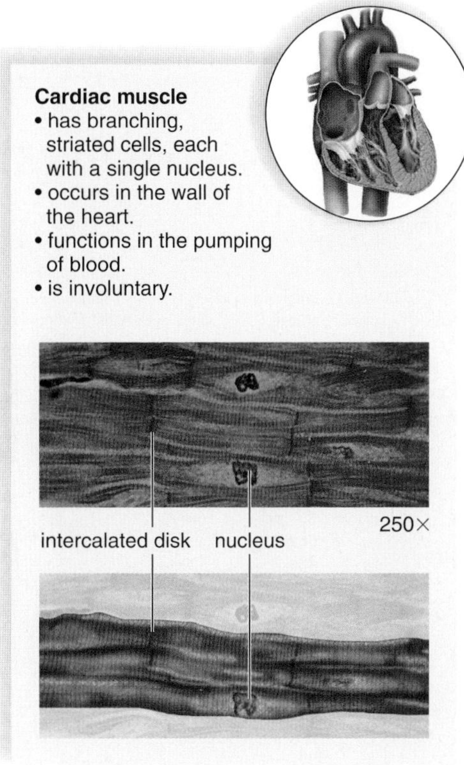

Cardiac muscle
• has branching, striated cells, each with a single nucleus.
• occurs in the wall of the heart.
• functions in the pumping of blood.
• is involuntary.
intercalated disk
nucleus
250×

Observation: Smooth Muscle

Smooth muscle is sometimes called **visceral muscle** because it makes up the walls of the internal organs, such as the intestines and the blood vessels. Smooth muscle is involuntary because its contraction does not require conscious effort.

1. Study a model or diagram of smooth muscle (Fig. 11.14), and note the shape of the cells and the centrally placed nucleus. Smooth muscle has spindle-shaped cells. What does *spindle-shaped* mean? ____

   ____

2. Examine a prepared slide of smooth muscle. Distinguishing the boundaries between the different cells may require you to bring the slide in and out of focus.

Figure 11.14 Smooth muscle.
Smooth muscle is nonstriated and involuntary. This type of muscle is composed of spindle-shaped cells.

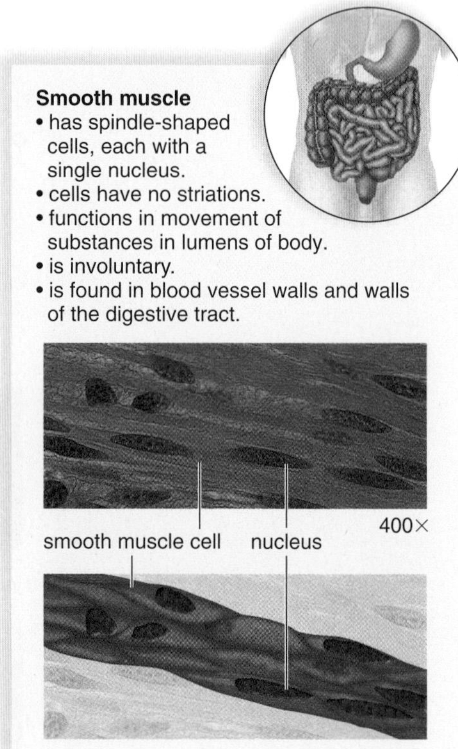

Smooth muscle
• has spindle-shaped cells, each with a single nucleus.
• cells have no striations.
• functions in movement of substances in lumens of body.
• is involuntary.
• is found in blood vessel walls and walls of the digestive tract.
400×
smooth muscle cell
nucleus

Summary of Muscular Tissue

Complete Table 11.3 to summarize your study of muscular tissue.

Table 11.3 Muscular Tissue

Type	Striations (Yes or No)	Branching (Yes or No)	Conscious Control (Yes or No)
Skeletal			
Cardiac			
Smooth			

11.4 Nervous Tissue

Nervous tissue is found in the brain, spinal cord, and nerves. Nervous tissue is composed of two types of cells: **neurons** that transmit messages and **neuroglia** that largely service the neurons (see Fig. 11.1). Motor neurons, which take messages from the spinal cord to the muscles, are often used to exemplify typical neurons (Fig. 11.15). Motor neurons have several **dendrites,** processes that take signals to a **cell body,** where the nucleus is located, and an **axon** that takes nerve impulses away from the cell body.

Observation: Nervous Tissue

1. Study a model or diagram of a neuron, and then examine a prepared slide. You will not be able to see neuroglial cells because they are small and cannot be seen at this magnification.
2. Identify the dendrites, cell body, and axon in Figure 11.15 and label the micrograph. Long axons are called nerve fibers.
3. Explain the appearance and function of the parts of a motor neuron:
   a. Dendrites ______________________________
   b. Cell body ______________________________
   c. Axon ______________________________

Figure 11.15 Motor neuron anatomy.

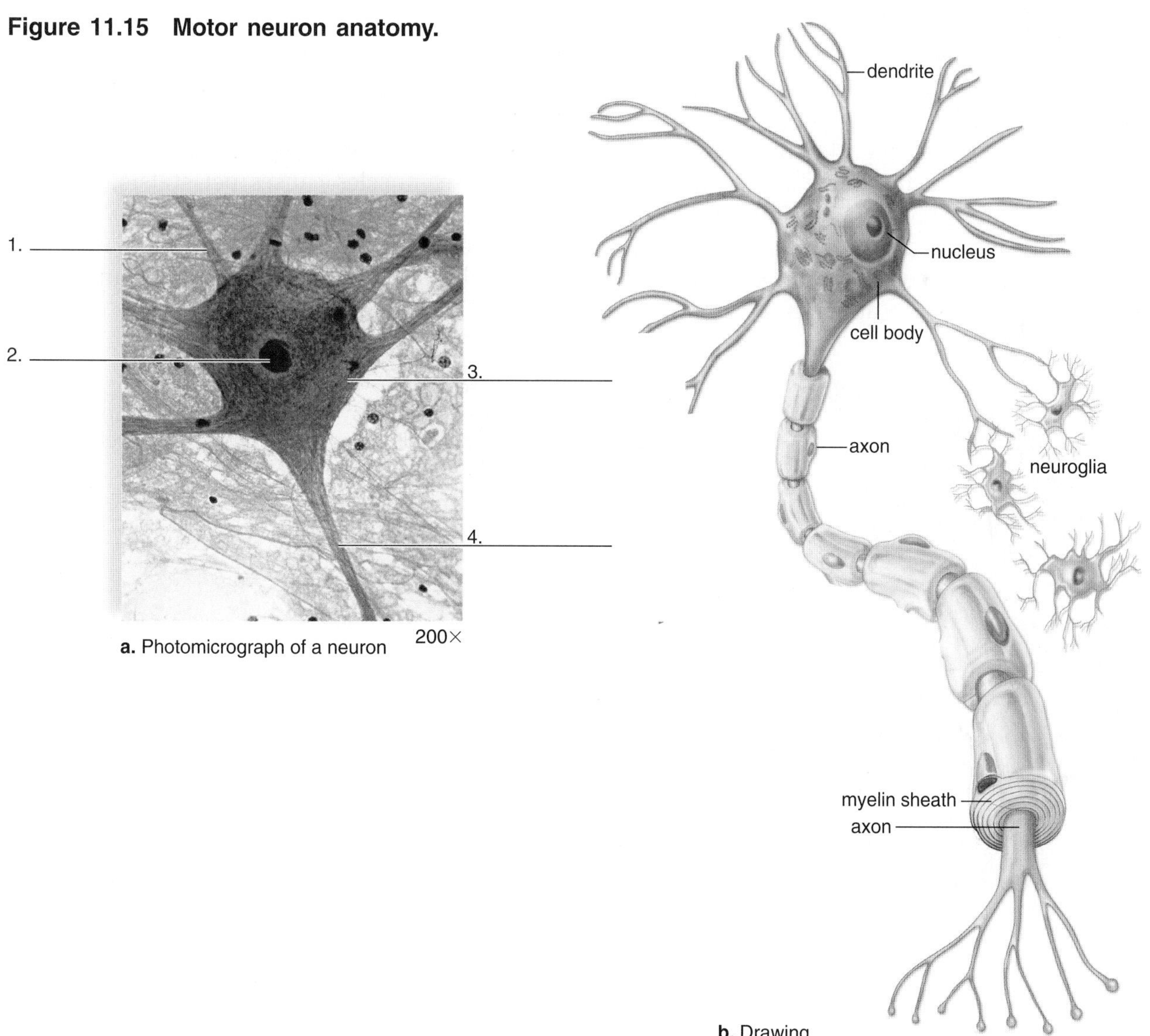

a. Photomicrograph of a neuron

b. Drawing

11.5 Tissues Form Organs

Organs are structures composed of two or more types of tissue that work together to perform particular functions. You may tend to think that a particular organ contains only one type of tissue. For example, muscular tissue is usually associated with muscles and nervous tissue with the brain. However, muscles and the brain also contain other types of tissue—for example, loose connective tissue and blood. Here we will study the compositions of two organs—the intestine and the skin.

Intestine

The **intestine,** a part of the digestive system, processes food and absorbs nutrient molecules.

Observation: Intestinal Wall

Study a slide of a cross section of intestinal wall. With the help of Figure 11.16, identify the following layers:

1. **Mucosa** (mucous membrane layer): This layer, which lines the central lumen (cavity), is made up of columnar epithelium overlying a layer of connective tissue. The epithelium is glandular—that is, it secretes mucus from goblet cells and digestive enzymes from the rest of the epithelium. The membrane is arranged in deep folds (fingerlike projections) called **villi,** which increase the small intestine's absorptive surface.
2. **Submucosa** (submucosal layer): This connective tissue layer contains nerve fibers, blood vessels, and lymphatic vessels. The products of digestion are absorbed into these blood and lymphatic vessels.
3. **Muscularis** (smooth muscle layer): Circular muscular tissue and then longitudinal muscular tissue are found in this layer. Rhythmic contraction of these muscles causes **peristalsis,** a wavelike motion that moves food along the intestine.
4. **Serosa** (serous membrane layer): In this layer, a thin sheet of connective tissue underlies a thin, outermost sheet of squamous epithelium. This membrane is part of the **peritoneum,** which lines the entire abdominal cavity.

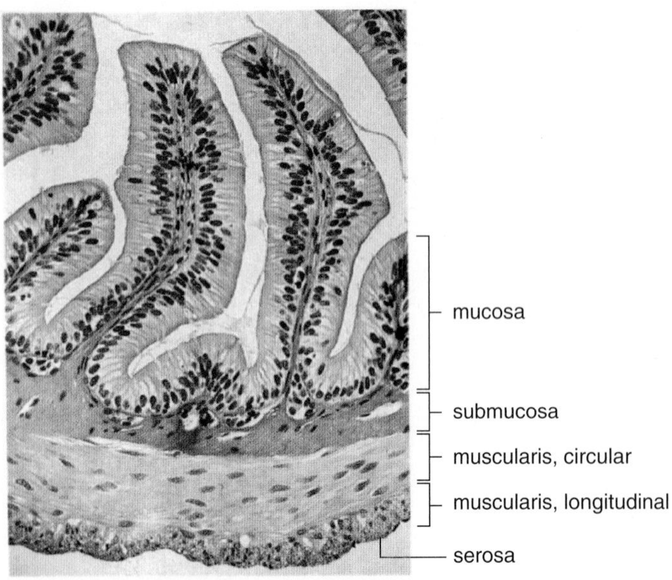

Figure 11.16 The intestinal wall. A cross section reveals the various layers of the intestinal wall, noted to the right of this photomicrograph. (Magnification ×25)

Skin

The skin covers the entire exterior of the human body. Skin functions include protection, water retention, sensory reception, body temperature regulation, and vitamin D synthesis.

Observation: Skin

Study a model or diagram and a prepared slide of the skin. With the help of Figure 11.17, identify the two skin regions and the subcutaneous layer from the exterior surface down:

1. **Epidermis:** This region is composed of stratified squamous epithelial cells. The outer cells of the epidermis are nonliving and create a waterproof covering that prevents excessive water loss. These cells are always being replaced because an inner layer of the epidermis is composed of living cells that constantly produce new cells.
2. **Dermis:** This region is a connective tissue containing blood vessels, nerves, sense organs, and the expanded portions of oil (sebaceous) and sweat glands and hair follicles.

   List the structures you can identify on your slide: ______________________________

   ____
3. **Subcutaneous layer:** This is a layer of loose connective tissue and adipose tissue that lies beneath the skin proper and insulates and protects inner body parts.

Figure 11.17 Human skin.
Human skin contains two regions, the epidermis and the dermis. (The dermis is blue in this image to distinguish it from the tan-colored epidermis.) The subcutaneous layer lies beneath the skin.

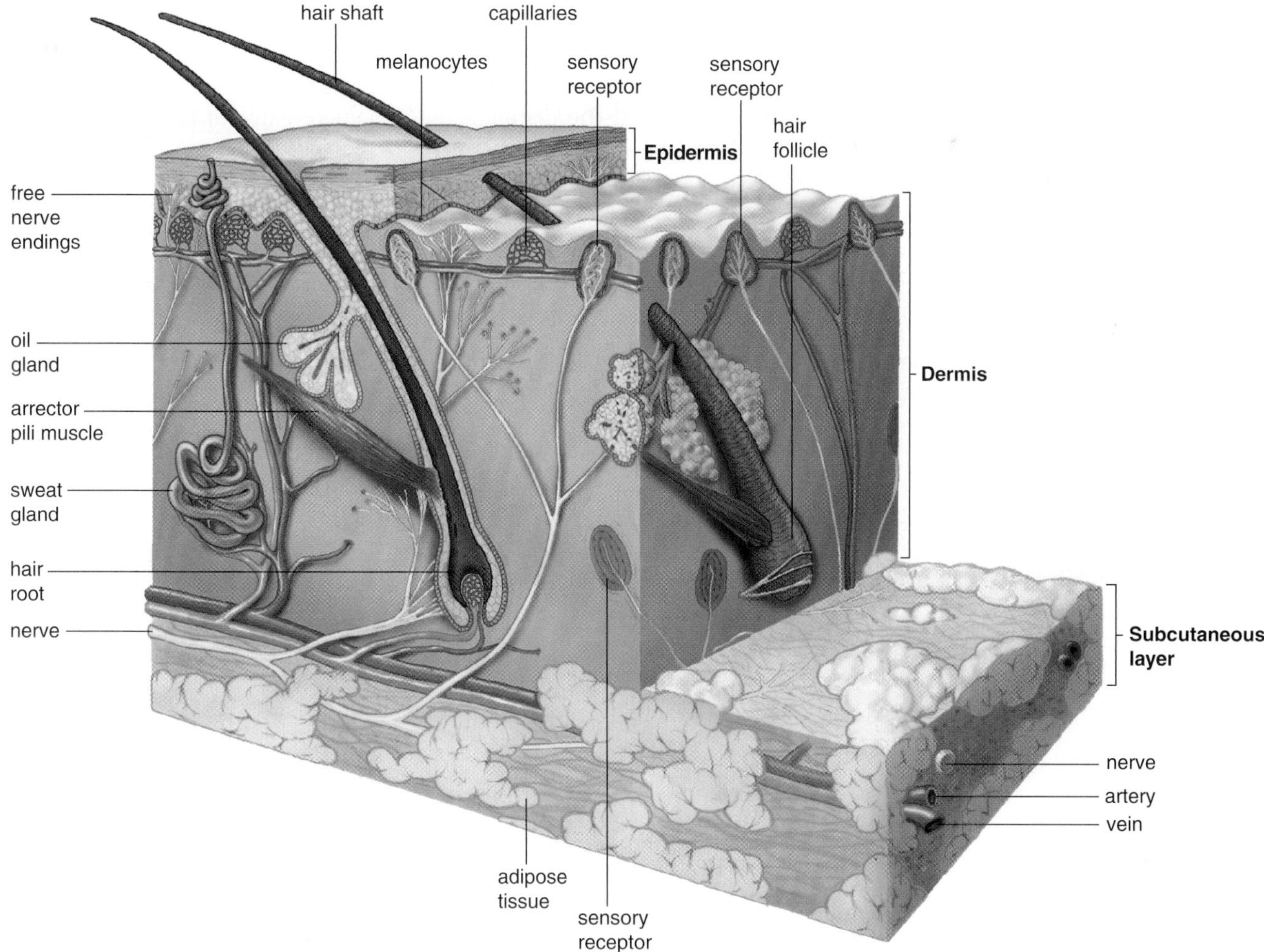

Laboratory Review 11

______ 1. What is the name for a group of cells that has the same structural characteristics and performs the same functions?

______ 2. Which type of epithelium has flattened cells?

______ 3. Name a body location for pseudostratified ciliated columnar epithelium.

______ 4. What is the function of goblet cells?

______ 5. What type of tissue occurs in the epidermis of the skin?

______ 6. Name a body location for hyaline cartilage.

______ 7. The cells of which tissue have a large, central, fat-filled vacuole?

______ 8. What type of muscular tissue is involuntary and striated?

______ 9. Name a body location for smooth muscle.

______ 10. What types of muscular tissue are striated?

______ 11. Name a body location for nervous tissue.

______ 12. Where is the nucleus located in a nerve cell?

______ 13. What type of tissue accounts for the movement of food along the digestive tract?

______ 14. Which skin layer contains blood vessels?

______ 15. What portion of a nerve cell transmits information away from the cell body?

Thought Questions

16. List the four major types of human body tissues and the distinguishing characteristics of each. Why does the human body contain all four types rather than a single type?

17. List the five types of epithelial tissue and the distinguishing characteristics of each.

18. Your lab instructor gives you a slide containing prepared muscle tissue. How would you identify the type of muscle tissue located on the slide?

19. Why might an injury to a bone have a faster recovery/healing time when compared to a muscle or nerve cell?

12

Chemical Aspects of Digestion

Learning Outcomes

12.1 Starch Digestion by Salivary Amylase

- Discuss the role of water in amylase activity.
- Relate the terms enzyme, substrate, and product to the digestive action of amylase.
- Describe the effect of length of time on the digestive action of amylase.
- Explain the effect of boiling on the digestive action of amylase.
- Generalize how results are evaluated to determine if starch digestion occurred.

12.2 Fat Digestion by Pancreatic Lipase

- Explain why the emulsification process was important prior to the action of lipase.
- Explain the function of a control in an experiment.
- Explain why a change in pH would indicate that fat digestion had occurred in the experimental procedure.

12.3 Protein Digestion by Pepsin

- Describe the effect of pH on the digestive action of pepsin.
- Describe the effect of environmental temperature on the digestive action of pepsin.
- Discuss the importance of accurate interpretation of color in evaluating this experiment.

Planning Ahead

Both the starch and protein digestion experiments (see pages 150 and 154) contain activities that require prior setup and taking subsequent readings at intervals. Your instructor may wish to have you set these experiments up in advance. In addition, a boiling water bath will be needed and should be started early.

Introduction

Digestion is the process by which food is degraded or broken down by **hydrolytic reactions**—that is, water is added to a large molecule, which splits into smaller, soluble molecules that can be absorbed into the bloodstream. For example, water is added until proteins are broken down into amino acids, starch is broken down into glucose, and fats are broken down into glycerol and fatty acids.

Enzymes are necessary for digestion, just as they are required for other chemical reactions in the body. The experimental procedures in this laboratory will demonstrate that **pepsin** speeds the hydrolysis of protein, that **pancreatic lipase** acts on fat, and that **salivary amylase** hydrolyzes starch. Enzymes are very specific and usually participate in only one type of reaction. According to the lock-and-key model of enzymatic action, this is because an enzyme fits its substrate just as a key fits a lock. Enzymes must maintain an appropriate shape or configuration to take part in a reaction. Enzymes have an optimum pH that allows them to maintain their usual three-dimensional shape. Enzymatic reactions speed up with increased temperature but are destroyed by excessive heat or boiling.

Each of the experimental procedures in this laboratory contain a **control,** a sample that goes through all the steps of the experiment except the one being tested (the **experimental variable**). If all of the chemicals (materials) are performing as expected, the results of this control should be negative. In experiments such as the ones in this laboratory, however, many factors, such as the purity of the chemical reagents, the age of the reagent, and the conditions of storage, can affect the outcome. In addition, organic compounds can deteriorate upon storage, particularly if they are stored under the wrong conditions. Finally, the accuracy of your results will be greatly enhanced if you carefully read each experimental procedure **before** beginning the experiment.

12.1 Starch Digestion by Salivary Amylase

Starch is present in bakery products and in potatoes, rice, and corn. Starch is digested by **salivary amylase** in the mouth, a process described by the following reaction:

$$\text{starch} + \text{water} \xrightarrow{\text{amylase (enzyme)}} \text{maltose}$$

The placement of an enzyme on top of a reaction arrow indicates that the enzyme is not consumed or changed in the process. Enzymes are recycled. Look for this type of notation throughout this laboratory.

1. Why is this reaction called a *hydrolytic reaction*? ______________________________
____
2. If digestion does *not* occur, which will be present—starch or maltose? ______________
3. If digestion *does* occur, which will be present—starch or maltose? ________________
4. Why is it advantageous for the enzyme *not* to be consumed during the reaction? ________

Tests for Starch Digestion

You will be using two tests for starch digestion:

1. If digestion has not taken place, the iodine test for starch will be positive. If starch is present, a blue-black color immediately appears after a few drops of iodine are added to the test tube.
2. If digestion has taken place, a test for sugar (maltose) will be positive, with red showing the highest concentration of maltose and green showing the lowest (see Table 3.4, page 35). To test for sugar, add an equal amount of Benedict's reagent to the test tube. Place the tube in a boiling water bath for two to five minutes, and note any color changes. A color change of blue ⟶ green ⟶ yellow ⟶ orange ⟶ red indicates the presence of maltose. Boiling the test tube is necessary for the Benedict's reagent to react. To which category of organic compounds (lipid, carbohydrate, or protein) do enzymes such as amylase belong? ______________________

   What happens when enzymes are boiled? ______________________________________

Experimental Procedure: Starch Digestion

Caution: Use protective eyewear when performing this experiment.

Preparation

1. With a wax pencil, number eight clean test tubes above the level of the boiling water bath, and mark at the 1 cm and 2 cm levels.
2. Fill tubes 1 through 6 to the 1 cm mark with *alpha-amylase solution.* Fill tubes 7 and 8 to the 1 cm mark with *water.*
3. Shake the *starch suspension* well each time before dispensing.
4. Fill tubes 3 and 4 to the 2 cm mark with *starch suspension,* and allow them to stand at room temperature for 30 minutes.
5. Place tubes 5 and 6 in a boiling water bath for 10 minutes. After boiling, fill to the 2 cm mark with *starch suspension,* and allow the tubes to stand for 20 minutes.

6. Fill tubes 7 and 8 to the 1 cm mark with *water* and to the 2 cm mark with *starch suspension.* Allow the tubes to stand for 30 minutes.

Tube 1 Fill to the 2 cm mark with *starch suspension,* and test for starch **immediately,** using the iodine test described previously. Record your results in Table 12.1.

Tube 2 Fill to the 2 cm mark with *starch suspension,* and test for sugar **immediately,** using *Benedict's reagent,* described earlier, which requires boiling. Record your results in Table 12.1.

Why do you expect tube 1 to have a positive test for starch and tube 2 to have a negative test for sugar? ____________________

Record your explanation in Table 12.1.

Tubes 3, 5, and 7 After 30 minutes, test for starch using the iodine test. Record your results in Table 12.1.

Tubes 4, 6, and 8 After 30 minutes, test for sugar using the Benedict's test. Record your results in Table 12.1.

Why do you expect tube 3 to have a negative test for starch and tube 4 to have a positive test for sugar? ____________________

Why do you expect tube 5 to have a positive test for starch and tube 6 to have a negative test for sugar? ____________________

Why do you expect tube 7 to have a positive test for starch and tube 8 to have a negative test for sugar? ____________________

Record your explanations for your results in Table 12.1.

Table 12.1 Starch Digestion by Amylase

Tube	Contents	Time	Type of Test	Results	Explanation
1	Alpha-amylase Starch				
2	Alpha-amylase Starch				
3	Alpha-amylase Starch				
4	Alpha-amylase Starch				
5	Alpha-amylase, boiled Starch				
6	Alpha-amylase, boiled Starch				
7	Water Starch				
8	Water Starch				

Conclusions

- This experiment demonstrated that, in order for an enzymatic reaction to occur, an active ________ must be present, and ________ must pass to allow the reaction to occur.
- Which test tubes served as a control in this experiment? ________
  Explain. ____

12.2 Fat Digestion by Pancreatic Lipase

Lipids include fats (e.g., butterfat) and oils (e.g., sunflower, corn, olive, and canola). Lipids are digested by **pancreatic lipase** in the small intestine (Fig. 12.1), a process described by the following two reactions:

(1) $$\text{fat} \xrightarrow{\text{bile (emulsifier)}} \text{fat droplets}$$

(2) $$\text{fat droplets} + \text{water} \xrightarrow{\text{lipase (enzyme)}} \text{glycerol} + \text{fatty acids}$$

Bile, a product of the liver, enters the small intestine by way of the bile duct. Since bile is not an enzyme, the first reaction is not enzymatic. It is an emulsification reaction in which fat is physically dispersed by the emulsifier (bile) into small droplets. The small droplets provide a greater surface area for enzyme actions. Lipids are hydrophobic and therefore insoluble, so they are hydrolyzed slowly unless an emulsifier is used.

Given the second reaction, would the pH of the solution be lower before or after the reaction? (*Hint:* Remember that an acid decreases pH and a base increases pH.) ____

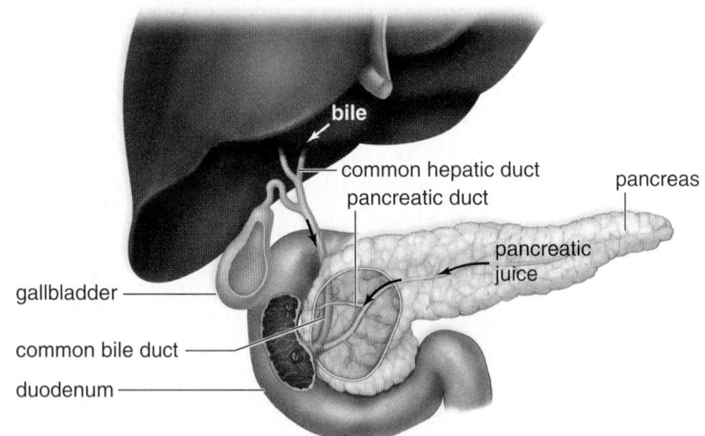

Figure 12.1 Emulsification and digestion of fat.
Bile from the liver (stored in gallbladder) enters small intestine, where lipase in pancreatic juice from the pancreas digests fat.

Test for Fat Digestion

In the test for fat digestion you will be using a pH indicator that changes color as the solution in the test tube goes from basic conditions to acidic conditions. Phenol red is a pH indicator that is red in basic solutions and yellow in acidic solutions. Bile salts will be used as an emulsifier.

Experimental Procedure: Fat Digestion

With a wax pencil, number three clean test tubes, and mark at the 1 cm, 3 cm, and 5 cm levels. Fill all the tubes to the 1 cm mark with *vegetable oil* and to the 3 cm mark with *phenol red.*

Tube 1 Fill to the 5 cm mark with *pancreatin solution* (pancreatic lipase). Add a pinch of *bile salts.* Invert gently to mix, and record the initial color in Table 12.2. Incubate at 37°C, and check every 20 minutes. Record the length of time for any color change.

Tube 2 Fill to the 5 cm mark with *pancreatin solution.* Invert gently to mix, and record the initial color in Table 12.2. Incubate at 37°C, and check every 20 minutes. Record the length of time for any color change.

Tube 3 Fill to the 5 cm mark with *water.* Invert gently to mix, and record the initial color in Table 12.2. Incubate at 37°C, and check every 20 minutes. Record the length of time for any color change.

Table 12.2 Fat Digestion by Pancreatic Lipase

Tube	Contents	Time	Color Change		Explanation
			Initial	*Final*	
1	Vegetable oil Phenol red Pancreatin Bile salts				
2	Vegetable oil Phenol red Pancreatin				
3	Vegetable oil Phenol red Water				

Conclusions

- Explain your results in Table 12.2 by giving a reason digestion did or did not occur.
- What role did bile play in this experiment? ______
- What role did phenol red play in this experiment? ______
- Which test tube in this experiment could be considered a control? ______

12.3 Protein Digestion by Pepsin

Certain foods, such as meat and egg whites, are rich in protein. Egg whites contain albumin, which is the protein used in this exercise. Protein is digested by **pepsin** in the stomach (Fig. 12.2), a process described by the following reaction:

$$\text{protein} + \text{water} \xrightarrow{\text{pepsin (enzyme)}} \text{peptides}$$

The stomach has a very low pH. Does this indicate that pepsin works effectively in an acidic or a basic environment? ____________________

Figure 12.2 Digestion of protein.
Pepsin, produced by the gastric glands of the stomach, helps digest protein.

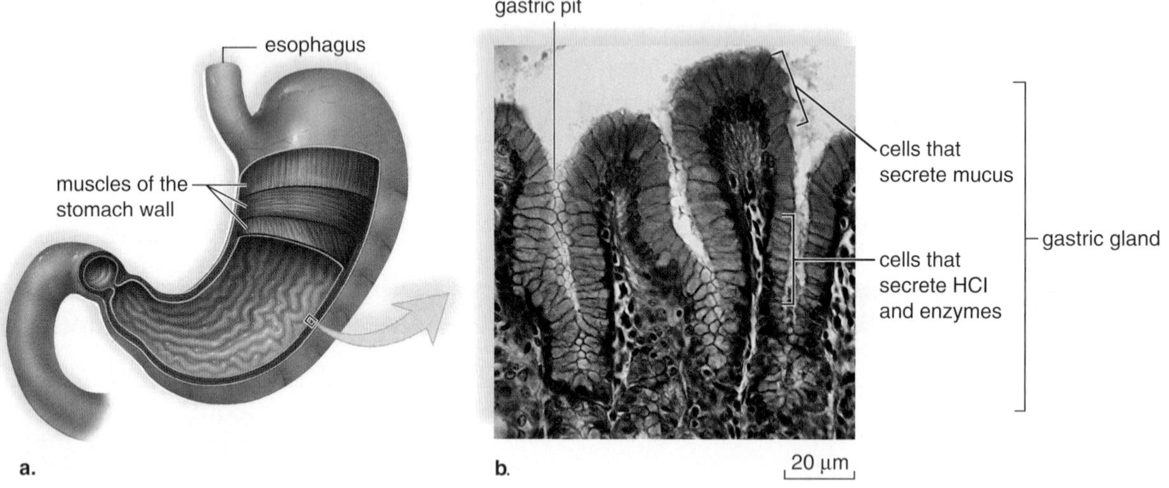

Test for Protein Digestion

Biuret reagent is used to test for protein digestion. If digestion has not occurred, biuret reagent turns purple, indicating that protein is present. If digestion has occurred, biuret reagent turns pinkish-purple, indicating that peptides are present.

> **Caution:** Use protective eyewear when performing this experiment. Biuret reagent contains a strong solution of sodium or potassium hydroxide. Exercise care in using this reagent, and follow your instructor's directions for disposing of these tubes. If any biuret reagent should spill on your skin, rinse immediately with water.

Experimental Procedure: Protein Digestion

With a wax pencil, number four test tubes, and mark at the 2 cm, 4 cm, 6 cm, and 8 cm levels. Fill all tubes to the 2 cm mark with *albumin solution.*

Tube 1 Fill to the 4 cm mark with *pepsin solution* and to the 6 cm mark with *0.2% HCl.* Swirl to mix, and incubate at 37°C. After 1½ hours, fill to the 8 cm mark with *biuret reagent.* Record your results in Table 12.3.

Tube 2 Fill to the 4 cm mark with *pepsin solution* and to the 6 cm mark with *0.2% HCl.* Swirl to mix, and keep at room temperature. After 1½ hours, fill to the 8 cm mark with *biuret reagent.* Record your results in Table 12.3.

Tube 3 Fill to the 4 cm mark with *pepsin solution* and to the 6 cm mark with *water.* Swirl to mix, and incubate at 37°C. After 1½ hours, fill to the 8 cm mark with *biuret reagent.* Record your results in Table 12.3.

Tube 4 Fill to the 6 cm mark with *water.* Swirl to mix, and incubate at 37°C. After 1½ hours, fill to the 8 cm mark with *biuret reagent.* Record your results in Table 12.3.

Table 12.3 Protein Digestion by Pepsin

Tube	Contents	Temperature	Results of Test	Explanation
1	Albumin Pepsin HCl Biuret reagent			
2	Albumin Pepsin HCl Biuret reagent			
3	Albumin Pepsin Water Biuret reagent			
4	Albumin Water Biuret reagent			

Conclusions

- Explain your results in Table 12.3 by giving a reason digestion did or did not occur.
- Which tube was the control? ______

  Explain. ______
- If this control tube had given a positive result for protein digestion, what could you conclude about this experiment? ______

Requirements for Digestion

Explain in Table 12.4 how each of the requirements listed influences effective digestion.

Table 12.4 Requirements for Digestion

Requirement	Explanation
Specific enzyme	
Specific substrate	
Warm temperature	
Time	
Specific pH	
Fat emulsifier	

Laboratory Review 12

1. When iodine (IKI) solution turns blue-black, what substance is present?
2. What color is Benedict's reagent originally?
3. What happens to an enzyme when it is boiled?
4. Saliva contains what enzyme?
5. As oil is digested, why does the first tube turn from red to yellow?
6. What temperature promotes enzymatic action?
7. What do you call a sample that goes through all the steps of an experiment but lacks the factor being tested?
8. What role do bile salts play in the digestion of fat?
9. What color does biuret reagent turn when peptides are present?
10. Is the optimal pH for pepsin acidic or basic?
11. Why would you predict that pepsin would not digest starch?
12. In addition to pepsin and water, what is needed to digest protein?
13. Name the enzyme responsible for the hydrolysis of starch.

Thought Questions

14. Which of the following two combinations is most likely to result in digestion?
    a. Pepsin, protein, water, body temperature
    b. Pepsin, protein, hydrochloric acid (HCl), body temperature
    Explain.

15. Which of the following two combinations is most likely to result in digestion?
    a. Amylase, starch, water, body temperature, testing immediately
    b. Amylase, starch, water, body temperature, waiting 30 minutes
    Explain.

(Review continues on reverse.)

16. What would you conclude if the tests for starch and maltose were both positive?

17. Amylase operates best in a neutral pH environment. What would you expect to happen to the digestion of starch once it enters the stomach? Why?

13

Basic Mammalian Anatomy I

Learning Outcomes

13.1 External Anatomy
- Compare the limbs of a pig to the limbs of a human.
- Identify the gender of a fetal pig.

13.2 Oral Cavity and Pharynx
- Using a preserved specimen, photo, or anatomical drawing of a fetal pig, find and identify the teeth, tongue, and hard and soft palates, including the uvula.
- Using a preserved specimen, photo, or anatomical drawing of a fetal pig, identify and state a function for the glottis, nasopharynx, and esophagus.
- Name the two pathways that cross in the mammalian pharynx and explain how their separate functions are maintained despite having a common or shared entrance.

13.3 Thoracic and Abdominal Incisions
- State the major organs of the thoracic cavity and distinguish these from the major organs of the abdominal cavity.

13.4 Neck Region
- Using a preserved specimen, photo, or anatomical drawing of a fetal pig, find, identify, and state a function for the thymus, the larynx, and the thyroid gland.

13.5 Thoracic Cavity
- Identify the three compartments and the organs of the thoracic cavity.
- Using a preserved specimen, photo, or anatomical drawing of a fetal pig, find and identify the diaphragm and describe its primary function.

13.6 Abdominal Cavity
- Using a preserved specimen, photo, or anatomical drawing of a fetal pig, find, identify, and state a function for the liver, stomach, spleen, gallbladder, pancreas, large intestine, and small intestine. Describe where these organs are positioned in relation to one another.

13.7 Human Anatomy
- Using a photograph or anatomical drawing or model of a human find, identify, and state a function for the human organs studied in this laboratory.
- Associate each organ with a particular system of the body.

Introduction

In this laboratory, you will dissect a fetal pig. Both pigs and humans are mammals; therefore, you will be studying mammalian anatomy. The period of pregnancy, or gestation, in pigs is approximately 17 weeks (compared with an average of 40 weeks in humans). The age of the piglets used in class will usually be within 1 to 2 weeks of birth.

The pigs will have a slash in the right neck region, indicating the site of blood drainage. A red latex solution was injected into the **arterial system,** and a blue latex solution was injected into the **venous system** of the pigs. Therefore, when a vessel appears red, it is an artery, and when a vessel appears blue, it is a vein. Do not confuse this color pattern with circulatory diagrams in your textbook, or like the one shown in Figure 14.1 of this manual, that differentiate oxygen-rich blood flow and oxygen-poor blood flow by using red- and blue-colored vessels, respectively.

> **Caution:** Wear protective latex gloves when handling preserved animal organs. Use protective eyewear and exercise caution when using sharp instruments during this experiment. Wash hands thoroughly upon completion of the experiment.

13.1 External Anatomy

Mammals are characterized by the presence of mammary glands and hair. Mammals also occur in two distinct sexes, called males and females, that are often distinguishable by their external **genitals,** the reproductive organs.

Both pigs and humans are placental mammals, which means that development occurs internally, within the uterus of the mother. The **placenta** is a layer composed of tissue from the embryo and the mother's uterine wall. As development proceeds, the **umbilical cord** forms, connecting the placenta and uterine wall to the embryo. The umbilical cord remains attached as the embryo develops into a fetus, providing a connection to the mother's system until shortly after birth. Oxygen and nutrients from the mother's blood supply are delivered to the fetus by the umbilical cord and in exchange, waste products and carbon dioxide are transported away by the mother's bloodstream.

Pigs and humans are tetrapods—that is, they have four limbs. Pigs walk on all four of their limbs; in fact, they walk on their toes, and their toenails have evolved into hooves. In contrast, humans walk only on the feet of their hind limbs.

Perspective is important in dissections. If an animal is lying face-down, *top* and *bottom* mean something different than when the animal is lying on its back. For this reason, the terms *dorsal, ventral, anterior, and posterior* are used in place of terms that you may commonly use. See Figure 13.1 to locate the appropriate positions of these orientations.

Observation: External Anatomy

Body Regions and Limbs

1. Place your animal in a dissecting pan, and observe the following body regions: the rather large head; the short, thick neck; the cylindrical trunk with two pairs of appendages (forelimbs and hind limbs); and the short tail (Fig. 13.1*a*). The tail is an extension of the vertebral column.
2. Examine the four limbs, and feel for the joints of the digits, wrist, elbow, shoulder, hip, knee, and ankle.
3. Determine which parts of the forelimb correspond to your upper arm, elbow, lower arm, wrist, and hand.
4. Do the same for the hind limb, comparing it with your leg.
5. The pig walks on its toenails, which would be like a ballet dancer on "tiptoe." Notice how your heel touches the ground when you walk. Where is the heel of the pig? ______________

Umbilical Cord

1. Locate the umbilical cord arising from the ventral (toward the belly) portion of the abdomen.
2. Note the cut ends of the umbilical blood vessels. If they are not easily seen, cut the umbilical cord near the end and observe this new surface.
3. What is the function of the umbilical cord? ______________________________

   ____

Nipples and Hair

1. Locate the small **nipples,** which are the external openings of the **mammary glands.** Note that the nipples are *not* an indication of sex, since both males and females possess them. How many nipples does your pig have? ________ Why is it advantageous for a pig to have so many nipples? ______________________________

Figure 13.1 External anatomy of the fetal pig.
a. Body regions and limbs. **b,c.** the sexes can be distinguished by the external genitals.

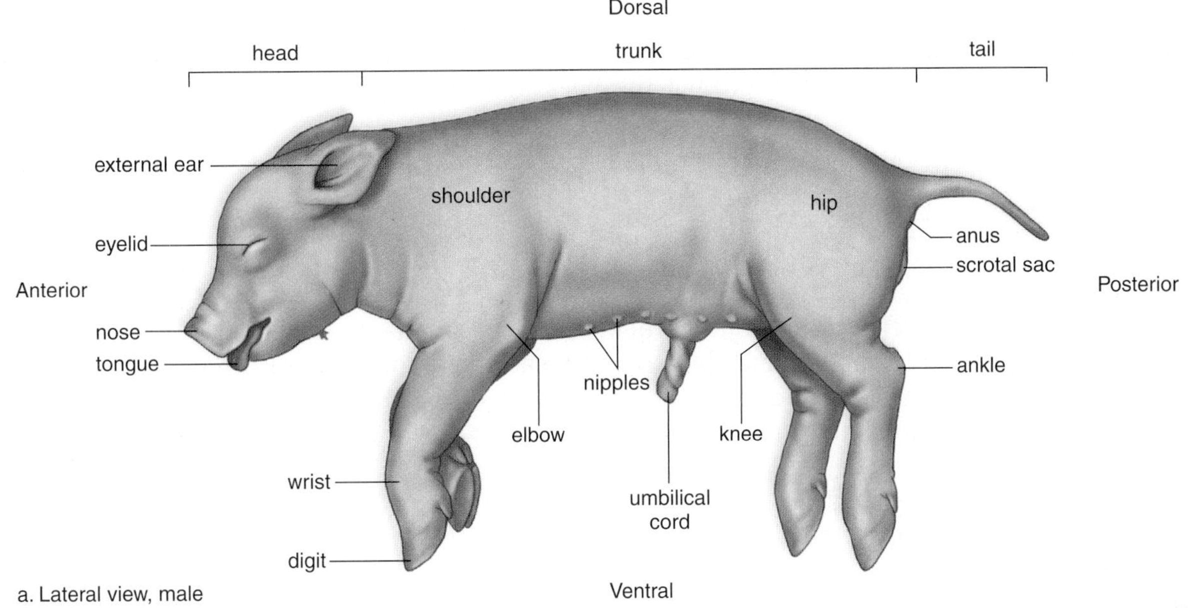

a. Lateral view, male

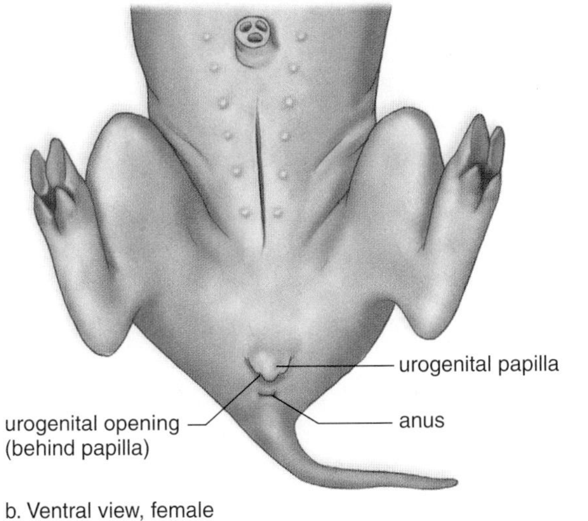

b. Ventral view, female

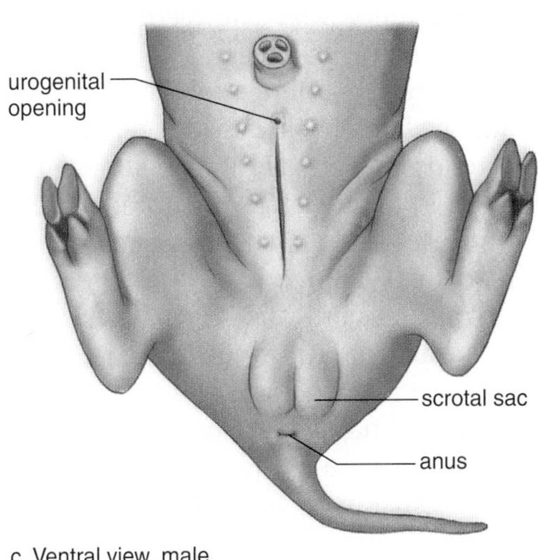

c. Ventral view, male

2. Can you find hair on your pig? ________ If so, where? ______________________________

____

Anus and External Genitals

1. Locate the **anus** under the tail. The anus is an opening for what system in the body? ________
2. In females, locate the **urogenital opening,** just anterior to the anus, and a small, fleshy **urogenital papilla** projecting from the urogenital opening (Fig. 13.1*b*).
3. In males, locate the urogenital opening just posterior to the umbilical cord. The duct leading to it runs forward from between the legs in a long, thick tube, the **penis,** which can be felt under the skin. In males, the urinary system and the genital system are always joined (Fig. 13.1*c*).
4. You are responsible for identifying pigs of both sexes. What sex is your pig? ______________ Be sure to look at a pig of the opposite sex that another group of students is dissecting.

13.2 Oral Cavity and Pharynx

The **oral cavity** is the space in the mouth that contains the tongue and the teeth. The **pharynx** is posterior and slightly dorsal to the oral cavity and has three openings: The **glottis** is an opening through which air passes on its way to the **trachea** (the windpipe) and lungs. The **esophagus** is a portion of the digestive tract that leads through the neck and thorax to the stomach. The **nasopharynx** leads to the nasal passages.

Observation: Oral Cavity and Pharynx

Oral Cavity

1. Insert a sturdy pair of scissors into one corner of the specimen's mouth, and cut posteriorly (toward the hind end) for approximately 4 cm. Repeat on the opposite side.
2. Place your thumb on the tongue at the front of the mouth, and gently push downward on the lower jaw. This will tear some of the tissue in the angles of the jaws so that the mouth will remain partly open (Fig. 13.2).
3. Note small, underdeveloped teeth in both the upper and lower jaws. Other embryonic, nonerupted teeth may also be found within the gums. The teeth are used to chew food.
4. Examine the tongue, which is partly attached to the lower jaw region but extends posteriorly and is attached to a bony structure at the back of the oral cavity (Fig. 13.2). The tongue manipulates food for swallowing.
5. Locate the hard and soft palates (Fig. 13.2). The **hard palate** is the ridged roof of the mouth that separates the oral cavity from the nasal passages. The **soft palate** is a smooth region posterior to the hard palate. An extension of the soft palate—the **uvula**—hangs down into the throat in humans. (A pig does not have a uvula.)

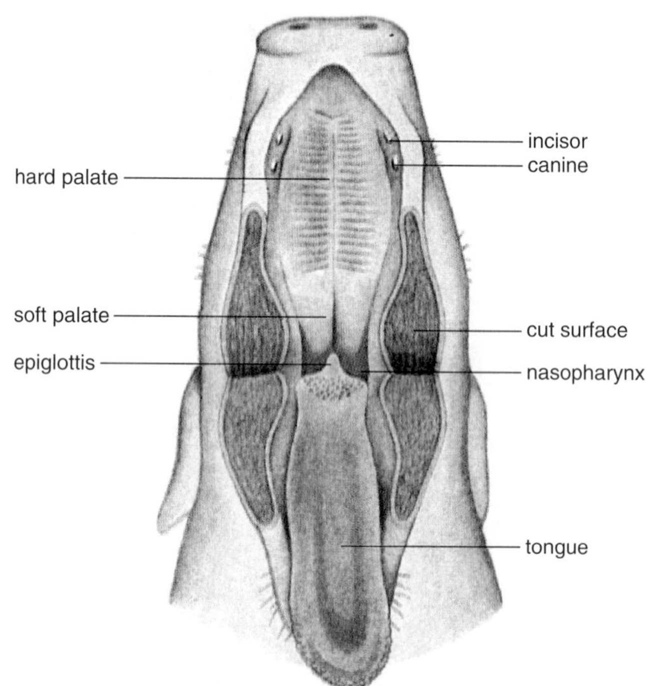

Figure 13.2 Oral cavity of the fetal pig.
The roof of the oral cavity contains the hard and soft palates, and the tongue lies above the floor of the oral cavity.

Pharynx

1. Push down on the tongue until you open the jaws far enough to see a slightly pointed flap of tissue pointing dorsally (toward the back) (Fig. 13.2). This flap is the **epiglottis,** which covers the glottis. The glottis leads to the trachea (Fig. 13.3*a*).
2. Posterior and dorsal to the glottis, find the opening into the esophagus. Note the proximity of the glottis and the opening to the esophagus. Each time the pig—or a human—swallows, the epiglottis shifts position, closing the glottis to keep food and fluids from going into the lungs via the trachea.
3. Insert a blunt probe into the glottis, and note that it enters the trachea. Remove the probe, insert it into the esophagus, and note the position of the esophagus beneath, or dorsal to, the trachea.
4. Make a midline cut in the soft palate from the epiglottis to the hard palate. Then make two lateral cuts at the edge of the hard palate.
5. Posterior to the soft palate, locate the openings to the nasal passages.
6. Explain why it is correct to say that the air and food passages cross in the pharynx.

____

____

Figure 13.3 Air and food passages in the fetal pig.
The air and food passages cross in the pharynx. **a.** Drawing. **b.** Dissection of specimen.

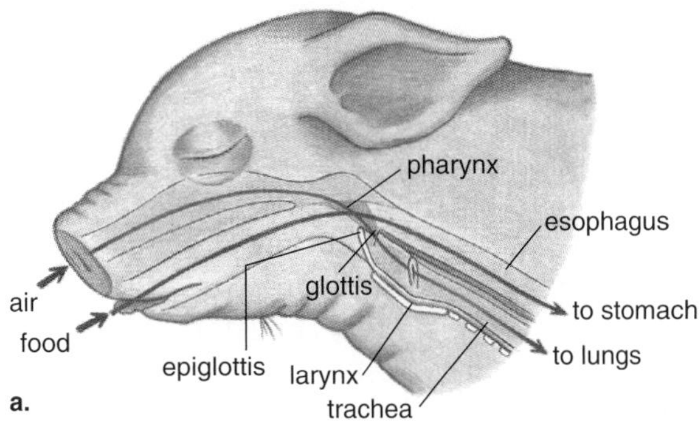

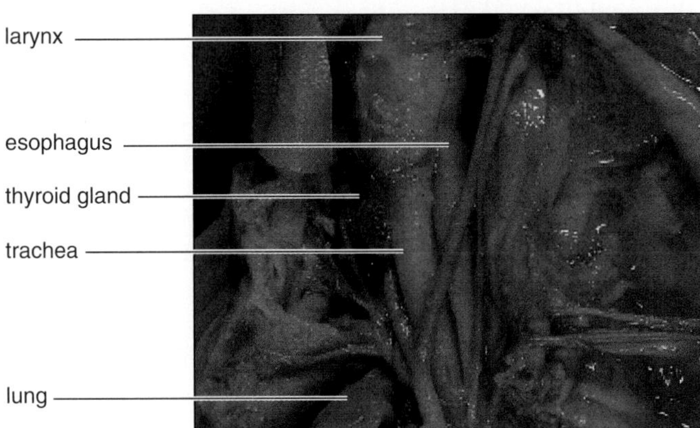

13.3 Thoracic and Abdominal Incisions

First, prepare your pig according to the following directions, and then make thoracic and abdominal incisions so that you will be able to study the internal anatomy of your pig.

Preparation of Pig for Dissection

1. Place the fetal pig on its back in the dissecting pan.
2. Tie a cord around one forelimb, and then bring the cord around underneath the pan to fasten back the other forelimb.
3. Spread the hind limbs in the same way.
4. With scissors always pointing up (never down), make the following incisions to expose the thoracic and abdominal cavities. The incisions are numbered on Figure 13.4 to correspond with the following steps.

Thoracic Incisions

1. Cut anteriorly up from the **diaphragm,** a structure that separates the thoracic cavity from the abdominal cavity, until you reach the hairs in the throat region.
2. Make two lateral cuts, one on each side of the midline incision anterior to the forelimbs, taking extra care not to damage the blood vessels around the heart.
3. Make two lateral cuts, one on each side of the midline just posterior to the forelimbs and anterior to the diaphragm, following the ends of the ribs. Pull back the flaps created by these cuts to expose the **thoracic cavity.** List the organs you find in the thoracic cavity.

   ____

Abdominal Incisions

4. With scissors pointing up, cut posteriorly from the diaphragm to the umbilical cord.
5. Make a flap containing the umbilical cord by cutting a semicircle around the cord and by cutting posteriorly to the left and right of the cord.
6. Make two cuts, one on each side of the midline incision posterior to the diaphragm. Examine the diaphragm, which is attached to the chest wall by radially arranged muscles. The central region of the diaphragm, called the **central tendon,** is a membranous area.
7. Make two more cuts, one on each side of the flap containing the umbilical cord and just anterior to the hind limbs. Pull back the side flaps created by these cuts to expose the **abdominal cavity.**
8. Lifting the flap with the umbilical cord requires cutting the **umbilical vein.** Before cutting the umbilical vein, tie a thread on each side of where you will cut to mark the vein for future reference.
9. Rinse out your pig as soon as you have opened the abdominal cavity. If you have a problem with excess fluid, obtain a disposable plastic pipette to suction off the liquid.
10. Anatomically, the diaphragm separates what two cavities?

    ____
11. List the organs you find in the abdominal cavity.

    ____

Figure 13.4 Ventral view of the fetal pig indicating incisions.
These incisions are to be made preparatory to dissecting the internal organs. They are numbered here in the order they should be done.

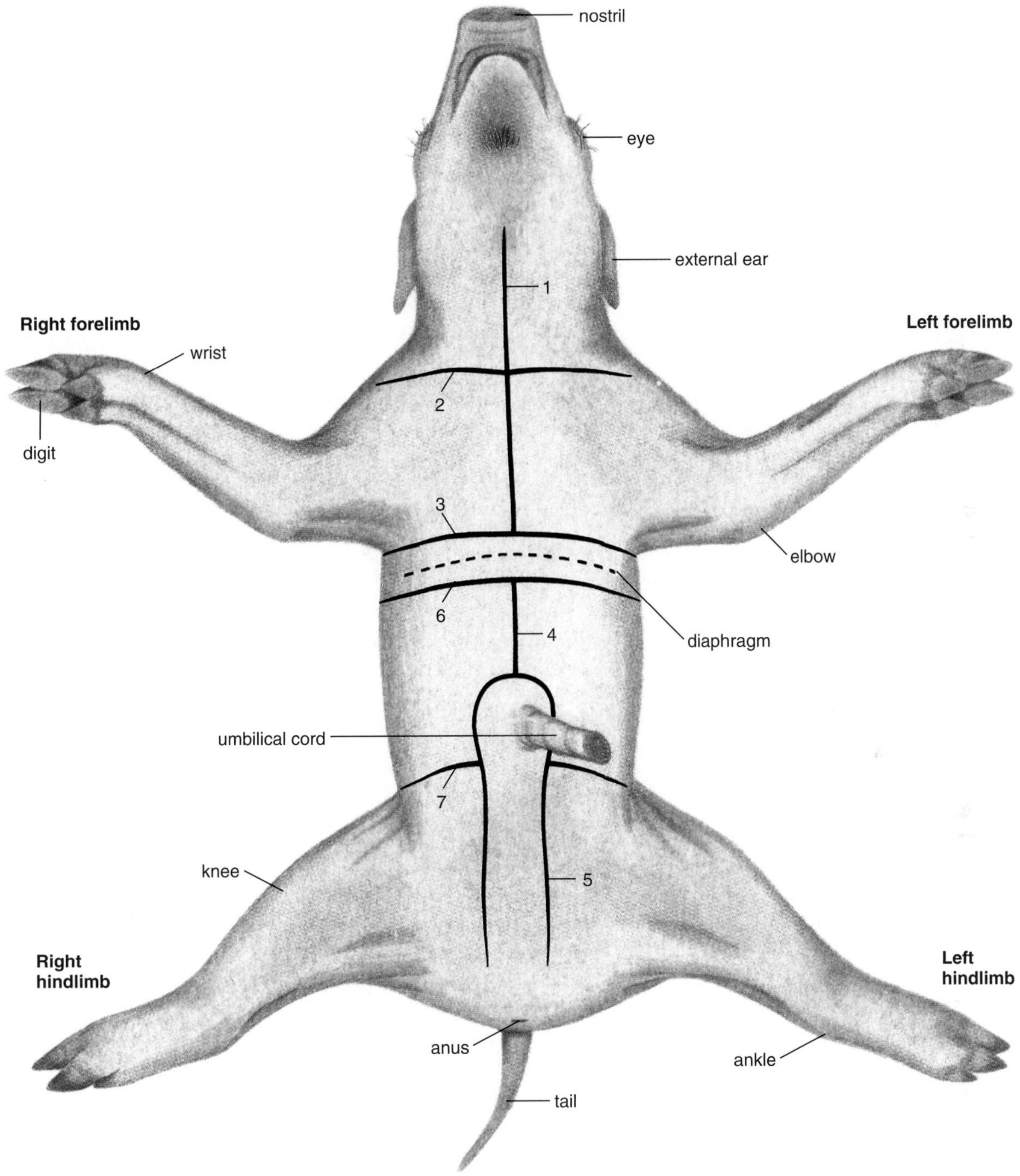

13.4 Neck Region

You will locate several organs in the neck region. Use Figures 13.3*b* and 13.5 as a guide, but **keep all the flaps on your pig** so you can close the thoracic and abdominal cavities at the end of the laboratory session.

The **thymus gland** is a part of the lymphatic system. Certain white blood cells called T (for thymus) lymphocytes mature in the thymus gland and help fight disease. The **larynx,** or voice box, sits atop the **trachea,** or windpipe. The **esophagus** is a portion of the digestive tract that leads to the stomach. The **thyroid gland** secretes hormones that travel in the blood and act upon other body cells. These hormones (e.g., thyroxine) regulate the rate at which metabolism occurs in cells.

Observation: Neck Region

Thymus Gland

1. Move the skin apart in the neck region just below the hairs mentioned earlier. If necessary, cut the body wall laterally to make flaps. You will most likely be viewing exposed muscles.
2. *Cut through and clear away muscle* to expose the *thymus gland,* a diffuse gland that lies among the muscles. The thymus is particularly large in fetal pigs, since their immune systems are still developing.

Larynx, Trachea, and Esophagus

1. Probe down into the deeper layers of the neck. Medially (toward the center), beneath several strips of muscle, you will find the hard-walled larynx and the trachea, parts of the respiratory passage to be examined later. Dorsal to the trachea, find the esophagus.
2. Open the mouth and insert a probe into the glottis and esophagus from the pharynx to better understand the orientation of these two organs.

Thyroid Gland

Locate the thyroid gland just posterior to the larynx, lying ventral to (on top of) the trachea.

13.5 Thoracic Cavity

As previously mentioned, the body cavity of mammals, including human beings, is divided by the diaphragm into the thoracic cavity and the abdominal cavity. The heart and lungs are in the thoracic cavity (Figs. 13.5 and 13.6). The **heart** is a pump for the cardiovascular system, and the **lungs** are organs of the respiratory system where gas exchange occurs.

Observation: Thoracic Cavity

Heart and Lungs

1. If you have not yet done so, fold back the chest wall flaps. To do this, you will need to tear the thin membranes that divide the thoracic cavity into three compartments: the left **pleural cavity** containing the left lung, the right pleural cavity containing the right lung, and the **pericardial cavity** containing the heart.
2. Examine the lungs. Locate the four lobes of the right lung and the three lobes of the left lung. The trachea, dorsal to the heart, divides into the **bronchi,** which enter the lungs. Later, when the heart is removed, you will be able to see the trachea and bronchi.
3. Trace the path of air from the nasal passages to the lungs.

   ____

   ____

Figure 13.5 Internal anatomy of the fetal pig.
The major organs are featured in this drawing. In the fetal pig, a red color tells you a vessel is an artery, and a blue color tells you it is a vein. (It does not tell you whether this vessel carries O_2-rich or O_2-poor blood.) Contrary to this drawing, *keep all the flaps on your pig* so you can close the thoracic and abdominal cavities at the end of the laboratory session.

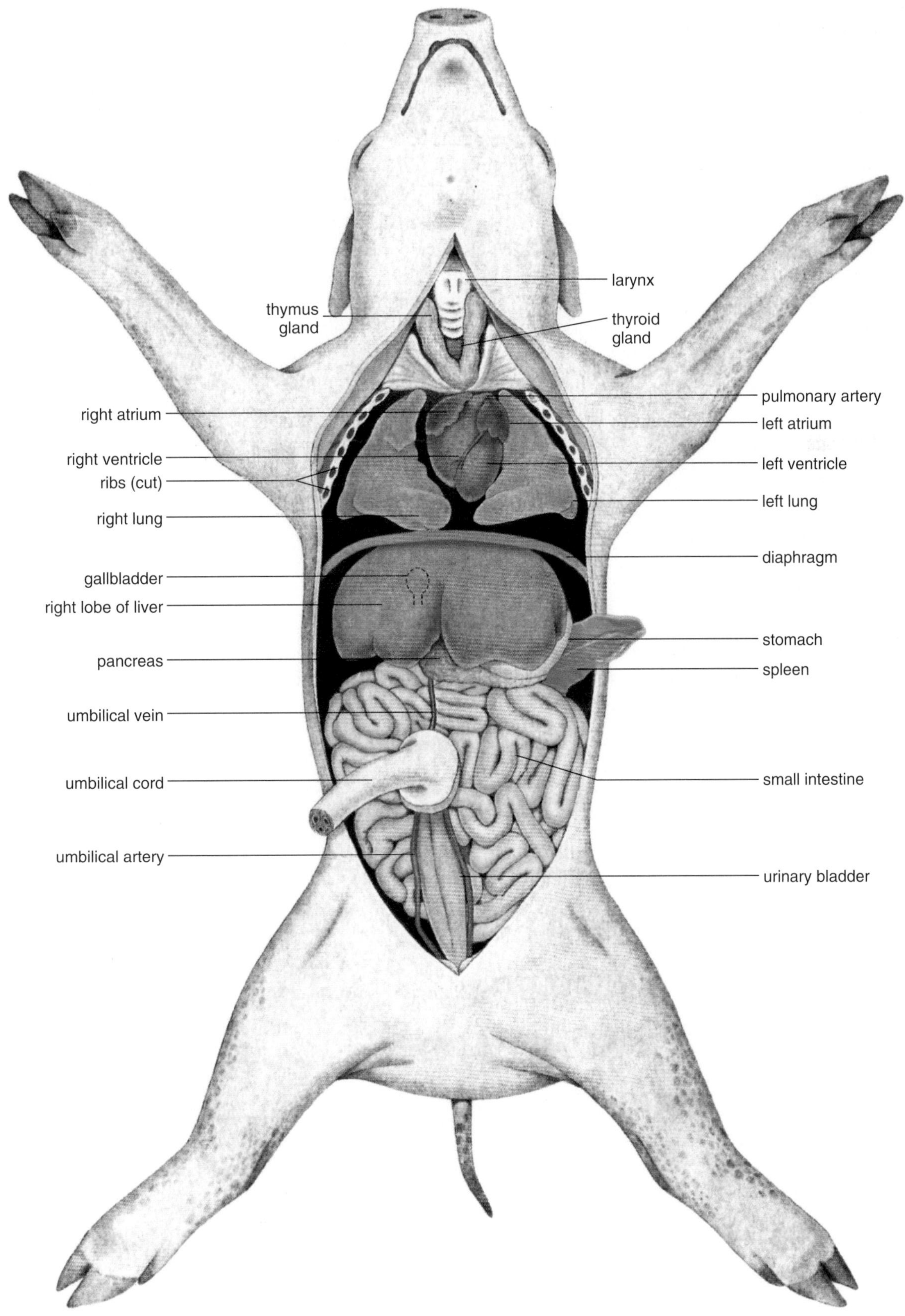

13.6 Abdominal Cavity

The abdominal wall and organs are lined by a membrane called **peritoneum,** consisting of epithelium supported by connective tissue. Double-layered sheets of peritoneum, called **mesenteries,** project from the body wall and support the organs.

The **liver,** the largest organ in the abdomen (Fig. 13.6), performs numerous vital functions, including (1) disposing of worn-out red blood cells, (2) producing bile, (3) storing glycogen, (4) maintaining the blood glucose level, and (5) producing blood proteins.

The abdominal cavity also contains organs of the digestive tract, such as the stomach, small intestine, and large intestine. The **stomach** (see Fig. 13.5) stores food and has numerous gastric glands that secrete gastric juice, which digests protein. The **small intestine** is the part of the digestive tract that receives secretions from the pancreas and gallbladder. Besides being an area for the digestion of all components of food—carbohydrate, protein, and fat—the small intestine absorbs the products of digestion: glucose, amino acids, glycerol, and fatty acids. The **large intestine** is the part of the digestive tract that absorbs water and prepares feces for defecation at the anus.

The **gallbladder** stores and releases bile, which aids the digestion of fat. The **pancreas** (see Fig. 13.5) is both an exocrine and an endocrine gland. As an exocrine gland, it produces and secretes pancreatic juice, which digests all the components of food in the small intestine. Both bile and pancreatic juice enter the duodenum by way of ducts. As an endocrine gland, the pancreas secretes the hormones insulin and glucagon into the bloodstream. Insulin and glucagon regulate blood glucose levels.

The **spleen** (see Fig. 13.5) is a lymphoid organ in the lymphatic system that contains both white and red blood cells. It purifies blood and disposes of worn-out red blood cells.

Observation: Abdominal Cavity

Liver

1. If your particular pig is partially filled with dark, brownish material, take your animal to the sink and rinse it out. This material is clotted blood. Consult your instructor before removing any red or blue latex masses, since they may enclose organs you will need to study.
2. Locate the liver, a large, brown organ, and note that its anterior surface is smoothly convex and fits snugly into the concavity of the diaphragm.
3. Name several functions of the liver. ______________________________

____

Stomach and Spleen

1. Push aside and identify the stomach, a large sac dorsal to the liver on the left side.
2. Locate the point near the midline of the body where the **esophagus** penetrates the diaphragm and joins the stomach.
3. Find the spleen, a long, flat, reddish organ attached to the stomach by mesentery.
4. The stomach is a part of what system? ______________________________

   What is its function? ______________________________
5. The spleen is a part of what system? ______________________________

   What is its function? ______________________________

Figure 13.6 Internal anatomy of the fetal pig.
Most of the major organs are shown in this photograph. The stomach has been removed. The spleen, gallbladder, and pancreas are not visible. *Do not* remove any organs or flaps from your pig.

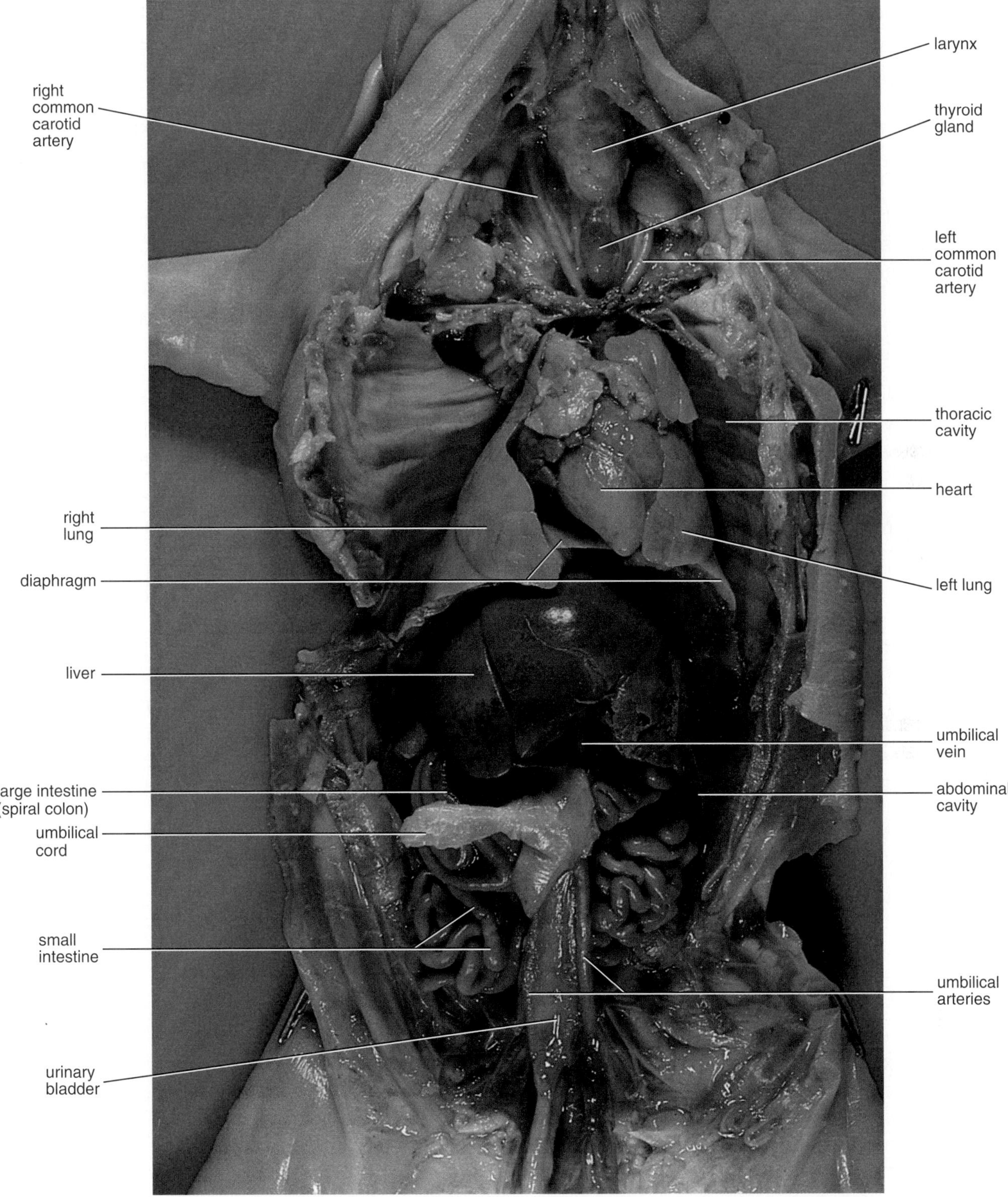

Small Intestine

1. Look posteriorly where the stomach makes a curve to the right and narrows to join the anterior end of the small intestine called the **duodenum.**
2. From the duodenum, the small intestine runs posteriorly for a short distance and is then thrown into an irregular mass of bends and coils held together by a common mesentery.
3. The small intestine is a part of what system? ______________________________

   What is its function? ______________________________

Gallbladder and Pancreas

1. Locate the **bile duct,** which runs in the mesentery stretching between the liver and the duodenum. Find the gallbladder, embedded in the liver on the underside of the right lobe. It is a small, greenish sac.
2. Lift the stomach and locate the pancreas, the light-colored, diffuse gland lying in the mesentery between the stomach and the small intestine. The pancreas has a duct that empties into the duodenum of the small intestine.
3. What is the function of the gallbladder? ______________________________
4. What is the function of the pancreas? ______________________________

   ____

Large Intestine

1. Locate the distal (far) end of the small intestine, which joins the large intestine posteriorly, in the left side of the abdominal cavity (right side in humans). At this junction, note the **cecum,** a blind pouch.
2. Compare the large intestine of your pig to Figure 13.7, and note that the organ does not have the same appearance in humans.
3. Follow the main portion of the large intestine, known as the **colon,** as it runs from the point of juncture with the small intestine into a tight coil (spiral colon), then out of the coil anteriorly, then posteriorly again along the midline of the dorsal wall of the abdominal cavity. In the pelvic region, the **rectum** is the last portion of the large intestine. The rectum leads to the **anus.**
4. The large intestine is a part of what system? ______________________________
5. What is the function of the large intestine? ______________________________
6. Trace the path of food from the mouth to the anus. ______________________________

   ____

Storage of Pigs

1. Before leaving the laboratory, place your pig in the plastic bag provided.
2. Expel excess air from the bag, and tie it shut.
3. Write your *name* and *section* on the tag provided, and attach it to the bag. Your instructor will indicate where the bags are to be stored until the next laboratory period.
4. Clean the dissecting tray and tools, and return them to their proper location.
5. Wipe off your goggles.
6. Wash your hands.

13.7 Human Anatomy

Humans and pigs are both mammals, and their organs are similar. A human torso model shows the exact location of the organs in human beings (Fig. 13.7). You should learn to associate each human organ with its particular system.

Observation: Human Torso

1. Examine a human torso model, and using Figure 13.7 as a guide, locate the same organs you have just dissected in the fetal pig.
2. From your studies so far, list any major differences between pig internal anatomy and human internal anatomy that you have identified. ______________________________

Figure 13.7 Human internal organs.
The dotted lines indicate the full shape of an organ that is partially covered by another organ.

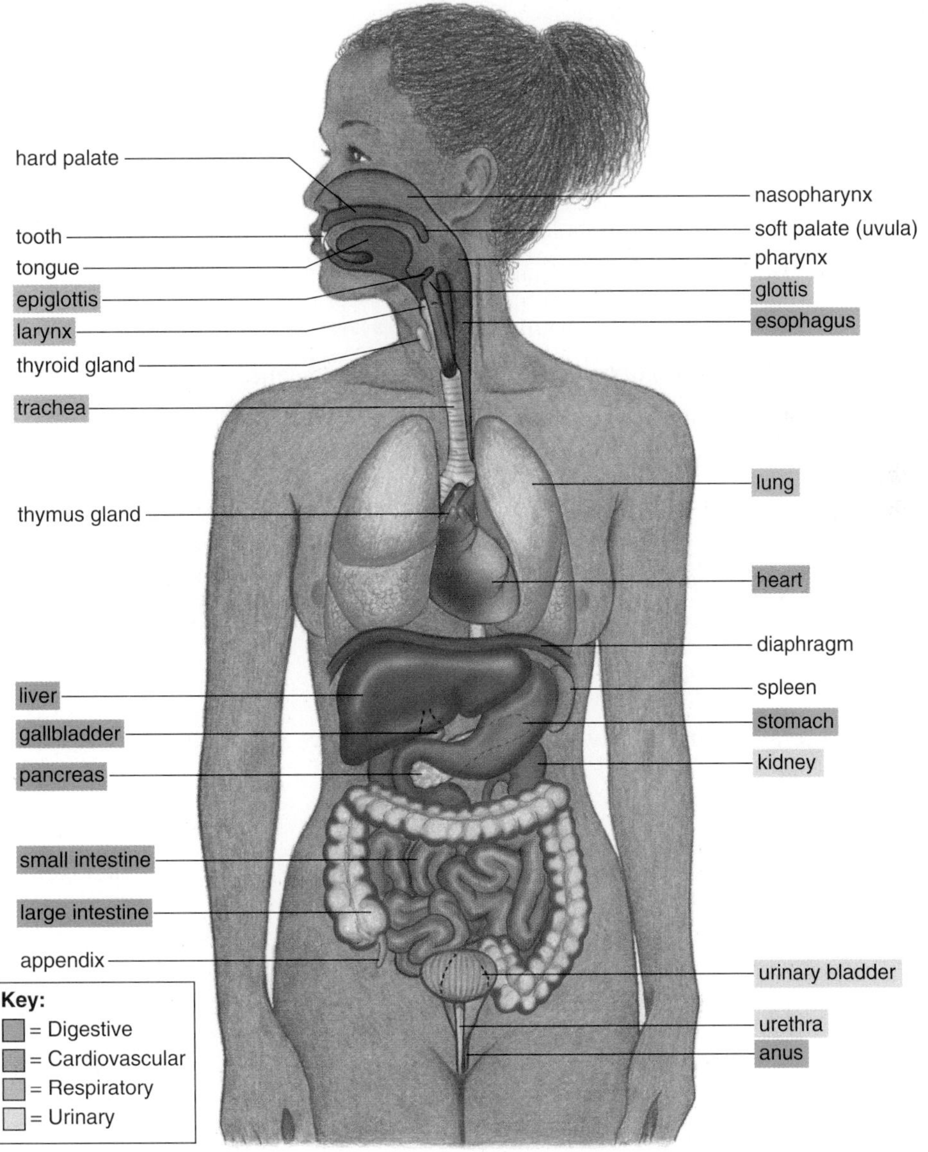

Laboratory Review 13

_______________ 1. In the fetal pig, what sex has a urogenital opening beneath the papilla just superior to the anus?

_______________ 2. What two characteristics do all mammals have?

_______________ 3. The esophagus connects the pharynx with which organ?

_______________ 4. What is the hard portion of the roof of the mouth called?

_______________ 5. What is the opening to the trachea called?

_______________ 6. Name the largest organ in the abdominal cavity.

_______________ 7. What structure separates the thoracic cavity from the abdominal cavity?

_______________ 8. Name the structure just dorsal to the thyroid gland.

_______________ 9. What structure covers the glottis?

_______________ 10. If a probe is placed through the glottis, it will enter what structure?

_______________ 11. The heart is located in what cavity?

_______________ 12. What organs are in the pleural cavity?

_______________ 13. The stomach connects to what part of the small intestine?

_______________ 14. Identify the gland that is located by lifting the stomach.

_______________ 15. Name a lymphoid organ in the abdominal cavity.

_______________ 16. Is the spleen located on the right or left side of the abdominal cavity?

_______________ 17. The pancreas belongs to what system of the body?

_______________ 18. Where do air and food passages cross one another?

_______________ 19. What organ releases bile?

Thought Questions

20. What difficulty would probably arise if a person were born without an epiglottis?

21. A large portion of the abdominal cavity is taken up by digestive organs. Which organs are these?

22. The small intestine exists as a series of folds and coils. What might be the advantage of such a configuration?

23. Difficulties maintaining blood glucose level, bile production, and the production of blood proteins might be associated with problems in what organ?

LABORATORY

14

Cardiovascular System

Learning Outcomes

14.1 Path of Blood in an Adult Versus a Fetus
- Distinguish the role of the pulmonary circuit from that of the systemic circuit.
- Illustrate the two pathways that blood can take through the fetal heart.
- Describe the path of blood from the heart to and from the placenta in the fetus.
- Compare the circulation of blood in a fetus with the circulation of blood in an adult.

14.2 Pulmonary Circuit
- Using a fetal pig, photographs, or charts, identify the pulmonary blood vessels and trace the path of blood from the heart to and from the lungs.

14.3 Systemic Circuit
- Using a fetal pig, photographs, or charts, identify the coronary arteries and cardiac veins.
- Using a fetal pig, photographs, or charts, identify the arteries and veins in the thoracic cavity.
- Using a fetal pig, photographs, or charts, trace the path of blood to and from the heart to the head; trace the path of blood to and from the forelimbs.
- Using a fetal pig, photographs, or charts, identify the arteries and veins in the abdominal cavity.
- Using a fetal pig, photographs, or charts, trace the path of blood from the heart to and from the hind limbs; trace the path of blood to and from the kidneys.

14.4 Blood Vessel Comparison
- Describe each of the layers of a blood vessel wall.
- Contrast the wall of an artery to that of a vein.

Introduction

Blood must circulate to serve the body. The heart pumps the blood, which moves away from the heart in **arteries** and **arterioles** and returns to the heart in **venules** and **veins. Capillaries** connect arterioles to venules.

Arteries have thick elastic walls that can expand when blood enters under pressure. Blood pressure moves blood in arteries and arterioles. Veins have weak walls and valves that prevent blood from falling back. When skeletal muscles contract, they press against the venules and veins, keeping the blood moving toward the heart. The thin walls of the capillaries allow exchange of materials with the tissue fluid that surrounds the cells.

In today's laboratory, you will first learn to trace the path of blood in human adult and fetal hearts. There are differences in the circulatory patterns of human adults and fetuses. In adults, oxygen enters the blood at the lungs, and nutrient molecules enter the blood at the digestive tract. In fetuses, the lungs and digestive tract do not function. Blood is transported to the placenta, where exchange with the mother's blood takes place, and fetal blood receives oxygen and nutrient molecules. You will dissect some of the major arteries and veins in the body of a fetal pig. As you see, not all arteries carry O_2-rich blood, and not all veins carry O_2-poor blood.

Caution: Wear protective eyewear and latex gloves when handling preserved animal organs. Exercise caution when using scalpel and wash your hands thoroughly upon completion of this exercise.

14.1 Path of Blood in an Adult Versus a Fetus

Your primary goal is to know the path of blood in an adult; therefore, we will begin by learning to trace the path of blood in adults. Your pig is a fetal pig; therefore, we have an opportunity to also learn the path of blood in a fetus.

Path of Blood in Adult Humans

In adult mammals, including humans, the heart is a double pump. The right side of the heart, consisting of two chambers (the right atrium and right ventricle), pumps blood into the **pulmonary circuit**—that is, to the lungs and back to the heart (Fig. 14.1). While the blood is in the lungs, it gives up carbon dioxide and gains oxygen. The left side of the heart, consisting of two chambers (the left atrium and left ventricle), pumps blood to the **systemic circuit**—that is, throughout the whole body except the lungs. Blood in the systemic circuit gives up oxygen and gains carbon dioxide.

Figure 14.1 shows how to trace the path of blood in both the pulmonary and systemic circuits in adult humans. It will also assist you in learning the names of some of the major blood vessels.

Figure 14.1 Diagram of the human cardiovascular system.
In the pulmonary circuit, the pulmonary arteries take O_2-poor blood to the lungs, and the pulmonary veins return O_2-rich blood to the heart. In the systemic circuit, the aorta branches into the various arteries that go to all other parts of the body. After blood passes through arterioles, capillaries, and venules, it enters various veins and then the superior and inferior venae cavae, which return it to the heart.

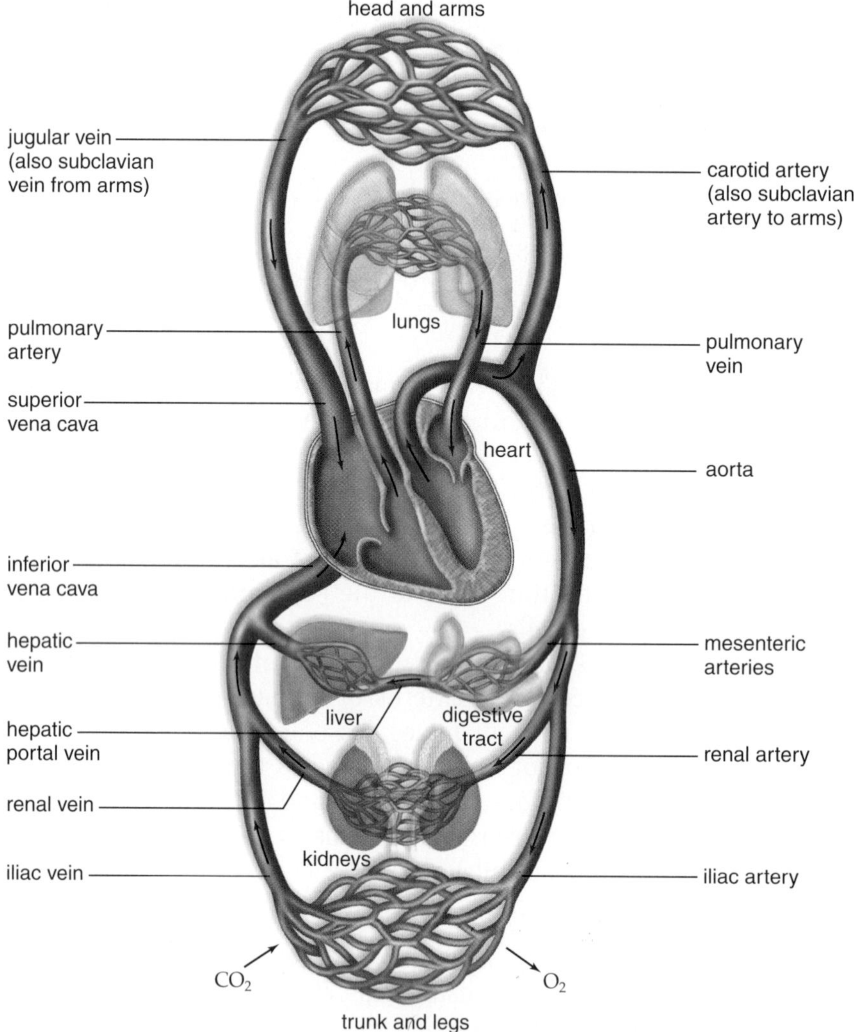

Pulmonary Circuit

1. Trace the path of blood in the pulmonary circuit from the heart to the lungs, and then from the lungs to the heart. Follow the arrows in Figure 14.1, and use the label names provided there.

right ventricle of heart → ______________ → lungs ——→ ______________ → ______________ of heart

Notice that in the adult there is no connection between the right and left sides of the heart except via the lungs.

Systemic Circuit

2. Trace the path of blood in the systemic circuit from the heart to the kidneys, and then from the kidneys to the heart.

left ventricle of heart → _______ → _______ → kidneys → _______ → _______ → _______ of heart

Names of Blood Vessels

3. Use Figure 14.1 to complete Table 14.1.

Table 14.1 Major Blood Vessels in the Systemic Circuit

Body Part	Artery	Vein
Heart		
Head		
Arms		
Kidney		
Legs		
Intestines		

Path of Blood in Fetal Humans

Human fetal circulation has five features that are absent in adults. These features are listed in Table 14.2 and shown in Figure 14.2. Notice that the blood passes through the heart to the aorta without going to the lungs and that the umbilical vessels take blood to and from the **placenta,** the organ that develops on the inner wall of the uterus where exchange takes place between fetal blood and the mother's blood. The respiratory and digestive systems are not operative in the fetus; oxygen and nutrients are acquired at the placenta.

Table 14.2 Unique Features of Human Fetal Circulation

1. Oval opening (foramen ovale): an opening between the atria
2. Arterial duct (ductus arteriosus): a short, stout vessel leading directly from the pulmonary trunk to the aorta
3. Umbilical arteries: take blood from the iliac arteries to the placenta
4. Umbilical vein: returns blood from the placenta to the liver
5. Venous duct (ductus venosus): a continuation of the umbilical vein that takes blood to the inferior vena cava

Figure 14.2 Human fetal circulation.

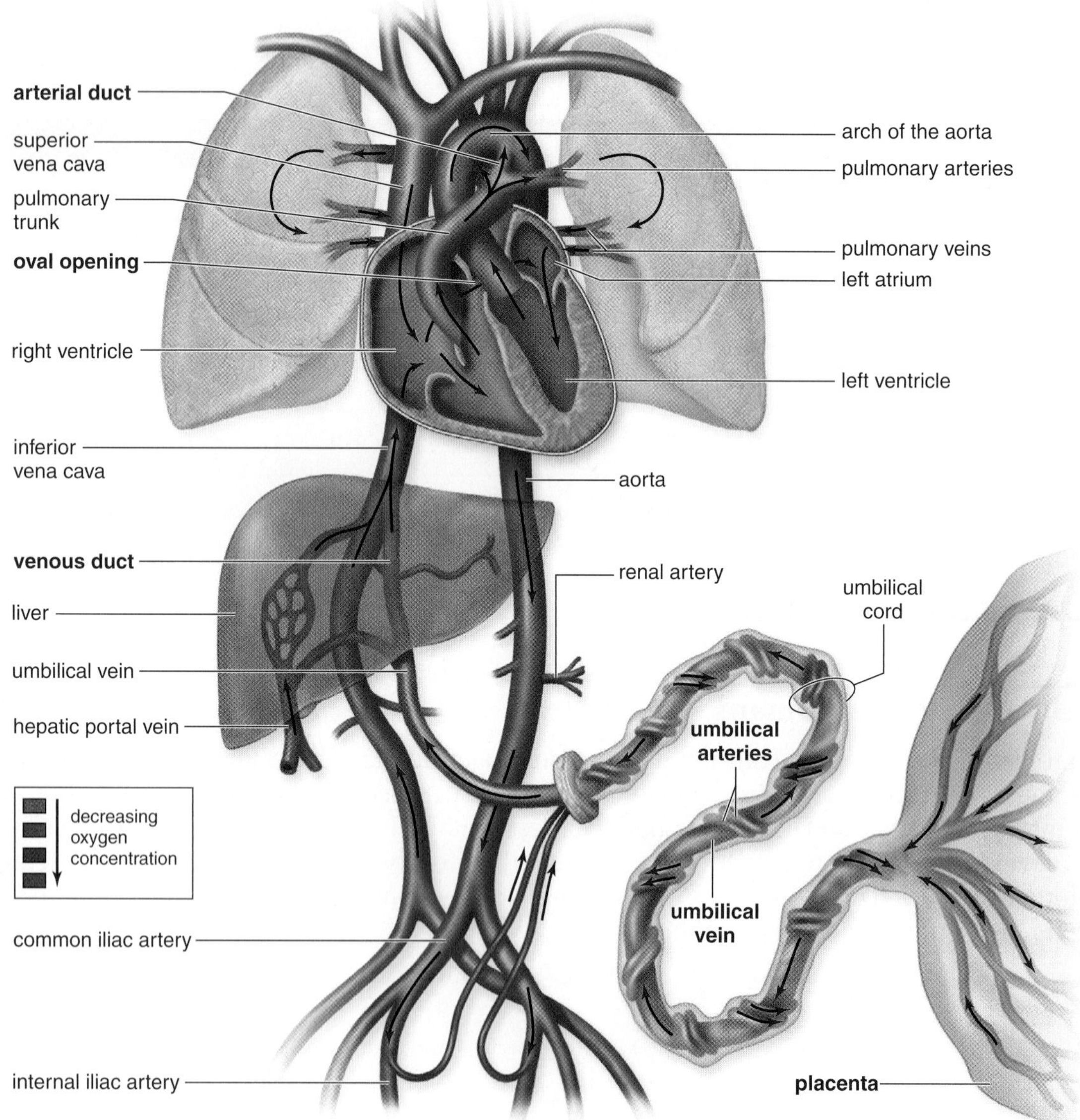

Through the Heart

Trace the path of blood from the right atrium to the aorta, going through the oval opening, and then trace the path of blood from the right atrium to the aorta, going through the arterial duct.

First pathway: right atrium → ________________ → ________________ → ________________ → aorta

Second pathway: right atrium → ________________ → ________________ → ________________ → aorta

To the Placenta and Return

Trace the path of blood from the aorta to the placenta and from the placenta to the inferior vena cava.

aorta → ________________ → ________________ → placenta

placenta → ________________ → ________________ → inferior vena cava

Comparison

Complete Table 14.3 to compare human fetal circulation with human adult circulation.

Table 14.3 Comparison of Human Fetal and Adult Circulation

	Fetus	Adult
Vessel with the highest oxygen concentration		
Passage from right to left side of heart		
Entrance of blood into aorta 1. 2.		
Area of gas exchange		

14.2 Pulmonary Circuit

In this section, we will use the fetal pig to examine the pulmonary arteries and veins that occur in mammals, including humans. The fetal pig will have pulmonary arteries and veins, even though they are not functional until after the pig is born. In the pulmonary circuit of adult mammals, pulmonary arteries take blood away from the heart to the lungs, and pulmonary veins take blood from the lungs to the heart. Remember that in your pig, *all arteries* have been injected with red latex, and *all veins* have been injected with blue latex.

Observation: Pulmonary Circuit

Pulmonary Trunk and Pulmonary Arteries

1. Locate the **pulmonary trunk,** which arises from the ventral side of the heart (Fig. 14.3). It may appear white because the thick wall prevents the color of the red latex from showing through.
2. If you have not already done so, remove the pericardial sac from the heart to reveal the blood vessels entering the heart. Veins take blood to the heart.

Figure 14.3 Ventral view of fetal pig heart.
In the fetal pig, two major arteries branch off the aorta: the brachiocephalic artery and the right subclavian artery. The brachiocephalic artery branches, in turn, into the carotid trunk and the right subclavian artery (see Fig. 14.5).

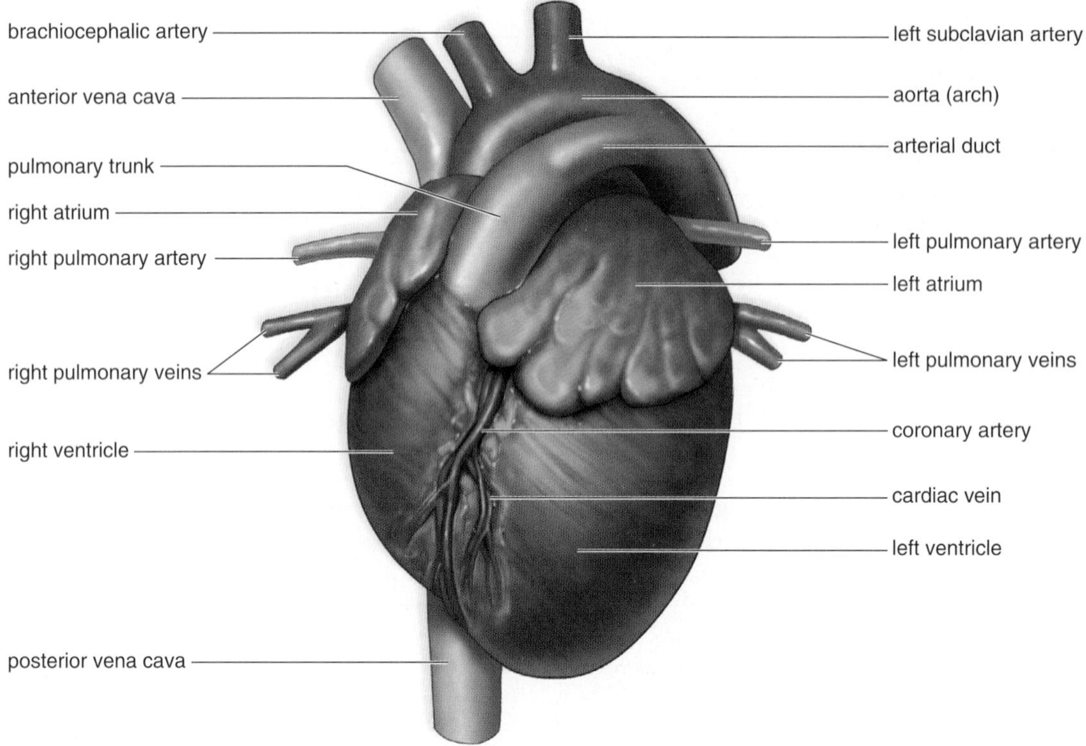

3. Trace the pulmonary trunk, and notice that it seems to connect directly with the aorta, the major artery. This connection is the **arterial duct** (ductus arteriosus), discussed in the previous section.
4. In addition to this duct, look closely and you will find the **pulmonary arteries,** which leave the pulmonary trunk and go to the lungs.

Pulmonary Veins

1. The pulmonary veins are hard to find. If you wish, clean away the membrane dorsal to the heart, and carefully note the vessels (pulmonary veins) that leave the lungs. Trace these to the left atrium of the heart.
2. In the adult, which blood vessels—pulmonary arteries or pulmonary veins—carry O_2-rich blood? ______________________________

____

14.3 Systemic Circuit

In adult mammals, the **systemic circuit** serves all parts of the body except the alveoli of the lungs. Arteries take O_2-rich blood from the heart to the organs, and veins take O_2-poor blood from the organs to the heart. The aorta is the major artery, and the **venae cavae** are the major veins. It will be possible for you to identify the arteries branching from the aorta and the corresponding veins branching from the venae cavae. In the fetal pig, the venae cavae are called the *anterior* vena cava and the *posterior* vena cava because the normal position of the body is horizontal rather than vertical. Use Figure 14.4 to trace the blood vessels, but **do not remove any organs. You will need them in Laboratory 16.**

Figure 14.4 Ventral view of fetal pig arteries and veins.
Use this diagram to trace blood vessels, **but do not remove any organs. You will need them in laboratory 16.** (a. = artery; v. = vein.)

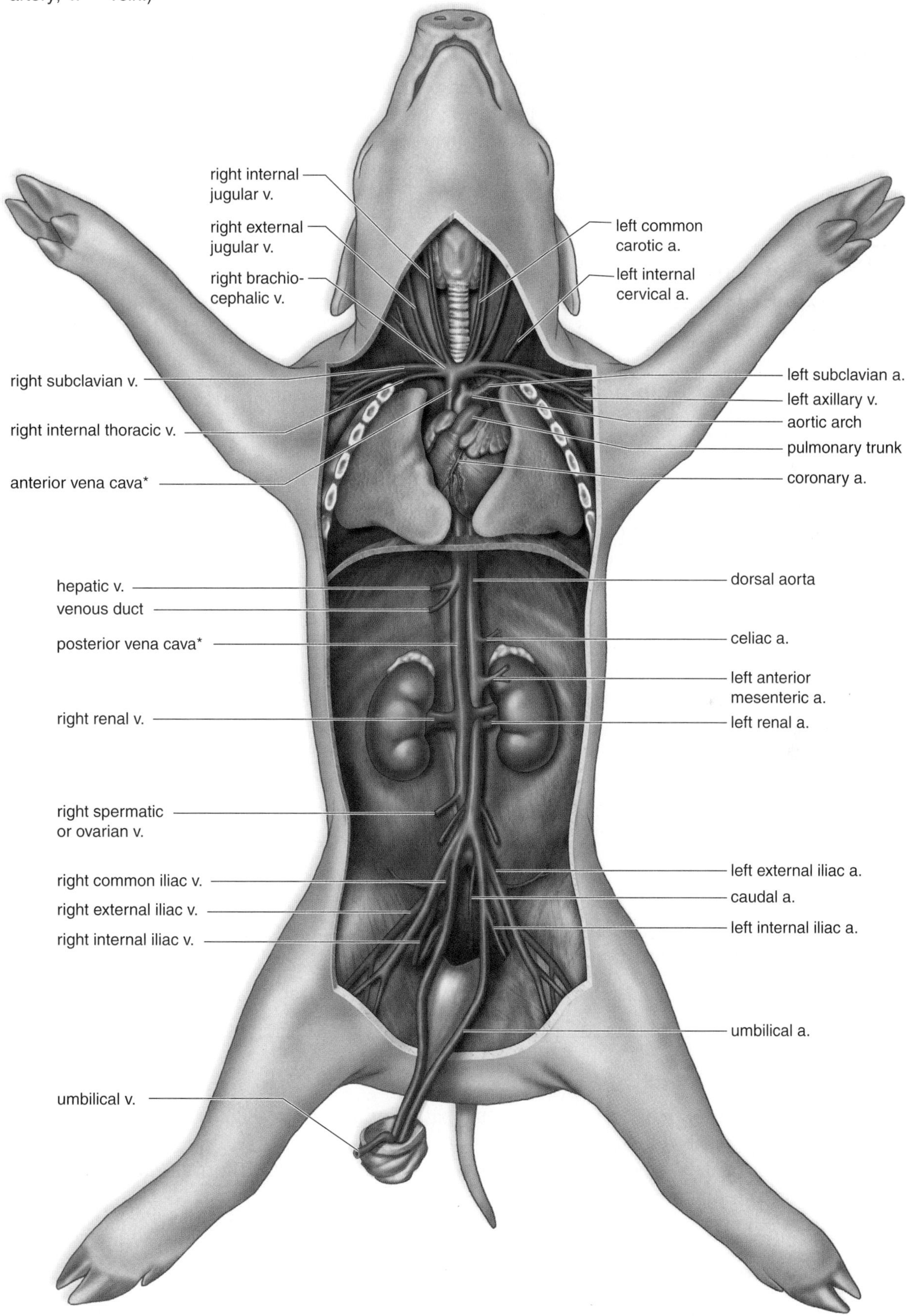

* Because the pig walks on all four limbs, the anterior vena cava in pigs is called the superior vena cava in humans, and the posterior vena cava in pigs is called the inferior vena cava in humans.

Observation: Aorta and Venae Cavae

1. Follow the aorta as it extends through the thoracic cavity. To do this, gently move the lungs and heart to the right side of the thoracic cavity. Notice how the thoracic or dorsal aorta is a large, whitish vessel extending through the thoracic cavity and then through the diaphragm to become the abdominal aorta. Also notice the esophagus, a smaller, flattened tube more toward the midline. The esophagus goes through the diaphragm to join with the stomach. Locate the anterior vena cava coming off the top of the heart and the posterior vena cava just to the right of the midline. The posterior vena cava also passes through the diaphragm.
2. Name three structures that pass through the diaphragm. ______________________

Blood Vessels of the Upper Body

The **coronary arteries** and **cardiac veins** lie on the surface of the heart. The **carotid arteries** and **jugular veins** serve the neck and head regions of the body.

Observation: Blood Vessels of the Upper Body

Coronary Arteries and Cardiac Veins

1. Locate the coronary arteries and cardiac veins, which are easily visible on the heart's surface (see Fig. 14.3).
2. The coronary arteries arise from the aorta just as it leaves the heart, and the cardiac veins go directly into the right atrium.

Carotid Arteries and Jugular Veins

1. Find the aorta and its first branch, the **brachiocephalic artery,** which divides almost immediately into the **right subclavian** (to the pig's right shoulder) **artery** and the **carotid trunk.** The carotid trunk divides into the **right** and **left common carotid arteries** (Fig. 14.5). The aorta makes an arch and then continues as the **dorsal aorta.**
2. Find the second branch of the aorta, which is the **left subclavian artery.**
3. Both the right and left subclavian arteries have branches called the right and left brachial arteries, which serve the upper limbs.
4. Find the **jugular veins** alongside the carotid arteries (Fig. 14.6). Trace the carotid arteries and jugular veins as far as possible toward the head. What part of the body is serviced by the carotid arteries and the jugular veins? ______________________

   ____
5. Trace the jugular veins until they join the anterior vena cava.
6. Name and locate all the veins that join to form the anterior vena cava (Fig. 14.6). ____________

   ____

Subclavian Arteries and Veins

1. Locate the **subclavian arteries** (Fig. 14.5) and **subclavian veins** (Fig. 14.6), which serve the forelimbs and are easily identified.
2. Note that while the left subclavian artery branches from the aorta, the right subclavian artery branches from the brachiocephalic artery.
3. Note that the jugular veins and the subclavian veins join to form the brachiocephalic veins.

Internal Thoracic Artery and Vein

Blood reaches and leaves the thoracic wall through several blood vessels, including the internal thoracic artery and vein. The internal thoracic vein is often cut when opening the thoracic cavity.

Figure 14.5 Arteries of the upper body of the fetal pig.
The subscapular artery supplies the shoulder muscles. (a. = artery.)

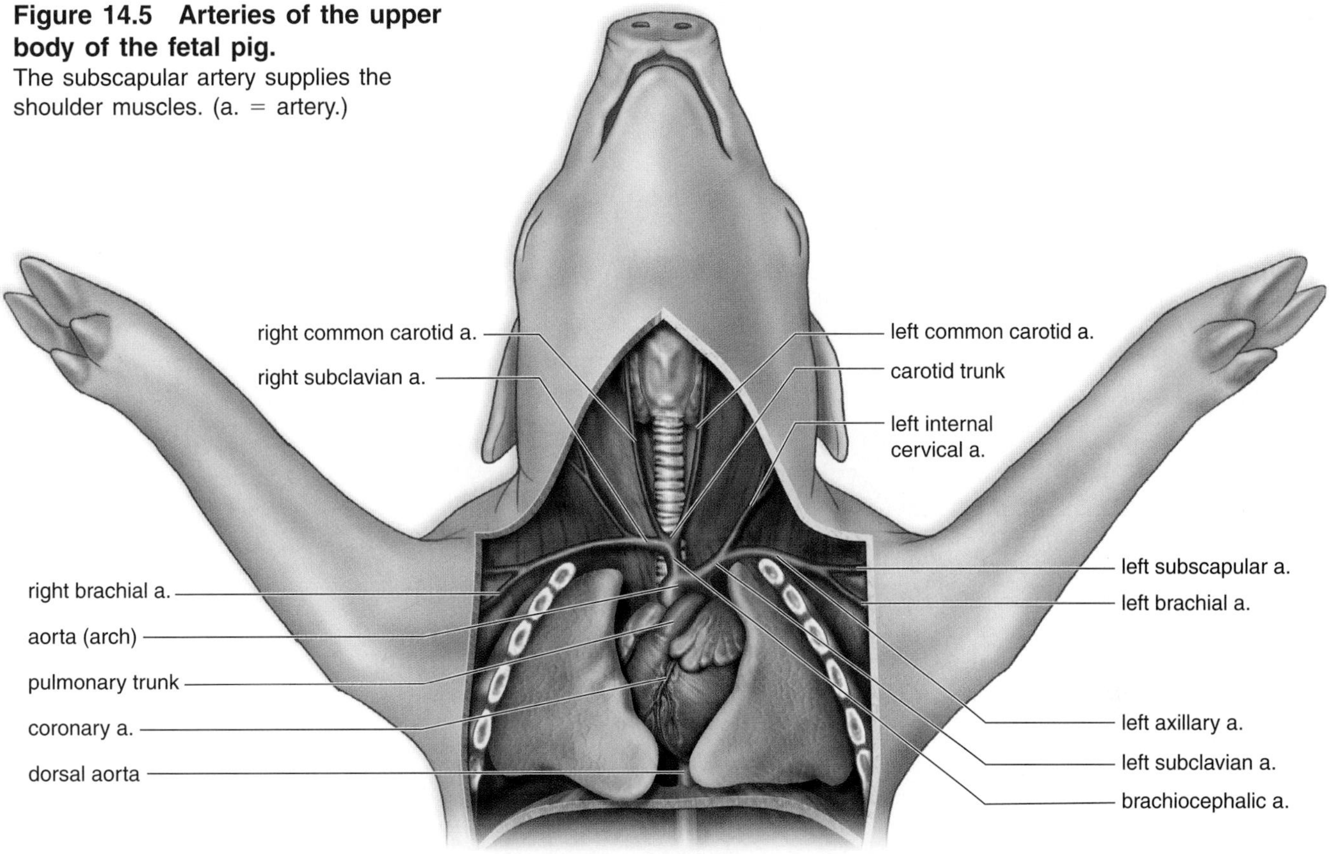

Figure 14.6 Veins in the thoracic cavity of the fetal pig.
The subclavian veins enter a brachiocephalic vein. (v. = vein.)

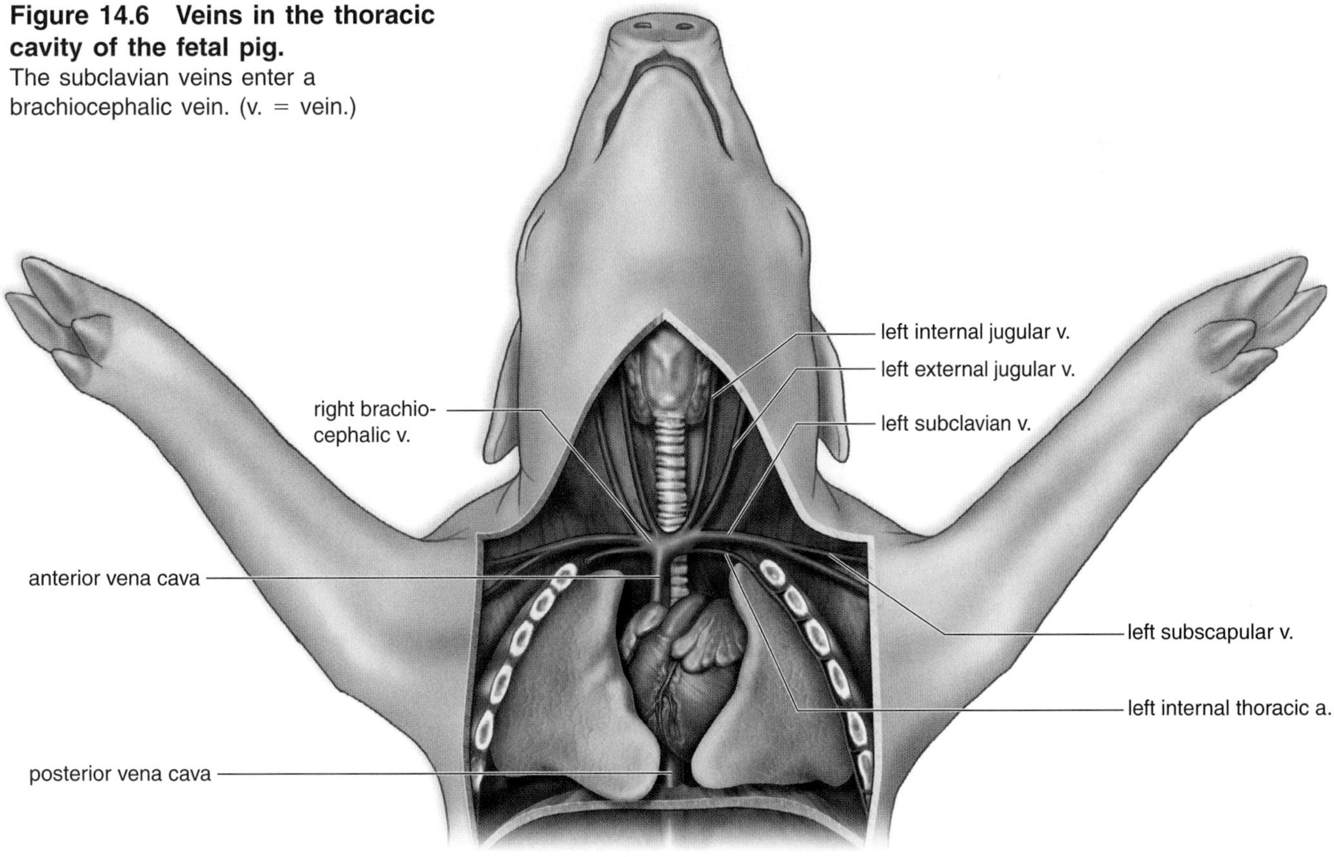

Blood Vessels of the Abdominal Cavity

You will find several major blood vessels in the abdominal cavity. Use Figures 14.7 and 14.8 to trace the blood vessels, but **do not remove any organs. You will need them in Laboratory 16.**

Figure 14.7 Arteries in the abdominal cavity of the fetal pig.
The external iliac arteries proceed into the hind limbs from the caudal end of the dorsal aorta. The internal iliacs also branch from the dorsal aorta. They give rise to the umbilical arteries and continue as much smaller vessels. (aa, = arteries.) Use this diagram to trace blood vessels, but **do not remove any organs. You will need them in Laboratory 16.**

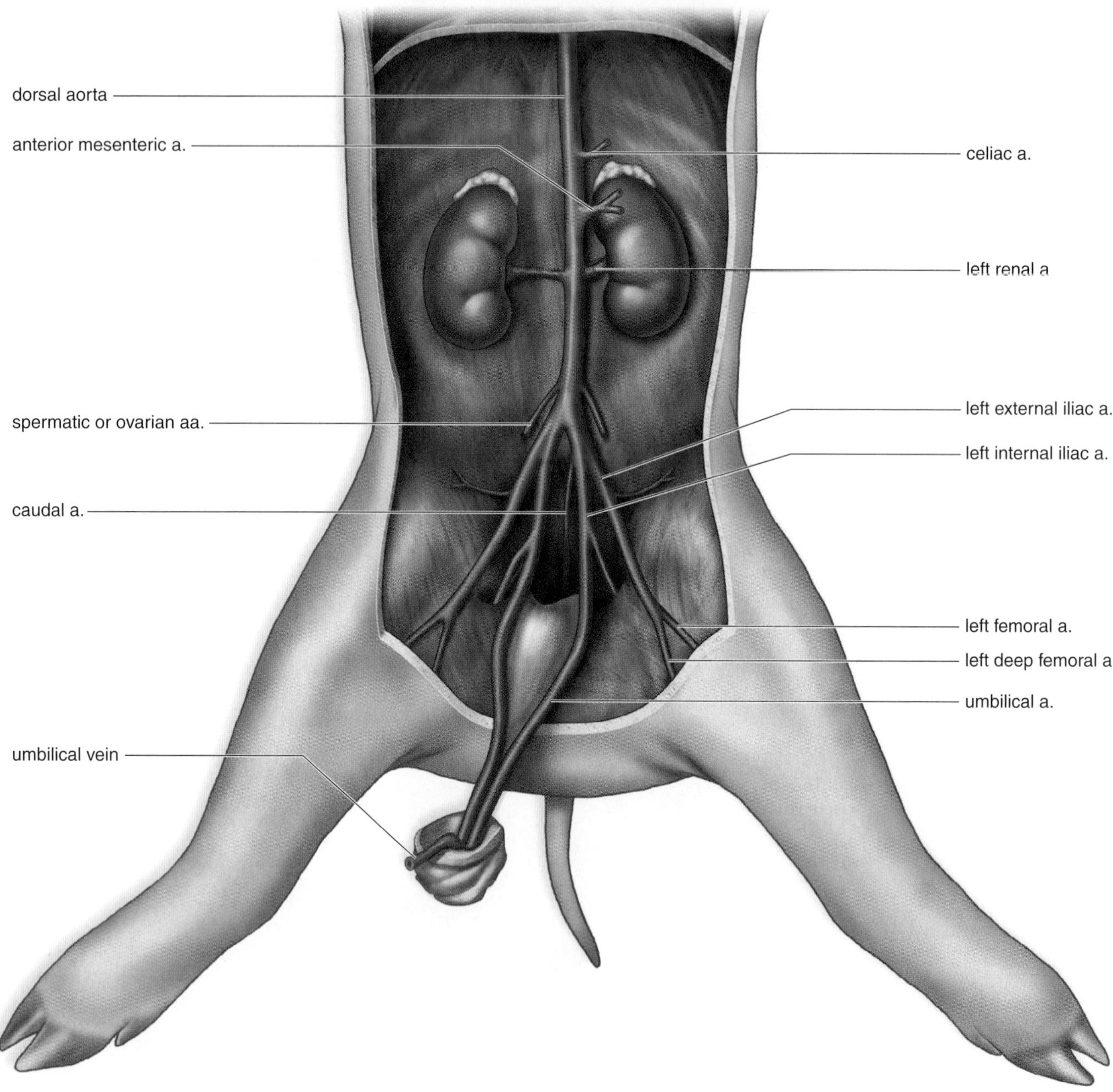

Figure 14.8 Veins in the abdominal cavity of the fetal pig.
The posterior vena cava divides into the common iliac veins. The common iliac veins branch into the external and internal iliac veins. (vv. = veins.) Use this diagram to trace blood vessels, but **do not remove any organs. You will need them in Laboratory 16.**

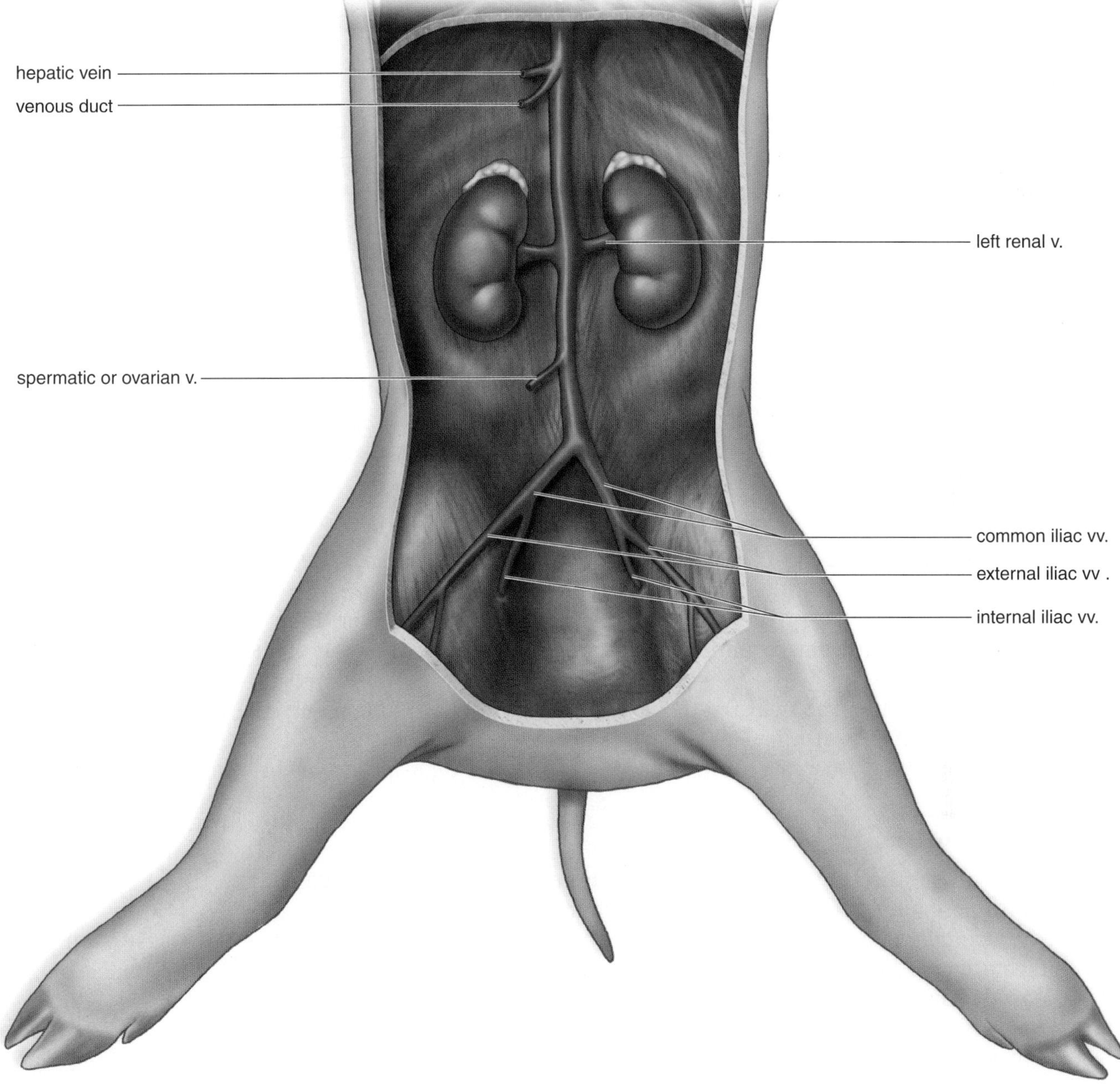

Observation: Blood Vessels of the Abdominal Cavity

Celiac and Mesenteric Arteries

1. Carefully lift up the liver and stomach, and put them aside to your left. Dissect the dorsal mesentery to see the **celiac artery** as it leaves the aorta. Tributaries from this vessel eventually reach the stomach, duodenum, liver, and spleen.
2. Branching from the aorta just posterior to the celiac artery is a long, unpaired trunk, called the **anterior mesenteric artery,** which has tributaries to the pancreas and small intestine (see Fig. 14.7).
3. The celiac and mesenteric arteries take blood to the intestines. Thereafter, the hepatic portal vein takes blood from the intestinal capillaries to capillaries in the liver. A portal system is defined as a circulatory unit that goes from one capillary bed to another without passing through the heart. You will locate the hepatic portal system in Laboratory 16.

Renal Arteries and Veins

1. Locate the **renal arteries** as they branch from the aorta, and trace these arteries as they go into the kidneys.
2. Locate the renal veins as they leave the kidneys (Fig. 14.8), and trace these veins as they join the posterior vena cava.

Iliac Arteries and Veins

1. At its posterior end, the aorta branches into the paired **iliac arteries.** Locate the iliac arteries at the posterior end of the aorta, and trace these arteries into the hind limbs (see Fig. 14.7).
2. Find the **iliac veins** alongside the iliac arteries, and trace these veins as they join the posterior vena cava (see Fig. 14.8).

Umbilical Arteries and Veins

1. Locate the **umbilical arteries** on either side of the bladder. Trace these arteries as they branch from the iliac arteries and as they pass into the **umbilical cord** (see Fig. 14.7).
2. When you were exposing the abdominal cavity, you cut the **umbilical vein.** Trace the umbilical vein from the umbilical cord to the liver. The umbilical vein is joined to the posterior vena cava by the **venous duct** (ductus venosus), which passes through the liver.

Posterior Vena Cava

1. Locate the **posterior vena cava,** which is easily seen as a large, blue vessel just ventral to the dorsal aorta (see Fig. 14.8).
2. Note that this vessel seems to disappear in the region of the liver. Here the posterior vena cava receives the hepatic veins coming from the liver. Scrape away some of the liver tissue in order to see these veins.
3. Locate the posterior vena cava as it passes through the diaphragm into the thoracic cavity and as it enters the right atrium.
4. Trace the posterior vena cava from the iliac veins to the right atrium of the heart (see Fig. 14.4).

Storage of Pigs

1. Before leaving the laboratory, place your pig in the plastic bag provided.
2. Expel excess air from the bag, and tie it shut.
3. Write your *name* and *section* on the tag provided, and attach it to the bag. Your instructor will indicate where the bags are to be stored until the next laboratory period.
4. Clean the dissecting tray and tools, and return them to their proper location.
5. Wipe off your goggles.
6. Wash your hands.

14.4 Blood Vessel Comparison

Blood pressure keeps blood moving in arteries away from the heart. The thick walls of arteries can expand when blood pours into them. Skeletal muscle contraction pushing in on veins keeps the blood in veins moving toward the heart. Do you predict that arteries or veins are generally more superficial in the body? ______________________________

Wall of an Artery Compared with Wall of a Vein

Arteries have thicker walls than veins (Fig. 14.9). Their thick walls can expand when the heart contracts and blood pours into them.

Both arteries and veins have three distinct layers, or **tunicas,** forming a wall around the lumen, the space through which blood flows. The three tunicas are called the inner layer (tunica intima), the middle layer (tunica media), and the outer layer (tunica externa).

Figure 14.9 Photomicrograph of an artery and a vein.
(Magnification ×100)

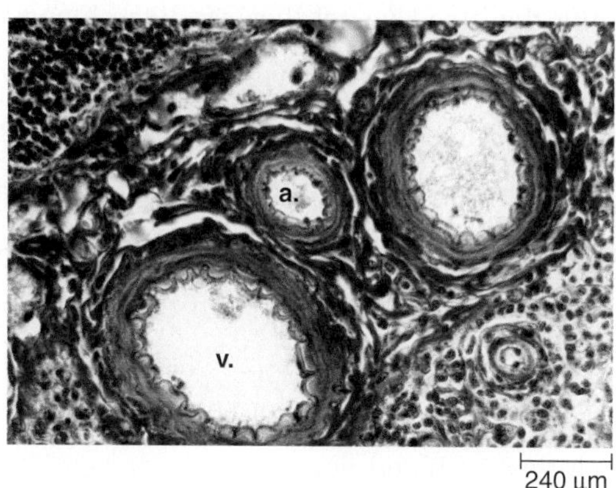

Observation: Blood Vessel Comparison

1. Obtain a microscope slide that shows an artery and a vein in cross section.
2. View the slide, under both low and high power, and with the help of Figure 14.9, determine which is the artery and which is the vein.
3. Identify the **outer layer,** which contains many collagen and elastic fibers and often appears white in specimens.
4. Identify the **middle layer,** the thickest layer, which is composed of smooth muscle and elastic tissue. Does this layer appear thicker in arteries than in veins? ______________________
5. Identify the **inner layer,** a smooth lining of simple squamous epithelial cells called the endothelium. In veins, the endothelium forms valves that keep the blood moving toward the heart. Arteries do not have valves. Considering the relationship of arteries and veins to the heart, why do veins have valves, while arteries do not? ______________________

   ____

Conclusions

- Which type of blood vessel (arteries or veins) has thicker walls? Why? ______________________

  ____
- Which type of blood vessel has thinner walls? ______________________

  ____
- Which type of blood vessel is more apt to lose its elasticity, leading to a discoloration that can be externally observed? ______________________

  ____

  What is this condition called? ______________________

Laboratory Review 14

_______________ 1. What fetal structure connects the pulmonary trunk to the aorta?

_______________ 2. What fetal blood vessel contains the most oxygen?

_______________ 3. What structure allows the blood to pass from the right to the left side of the heart into the fetus?

_______________ 4. Does the pulmonary artery in adults carry O_2-rich or O_2-poor blood?

_______________ 5. The coronary arteries and cardiac veins serve what organ?

_______________ 6. Identify the blood vessel that conducts blood to the head.

_______________ 7. Identify the artery that serves the kidney.

_______________ 8. Identify the large artery that runs dorsally along the wall of the abdominal cavity.

_______________ 9. Identify the arteries that take blood from the aorta to the hind limbs.

_______________ 10. What part of the human body is served by the subclavian vessels?

_______________ 11. Identify the large abdominal vein that runs alongside the aorta and enters the right atrium.

_______________ 12. What part of the body is not served by the systemic circuit?

_______________ 13. Which type of blood vessel (artery or vein) has thicker walls?

_______________ 14. Which type of blood vessel (artery or vein) has valves?

_______________ 15. Identify the artery whose tributaries serve the stomach, duodenum, liver, and spleen.

Thought Questions

16. During a heart attack, cardiac muscle cells are deprived of their blood supply, yet the ventricles are still full of blood. Explain.

17. Trace the path of blood from the left ventricle to the kidneys and back to the right atrium.

18. Evaluate the following statement:

    All arteries carry oxygenated blood and all veins carry deoxygenated blood.
    Based on what you have learned in this laboratory, is this statement correct? Explain your answer.

LABORATORY

15

Features of the Cardiovascular System

Learning Outcomes

15.1 The Blood

- Distinguish red blood cells from white blood cells using images or prepared slides.
- Contrast the structure and function of red blood cells and the major categories of white blood cells.

15.2 The Heart

- Using a fetal pig or photographs, locate and identify the chambers of the heart and their attached blood vessels.
- Using a preserved specimen or photographs, name and locate the valves of the heart.
- Using a preserved specimen or photographs, trace the path of blood through the heart.
- Discuss and explain the conduction system of the heart.

15.3 Heartbeat

- Determine the heartbeat by using a stethoscope or by taking the pulse.
- Explain why increased heartbeat is expected as a result of exercise.

15.4 Blood Pressure

- Explain why an increase in blood pressure is expected as a result of exercise.

Introduction

The circulatory fluid is called **blood.** Blood is composed of two parts: formed elements and plasma. Red blood cells transport oxygen, and white blood cells fight infection. The nutrients needed by cells are carried in the plasma. These nutrients exit the blood at the thin walls of the capillaries, which also allow wastes to enter the blood.

The heart and blood vessels form the cardiovascular system. The heart is a cone-shaped, muscular organ, about the size of a fist. It is located between the lungs, directly behind the sternum (breastbone), and is tilted so that the apex is directed to the left. The heart has right and left sides divided by the septum. There are four **chambers:** two upper, thin-walled atria and two lower, thick-walled ventricles. The heart has valves that keep the blood flowing in one direction; a backward flow closes the valves. The right side of the heart sends blood through the lungs, and the left side sends blood into the body. Therefore, the right side of the heart is the pump for the pulmonary circuit, and the left side of the heart is the pump for the systemic circuit.

During a heartbeat, the two atria contract simultaneously; then the two ventricles contract at the same time. When the heart beats, the familiar lub-dub sound may be heard as the valves of the heart close. The lub is caused by vibrations of the heart when the atrioventricular valves close, and the dub is heard when vibrations occur due to the closing of the semilunar valves. The beat of the heart is intrinsic; when the **SA (sinoatrial) node** initiates the heartbeat, the atria contract, and when the **AV (atrioventricular) node** is stimulated, it causes the ventricles to contract. **Blood pressure,** which keeps blood moving in the arteries, is created by the pumping of the heart. **Pulse** is due to the expansion and recoil of the arterial walls.

15.1 The Blood

The blood that flows through capillaries exchanges molecules with tissue fluid, and in this way the cells receive the nutrients and oxygen they need and rid themselves of waste molecules. Blood appears to be a somewhat viscous, homogeneous fluid. However, analysis shows it to be a fluid tissue, composed of plasma (the fluid portion) and formed elements (including the cells and platelets). The plasma also contains soluble proteins, such as clotting factors and hormones. Platelets are small cell fragments that help in clotting.

Red and White Blood Cells

There are two categories of blood cells: red blood cells (erythrocytes) and white blood cells (leukocytes). Red blood cells are smaller than white blood cells, and they lack a nucleus, which enables them to carry the maximum amount of oxygen. Red blood cells appear red because they contain the respiratory pigment hemoglobin. Each red blood cell lasts about 120 days in circulation.

White blood cells are larger than red blood cells, and they have a nucleus. The white blood cells are translucent if not stained; there are five different kinds (Fig. 15.1). White blood cells fight infection, and the white blood cell count is used by doctors to help diagnose diseases. The time white blood cells are in circulation is variable.

Observation: Blood Slide

1. Observe a prepared blood smear slide on high power, and note the biconcave (concave on both sides) red blood cells and the less numerous white blood cells. Also differentiate the two types of cells according to their size and the presence or absence of a nucleus.
2. Complete Table 15.1.

Table 15.1 Slide of Blood

	Relative Number	Relative Size	Presence of Nucleus	Function
Red blood cells				
White blood cells				

Observation: White Blood Cell Slide

1. Observe the demonstration slides under oil immersion. With the help of Figure 15.1, identify the different types of white blood cells, which have been stained with Wright stain.
2. Identify the **granular leukocytes:**

   Neutrophils have cytoplasmic granules; the nucleus is multilobed with two to five connected parts. Neutrophils are, therefore, called *polymorphonuclear leukocytes.*
   Eosinophils have granules that stain deep red. The nucleus is bilobed.
   Basophils have granules that stain deep blue. The nucleus is bilobed.

3. Identify the **agranular leukocytes:**

   Monocytes are the largest of the white blood cells. The nucleus varies in shape.
   Lymphocytes are usually only slightly larger than red blood cells and typically have a relatively large, round nucleus surrounded by a thin rim of cytoplasm.

Figure 15.1 The white blood cells.
(*a*) A neutrophil has a lobed nucleus with two to five lobes. (*b*) An eosinophil has red-staining granules. (*c*) A basophil has deep-blue-staining granules. (*d*) A monocyte is the largest of the blood cells. (*e*) A lymphocyte contains a large, round nucleus. (*a–c, e:* Magnification ×400; *d:* Magnification ×500)

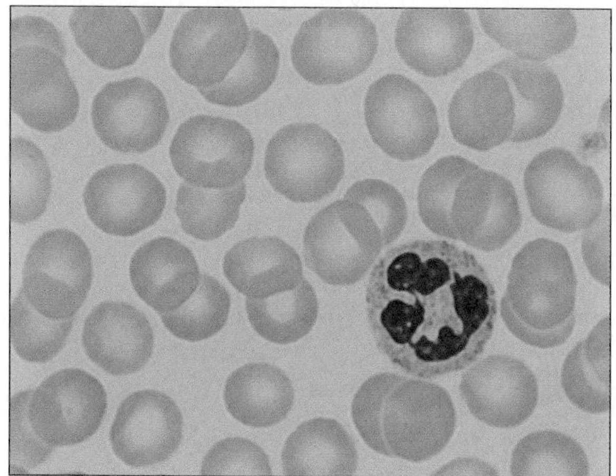

a. Neutrophil

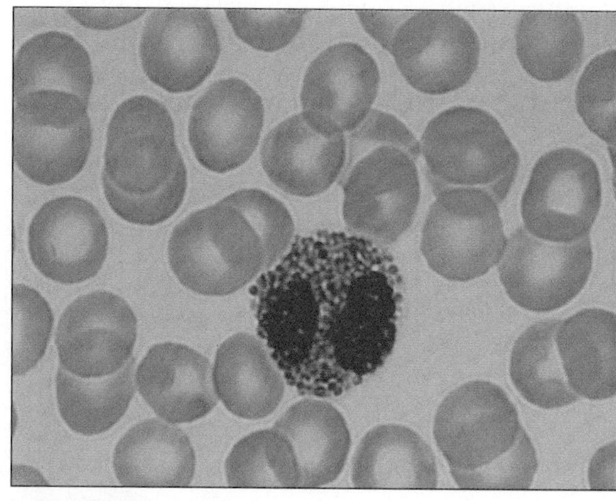

b. Eosinophil

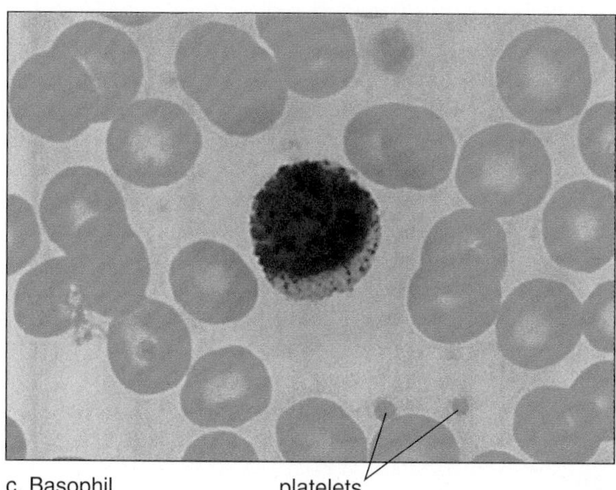

c. Basophil

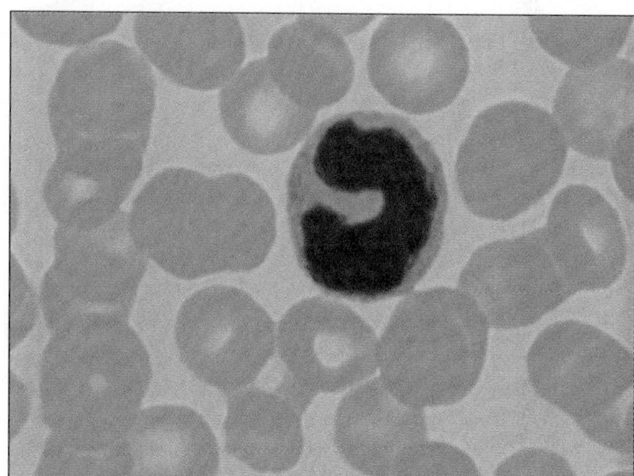

d. Monocyte

platelets

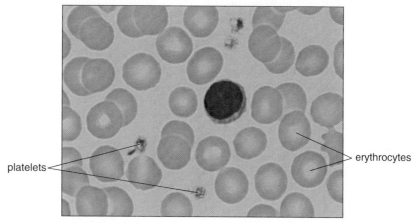

erythrocytes

e. Lymphocyte

Blood Flow

As shown in Figure 15.2, blood flows from an artery to an arteriole, then through a capillary to a venule, and finally to a vein. The precapillary sphincter is a muscle that can contract to shunt blood directly to the venule, bypassing the capillary bed.

Figure 15.2 Diagrammatic representation of a capillary bed.

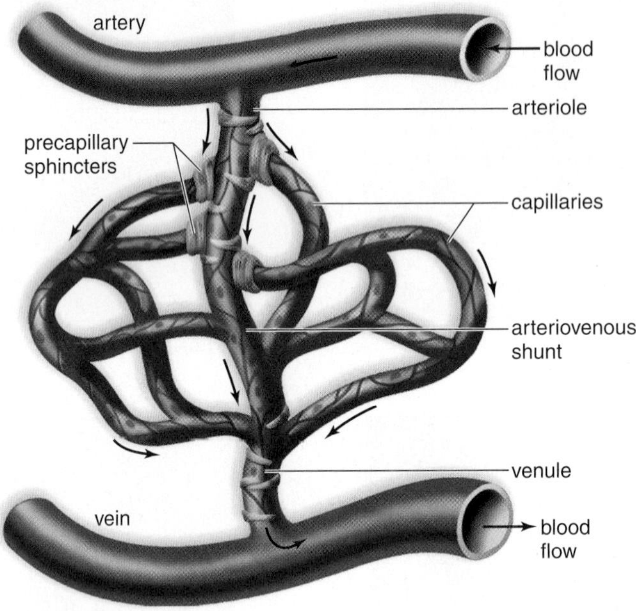

Experimental Procedure: Blood Flow

1. Observe blood flow through capillaries, arterioles, and venules, either in the tail of a goldfish or in the webbed skin between the toes of a frog, as prepared by your instructor. Refer to Figure 15.2 for help in following the path of blood.
2. Examine under low and high power of the microscope.
3. Watch the pulse and the swiftly moving blood in the arterioles.
4. Contrast this with the more slowly moving blood that circulates in the opposite direction in the venules. Many crisscrossing capillaries are visible.
5. Look for blood cells floating in the bloodstream. Don't confuse blood cells with chromatophores, irregular, black patches of pigment that may be visible in the skin.
6. The capillaries are thin-walled vessels. In what way is this feature useful to the organism?

____

15.2 The Heart

The mammalian heart has a right and left atrium and a right and left ventricle. *In order to tell the left from the right side, mentally position the heart so it corresponds to your own body*. There is at least one blood vessel attached to each of the chambers. The heart valves keep the blood moving forward because backward flow closes the valves. The heart nodes cause the heart to contract every 0.85 second. The contraction of the heart pumps the blood through the heart and out into the arteries. The right ventricle pumps blood into the pulmonary trunk, which leads to the pulmonary arteries, and the left ventricle pumps blood into the aorta, which is the major artery in the body. The heart is a muscle that has its own blood supply. The coronary arteries, which lie on the surface of the heart, arise from the aorta. The cardiac veins drain into the right atrium of the heart.

Caution:	Wear protective eyewear and latex gloves when handling preserved animal organs. Exercise caution when using the scalpel and wash your hands thoroughly upon completion of this exercise.

Heart Model or Preserved Sheep Heart

The heart model or preserved sheep heart will be used to study the anatomy of the heart (Fig. 15.3).

Figure 15.3 External view of mammalian heart.
Externally, notice the coronary arteries and cardiac veins that serve the heart itself.

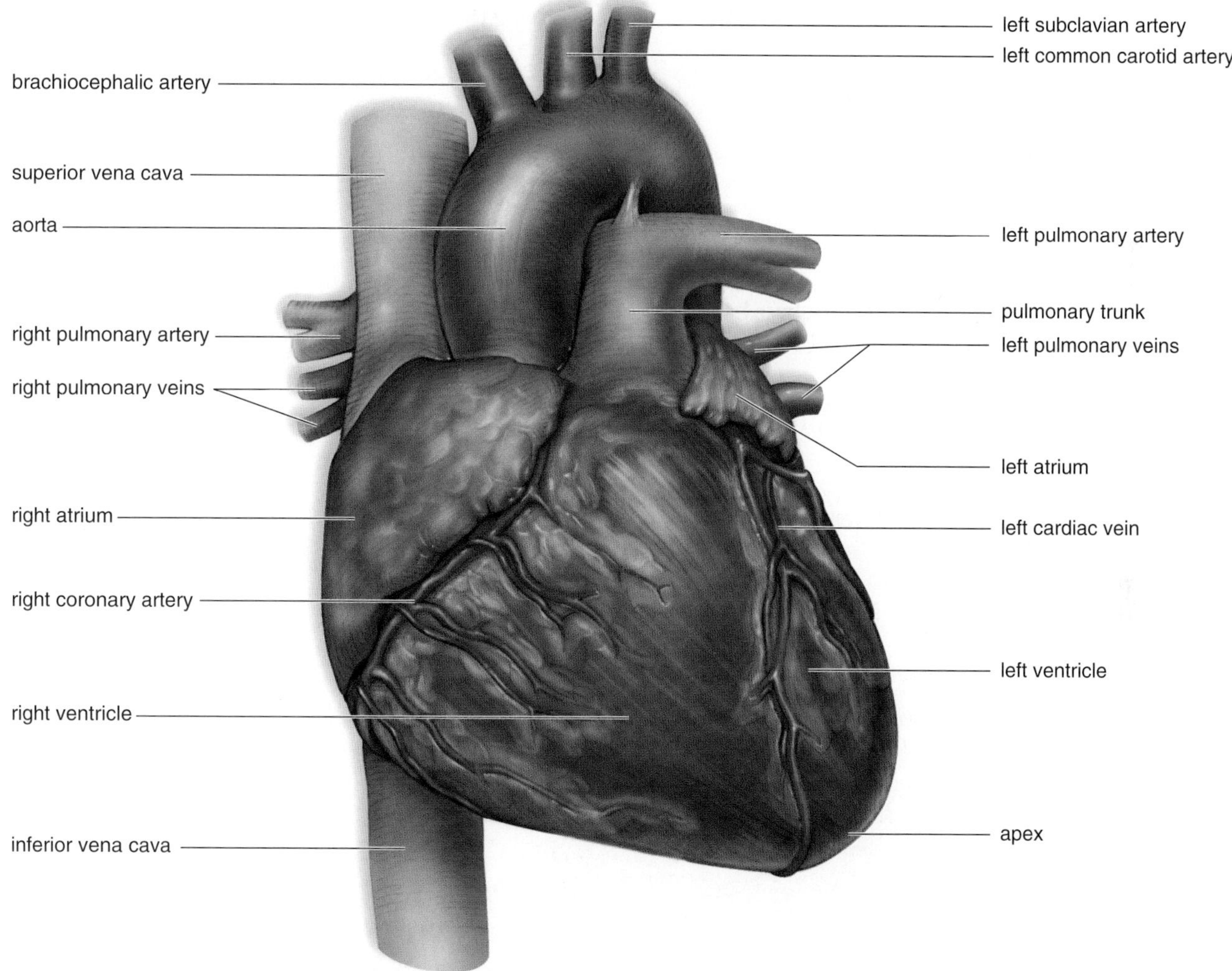

Observation: External Anatomy

1. Identify the **right atrium,** and its attached blood vessels, the superior and inferior **venae cavae.** The superior vena cava and the inferior vena cava return blood from the head and body, respectively, to the right atrium.
2. Identify the **right ventricle** and its attached blood vessel, the **pulmonary trunk.** The pulmonary trunk leaves the ventral side of the heart from the top of the right ventricle and then passes forward diagonally before branching into the **right and left pulmonary arteries.**

3. Identify the **left atrium** and its attached blood vessels, the left and right **pulmonary veins.** The pulmonary veins return blood from the lungs to the left atrium.
4. Identify the **left ventricle** and its attached blood vessel, the **aorta,** which arises from the anterior end of the left ventricle, just dorsal to the origin of the pulmonary trunk. The aorta soon bends to the animal's left as the aortic arch. The aorta carries blood to the body proper.
5. Identify the **coronary arteries** and the **cardiac veins,** which service the needs of the heart wall. The coronary arteries branch off the aorta as soon as it leaves the heart and appear on the surface of the heart. The cardiac veins, also on the surface of the heart, join and then enter the right atrium through the coronary sinus.

Observation: Internal Anatomy

Remove the ventral half of the human heart model (Fig. 15.4).

1. Identify the four chambers of the heart in longitudinal section: **right atrium, right ventricle, left atrium,** and **left ventricle.**
2. Which ventricle is more muscular? ______________________________
3. Why is this appropriate? ______________________________

Figure 15.4 Internal view of mammalian heart.
Internally, the heart has four chambers and there is a septum that separates the left side from the right side.

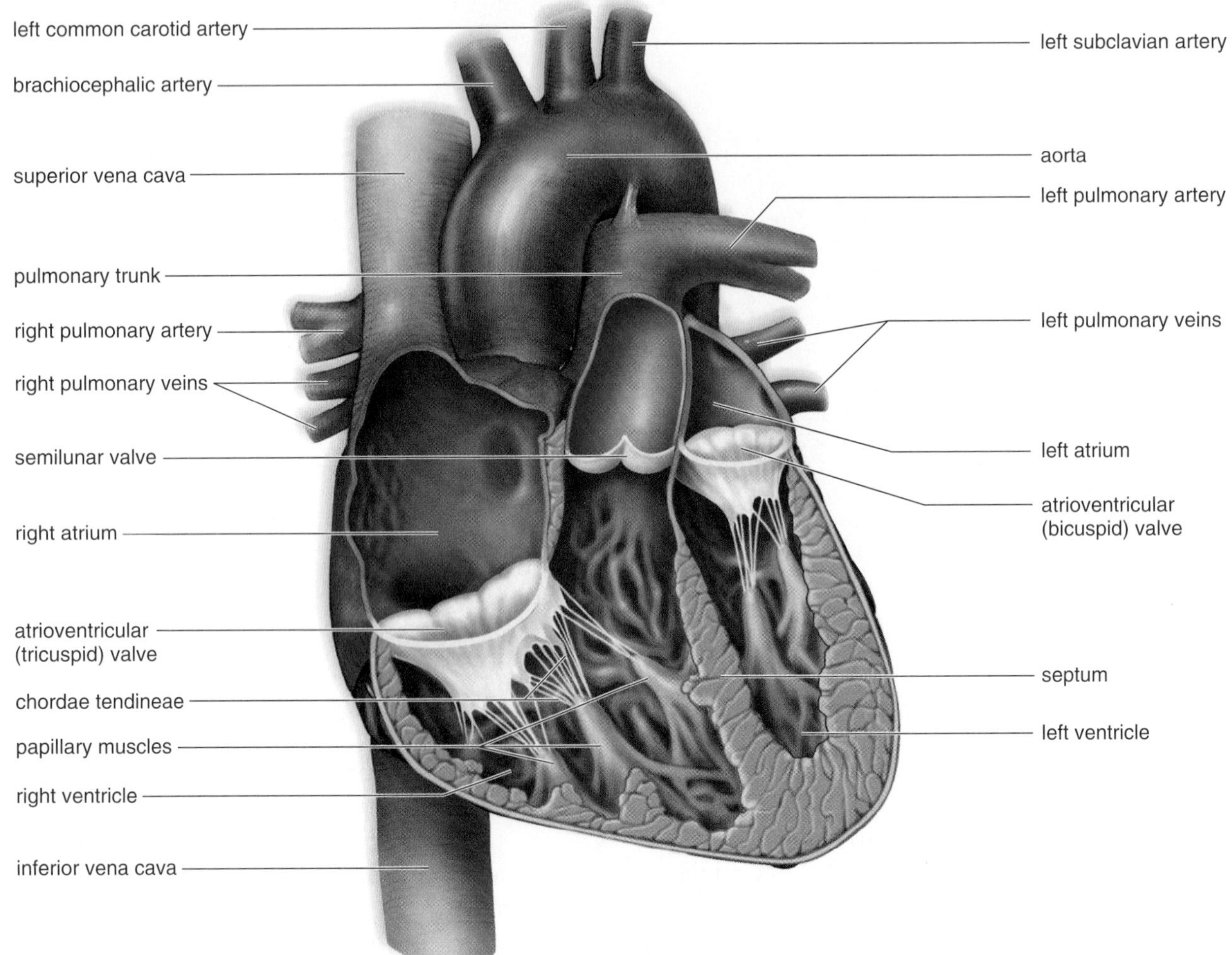

Valves of the Heart

1. Locate the four valves of the heart. Find the **right atrioventricular** (tricuspid) valve, which is located between the right atrium and the right ventricle.
2. Find the **left atrioventricular** (bicuspid or mitral) valve, which is located between the left atrium and the left ventricle.
3. Find the **pulmonary semilunar** valve, which is located in the base of the pulmonary trunk.
4. Find the **aortic semilunar** valve, which is located in the base of the aorta. What is the function of the heart valves? ______________________________
5. Note the **chordae tendineae** ("heartstrings") that hold the atrioventricular valves in place while the heart contracts. These extend from the papillary muscles. The chordae tendineae prevent the atrioventricular valves from inverting into the atria when the ventricles contract.

Path of Blood Through the Heart

To demonstrate that O_2-poor blood is kept separate from O_2-rich blood, trace the path of blood from the right side of the heart to the aorta by filling in the following blanks. Notice that after birth, the blood passes through the lungs in order to go from the right to the left sides of the heart.

Venae Cavae	**Lungs**
____________	____________
____________ valve	____________
____________	____________ valve
____________ valve	____________
____________	____________ valve
Lungs	Aorta

Conduction System of the Heart

The SA node is the pacemaker of the heart because it initiates the heartbeat and sends an excitation impulse every 0.85 second, causing the atria to contract. After the impulse reaches the **AV node,** it passes into large fibers and thereafter spreads out by way of the smaller **Purkinje fibers** (Fig. 15.5*a*). These fibers lie within the myocardium and signal the ventricles to contract.

With the contraction of any muscle, including the heart, electrolyte changes occur that can be detected by electrical recording devices. Therefore, it is possible to study the heartbeat by recording voltage changes that occur when the heart contracts. The record that results is called an *electrocardiogram* (ECG). The first wave in the electrocardiogram, called the P wave, occurs prior to the excitation and contraction of the atria. The second wave, or QRS wave, occurs prior to ventricular excitation and contraction. The third wave, or T wave, is caused by the recovery of the ventricles. (Atrial relaxation is not apparent in an ECG.) Examination of an ECG indicates whether the heartbeat has a normal or an irregular pattern (Fig. 15.5*b*).

Figure 15.5 Control of the heartbeat.
a. The SA node sends out a stimulus that causes the atria to contract. When this stimulus reaches the AV node, it signals the ventricles to contract by way of the atrioventricular bundle and Purkinje fibers. **b.** A normal ECG indicates that the heart is functioning properly. The P wave indicates that the atria are about to contract; the QRS wave indicates that the ventricles are about to contract; and the T wave indicates that the ventricles are recovering from contraction.

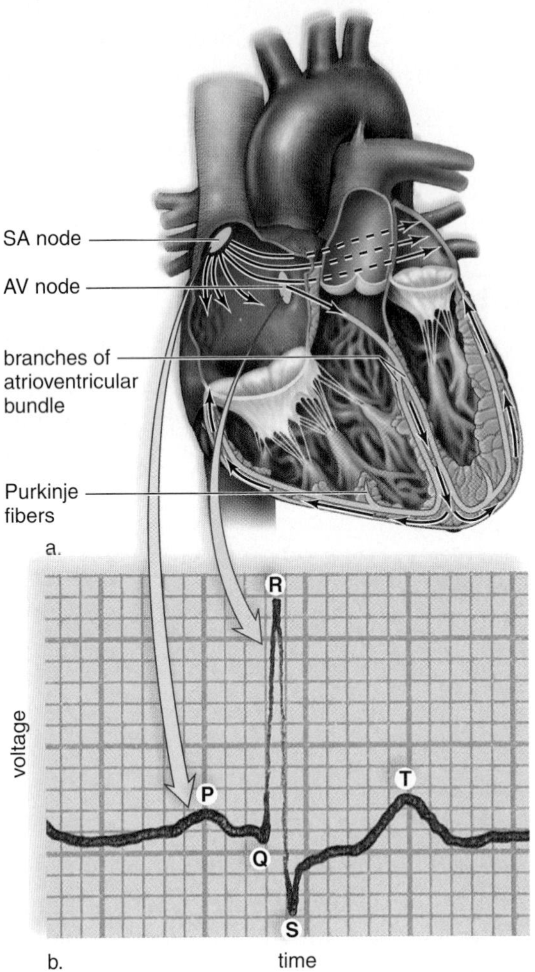

Observation: Nodes

1. Note the SA (sinoatrial) node in Figure 15.5*a*.
2. Explain Figure 15.5 by answering these questions:

   Why is an arrow drawn between the SA node and the P wave? ______________________

   ____

   Why is an arrow drawn between the AV node and the QRS wave? ______________________

   ____

   The voltage changes in an ECG are related to the ______________________

   ______________________.

15.3 Heartbeat

Recall that during a heartbeat, first the atria contract and then the ventricles contract. Usually, then, there are two heart sounds with each heartbeat. The first sound (lub) is low and dull and lasts longer than the second sound. It is caused by the closure of the atrioventricular valves following atrial contraction. The second sound (dub) follows the first sound after a brief pause. The sound has a snapping quality of higher pitch and shorter duration. The sound is caused by the closure of the semilunar valves. The beat of the heart creates blood pressure. When a chamber contracts, it is called **systole,** and when a chamber relaxes, it is called **diastole.**

Heartbeat at Rest

In the following procedure, you will employ two different methods to determine the heartbeat at rest: (1) using the stethoscope to listen directly to the heartbeat and (2) obtaining the pulse rate by detecting the pulse. Both of these exercises require that you work with a laboratory partner. The normal resting heartbeat rate in a young adult is between 60 and 80 beats per minute.

Experimental Procedure: Heartbeat at Rest

Stethoscope Method

1. Obtain a stethoscope, and properly position the earpieces. They should point forward. Place the bell of the stethoscope on the left side of your partner's chest between the fourth and fifth ribs. This is where the apex (tip) of the heart is closest to the body wall.
2. Which of the two sounds (lub or dub) is louder? ____________________
3. Count the heartbeat for 15 seconds, and then multiply by 4. __________ × 4 = __________
   Record your partner's heartbeat in Table 15.2.
4. Now switch, and your partner will determine your heartbeat. __________ × 4 = __________
   Record your heartbeat in Table 15.2.

Pulse-Rate Method

1. Use a digital monitor, if available, to measure the pulse (see Fig. 15.6), or position the fingers of one hand over the large artery near the outer (thumb) side of your partner's arm so that the little finger is slightly beyond the wrist. Count the pulse rate for 15 seconds, and then multiply by 4.

   __________ × 4 = __________

   Record your partner's pulse rate in Table 15.2.
2. Now switch, and your partner will determine your pulse rate.

   __________ × 4 = __________

   Record your pulse rate in Table 15.2.
3. Are the number of heartbeats the same regardless of the method used to determine the rate?
   __________ Explain. ____________________

Table 15.2 Heartbeat at Rest

Method	Partner	Yourself
Stethoscope		
Pulse rate		

Heartbeat After Exercise

You will employ only the pulse-rate method to determine the heartbeat after exercise.

Experimental Procedure: Heartbeat After Exercise

1. Jump on each foot 20 times. Then, using a digital monitor or the pulse method, determine your heartbeat after exercise, and complete Table 15.3.

Table 15.3 Heartbeat at Rest and After Exercise

Before Exercise (from Table 15.2)		After Exercise	
Partner	Yourself	Partner	Yourself

2. Why is it advantageous to have an increased heartbeat during exercise?
____

15.4 Blood Pressure

Blood pressure is highest just after ventricular systole, and it is lowest during ventricular diastole. Why? _______________

____

We would expect a person to have lower blood pressure readings at rest than after exercise. Why? _______________

Experimental Procedure: Blood Pressure at Rest and After Exercise

A number of different types of digital blood pressure monitors are now available, and your instructor will instruct you how to use the type you will be using for this Experimental Procedure. The normal resting blood pressure readings for a young adult are 120/80 (systolic/diastolic), as displayed on the monitor shown in Figure 15.6.

You may work with a partner or by yourself. If working with a partner, each of you will assist the other in taking blood pressure readings. After you have noted the blood pressure readings, also note the pulse reading.

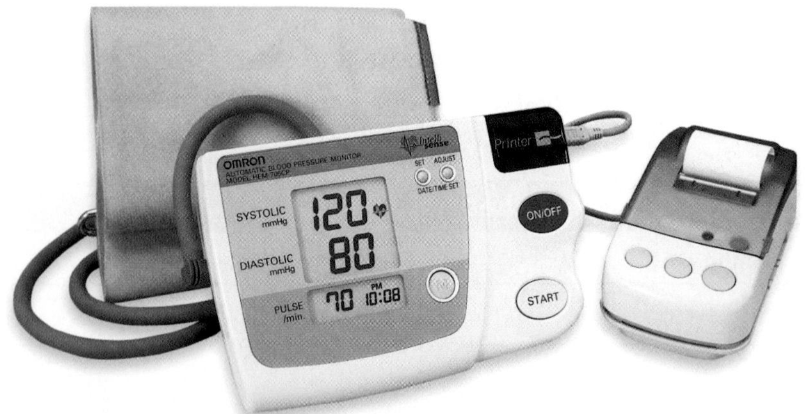

Figure 15.6 Measurement of blood pressure and pulse. There are many different types of digital blood pressure/pulse monitors now available. The one shown here uses a cuff to be placed on the arm. Others use a cuff for the wrist.

Blood Pressure at Rest

1. Reduce your activity as much as possible.
2. Use the blood pressure monitor to obtain blood pressure readings, and record them in Table 15.4. Are the pulse readings consistent with the blood pressure readings? Offer an explanation in Table 15.4.

Table 15.4 Blood Pressure At Rest

	Blood Pressure	Pulse	Explanation
Partner			
Yourself			

Blood Pressure After Exercise

1. Run in place for one minute.
2. Use the blood pressure monitor to obtain blood pressure readings, and record them in Table 15.5. Are the pulse readings consistent with the blood pressure readings? Offer an explanation in Table 15.5.

Table 15.5 Blood Pressure After Exercise

	Blood Pressure	Pulse	Explanation
Partner			
Yourself			

Conclusion

- Why would you expect the pulse readings to be consistent with the blood pressure readings in Tables 15.4 and 15.5?

___

___

Laboratory Review 15

_________________ 1. What type of blood cells are lymphocytes and monocytes?

_________________ 2. What is the function of red blood cells?

_________________ 3. What is the function of white blood cells?

_________________ 4. Nutrients exit and wastes enter which type of blood vessel?

_________________ 5. Which blood cells contain a respiratory pigment?

_________________ 6. Which chamber of the heart receives venous blood from the systemic circuit?

_________________ 7. Identify the vessel that conducts blood from the left ventricle.

_________________ 8. The pulmonary trunk leaves which chamber?

_________________ 9. Identify the artery that nourishes the heart tissue.

_________________ 10. Which heart chamber pumps blood throughout the body?

_________________ 11. How many times a minute does the heart normally beat in a young adult?

_________________ 12. What is it called when a heart chamber contracts?

_________________ 13. Which is higher—systolic or diastolic pressure?

_________________ 14. What is the normal resting blood pressure of a young adult?

_________________ 15. Identify the pacemaker region of the heart.

Thought Questions

16. Why is the body better served by having several different types of white blood cells?

17. Under what conditions in everyday life would you expect the heartbeat and the blood pressure to increase? When might this be an advantage? A disadvantage?

18. Why is it important to maintain an adequate blood pressure?

19. What might be the significance of nonfunctional or defective chordae tendinae?

LABORATORY

16

Basic Mammalian Anatomy II

Learning Outcomes

16.1 Urinary System
- Using preserved specimens, images, or charts, locate and identify the organs of the urinary system.
- State a function for each of the major organs of the urinary system.

16.2 Male Reproductive System
- Using preserved specimens, images, or charts, locate and identify the organs of the male reproductive system.
- State a function for each of the major organs of the male reproductive system.
- Compare the male reproductive system of the pig with that of the human male.

16.3 Female Reproductive System
- Using preserved specimens, images, or charts, locate and identify the organs of the female reproductive system.
- State a function for each of the major organs of the female reproductive system.
- Compare the female reproductive system of the pig with that of the human female.

16.4 Anatomy of Testis and Ovary
- Identify the major regions and structures from a cross-section slide of the testis and seminiferous tubules.
- Identify the major regions and structures from a slide of the ovary and the follicles.

16.5 Review of the Respiratory, Digestive, and Cardiovascular Systems
- Using preserved specimens, images, or charts, locate and identify the individual organs of the respiratory, digestive, and cardiovascular systems.
- Distinguish each organ by function.
- Using preserved specimens, images, or charts, locate and identify the hepatic portal system. State a function for this system. Distinguish each organ by function.

Introduction

The **urinary system** and the **reproductive system** are so closely associated in mammals that they are often considered together as the **urogenital system.** They are particularly associated in males, where certain structures function in both systems. In this laboratory, we will focus first on dissecting the urinary and reproductive systems in the fetal pig. We will then compare the anatomy of the reproductive systems in pigs with those in humans.

In mammalian reproductive systems, the testes (sing., *testis*) are the male gonads, and the ovaries (sing., *ovary*) are the female gonads. The testes produce sperm, and the ovaries produce oocytes (eggs). Examining prepared slides in this laboratory will allow you to observe the location of spermatogenesis in the testis and oogenesis in the ovary.

Finally, we will examine parts of the respiratory, digestive, and cardiovascular systems in the fetal pig. You will view the organs of the respiratory system in some detail; remove and examine the heart, stomach, and intestine; and view the exposed hepatic portal system.

Caution: Wear protective eyewear and latex gloves when handling preserved animal organs. Exercise care when using the scalpel and wash your hands thoroughly upon completion of the experiment.

Figure 16.1 Urinary system of the fetal pig.
In **(a)** females and **(b)** males, urine is made by the kidneys, transported to the bladder by the ureters, stored in the bladder, and then excreted from the body through the urethra.

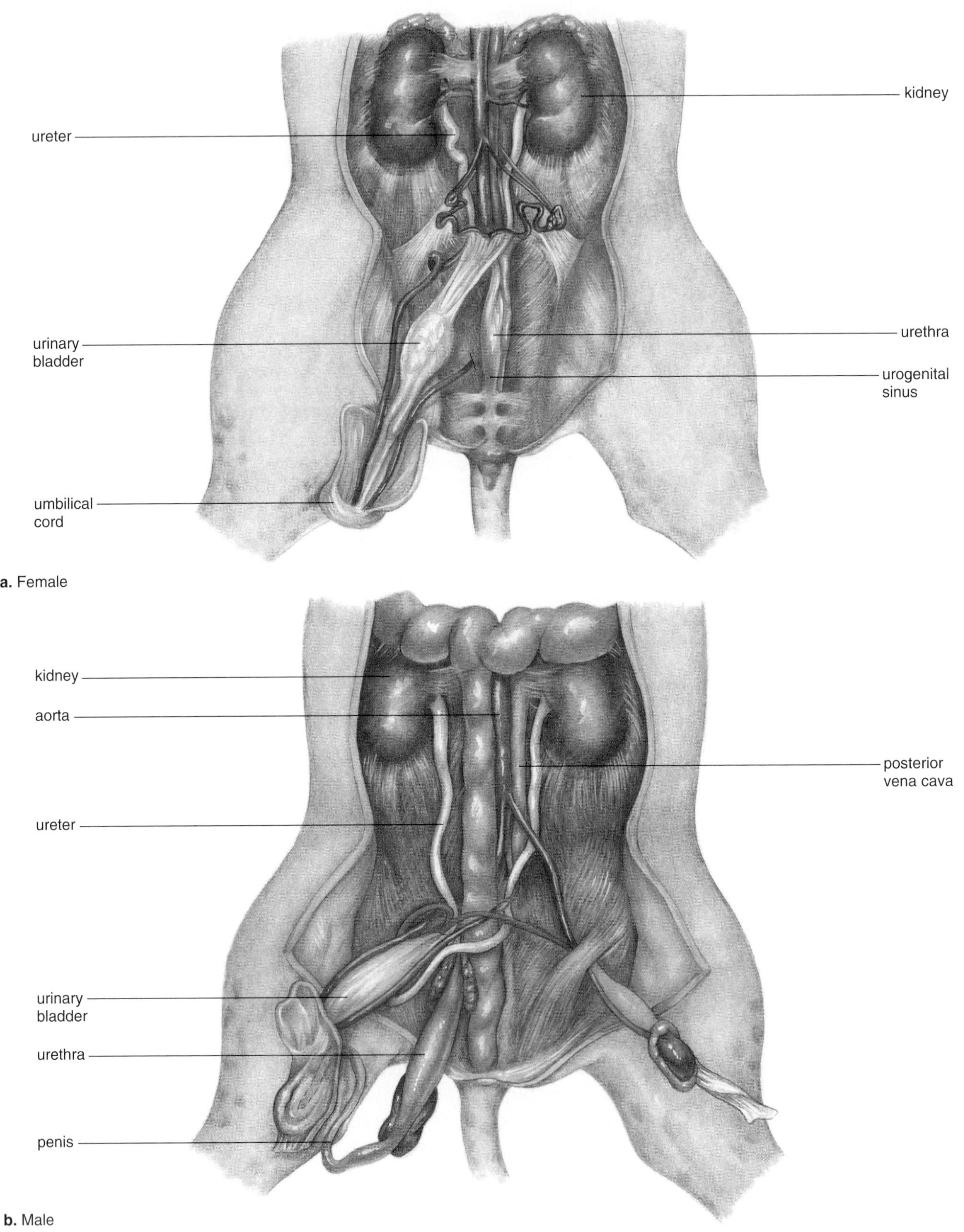

16.1 Urinary System

The urinary system consists of the **kidneys,** which produce urine; the **ureters,** which transport urine to the **urinary bladder,** where urine is stored; and the **urethra,** which transports urine to the outside. In males, the urethra also transports sperm during ejaculation.

During the upcoming dissection, compare the urinary system structures of both sexes of fetal pigs. Later in this laboratory period, exchange specimens with a neighboring team for a more thorough inspection.

Observation: Urinary System in Pigs

1. The large, paired kidneys (Fig. 16.1) are reddish organs covered by **peritoneum,** a membrane that anchors them to the dorsal wall of the abdominal cavity, which is sometimes called the **peritoneal cavity.** Clean the peritoneum away from one of the kidneys, and study it more closely.
2. Using a razor blade or scalpel, section one of the kidneys in place, cutting it lengthwise (Fig. 16.2). Note that at the center of the medial portion of the kidney is an irregular, cavity-like reservoir, the **renal pelvis.** The outermost portion of the kidney (the **renal cortex**) shows many small striations perpendicular to the outer surface. This region and the more even-textured **renal medulla** region inward from the renal cortex is composed of the conical *renal pyramids.* The renal pyramids are composed of **nephrons** (excretory tubules).
3. Locate the **ureters,** which leave the kidneys and run posteriorly under the peritoneum.
4. Clean the peritoneum away, and follow a ureter to the **urinary bladder,** which normally lies in the posterior ventral portion of the abdominal cavity. The urinary bladder is on the inner surface of the flap of tissue to which the umbilical cord was attached.
5. The **urethra,** which arises from the bladder posteriorly, runs parallel to the rectum. Follow the urethra until it passes from view into the ring formed by the pelvic girdle.
6. Trace the path of urine. ____

____

Figure 16.2 Anatomy of the kidney.
A kidney has a renal cortex, renal medulla, renal pelvis, and microscopic tubules called nephrons.

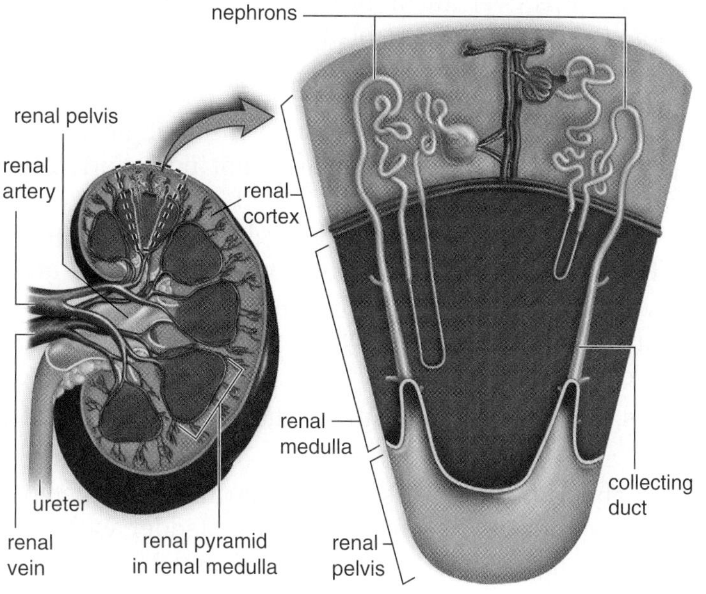

16.2 Male Reproductive System

Table 16.1 Male Reproductive Organs and Functions

Organ	Function
Testes	Produce sperm and sex hormones
Epididymis	Stores sperm as they mature
Vas deferens	Conducts and stores sperm
Seminal vesicle	Contributes fluid to semen
Prostate gland	Contributes secretions to semen
Urethra	Conducts sperm
Bulbourethral glands	Contribute mucoid fluid to semen
Penis	Organ of copulation

The **male reproductive system** consists of the **testes** (sing., *testis*), which produce sperm, and the **epididymis** (pl., *epididymides*), which stores sperm before they enter the **vas deferens** (pl., *vasa deferentia*). Just prior to ejaculation, sperm leave the vas deferens and enter the **urethra,** located in the penis. The **penis** is the male organ of copulation. **Seminal vesicles,** the **prostate gland,** and the **bulbourethral glands** add fluid to semen after sperm reach the urethra. Table 16.1 summarizes the male reproductive organs.

The testes begin their development in the abdominal cavity, just anterior and dorsal to the kidneys. Before birth, however, they gradually descend into paired scrotal sacs within the **scrotum,** which is suspended anterior to the anus. Each scrotal sac is connected to the body cavity by an **inguinal canal,** the opening of which can be found in your pig. The passage of the testes from the body cavity into the scrotal sacs is called the descent of the testes and it occurs in human males. The testes in most of the male fetal pigs being dissected will probably be partially or fully descended.

Observation: Male Reproductive System in Pigs

Inguinal Canal, Testis, Epididymis, and Vas Deferens

1. Locate the opening of the left inguinal canal, which leads to the left scrotal sac (Fig. 16.3).
2. Expose the canal and sac by making an incision through the skin and muscle layers from a point over this opening back to the left scrotal sac.
3. Open the sac, and find the testis. Note the much-coiled tubule—the epididymis—that lies alongside the testis. This is continuous with the vas deferens, which passes back toward the abdominal cavity.
4. Trace a vas deferens as it loops over an umbilical artery and ureter and unites with the urethra dorsally at the posterior end of the urinary bladder.

Penis, Urethra, and Accessory Glands

1. Cut through the ventral skin surface just posterior to the umbilical cord. This will expose the rather undeveloped penis, which extends from this point posteriorly toward the anus. The central duct of the penis is the urethra.
2. Lay the penis to one side, and then cut down through the ventral midline, laying the legs wide apart in the process (Fig. 16.4). The cut will pass between muscles and through pelvic cartilage (bone has not developed yet). Do not cut any of the ducts or tracts in the region.
3. You will now see the urethra passing ventrally above the rectum. It is somewhat heavier in the male due to certain accessory glands:
   a. Bulbourethral glands, about 1 cm in diameter, lie laterally and well back toward the anal opening.
   b. The prostate gland, about 4 mm across and 3 mm thick, is located on the dorsal surface of the urethra, just posterior to the juncture of the bladder with the urethra. It is often difficult to locate and is not shown in Figures 16.3 and 16.4.
   c. Small, paired seminal vesicles may be seen on either side of the prostate gland.

Figure 16.3 Male reproductive system of the fetal pig.
In males, the urinary system and the reproductive system are joined. The vasa deferentia (sing., vas deferens) enter the urethra, which also carries urine.

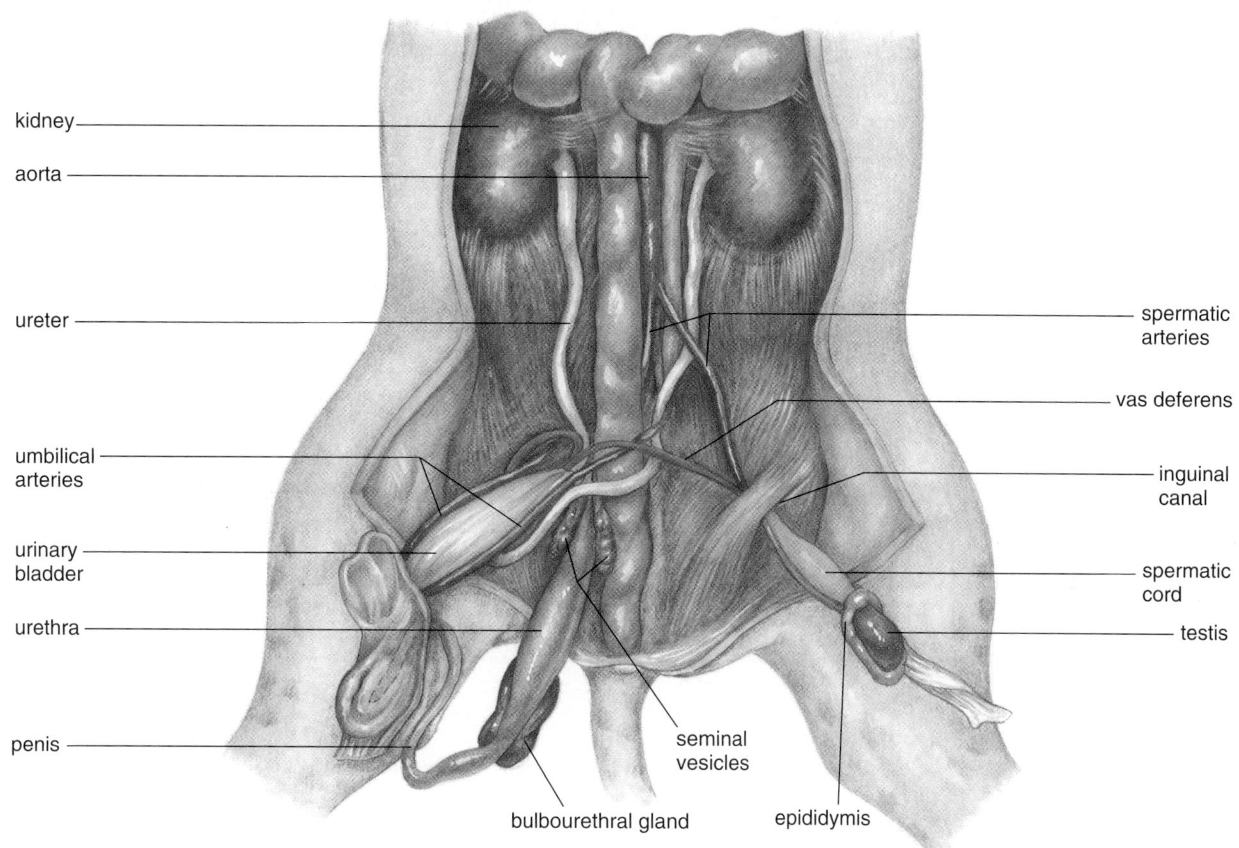

Figure 16.4 Photograph of the male reproductive system of the fetal pig.
Compare the diagram in Figure 16.3 with this photograph to help identify the structures of the male urogenital system.

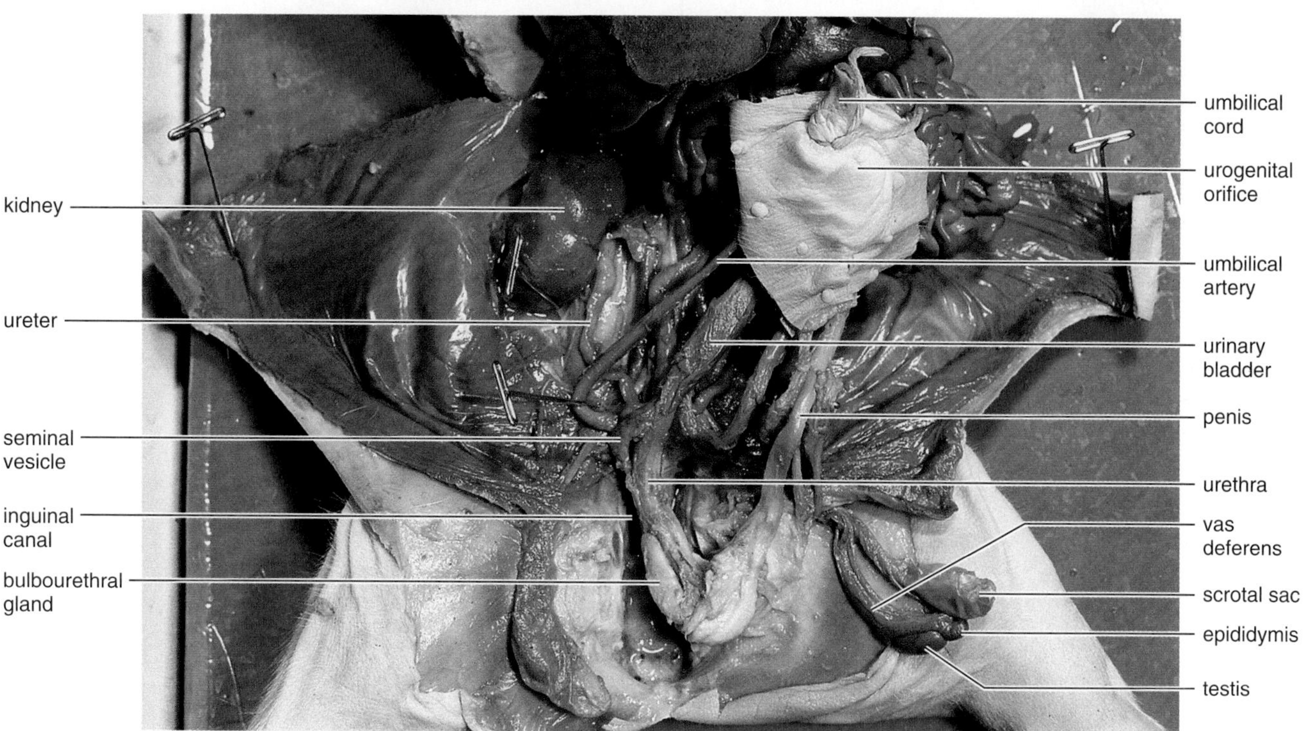

4. Trace the urethra as it leaves the bladder. It proceeds posteriorly, but when it nears the posterior end of the body, it turns rather abruptly anterioventrally and runs forward just under the skin of the midventral body wall, where you have just dissected it. This latter portion of the urethra is, then, within the penis.
5. Now you should also be able to see the entrance of the vasa deferentia into the urethra. If necessary, dissect these structures free from surrounding tissue, and expose the point of entrance of these ducts into the urethra near the location of the prostate gland. In males, the urethra transports sperm, as well as urinary wastes from the bladder.
6. Trace the path of sperm in the male. ______________________________

____

Comparison of Male Fetal Pig and Human Male

Use Figure 16.5 to help you compare the male pig reproductive system with the human male reproductive system. Complete Table 16.2, which compares the location of the penis in these two mammals.

Table 16.2 Location of Penis in Male Fetal Pig and Human Male

	Fetal Pig	Human
Penis		

Figure 16.5 Human male urogenital system.
In the fetal pig, but not in the human male, the penis lies beneath the skin and exits at the urogenital opening.

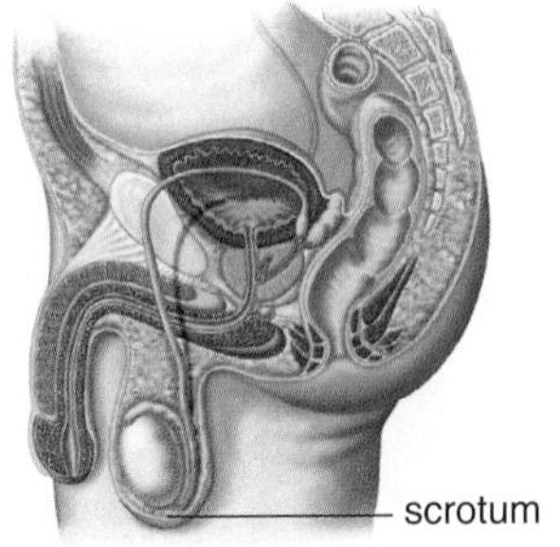

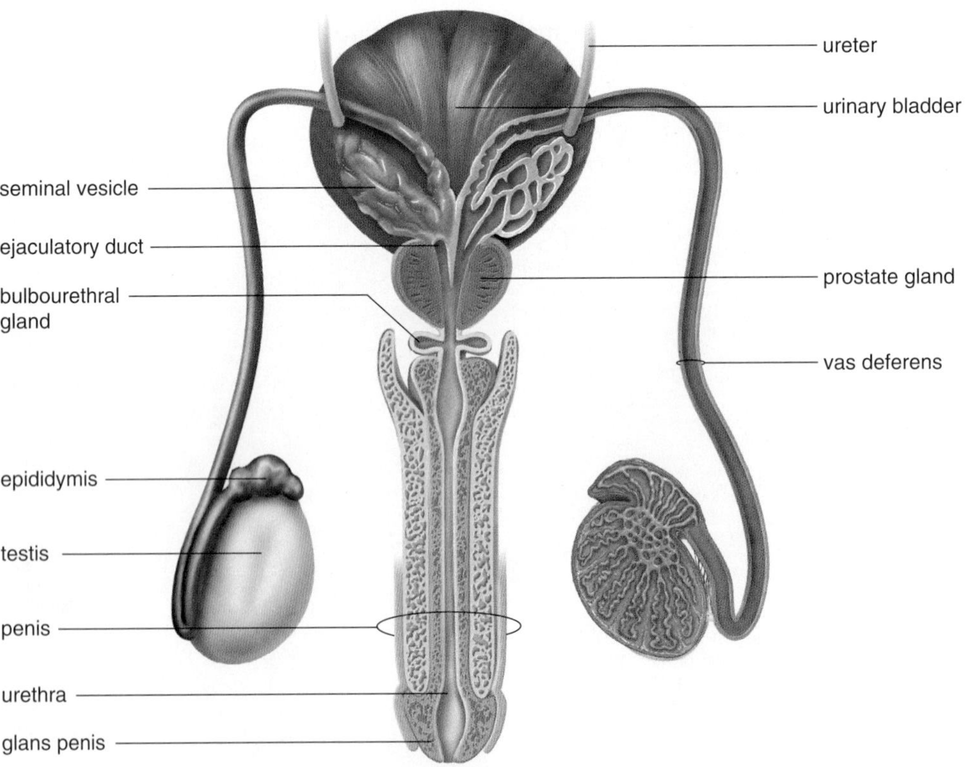

16.3 Female Reproductive System

The **female reproductive system** (Table 16.3) consists of the **ovaries,** which produce eggs, and the **oviducts,** which transport eggs to the **uterus,** where development occurs. In the fetal pig, the uterus does not form a single organ, as in humans, but is partially divided into external structures called **uterine horns,** which connect with the oviduct. The **vagina** is the birth canal and the female organ of copulation.

Table 16.3 Female Reproductive Organs and Functions

Organ	Function
Ovary	Produces egg and sex hormones
Oviduct (fallopian tube)	Conducts egg toward uterus
Uterus	Houses developing fetus
Vagina	Receives penis during copulation and serves as birth canal

Observation: Female Reproductive System in Pigs

Ovaries and Oviducts

1. Locate the paired ovaries, small bodies suspended from the peritoneal wall in mesenteries, posterior to the kidneys (Figs. 16.6 and 16.7).
2. Closely examine one ovary. Note the small, short, coiled oviduct, sometimes called the fallopian tube. The oviduct does not attach directly to the ovary but ends in a funnel-shaped structure with fingerlike processes (fimbriae) that partially encloses the ovary.

Uterine Horns

1. Locate the **uterine horns.** (Do not confuse the uterine horns with the oviducts; the latter are much smaller and are found very close to the ovaries.)
2. Find the median body of the uterus located at the joined posterior ends of the uterine horns.

Vagina

1. Separate the hind limbs of your specimen, and cut down along the midventral line. The cut will pass through muscle and the cartilaginous pelvic girdle. With your fingers, spread the cut edges apart, and use blunt dissecting instruments to separate connective tissue.
2. Note three ducts passing from the body cavity to the animal's posterior surface. One of these is the urethra, which leaves the bladder and passes into the **urogenital sinus.** The urethra is a part of the urinary system. The most dorsal of the three ducts is the **rectum,** which passes to its own opening, the **anus.** The rectum and anus are, of course, part of the digestive system, not the reproductive system.
3. Find the vagina, located dorsally to the urethra. The vagina is the birth canal and is the organ of copulation. Anteriorly, it connects to the uterus, and posteriorly it enters the urogenital sinus. This sinus is absent in adult humans and several other female mammals.

Figure 16.6 Female reproductive system of the fetal pig.
In the adult female, the urinary system and the reproductive system are separate. In the fetus, the vagina joins the urethra just before the urogenital sinus.

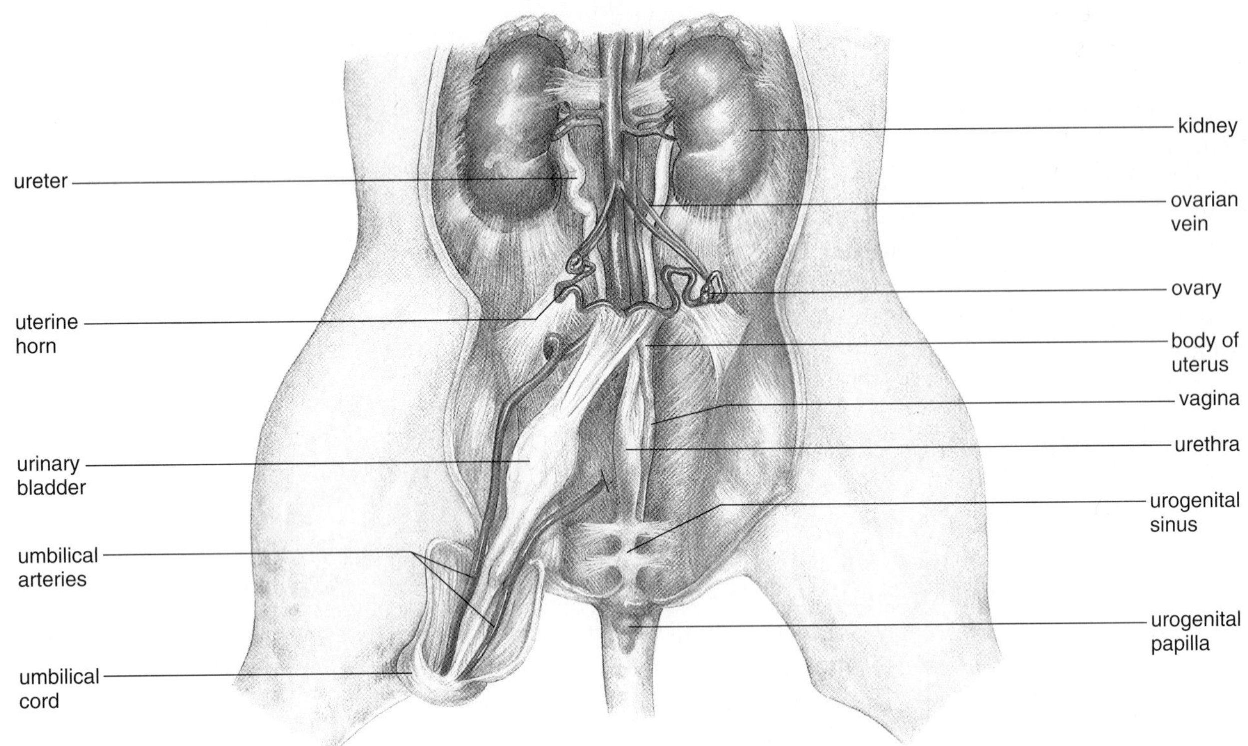

Figure 16.7 Photograph of the female reproductive system of the fetal pig.
Compare the diagram in Figure 16.6 with this photograph to help identify the structures of the female urinary and reproductive systems.

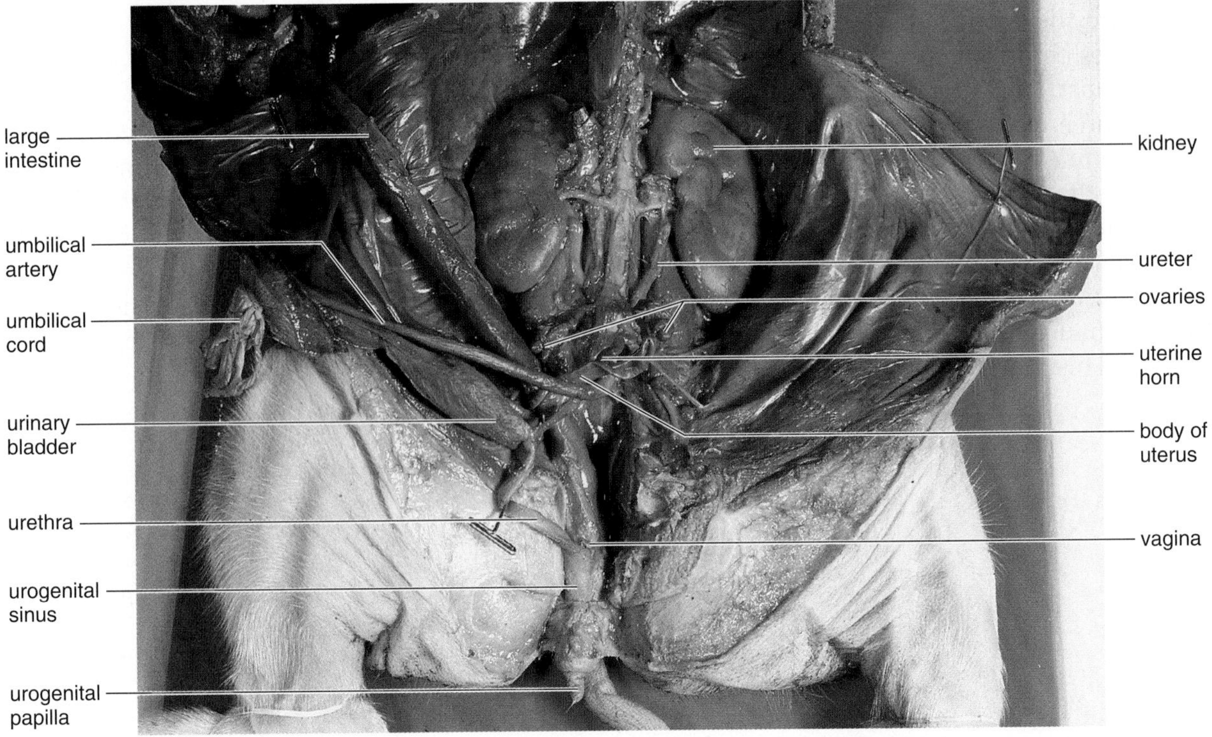

Comparison of Female Fetal Pig with Human Female

Use Figure 16.8 to compare the female pig reproductive system with the human female reproductive system. Complete Table 16.4, which compares the appearance of the oviducts and the uterus, as well as the presence or absence of a urogenital sinus in these two mammals.

Figure 16.8 Human female reproductive system.
Especially compare the anatomy of the oviducts in humans with that of the uterine horns in a pig. In a pig, the fetuses develop in the uterine horns; in a human female, the fetus develops in the body of the uterus.

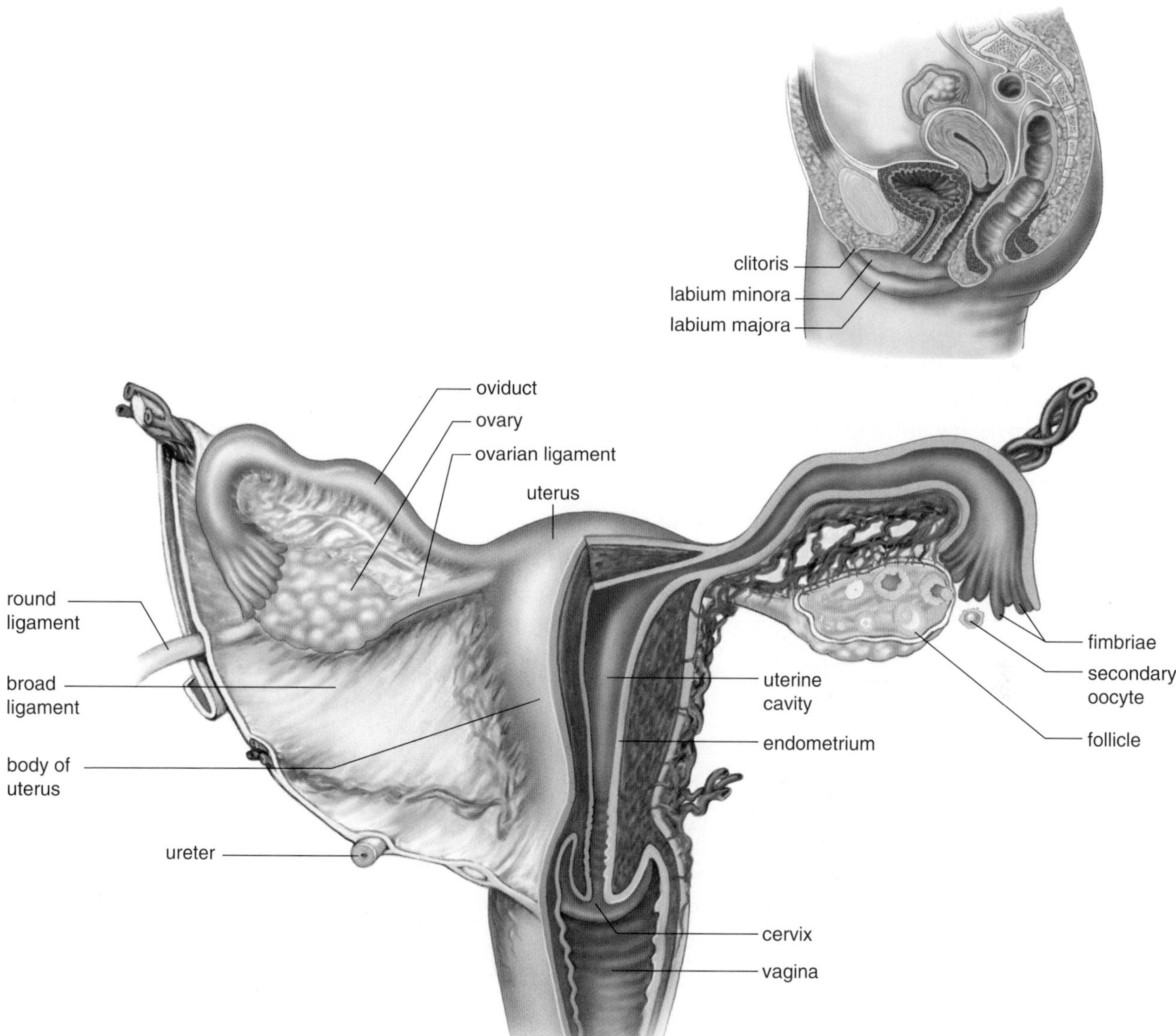

Table 16.4 Comparison of Female Fetal Pig with Human Female

	Fetal Pig	Human
Oviducts		
Uterus		
Urogenital sinus		

16.4 Anatomy of Testis and Ovary

Recall that the testes produce sperm (the male gametes) and that the ovaries produce oocytes (the female gametes). A testis contains **seminiferous tubules,** where sperm formation takes place, and **interstitial cells** scattered in the spaces between seminiferous tubules. Interstitial cells produce the male sex hormone testosterone. An ovary contains **follicles** in various stages of maturation. Ovarian follicles produce the female sex hormones estrogen and progesterone. One or more follicles complete maturation during each cycle and produce an oocyte.

Observation: Testis and Ovary

Testis

1. Examine a prepared slide of the testis. Note under low power the many circular structures—the seminiferous tubules.
2. Switch to high power, and observe one tubule in particular. With the help of Figure 16.9, find mature sperm, which look like thin, fine, dark lines, in the middle of the tubule. Interstitial cells are between the tubules.

Ovary

1. Examine a prepared slide of an ovary, and refer to Figure 16.10 for help in identifying the structures. Note that the female gonad contains an inner core of loose, fibrous tissue. The outer part contains the follicles that produce eggs.
2. Locate a **primary follicle,** which appears as a circle of cells surrounding a somewhat larger cell.

Figure 16.9 Photomicrograph of a seminiferous tubule.
A testis contains seminiferous tubules separated by interstitial cells. The tubules produce sperm, and the interstitial cells produce male sex hormones.

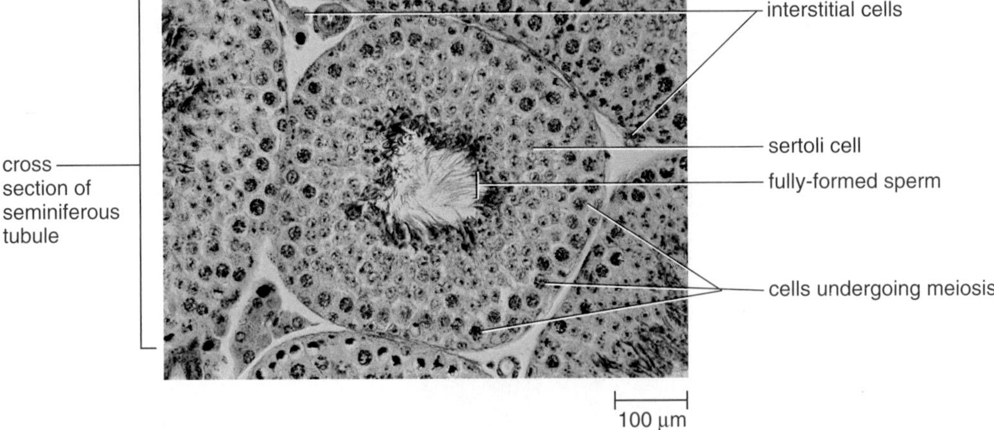

Figure 16.10 Photomicrograph of ovarian tissue.
An ovary contains follicles that are in different stages of maturity. A secondary follicle contains a secondary oocyte, which will burst from the ovary during ovulation. A follicle also produces the female sex hormones.

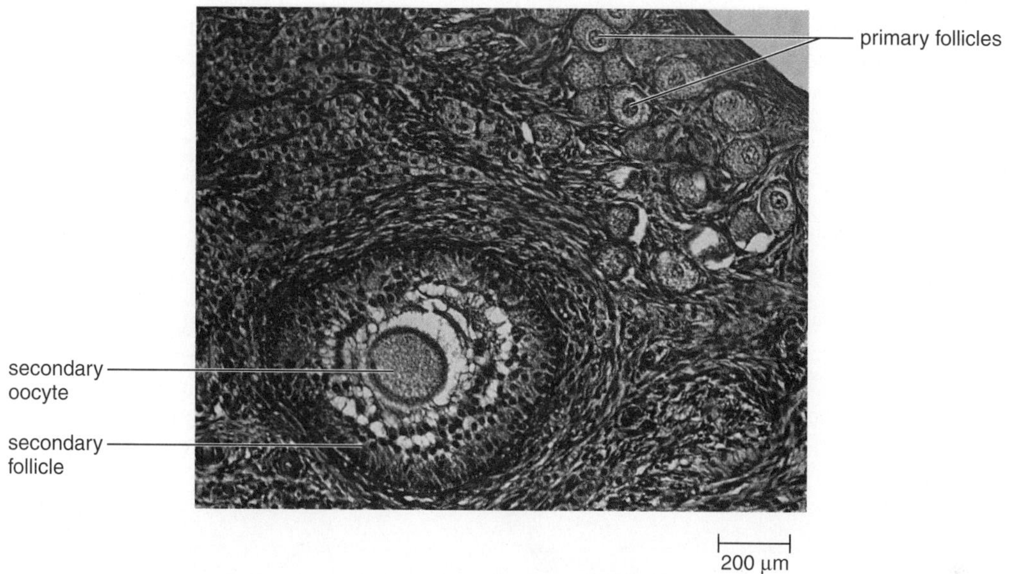

3. Find a **secondary follicle,** and switch to high power. Note the **secondary oocyte** (egg), surrounded by numerous cells, to one side of the liquid-filled follicle.
4. Also look for a large, fluid-filled vesicular (Graafian) follicle, which contains a mature secondary oocyte to one side. This follicle will be next to the outer surface of the ovary because it is the type of follicle that releases the egg during ovulation.
5. Also look for the remains of the corpus luteum, which will look like scar tissue. The corpus luteum develops after the vesicular follicle has released its egg, and then later it deteriorates. Not all slides will contain a vesicular follicle and corpus luteum because they may not have been present when the slide was made.

Comparison of Reproductive Systems

Complete Table 16.5 to describe the differences between the male and female mammalian reproductive systems.

Table 16.5 Comparison of Human Male and Female Reproductive Systems

	Male	Female
Gonad		
Duct from gonad		
Structure connected to gonad by duct		
Copulatory organ		

Figure 16.11 Internal anatomy of the fetal pig.
Most of the major organs are shown in this photograph. The stomach has been removed. The spleen, gallbladder, and pancreas are not visible.

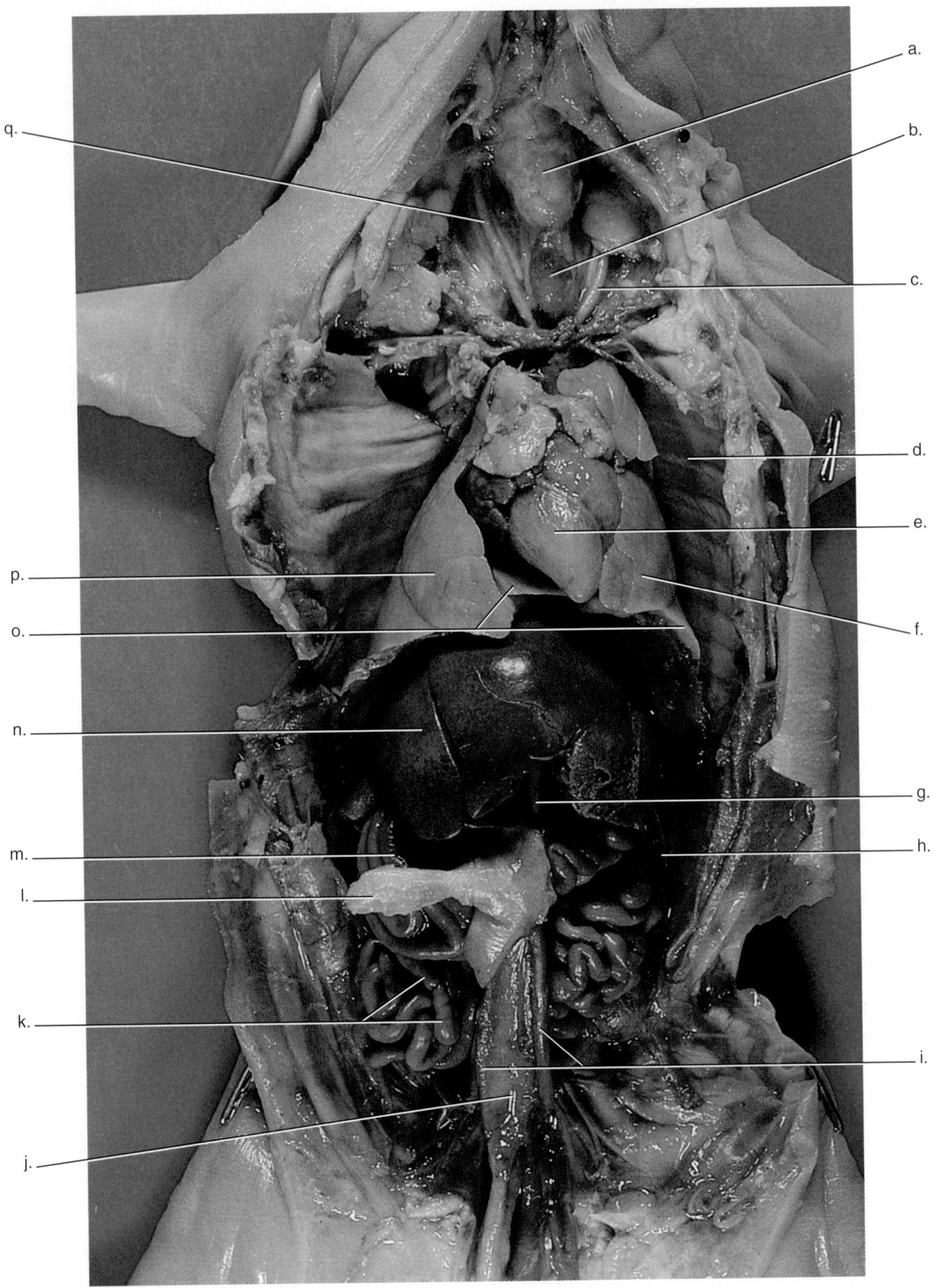

16.5 Review of the Respiratory, Digestive, and Cardiovascular Systems

In Laboratory 13, you dissected the respiratory, digestive, and cardiovascular systems of the fetal pig. *Review your knowledge of these systems by reexamining your dissection of the fetal pig and by labeling Figure 16.11.* In this portion of today's lab, you will review each system and examine some organs in more detail. **Do not remove any organs** unless told to do so by your instructor.

Observation: Respiratory System in Pigs

1. With the help of Figure 16.12, trace the path of air from the nasal passages to the lungs. List the first three organs in the left column and the last three organs in the right column.

   nasal passages

   ________________ ________________

   ________________ ________________

   ________________ ________________

   lungs
2. Make sure you have cut the corners of the mouth as directed in Lab 13, page 162. In the **pharynx,** you should be able to locate the **glottis,** an opening to the ________________.
3. If necessary, make a midventral incision in the neck to expose the **larynx.**
4. Clear away the "straplike" muscles covering the **trachea.** Now you should be able to feel the cartilaginous rings that hold the trachea open. Locate the esophagus, which lies below the trachea.
5. If available, observe a slide on display showing a section through the trachea and esophagus. Notice in Figure 16.12 that the air and food pathways cross in the pharynx.
6. Open the pig's mouth, insert a blunt probe into the glottis, and carefully work the probe down through the larynx to the level of the **bronchi.**
7. Observe the **lungs,** and if available, observe a prepared slide of lung tissue.
8. If so directed by your instructor, remove a portion of the trachea, the bronchi, and the lungs, keeping them all in one piece. Place this specimen in a small container of water. Holding the trachea with your forceps, gently but firmly stroke the lung repeatedly with the blunt wooden base of one of your probes. If you work carefully, the alveolar tissue will be fragmented and rubbed away, leaving the branching system of air tubes and blood vessels.

Figure 16.12 Air and food passages in the fetal pig.
A probe can pass from the mouth to the larynx to the esophagus.

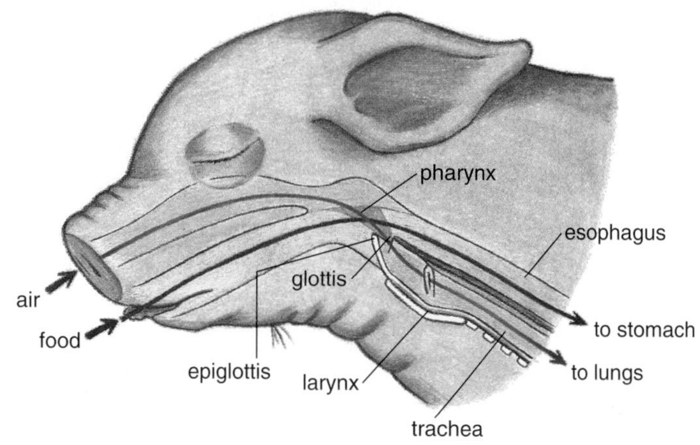

Observation: Digestive System in Pigs

1. Trace the path of food from the mouth to the anus:

   mouth

___________	___________
___________	___________
___________	___________
	anus

2. Open the **mouth** again, and insert a blunt probe into the esophagus (see Fig. 16.12). Then trace the **esophagus** to the stomach.
3. Open one side of the **stomach,** and examine its interior surface. Does it appear smooth or rough? ________
4. Find the pyloric sphincter, the muscle that surrounds the entrance to the duodenum, the first part of the **small intestine.** Record the length of the small intestine. ________
   If you have not done so before, find the bile duct that empties into the duodenum. The bile duct comes from the ________.
5. Find the **cecum,** a projection where the small intestine enters the large intestine. How does the appearance of the pig's large intestine differ from that of a human?

   ____

6. Carefully cut the mesenteries holding the colon of the **large intestine** in place, and uncoil the large intestine. Record the length of the large intestine. ________ How does the length of the large intestine compare with that of the small intestine? ____________________
7. Locate again the liver, pancreas, and gallbladder, three accessory organs of digestion.

Observation: Cardiovascular System in Pigs

Heart

1. Trace the path of blood through the heart, starting with the vena cava and ending with the aorta. Mention all the chambers of the heart and the valves (see Fig. 15.3).

To the heart:	From the lungs:
vena cava	______________
______________	______________
______________ valve	______________ valve
______________	______________
______________ valve	______________ valve
______________	aorta
______________	

2. Keeping the heart inside the pig, cut the pericardial sac (the tissue that surrounds the heart).
3. Look for and identify the vessels attached to the heart.
4. Section the heart, and look for its four chambers. Remnants of the atrioventricular valves can be seen as thin sheets of whitish tissue attached to fine, white, tendinous strands.
5. With your blunt probe, find the oval opening in the wall between the two atrial chambers. Recall that this is a shunt that allows blood to bypass lung circulation prior to birth.

Blood Vessels

1. In general, arteries take blood ________ the heart, and veins take blood ________ the heart.
2. Locate the following blood vessels in your pig. State their origin and destination. (For simplicity, use the name of the structure and either the aorta, the anterior vena cava, or the posterior vena cava before or after the name of the structure, as appropriate.)
   a. **coronary artery:** takes blood from the ________ to the ________
   b. **cardiac vein:** takes blood from the ________ to the ________
   c. **carotid artery:** takes blood from the ________ to the ________
   d. **jugular vein:** takes blood from the ________ to the ________
   e. **subclavian artery:** takes blood from the ________ to the ________
   f. **subclavian vein:** takes blood from the ________ to the ________
   g. **renal artery:** takes blood from the ________ to the ________
   h. **renal vein:** takes blood from the ________ to the ________
   i. **iliac artery:** takes blood from the ________ to the ________
   j. **iliac vein:** takes blood from the ________ to the ________

Hepatic Portal System and Associated Vessels

A **portal system** is a circulatory unit that goes from one capillary bed to another without passing through the heart. For example, in mammals, the mesenteric arteries take blood to the intestines. Thereafter, the hepatic portal vein takes blood from the intestinal capillaries to capillaries in the liver (Fig. 16.13). The significance of this circulatory arrangement is apparent, given the liver's important role in processing and storing materials absorbed from the intestine. The hepatic veins take blood from the liver to the inferior (posterior) vena cava.

1. To find the **hepatic portal vein** in your pig, carefully break the mesenteries in the region of the bile duct. The hepatic portal vein is dorsal to the duct and will not be blue if the latex did not enter it.
2. To see the **hepatic veins,** scrape away the liver substance with the edge of the forceps or the blunt side of the scalpel until all of the soft liver material has been removed and only a mass of cords remains.
3. Identify the **umbilical vein** leading into the liver; the **venous duct,** the main channel through the liver; and the **hepatic veins,** consisting of three or four vessels from the liver to the posterior vena cava. The great majority of the cords you have exposed consist of branches of veins and bile ducts.

Storage of Pigs

1. Before leaving the laboratory, place your pig in the plastic bag provided.
2. Expel excess air from the bag, and tie it shut.
3. Write your *name* and *section* on the tag provided, and attach it to the bag. Your instructor will indicate where the bags are to be stored until the next laboratory period.

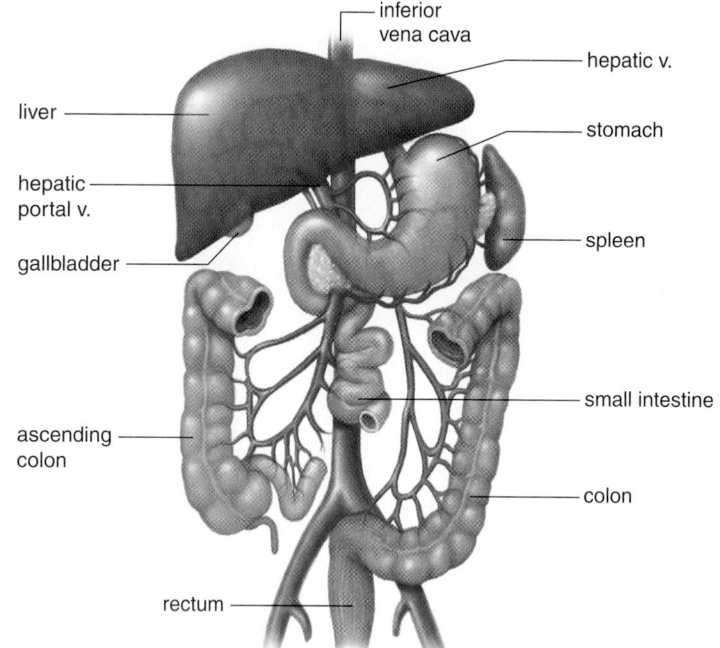

Figure 16.13 Hepatic portal system.
The hepatic portal vein takes the products of digestion from the digestive system to the liver, where they are processed. The hepatic veins take blood from the liver to the posterior vena cava. (v. = vein.)

4. Clean the dissecting tray and tools, and return them to their proper location.
5. Wipe off your goggles.
6. Wash your hands.

Laboratory Review 16

_______________ 1. Which structure in the urinary system carries urine to the bladder?

_______________ 2. Which structure in the urinary system receives urine from the bladder?

_______________ 3. What is the outermost portion of the kidney?

_______________ 4. Where are the testes located in human males?

_______________ 5. What is the function of the vas deferens?

_______________ 6. What is the function of the prostate gland?

_______________ 7. Where are the ovaries located?

_______________ 8. What is the function of the uterus?

_______________ 9. What is the function of the ovaries?

_______________ 10. The vas deferens in males compares with which structure in females?

_______________ 11. Which type of mammal, a pig or a human, has uterine horns?

_______________ 12. What organ in males is analogous to the vagina?

_______________ 13. Where are sperm produced in the testes?

_______________ 14. What structure in the ovary contains the developing oocyte?

_______________ 15. The lungs and heart are located in which body cavity?

_______________ 16. Name a circulatory pathway that goes from one capillary bed to another without passing through the heart.

_______________ 17. Name two glands that add fluid to semen after sperm reach the urethra.

Thought Questions

18. On the basis of anatomy, explain why the urethra is part of both the urinary and reproductive systems in males.

19. A vasectomy is a procedure in which the vas deferens are severed. Why would such a procedure cause sterility?

LABORATORY

17

Homeostasis

Learning Outcomes

17.1 Lungs
- Explain how the anatomy of the lungs contributes to homeostasis.
- Describe the microscopic anatomy of the lungs and the role of the alveoli in gas exchange.

17.2 Liver
- Explain how the anatomy of the liver contributes to homeostasis.
- Discuss the path of blood from the intestines, through the liver, and to the heart.

17.3 Kidneys
- Explain how the anatomy of the kidneys contributes to homeostasis.
- Discuss the path of blood from the heart to the kidneys, around a nephron, and back to the heart.
- Compare the three steps in urine formation and relate these events to the parts of a nephron.
- Predict whether substances will be in the filtrate and/or urine, and justify your conclusions.

17.4 Capillary Exchange in the Tissues
- Describe the exchange of molecules across a capillary wall and the mechanisms involved in this exchange.

Planning Ahead

The regulation of glucose level discussed in Section 17.2 (see page 220) requires a boiling water bath. Your instructor may advise you to begin the boiling water bath early.

Introduction

Homeostasis refers to the dynamic equilibrium of the body's internal environment. The internal environment of vertebrates, including humans, consists of blood and tissue fluid. The cells take nutrients from tissue fluid and return their waste molecules to it. Tissue fluid, in turn, exchanges molecules with the blood. This is called capillary exchange. All internal organs contribute to homeostasis, but this laboratory specifically examines the contributions of the lungs, liver, and kidneys (Fig. 17.1).

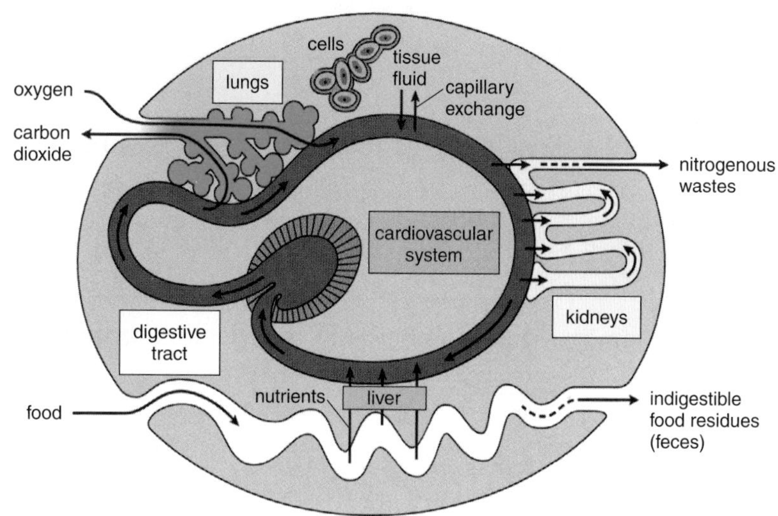

Figure 17.1 Contributions of organs to homeostasis.
The lungs contribute to homeostasis because they carry on gas exchange. The kidneys excrete nitrogenous wastes, and the liver, in association with the digestive tract, adds nutrients to the blood.

17.1 Lungs

Air moves from the nasal passages to the trachea, bronchi, bronchioles, and finally, lungs. The right and left lungs lie in the thoracic cavity on either side of the heart.

Lung Structure

A **lung** is a spongy organ consisting of irregularly shaped air spaces called **alveoli** (sing., *alveolus*). The alveoli are lined with a single layer of squamous epithelium and are supported by a mesh of fine, elastic fibers. The alveoli are surrounded by a rich network of tiny blood vessels called **pulmonary capillaries.**

Observation: Lung Structure

1. Observe a prepared slide of a stained section of a lung. In stained slides, the nuclei of the cells forming the thin alveolar walls appear purple or dark blue (Fig. 17.2*a*).
2. Look for areas with groups of red- or orange-colored, disc-shaped **erythrocytes** (red blood cells). When these appear in strings, you are looking at capillary vessels in side view.
3. In some part of the slide, you may even observe an artery. Thicker, circular or oval structures with a lumen (cavity) are cross sections of **bronchioles,** tubular pathways through which air reaches the air spaces.

Lung Function

Oxygen concentration in the air in alveoli is *greater* than in the blood in pulmonary capillaries. By the same token, carbon dioxide concentration in the air in alveoli is *less* than in the blood in pulmonary capillaries. Gas exchange in the lungs takes place by diffusion as gases move along a **concentration gradient** from greater to lesser concentration.

During gas exchange in the lungs, carbon dioxide (CO_2) leaves the blood and enters the alveoli, and oxygen (O_2) leaves the alveoli and enters the blood. Label Figure 17.2*b* to show gas exchange in the lungs. State one way the lungs contribute to homeostasis. ______________________________

Carbon Dioxide Transport and Release

Carbon dioxide is carried in the blood as bicarbonate ions (HCO_3^-):

$$CO_2 + H_2O \rightarrow \underset{\text{carbonic acid}}{H_2CO_3} \rightarrow \underset{\text{bicarbonate ion}}{HCO_3^-} + H^+$$

1. H^+ increases the acidity of the blood. Is blood more acidic when it is carrying carbon dioxide? ________ Explain. ____

   ____

2. As carbon dioxide leaves the blood, the following reaction is driven to the right:

   $$HCO_3^- + H^+ \rightarrow H_2CO_3 \rightarrow H_2O + CO_2$$

   Is blood less acidic when the carbon dioxide exits? ________ Explain. ____________________

   ____

3. In summary, state another way the lungs contribute to homeostasis. ____________________

   ____

Figure 17.2 Gas exchange in the lungs.
a. A photomicrograph shows that the lungs contain many air sacs called alveoli. The alveoli are surrounded by blood capillaries. **b.** During gas exchange, carbon dioxide leaves the blood and enters the alveoli; oxygen leaves the alveoli and enters the blood. Label the arrows with O_2 and CO_2 to show gas exchange in the lungs.

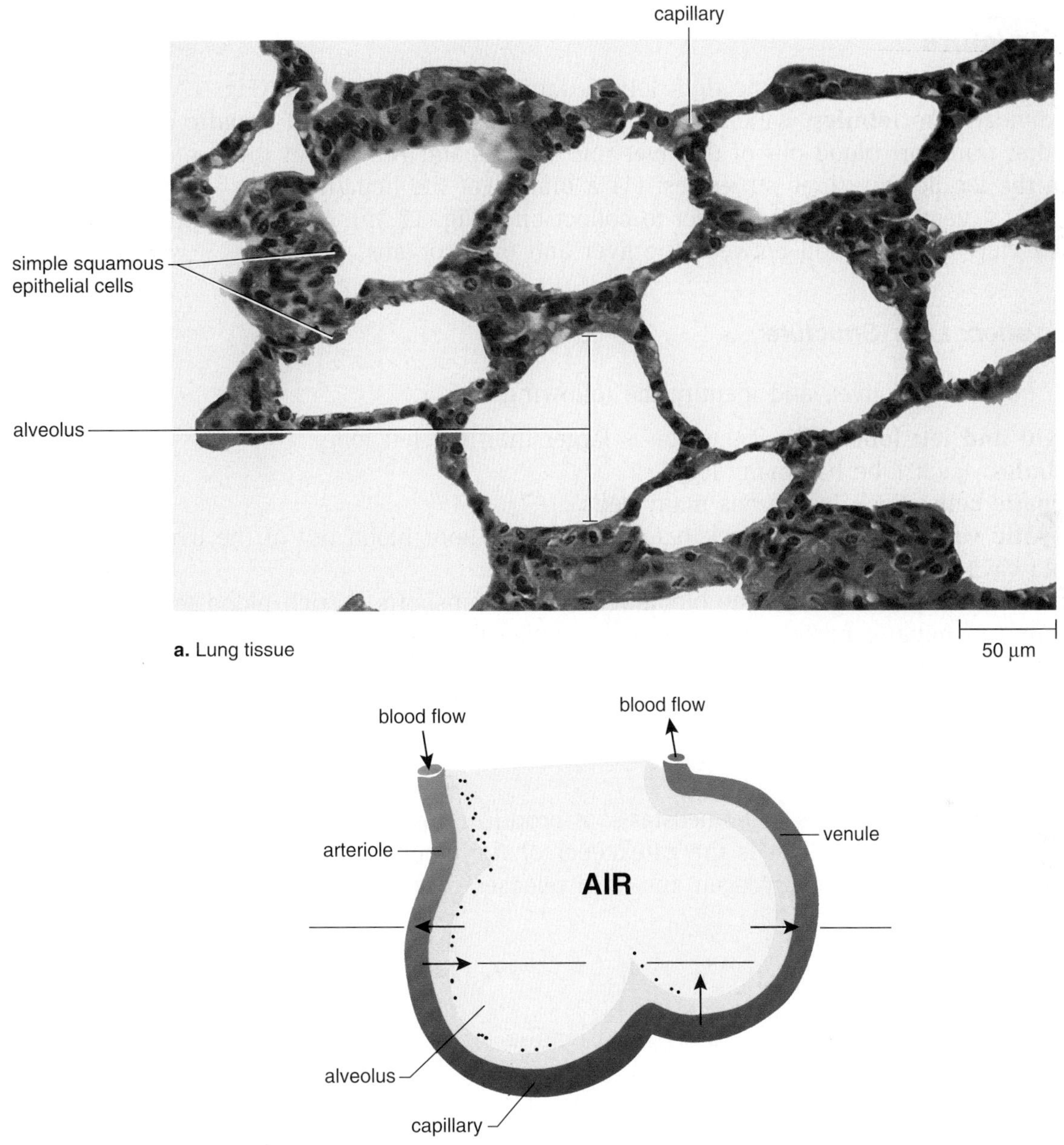

a. Lung tissue

b. Alveolus

17.2 Liver

The **liver,** which is the largest organ in the body, lies mainly in the upper right quadrant of the abdominal cavity under the diaphragm.

Liver Structure

The liver has two main **lobes;** the right lobe is larger than the left lobe (Fig. 17.3*a*). The lobes are further divided into **lobules,** which contain the cells of the liver, called **hepatic cells.** Small blood vessels that transport blood out of the liver into the hepatic vein are at the center of each lobule. Between the lobules are three structures: (1) a branch of the hepatic artery, (2) a branch of the hepatic portal vein, and (3) a bile duct to collect bile (Fig. 17.3*b*). Refer to Figure 17.3*c* for an expanded view of circulation between the liver and other organs.

Observation: Liver Structure

Study a model of the liver, and identify the following:

1. **Right and left lobes:** The right lobe is larger than the left lobe.
2. **Lobules:** Each lobe has many lobules.
3. **Hepatic cells:** Each lobule has many cells.
4. **Hepatic vein:** The blood vessel that transports O_2-poor blood out of the liver to the inferior vena cava.
5. **Branch of hepatic artery:** The blood vessel that transports O_2-rich blood to the liver.
6. **Branch of hepatic portal vein:** The blood vessel that transports blood containing nutrients from the intestine to the liver.
7. **Bile duct:** The passageway for bile going to the gallbladder.

Liver Function

The liver has many functions in homeostasis. It produces **urea,** the primary nitrogenous end product of humans. In general, the liver is the gatekeeper of the blood—it regulates blood composition. For example, it stores glucose as glycogen and then releases glucose to keep the blood glucose concentration at about 0.1%.

Urea Formation

The liver removes amino groups ($—NH_2$) from amino acids and converts them to urea, a relatively nontoxic nitrogenous end product.

1. In the chemical formula for urea that follows, circle the portions that would have come from amino groups:

$$NH_2—\overset{\overset{\displaystyle O}{\|}}{C}—NH_2$$

2. State one way the liver contributes to homeostasis. ______________________________

____

Regulation of Blood Glucose Level

After you eat, the liver stores excess glucose as glycogen. In between eating, glycogen is broken down by liver cells to produce glucose, and this glucose enters the bloodstream. The hormone insulin, made by the pancreas, promotes the uptake and storage of glucose in the liver. The hormone glucagon promotes the breakdown of glycogen and the release of glucose.

Figure 17.3 Anatomy of the liver.
a. The liver, a large organ in the abdominal cavity, plays a primary role in homeostasis. **b**. Each lobule is served by branches of the hepatic artery that bring O_2-rich blood to the liver and by a branch of the hepatic portal vein that brings nutrients to the liver from the intestines. The central vein of each lobule takes blood to the hepatic vein, which enters the posterior vena cava. Bile canals enter bile ducts taking bile away from each lobule for storage in the gallbladder. **c.** The hepatic portal system consists of the hepatic portal vein and hepatic vein.

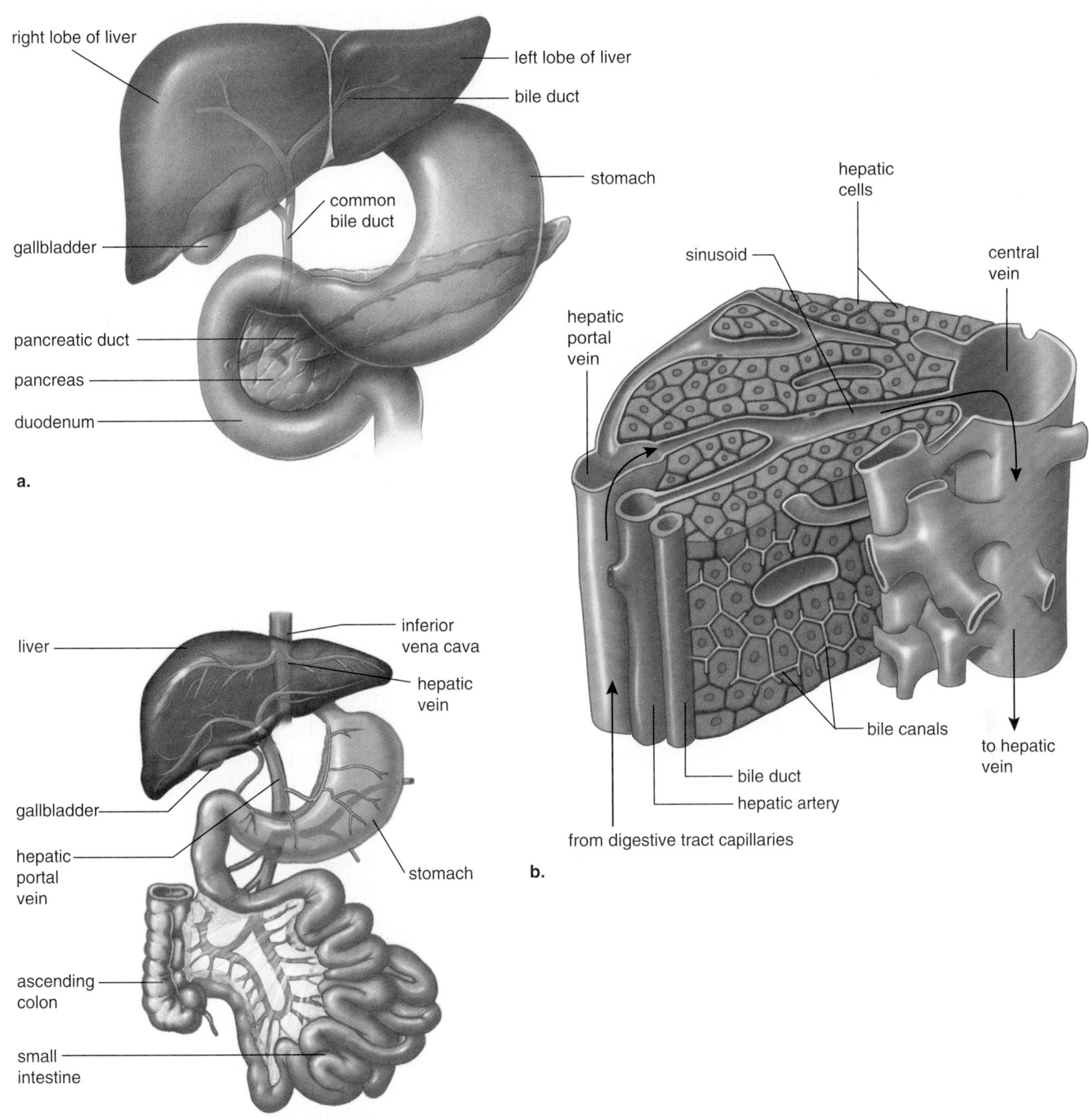

1. In the following equation, write the phrases *after eating* and *before eating* above or below the appropriate arrow:

$$\text{glucose} \xleftarrow{\hspace{3cm}} \text{glycogen}$$

2. Now add the words *insulin* and *glucagon* to the appropriate arrow in the equation. If glucose is excreted in the urine, instead of being stored, the individual has the medical condition called **diabetes mellitus,** commonly known simply as diabetes. In type 1 diabetes, the pancreas is no longer making insulin; in type 2 diabetes, the plasma membrane receptors are unable to bind properly to insulin. In type 1 diabetes, but not type 2, ketones (strong organic acids), which are a breakdown product of fat metabolism, also appear in the urine.
3. State another way in which the liver contributes to homeostasis. ______________________
4. Based upon the information presented in this section, how might type 1 diabetes be treated?

____

Experimental Procedure: Blood Glucose Level After Eating

Study the diagram of the human cardiovascular system in Figure 14.1, and trace the path of blood from the aorta to the vena cava via the intestine and liver. Simulated serum samples have been prepared to correspond to these blood vessels in a person who ate a short time ago:

A_1: Serum from a mesenteric artery. The mesenteric arteries take blood from the aorta to the intestine.

B_1: Serum from the hepatic portal vein, which lies between the intestine and the liver.

C_1: Serum from the hepatic vein, which takes blood from the liver to the inferior vena cava.

Caution: Use protective eyewear when performing this experiment.

1. With a wax pencil, label three test tubes A_1, B_1, and C_1, and mark them at 1 cm and 2 cm.
2. Fill test tube A_1 to the 1 cm mark with *serum A_1* and to the 2 cm mark with *Benedict's reagent.*
3. Fill test tube B_1 to the 1 cm mark with *serum B_1* and to the 2 cm mark with *Benedict's reagent.*
4. Fill test tube C_1 to the 1 cm mark with *serum C_1* and to the 2 cm mark with *Benedict's reagent.*
5. Place all three test tubes in the water bath *at the same time.* Heat the tubes in the same boiling water bath for five minutes.
6. Note any color change in the test tubes, and record the color and your conclusions in Table 17.1. The tube that shows color first has the most glucose, and so forth. Use the list to the right to assist you in making your conclusions:

Color change	Amount of glucose
Color is still blue	None
Green	Very low
Yellow-orange	Moderate
Orange	High
Orange-red	Very high

Table 17.1 Blood Glucose Level After Eating

Test Tubes	Color (after heating)	Conclusion
A_1 (mesenteric artery)		
B_1 (hepatic portal vein)		
C_1 (hepatic vein)		

Conclusions

- Which blood vessel—a mesenteric artery, the hepatic portal vein, or the hepatic vein—contains the most glucose after eating? ______
- Why do you suppose that the hepatic vein does not contain as much glucose as the hepatic portal vein after eating? ______

______

Experimental Procedure: Blood Glucose Level Before Eating

Simulated serum samples have been prepared to correspond to these blood vessels in a person who has not eaten for some time:

A_2: Serum from a mesenteric artery
B_2: Serum from the hepatic portal vein
C_2: Serum from the hepatic vein

Caution: Use protective eyewear when performing this experiment.

1. With a wax pencil, label three test tubes A_2, B_2, and C_2, and mark them at 1 cm and 2 cm.
2. Fill test tube A_2 to the 1 cm mark with *serum* A_2 and to the 2 cm mark with *Benedict's reagent*.
3. Fill test tube B_2 to the 1 cm mark with *serum* B_2 and to the 2 cm mark with *Benedict's reagent*.
4. Fill test tube C_2 to the 1 cm mark with *serum* C_2 and to the 2 cm mark with *Benedict's reagent*.
5. Heat the tubes in the same boiling water bath for five minutes.
6. Note any color change in the test tubes, and record the color and your conclusions in Table 17.2. Use the following list to assist you in making your conclusions:

Color change	Amount of glucose
Color is still blue	None
Green	Very low
Yellow-orange	Moderate
Orange	High
Orange-red	Very high

Table 17.2 Blood Glucose Level Before Eating

Test Tubes	Color (after heating)	Conclusion
A_2 (mesenteric artery)		
B_2 (hepatic portal vein)		
C_2 (hepatic vein)		

Conclusions

- Which blood vessel—a mesenteric artery, the hepatic portal vein, or the hepatic vein—contains the most glucose before eating? ______
- Why do you suppose that the hepatic vein now contains more glucose than the hepatic portal vein? ______

17.3 Kidneys

The **kidneys** are bean-shaped organs that lie along the dorsal wall of the abdominal cavity.

Kidney Structure

Figure 17.4 shows the macroscopic and microscopic structure of a kidney. The macroscopic structure of a kidney is due to the placement of over 1 million **nephrons.** Nephrons are tubules that do the work of producing urine.

Figure 17.4 Longitudinal section of a kidney.
a. The kidneys are served by the renal artery and renal vein. **b**. Macroscopically, a kidney has three parts: renal cortex, renal medulla, and renal pelvis. **c**. Microscopically, each kidney contains over a million nephrons.

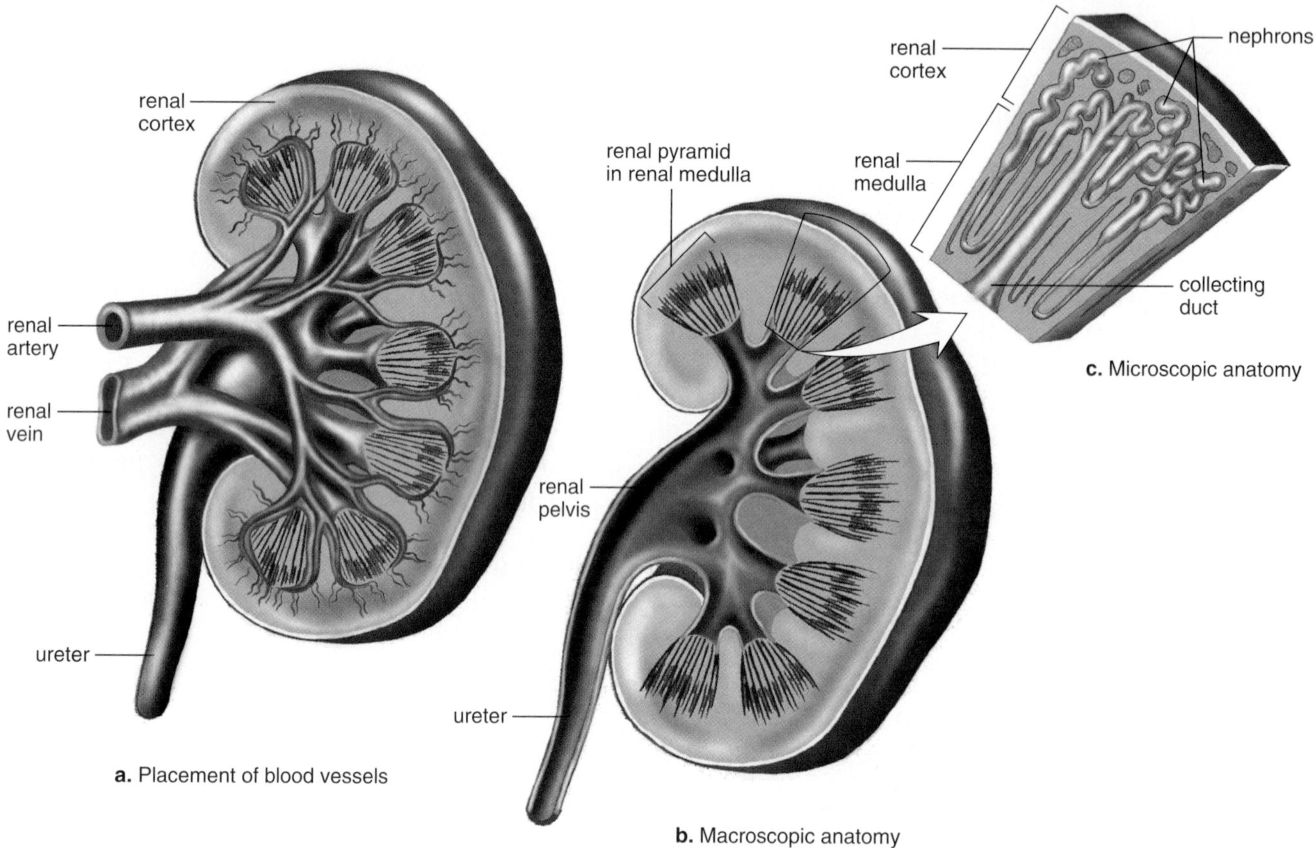

Observation: Kidney Model

Study a model of a kidney, and with the help of Figure 17.4, locate the following:

1. **Renal cortex:** a granular region
2. **Renal medulla:** contains the renal pyramids
3. **Renal pelvis:** where urine collects

Nephron Structure and Circulation

Figure 17.5 shows that the **afferent arteriole** enters the **glomerulus,** which is situated within the cup-shaped **glomerular capsule** (Bowman's capsule). The **efferent arteriole** leaves the glomerular capsule and enters the **peritubular capillary network** that surrounds the **proximal convoluted tubule,** the **loop of the nephron** (loop of Henle), and the **distal convoluted tubule.** Distal convoluted tubules from several nephrons enter one collecting duct.

Macroscopic and microscopic studies of kidney anatomy show that the glomerular capsule and convoluted tubules are in the renal cortex, while the loop of the nephron and the collecting ducts are in the renal medulla, accounting for the striated appearance of the renal pyramids. The collecting ducts enter the renal pelvis.

Figure 17.5 Structure of a nephron and its blood supply.
As the blood moves through the blood vessels about a nephron, substances exit and/or enter the blood from portions of the nephron.

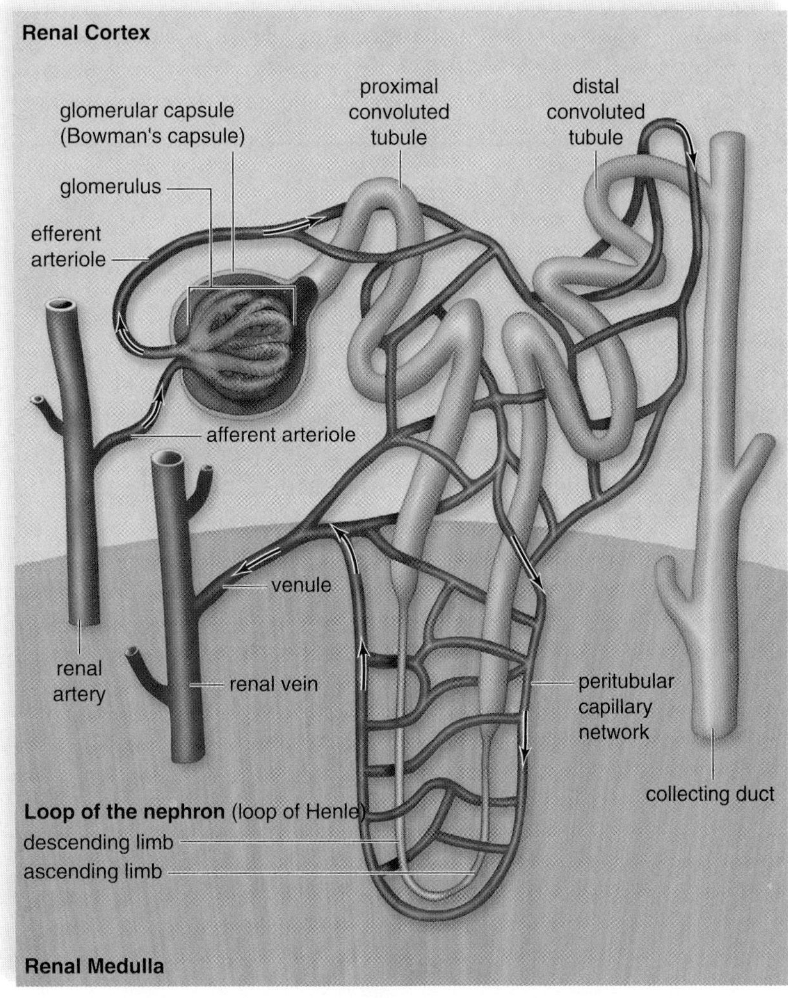

1. With the help of Figure 17.5, list the four parts of a nephron, and tell whether they are located in the renal cortex or the renal medulla (assume that the nephron has a long loop).

___

___

___

___

2. With the help of Figure 17.5 and Table 17.3, trace the path of blood toward, around, and away from an individual nephron. ____

____

____

Table 17.3 Circulation of Blood Around a Nephron

Name of Structure	Significance
Afferent arteriole	Brings arteriolar blood toward glomerular capsule
Glomerulus	Capillary tuft enveloped by glomerular capsule
Efferent arteriole	Takes arteriolar blood away from glomerular capsule
Peritubular capillary network	Capillary bed that envelops the rest of the nephron
Veins	Take venous blood away from the nephron

Kidney Function

The kidneys contribute to homeostasis by excreting nitrogenous wastes and by regulating blood volume, blood pressure, and pH.

Urine formation requires three steps:

1. **Glomerular filtration** requires the movement of molecules outward from the glomerulus to the inside of the glomerular capsule. Blood pressure forces small molecules into the glomerular capsule. Label the arrow in Figure 17.6 that marks the location of glomerular filtration.
2. **Tubular reabsorption** requires the movement of molecules primarily from the proximal convoluted tubule to the peritubular capillary network. Nutrient molecules and water in the nephron filtrate are returned to the blood. Label the arrow in Figure 17.6 that refers to tubular reabsorption.
3. **Tubular secretion** requires the movement of molecules from the peritubular capillary network to the nephron. Waste molecules remaining in the blood after glomerular filtration are moved into the nephron. Label the arrow in Figure 17.6 that refers to tubular secretion.

The two parts not considered in this discussion—the loop of the nephron and the collecting duct—are both active in water reabsorption. Regulation of water reabsorption maintains blood volume at the proper level.

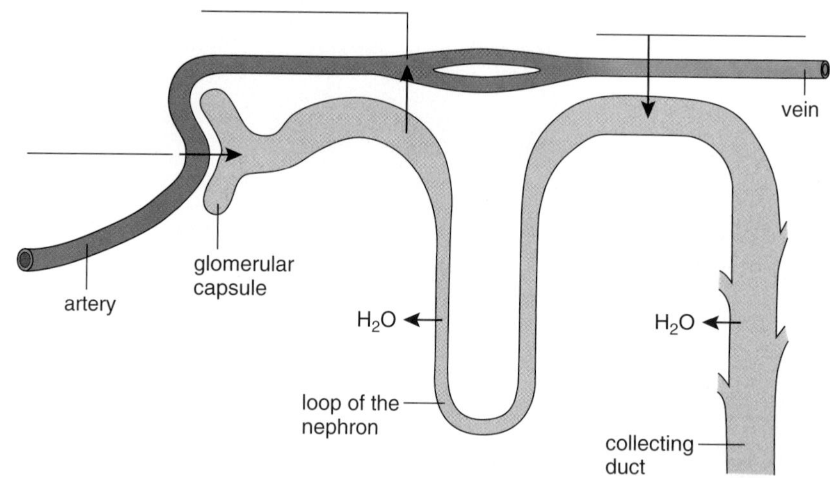

Figure 17.6 Urine formation. The three steps in urine formation are glomerular filtration, tubular reabsorption, and tubular secretion. Label the arrows as directed above.

Glomerular Filtration

Blood entering the glomerulus contains cells, proteins, glucose, amino acids, salts, urea, and water. Blood pressure causes small molecules of glucose, amino acids, salts, urea, and water to exit the blood and enter the glomerular capsule. The fluid in the glomerular capsule is called the **filtrate.**

1. In the list that follows, draw an arrow from left to right for all those molecules that leave the glomerulus and enter the glomerular capsule:

Glomerulus	**Glomerular Capsule (Filtrate)**
Cells	
Proteins	
Glucose	
Amino acids	
Salts	
Urea	
Water	

2. What substances are too large to leave the glomerulus and enter the glomerular capsule?

   ____

   These substances remain in the blood.

Tubular Reabsorption

When the filtrate enters the proximal convoluted tubule, it contains water, glucose, amino acids, urea, and salts. Enough water and salts are passively reabsorbed to maintain blood volume and pH.

1. What would happen to blood volume and blood pressure if water were not reabsorbed? (Consider what happens to water pressure when the volume of water moving through a hose is reduced.)

   ____

   ____

2. The cells that line the proximal convoluted tubule are also engaged in active transport and usually completely reabsorb nutrients (glucose and amino acids). What would happen to cells if the body lost all its nutrients by way of the kidneys? ______________________

   ____

3. In the list that follows, draw an arrow from left to right for all those molecules that are passively reabsorbed into the blood. Use darker arrows for those that are actively reabsorbed.

Proximal Convoluted Tubule	**Peritubular Capillary**
Water	
Glucose	
Amino acids	
Urea	
Salts	

4. What molecule is reabsorbed the least? __________.

Tubular Secretion

During tubular secretion, certain substances—for example, penicillin and histamine—are actively secreted from the peritubular capillary into the fluid of the tubule. Also, hydrogen ions and ammonia are secreted as necessary to maintain blood homeostasis.

Summary of Urine Formation

For each substance listed at the left in Table 17.4, place an X in the appropriate column(s) to indicate where you expect the substance to be present.

Table 17.4 Urine Constituents

Substance	In Blood of Glomerulus	In Filtrate	In Urine
Protein (albumin)			
Glucose			
Urea			
Water			

Answer the following questions.

1. What molecule is reabsorbed from the collecting duct so that urine becomes hypertonic?
____
2. Based on Table 17.4, state two ways the kidneys contribute to homeostasis. _______________
____
3. Which organ—the lung, liver, or kidney—makes urea? _______________
4. Which organ produces urea? _______________
5. Which organ excretes urine? _______________
6. If the blood is more acidic than normal, what pH do you suppose the urine will be? __________
7. If the blood is more basic than normal, what pH do you suppose the urine will be? __________
8. State another way the kidneys contribute to homeostasis. _______________
____

Urinalysis

Urinalysis can help diagnose a patient's illness. The procedure is easily performed with a Chemstrip test strip, which has indicator spots that produce specific color reactions when certain substances are present in the urine.

Experimental Procedure: Urinalysis

Suppose a patient complains of excessive thirst and urination, loss of weight despite an intake of sweets, and feelings of being tired and run-down. A urinalysis has been ordered, and you are to test the urine. (In this case, you will be testing simulated urine, just as you tested simulated blood sera earlier in this lab.)

1. Review "Regulation of Blood Glucose Level," step 2, on page 220. Obtain a Chemstrip urine test strip (Fig. 17.7) that tests for leukocytes, pH, protein, glucose, ketones, and blood, which are noted in the "Tests For" column of the figure.

2. The color key on the diagnostic color chart or on the Chemstrip vial label will explain what the color changes mean in terms of the pH level and amount of each substance present in the urine sample. You will use these color blocks to read the results of your test.
3. Obtain a "specimen container of the patient's urine."
4. Briefly (no longer than 1 second) dip the test strip into the urine. Ensure that the chemically treated patches on the test strip are totally immersed.
5. Draw the edge of the strip along the rim of the specimen container to remove excess urine.
6. Turn the test strip on its side, and tap once on a piece of absorbent paper to remove any remaining urine and to prevent the possible mixing of chemicals.
7. After 60 seconds, read the tests as follows: *Hold the strip close to the color blocks on the diagnostic color chart (Figure 17.7) or vial label, and match carefully,* ensuring that the strip is properly oriented to the color chart. For each test, circle the colors in Figure 17.7 that match those on your tested Chemstrip.

Figure 17.7 Urinalysis test.
A Chemstrip test strip can help determine illness in a patient by detecting substances in the urine.

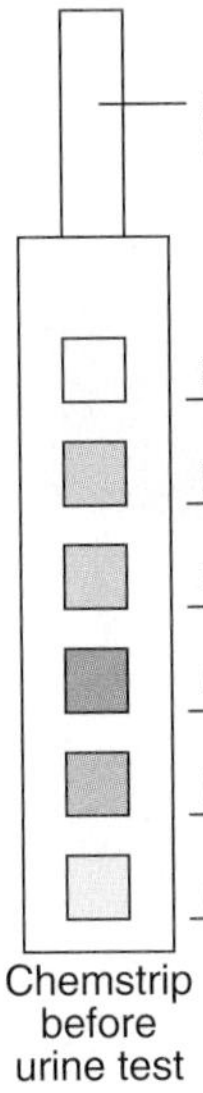

strip handle

Normal	Tests For:	Results
negative	leukocytes	
pH 5	pH	
negative	protein	
normal	glucose	
negative	ketones	
negative	blood	

Chemstrip before urine test

Conclusions

- According to your results, what condition might the patient have? ________
  Explain. ________
- Given that the patient's blood contains excess glucose, why is the patient suffering from excessive thirst and urination? ________
- Since neither the liver nor the body cells are taking up glucose, why is the patient tired? ________
- The metabolism of fat can explain the low pH of the urine. Why? ________

17.4 Capillary Exchange in the Tissues

Tissue fluid is continually created and refreshed at the capillaries when certain molecules leave the blood and others are picked up by the blood.

1. In Figure 17.8, write *oxygen* and *glucose* next to the appropriate arrow. Write *wastes* and *carbon dioxide* next to the appropriate arrow.
2. What type of pressure causes water to exit from the arterial side of the capillary? ______________
3. What type of pressure causes water to enter the venous side of the capillary? ______________

Figure 17.8 Capillary exchange.
A capillary, illustrating the exchange that takes place across a capillary wall.

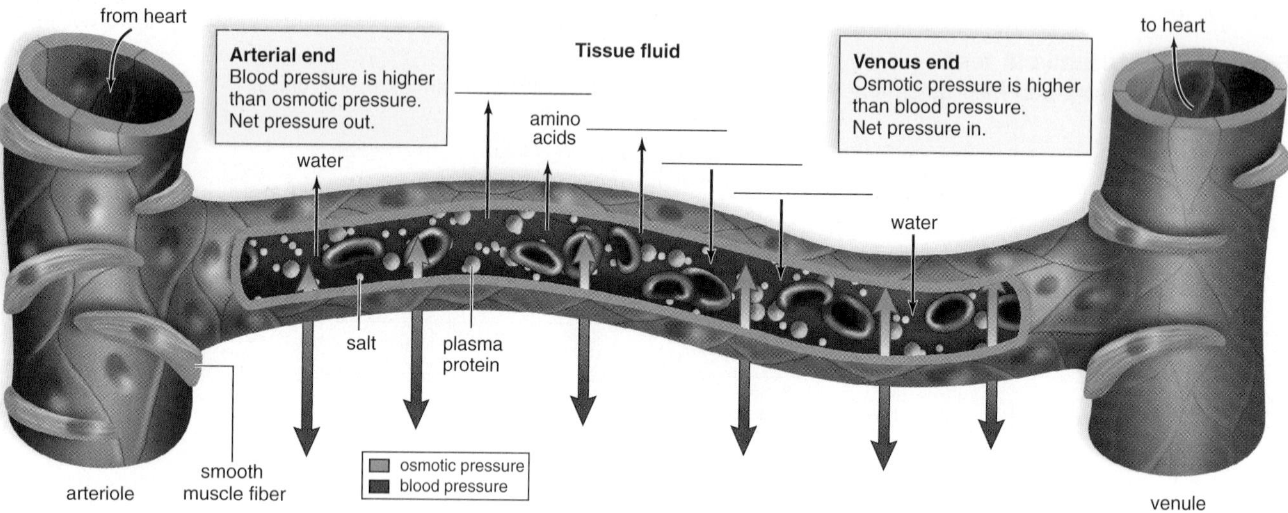

Conclusions

- What is the function of capillaries? ____
- Why are cells always in need of glucose and oxygen? ______________________________
- Why are cells always producing carbon dioxide? __________________________________

Summary of Homeostasis

As noted at the beginning of this laboratory, homeostasis is the dynamic equilibrium of the body's internal environment, which is the blood and tissue fluid surrounding tissue cells. The lungs and kidneys have boundaries that interact with the external environment in order to refresh blood. The liver also regulates blood content. *Fill in the table below to show the activities of these three organs.*

Processes	Lungs	Liver	Kidneys
Gas exchange	a.	______	______
pH maintenance	b.	______	f.
Glucose level	______	d.	g.
Waste removal	c.	e.	h.
Blood volume	______	______	i.

- Which of these organs contributes most to homeostasis? ______________

Laboratory Review 17

______________ 1. What process accounts for gas exchange in the lungs?

______________ 2. What molecule is removed by the lungs?

______________ 3. What are the air spaces in the lungs called?

______________ 4. What blood vessel lies between the intestines and the liver?

______________ 5. In what form is glucose stored in the liver?

______________ 6. The liver removes the amino group from amino acids to form what molecule?

______________ 7. The hepatic vein enters what blood vessel?

______________ 8. When molecules leave the glomerulus, they enter what portion of the nephron?

______________ 9. Name a substance that is in the glomerular filtrate but not in the urine.

______________ 10. Glucose in the urine indicates that a person may have what condition?

______________ 11. Name the process by which molecules move from the proximal convoluted tubule into the blood.

______________ 12. Where does urine collect before exiting the kidney?

______________ 13. Does venous blood in the tissues contain more or less carbon dioxide than arterial blood?

______________ 14. What type of pressure causes water to exit from the arterial side of the capillary?

______________ 15. What occurs during tubular reabsorption?

Thought Questions

16. Which systemic blood vessel would you expect to have a high glucose content immediately after eating? Explain.

17. In what ways do the kidneys aid homeostasis?

18. What might happen to the pH of the blood when a person hyperventilates?

LABORATORY

18

Nervous System and Senses

Learning Outcomes

18.1 The Mammalian Brain

- Using preserved specimens, photographs, other images, or models, identify the parts of the brain studied, and state the functions of each part.
- State several differences between a sheep brain and a human brain.

18.2 Spinal Nerves and Spinal Cord

- Describe the anatomy of the spinal cord and tell how the cord functions in relation to the brain and spinal nerves.
- Describe the anatomy and physiology of a spinal reflex arc.

18.3 The Human Eye

- Using preserved specimens, photographs, other images, or models, identify the parts of the eye and state a function for each part.

18.4 The Human Ear

- Using photographs, other images, or models, identify the parts of the ear and state a function for each part.

18.5 Sensory Receptors in Human Skin

- Describe the anatomy of the human skin and explain the distribution and function of sensory receptors.
- Relate the abundance of touch receptors to the ability to distinguish between two different touch points.

18.6 Human Chemoreceptors

- Relate the ability to distinguish tastes to the distribution of taste receptors on the human tongue.
- Relate the ability to distinguish foods to the senses of smell and taste.

Introduction

Nervous tissue, which is found in the organs of the nervous system, contains **neurons** (nerve cells) that conduct nerve impulses and glial cells that support them. The nervous system is divided into the **central nervous system** and the **peripheral nervous system.** The central nervous system includes the brain and the spinal cord. The peripheral nervous system includes the **cranial nerves,** which take nerve impulses (messages) to and from the brain, and the **spinal nerves,** which take nerve impulses to and from the spinal cord. (Neuron cell bodies are located in the central nervous system or in ganglia occurring along the nerves.)

Sense organs contain sensory receptors that are sensitive to a particular type of environmental stimulus. After receptors receive stimuli, they generate nerve impulses that travel to the spinal cord and/or brain via spinal or cranial nerves. The spinal cord and most regions of the brain interpret nerve impulses below the level of consciousness. The **cerebrum,** the highest level of the brain, is responsible for sensation. After interpretation, the spinal cord or brain sends nerve impulses to effectors (muscles or glands) that bring about a response to the stimulus.

Caution: Wear protective eyewear and latex gloves when handling preserved animal organs. Exercise caution when using scalpel and wash hands thoroughly upon completion of experiment.

18.1 The Mammalian Brain

The brain is the enlarged anterior end of the spinal cord containing parts and centers that receive input from, and can command other regions of, the nervous system.

Preserved Sheep Brain

The sheep brain (Fig. 18.1) is often used to study the mammalian brain. It is easily available and large enough that individual parts can be identified.

Observation: Preserved Sheep Brain

Examine the exterior and a midsaggital (longitudinal) section of a preserved sheep brain or a model of the human brain, and with the help of Figure 18.1, identify the following.

1. **Ventricles:** Interconnecting spaces that produce and serve as a reservoir for cerebrospinal fluid, which cushions the brain. Toward the anterior, note the large first ventricle (on one midsaggital section) and the second ventricle (on the other midsaggital section). Trace the second ventricle to the third and then the fourth ventricles.
2. **Medulla oblongata** (or simply **medulla**): The most posterior portion of the brain stem. It controls internal organs; for example, cardiac and breathing control centers are present in the medulla. Nerve impulses pass from the spinal cord through the medulla to higher brain regions.
3. **Pons:** The ventral, bulblike enlargement on the brain stem. It serves as a passageway for nerve impulses running between the medulla and the higher brain regions.
4. **Midbrain:** Anterior to the pons, the midbrain serves as a relay station for sensory input and motor output. It also contains a reflex center for eye muscles.
5. **Diencephalon:** The portion of the brain where the third ventricle is located. The hypothalamus and thalamus are also located here.
6. **Hypothalamus:** Forms the floor of the third ventricle and contains control centers for appetite, body temperature, and water balance. Its primary function is homeostasis. The hypothalamus also has centers for pleasure, reproductive behavior, hostility, and pain.
7. **Thalamus:** Two connected lobes located in the roof of the third ventricle. The thalamus is the highest portion of the brain to receive sensory impulses before the cerebrum. It is believed to control which of the received impulses is passed on to the cerebrum. For this reason, the thalamus sometimes is called the "gatekeeper to the cerebrum."
8. **Cerebellum:** Located just posterior to the cerebrum as you observe the brain dorsally, the cerebellum's two lobes make it appear rather like a butterfly. In cross section, the cerebellum has an internal pattern that looks like a tree. The cerebellum coordinates equilibrium and motor activity to produce smooth movements.
9. **Cerebrum:** The most developed area of the brain and responsible for higher mental capabilities. The cerebrum is divided into the right and left **cerebral hemispheres,** which are joined by the **corpus callosum,** a broad sheet of white matter. The outer portion of the cerebrum is highly convoluted and divided into the following surface lobes:
   a. **Frontal lobe:** Controls motor functions and permits voluntary muscle control. It also is responsible for abilities to think, problem solve, and speak.
   b. **Parietal lobe:** Receives information from sensory receptors located in the skin. It also helps in the understanding of speech. A groove called the **central sulcus** separates the frontal lobe from the parietal lobe.
   c. **Occipital lobe:** Interprets visual input and combines visual images with other sensory experiences. The optic nerves split and enter opposite sides of the brain at the optic chiasma, located in the diencephalon.
   d. **Temporal lobe:** Has sensory areas for hearing and smelling. The olfactory bulb contains nerve fibers that communicate with the olfactory cells in the nasal passages and take nerve impulses to the temporal lobe.

Figure 18.1 The sheep brain.

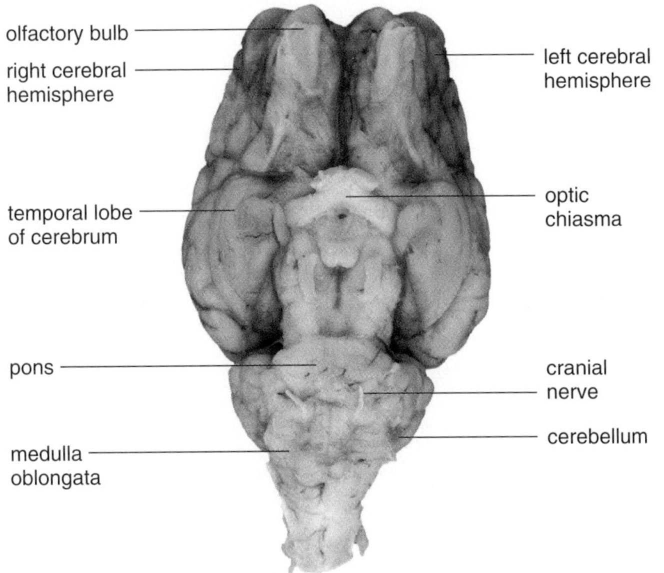

a. Ventral view

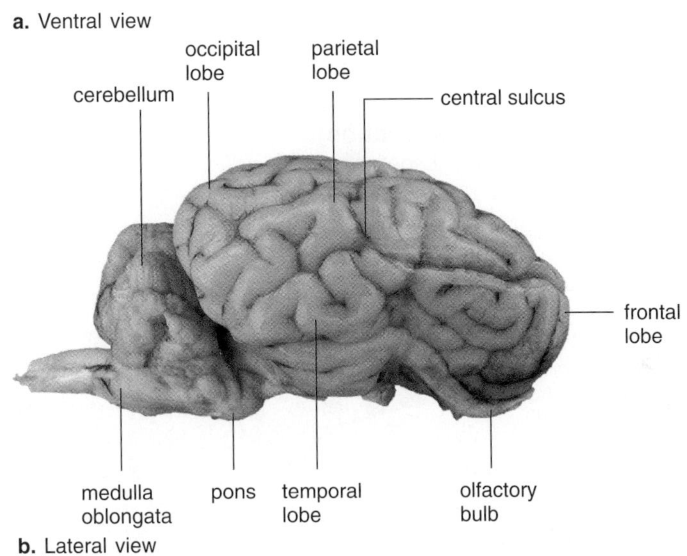

b. Lateral view

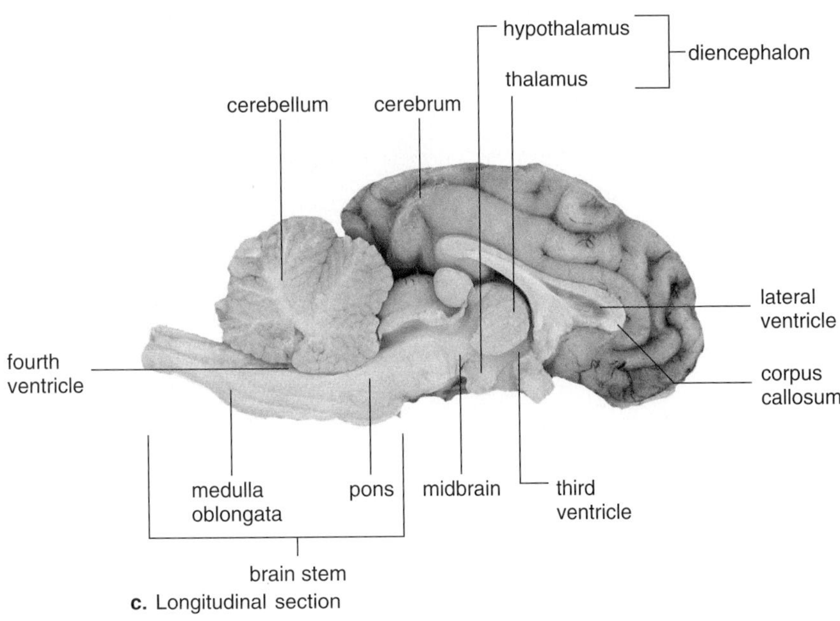

c. Longitudinal section

The Human Brain

Show that the human brain contains the same structures as the sheep brain by labeling Figure 18.2. However, you should also be able to detect some differences when comparing a sheep brain with a human brain.

Observation: The Human Brain

Examine a model of the human brain. State two similarities and two general differences in the structure of the brain between sheep and humans.

Similarities

1. ______________________________

2. ______________________________

Differences

3. ______________________________

4. ______________________________

Figure 18.2 The human brain (longitudinal section).
The cerebrum is larger in humans than in sheep. *Label where indicated.*

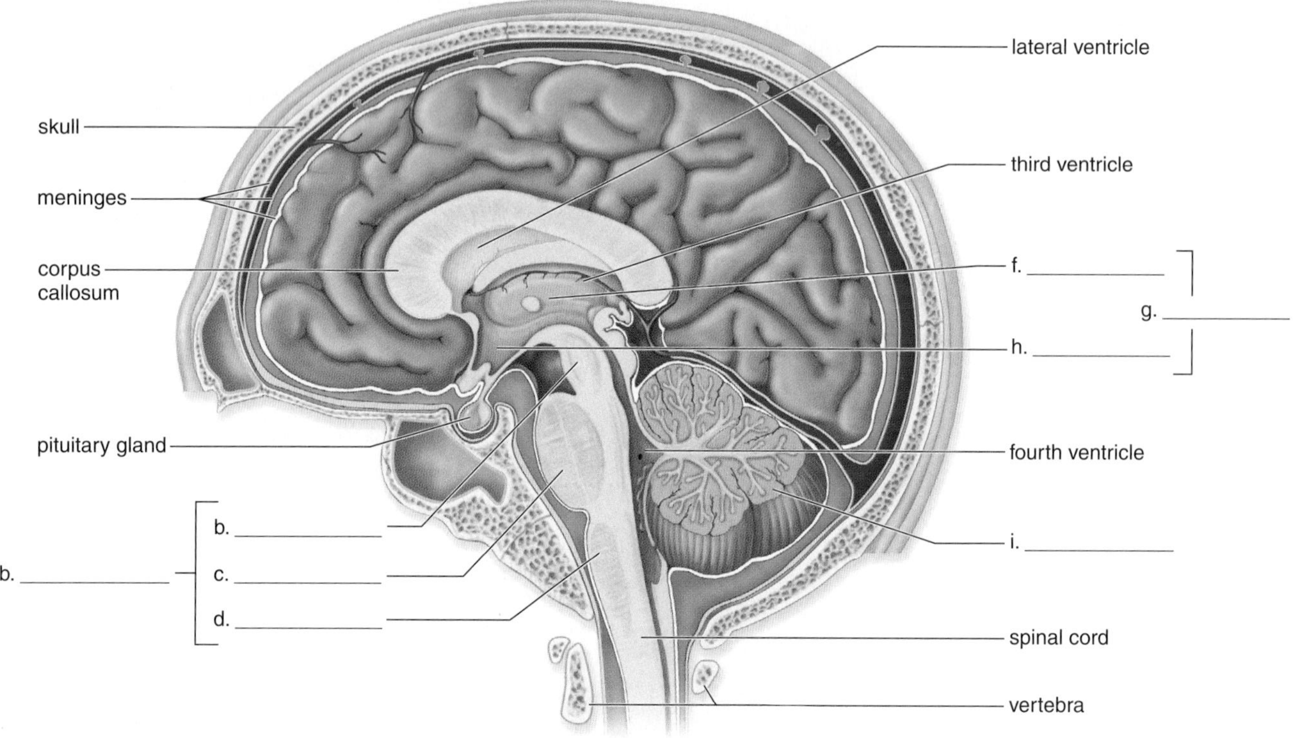

18.2 Spinal Nerves and Spinal Cord

The spinal nerves and spinal cord function below the level of consciousness, the reflex actions allowing quick responses to environmental stimuli without communicating with the brain.

Spinal Nerves

Pairs of spinal nerves are connected to the spinal cord, which lies in the middorsal region of the body and is protected by the vertebral column. Each spinal nerve contains long fibers of sensory neurons and long fibers of motor neurons.

1. **Sensory neuron,** whose long axon takes nerve impulses from a sensory receptor to the spinal cord. Note that the cell body of the sensory neuron is in the dorsal root ganglion. Why is this neuron called a sensory neuron? ______________________________

   ____
2. **Interneuron,** which lies completely within the spinal cord. Some interneurons have long fibers and take nerve impulses to and from the brain. The neuron in Figure 18.3 connects the sensory neuron to the motor neuron. Why is this neuron called an interneuron? ______________________________

   ____
3. **Motor neuron,** whose long axon takes nerve impulses from the spinal cord to an effector—in this case, a muscle. Muscle contraction often follows a response to a stimulus. Why is this neuron called a motor neuron? ______________________________

   ____

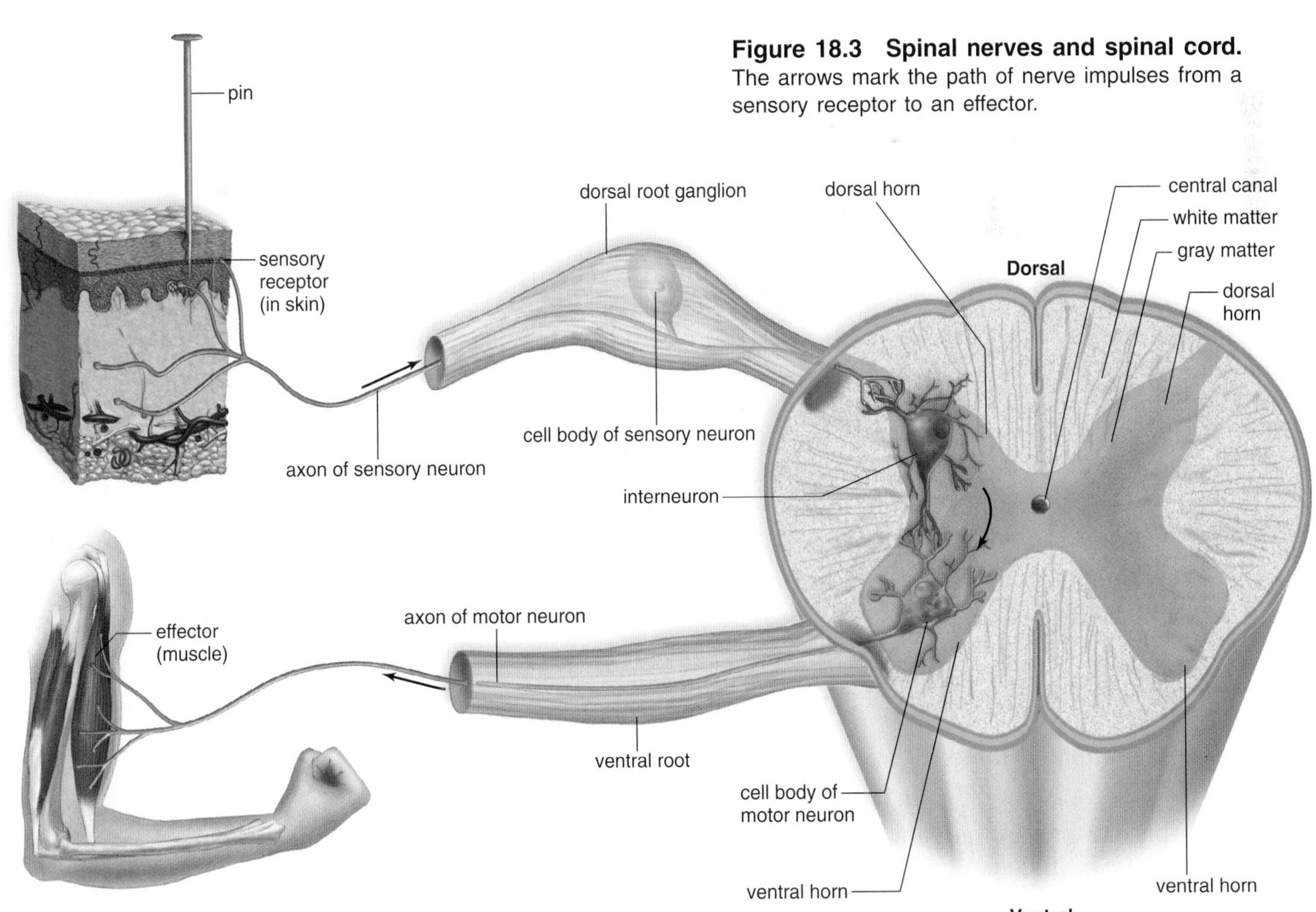

Figure 18.3 Spinal nerves and spinal cord. The arrows mark the path of nerve impulses from a sensory receptor to an effector.

The Spinal Cord

The spinal cord is a part of the central nervous system. It lies in the middorsal region of the body and is protected by the vertebral column.

Observation: The Spinal Cord

1. Examine a prepared slide of a cross section of the spinal cord under the lowest magnification possible. For example, some microscopes are equipped with a very short scanning objective that enlarges about 3.5×, with a total magnification of 35×. If a scanning objective is not available, observe the slide against a white background with the naked eye.
2. Identify the following with the help of Figure 18.4:
   a. **Gray matter:** A central, butterfly-shaped area composed of masses of short nerve fibers, interneurons, and motor neuron cell bodies.
   b. **White matter:** Masses of long fibers that lie outside the gray matter and carry impulses up and down the spinal cord. In living animals, white matter appears white because an insulating myelin sheath surrounds long fibers.

Figure 18.4 The spinal cord. Photomicrograph of spinal cord cross section.

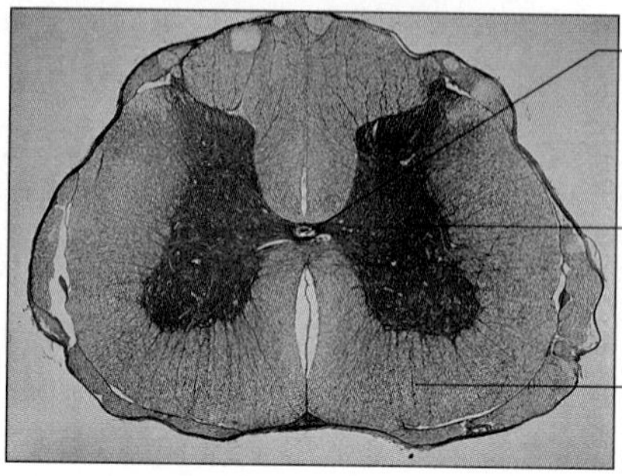

Spinal Reflexes

A **reflex** is an involuntary and predictable response to a given stimulus. When you touch a sharp tack, you immediately withdraw your hand (see Fig. 18.3). When a spinal reflex occurs, a sensory receptor is stimulated and generates nerve impulses that pass along the three neurons mentioned earlier—the sensory neuron, interneuron, and motor neuron—until the effector responds.

Experimental Procedure: Spinal Reflex

Although many reflexes occur in the body, only a tendon reflex is investigated in this Experimental Procedure. Two easily tested tendon reflexes involve the Achilles and **patellar tendons.** When these tendons are tapped with a reflex hammer (Fig. 18.5) or, in this experiment, with a meterstick, the attached muscle is stretched. The stretch receptor generates nerve impulses that are transmitted along sensory neurons to the spinal cord. Nerve impulses from the cord then pass along motor neurons and stimulate the muscle, causing it to contract. As the muscle contracts, it tugs on the tendon, causing movement of a bone opposite the joint. Receptors in other tendons, such as the Achilles tendon, respond similarly.

Ankle (Achilles) Reflex

1. Have the subject sit on a table so that his or her legs hang freely.
2. Tap the subject's Achilles tendon at the ankle with a meterstick.
3. Which way does the foot move? Does it extend (move away from the knee) or flex (move toward the knee)? ____________________

Knee-Jerk (Patellar) Reflex

1. Have the subject sit on a table so that his or her legs hang freely.
2. Sharply tap one of the patellar tendons just below the patella (kneecap) with a meterstick.
3. In this relaxed state, does the leg flex (move toward the buttocks) or extend (move away from the buttocks)? ____________________

Figure 18.5 Two human reflexes.
The quick response when either the **(a)** Achilles tendon or the **(b)** patellar tendon is stimulated by tapping with a rubber hammer indicates that a reflex has occurred.

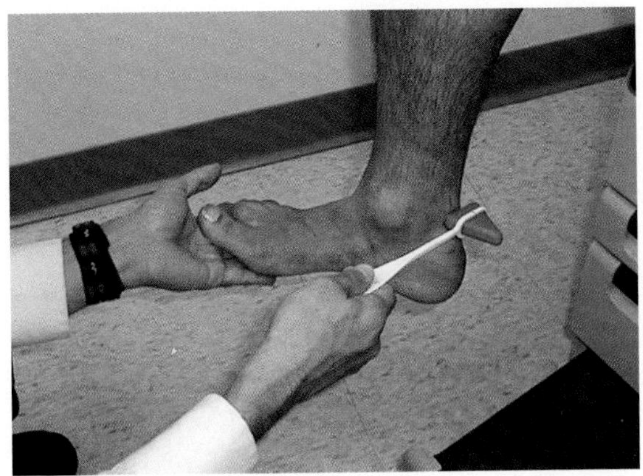

a. Ankle (Achilles) reflex

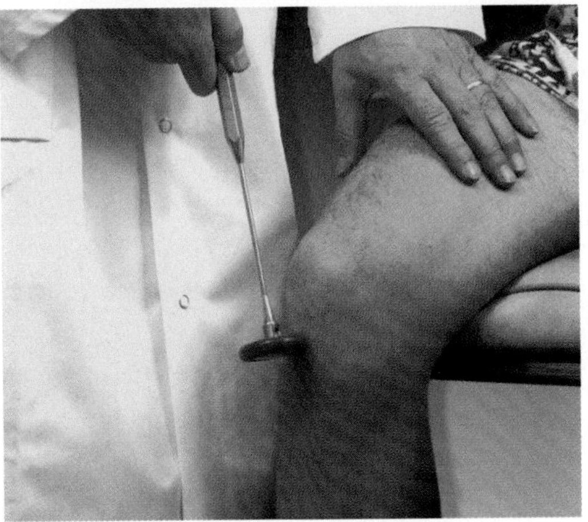

b. Knee-jerk (patellar) reflex

18.3 The Human Eye

The human eye is responsible for sight. Light rays enter the eye and strike the **rod cells** and **cone cells,** the photoreceptors for sight. The rods and cones generate nerve impulses that go to the brain via the optic nerve.

Observation: The Human Eye

1. Examine a human eye model, and identify the structures listed in Table 18.1.
2. Trace the path of light from outside the eye to the retina.

___

___

3. During **accommodation,** the lens rounds up to aid in viewing near objects or flattens to aid in viewing distant objects. Which structure holds the lens and is involved in accommodation?

___

4. **Refraction** is the bending of light rays so that they can be brought to a single focus. Which of the structures listed in Table 18.1 aid in refracting and focusing light rays?

___

5. Specifically, what are the sensory receptors for sight, and where are they located in the eye?

___

6. What structure takes nerve impulses to the brain from the rod cells and cone cells?

___

Table 18.1 Parts of the Human Eye

Part	Location	Function
Sclera	Outer layer of eye	Protects and supports eyeball
Cornea	Transparent portion of sclera	Refracts light rays
Choroid	Middle layer of eye	Absorbs stray light rays
Retina	Inner layer of eye	Contains receptors for sight
Rod cells	In retina	Make black-and-white vision possible
Cone cells	Concentrated in fovea centralis	Make color vision possible
Fovea centralis	Special region of retina	Makes acute vision possible
Lens	Interior of eye between cavities	Refracts and focuses light rays
Ciliary body	Extension from choroid	Holds lens in place; functions in accommodation
Iris	More anterior extension of choroid	Regulates light entrance
Pupil	Opening in middle of iris	Admits light
Humors (aqueous and vitreous)	Fluid media in anterior and posterior compartments, respectively, of eye	Transmit and refract light rays; support eyeball
Optic nerve	Extension from posterior of eye	Transmits impulses to brain

Figure 18.6 Anatomy of the human eye.
The sensory receptors for vision are the rod cells and cone cells present in the retina of the eye.

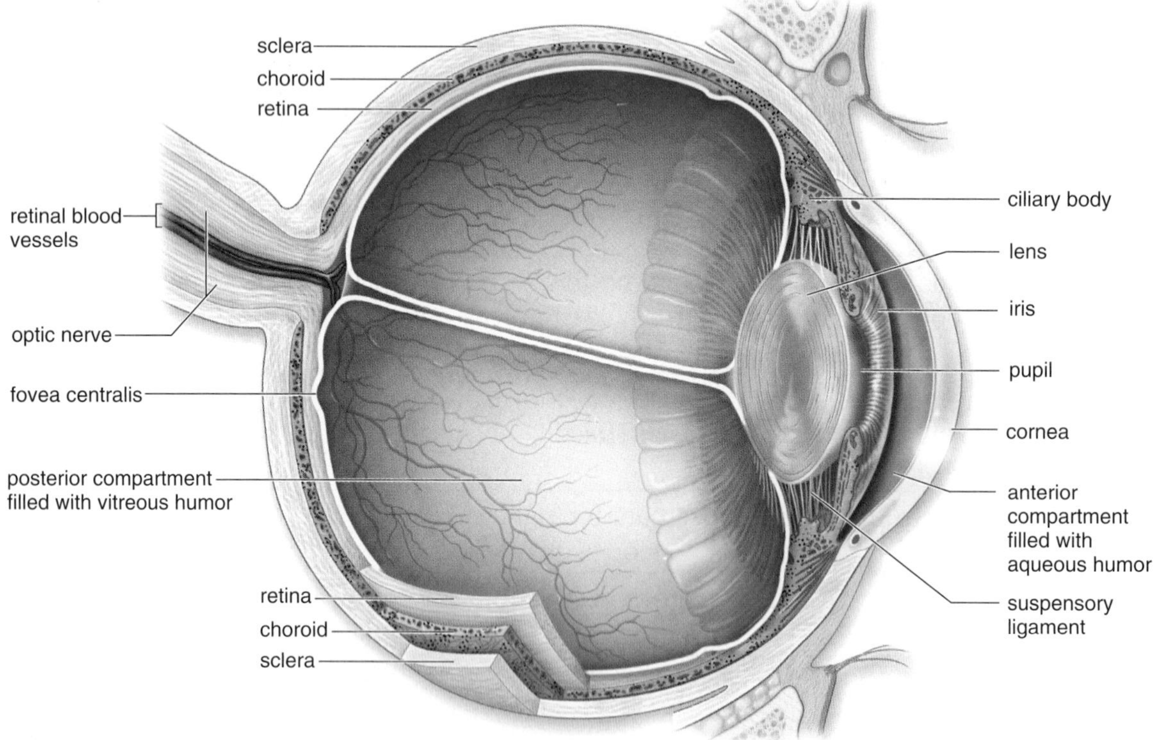

Figure 18.7 Blind spot.
This dark circle (or cross) will disappear at one location because there are no rod cells or cone cells at each eye's blind spot, where vision does not occur.

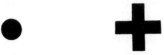

The Blind Spot of the Eye

The **blind spot** occurs where the optic nerve fibers exit the retina. No vision is possible at this location because of the absence of rod cells and cone cells.

Experimental Procedure: Blind Spot of the Eye

This Experimental Procedure requires a laboratory partner. Note that Figure 18.7 shows a small circle and a cross several centimeters apart.

Left Eye

1. Hold Figure 18.7 approximately 30 cm from your eyes. If you wear glasses, keep them on.
2. Close your right eye.
3. Stare only at the cross with your left eye. You should also be able to see the circle in the same field of vision. Slowly move the paper toward you until the circle disappears.
4. Repeat the procedure as many times as needed to find the blind spot.
5. Then slowly move the paper closer to your eyes until the circle reappears. Because only your left eye is open, you have found the blind spot of your left eye.
6. With your partner's help, measure the distance from your eye to the paper when the circle first disappeared. Left eye: ________ cm

Right Eye

1. Hold Figure 18.7 approximately 30 cm from your eyes. If you wear glasses, keep them on.
2. Close your left eye.
3. Stare only at the circle with your right eye. You should also be able to see the cross in the same field of vision. Slowly move the paper toward you until the cross disappears.
4. Repeat the procedure as many times as needed to find the blind spot.
5. Then slowly move the paper closer to your eyes until the cross reappears. Because only your right eye is open, you have found the blind spot of your right eye.
6. With your partner's help, measure the distance from your eye to the paper when the cross first disappeared. Right eye: ________ cm

Accommodation of the Eye

When the eye accommodates in order to see objects at different distances, the shape of the lens changes. The lens shape is controlled by the ciliary muscles attached to it. When you are looking at a distant object, the lens is in a flattened state. When you are looking at a closer object, the lens becomes more rounded. The elasticity of the lens determines how well the eye can accommodate. Lens elasticity decreases with increasing age, a condition called **presbyopia.** Because of presbyopia, many older people need bifocals to see near objects.

Experimental Procedure: Accommodation of the Eye

This Experimental Procedure requires a laboratory partner. It tests accommodation of either your left or right eye.

1. Hold a pencil upright by the eraser and at arm's length in front of whichever of your eyes you are testing (Fig. 18.8).
2. Close the opposite eye.
3. Move the pencil from arm's length toward your eye.
4. Focus on the end of the pencil.
5. Move the pencil toward you until the end is out of focus. Measure the distance (in centimeters) between the pencil and your eye: ________ cm
6. At what distance can your eye no longer accommodate for distance? ________ cm
7. If you wear glasses, repeat this experiment without your glasses, and note the accommodation distance of your eye without glasses: ________ cm. (Contact lens wearers need not make these determinations, and they should write the words *contact lens* in this blank.)
8. The "younger" lens can easily accommodate for closer distances. The nearest point at which the end of the pencil can be clearly seen is called the **near point.** The more elastic the lens, the "younger" the eye (Table 18.2). How "old" is the eye you tested? ________

Figure 18.8 Accommodation. When testing the ability of your eyes to accommodate in order to see a near object, always keep the pencil in this position.

Table 18.2 Near Point and Age Correlation

Age (Years)	10	20	30	40	50	60
Near Point (cm)	9	10	13	18	50	83

18.4 The Human Ear

The human ear, whose parts are listed in Table 18.3 and depicted in Figure 18.9, serves two functions: hearing and balance. When you hear, sound waves are picked up by the **tympanic membrane** and amplified by the **malleus, incus,** and **stapes.** This creates pressure waves in the canals of the **cochlea** that lead to stimulation of **hair cells,** the receptors for hearing. Nerve impulses travel by way of the **cochlear nerve** to the brain. Hair cells in the utricle and saccule of the vestibule and in semicircular canals are receptors for balance.

Observation: The Human Ear

Examine a human ear model, and find the structures depicted in Figure 18.9 based on the information given in Table 18.3.

Table 18.3 Parts of the Human Ear

Part	Medium	Function	Mechanoreceptor
Outer ear	Air		
Pinna		Collects sound waves	—
Auditory canal		Filters air	—
Middle ear	Air		
Tympanic membrane and ossicles		Amplify sound waves	—
Auditory tube		Equalizes air pressure	—
Inner ear	Fluid		
Semicircular canals		Rotational equilibrium	Stereocilia embedded in cupula
Vestibule (contains utricle and saccule)		Gravitational equilibrium	Stereocilia embedded in otolithic membrane
Cochlea (spiral organ)		Hearing	Stereocilia embedded in tectorial membrane

Figure 18.9 Anatomy of the human ear.
The outer ear extends from the pinna to the tympanic membrane. The middle ear extends from the tympanic membrane to the oval window. The inner ear encompasses the semicircular canals, the vestibule, and the cochlea.

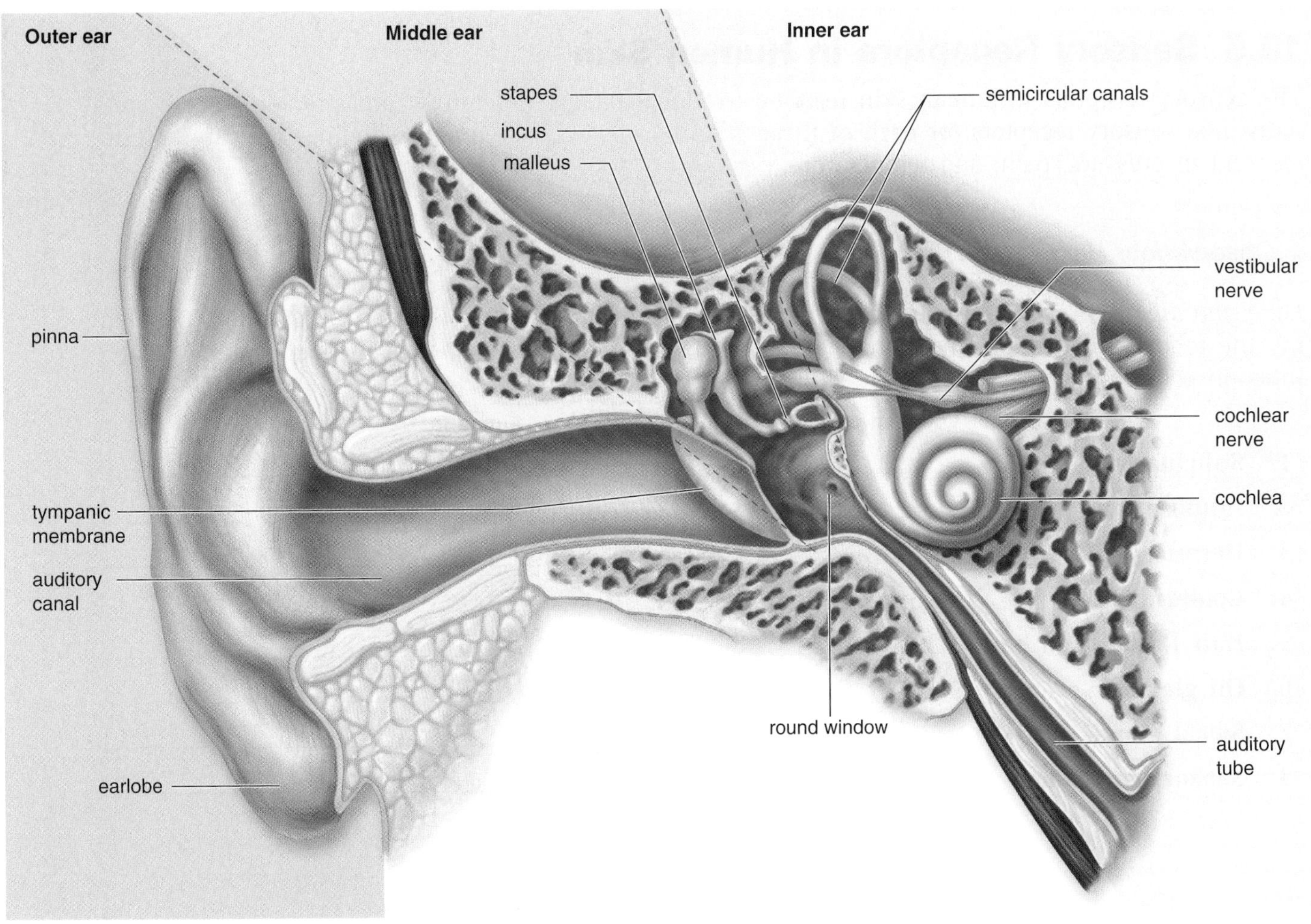

Experimental Procedure: Locating Sound

Humans locate the direction of sound according to how fast it is detected by either or both ears. A difference in the hearing ability of the two ears can lead to a mistaken judgment about the direction of sound. Both you and a laboratory partner should perform this Experimental Procedure on each other. Enter the data for *your* ears, not your partner's ears, in the spaces provided below.

1. Ask the subject to be seated, with eyes closed.
2. Then strike a tuning fork or rap two spoons together at the five locations listed in number 4. Use a random order.
3. Ask the subject to give the exact location of the sound in relation to his or her head.
4. Record the subject's perceptions when the sound is
   a. Directly below and behind the head ____________________
   b. Directly behind the head ____________________
   c. Directly above the head ____________________
   d. Directly in front of the face ____________________
   e. To the side of the head ____________________
5. Is there an apparent difference in hearing between your two ears? ____________________

____

18.5 Sensory Receptors in Human Skin

The sensory receptors in human skin respond to touch, pain, temperature, and pressure. There are individual sensory receptors for each of these stimuli, as well as free nerve endings that are able to respond to pressure, pain, and temperature.

Observation: Human Skin

Label the diagram of human skin in Figure 18.10 (refer to lab manual Figure 11.17, page 147). Identify the following structures in a model of the skin, if available. Below, indicate the location and the function of each of these parts of the skin.

	Location	Function
1. **Subcutaneous layer:**		
2. **Adipose tissue:**		
3. **Dermis:**		
4. **Epidermis:**		
5. **Hair follicle and hair:**		
6. **Oil gland:**		
7. **Sweat gland:**		
8. **Sensory receptors:**		

Figure 18.10 Sensory receptors in human skin.
The classical view is that each sensory receptor has the main function shown here. However, investigators report that matters are not so clear-cut. For example, microscopic examination of the skin of the ear shows only free nerve endings (pain receptors), and yet the skin of the ear is sensitive to all sensations. Therefore, it appears that the receptors of the skin are somewhat, but not completely, specialized.

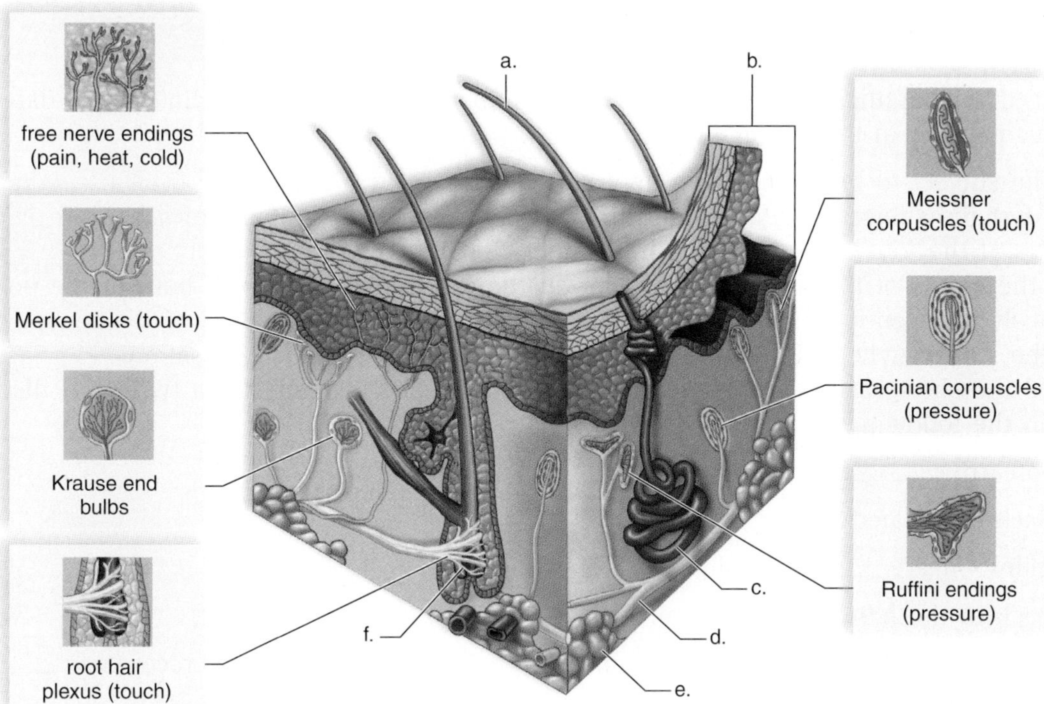

Sense of Touch

The dermis of the skin contains touch receptors, whose concentration differs in various parts of the body.

Experimental Procedure: Sense of Touch

You will need a laboratory partner to perform this Experimental Procedure. Enter *your* data, not the data of your partner, in the spaces provided below.

1. Ask the subject to be seated, with eyes closed.
2. Then test the subject's ability to discriminate between the two points of a hairpin or a pair of scissors at the four locations listed in number 5.
3. Hold the points of the hairpin or scissors on the given skin area, with both of the points simultaneously and gently touching the subject.
4. Ask the subject whether the experience involves one or two touch sensations.
5. Record the shortest distance between the hairpin or scissor points for a two-point discrimination in the following areas:

   a. Forearm: ________ mm

   b. Back of the neck: ________ mm

   c. Index finger: ________ mm

   d. Back of the hand: ________ mm

6. Which of these areas apparently contains the greatest density of touch receptors? ____________

   Why is this useful? ____

   ____
7. Do you have a sense of touch at every point in your skin? Explain. ____________________

   ____

Sense of Heat and Cold

Temperature receptors respond to a change in temperature.

Experimental Procedure: Sense of Heat and Cold

1. Obtain three 1,000 ml beakers, and fill one with *ice water,* one with *tap water* at room temperature, and one with *warm water* (45°–50°C).
2. Immerse your left hand in the ice-water beaker and your right hand in the warm-water beaker for 30 seconds.
3. Then place both hands in the beaker with room-temperature tap water.
4. Record the sensation in the right and left hands.

   a. Right hand: ____

   b. Left hand: ____

5. Explain your results: ____

   ____

18.6 Human Chemoreceptors

The taste receptors, located in the mouth, and the smell receptors, located in the nasal cavities, are the chemoreceptors that respond to molecules in the air and water.

Experimental Procedure: Sense of Taste

You will need a laboratory partner to perform the following procedures. It will not be necessary for all tests to be performed on both partners. You should take turns being either the subject or the experimenter. Dispose of used cotton swabs in a hazardous waste container, or as directed by your instructor.

Taste

1. The experimenter should be sure to use a clean cotton swab *each* time. The subject should be sure to rinse the mouth between applications.
2. For the sensation of sweet, apply *5% sucrose* to the tip, sides, and back of the tongue, and record on the first drawing in Figure 18.11 where the subject tastes the solution.
3. For the sensation of sour, apply *5% acetic acid* to the tip, sides, and back of the tongue, and record on the second drawing in Figure 18.11 where the subject tastes the solution.
4. For the sensation of salty, apply *10% NaCl* to the tip, sides, and back of the tongue, and record on the third drawing in Figure 18.11 where the subject tastes the solution.
5. For the sensation of bitter, apply *0.1% quinine sulfate* to the tip, sides, and back of the tongue, and record on the fourth drawing in Figure 18.11 where the subject tastes the solution.

Conclusions

- Do your results agree with those of other students in your laboratory? ______________________

  Explain: ____
- Give two general reasons the results might not agree, and explain your reasons based upon the technique used and the subject of the test.

  a. Technique: ____

  b. Subject: ____

Figure 18.11 Human tongue.
Recordings of where a subject tastes the solutions indicated.

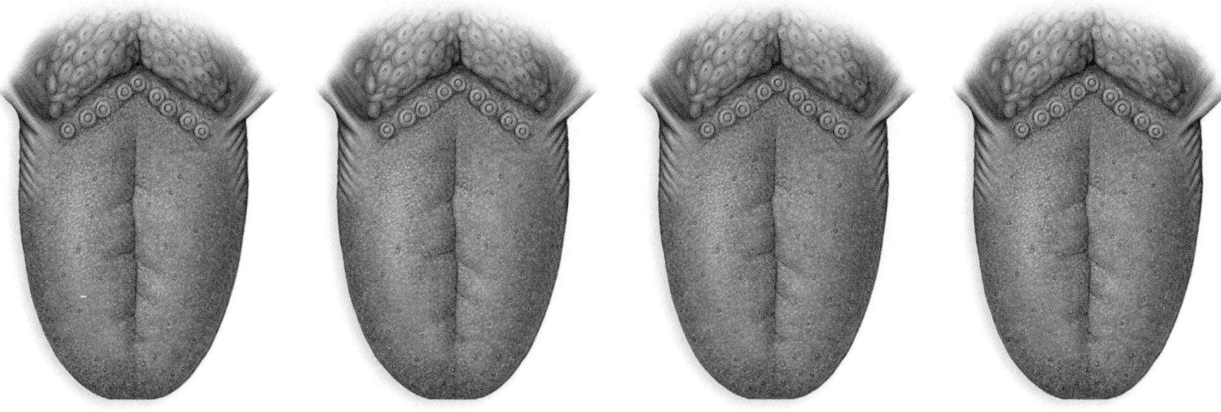

Taste and Smell

1. Students work in groups. Each group has one experimenter and several subjects.
2. The experimenter should obtain a LifeSavers candy from the various flavors available, without letting the subject know what flavor it is.
3. The subject closes both eyes and holds his or her nose.
4. The experimenter gives the LifeSavers candy to the subject, who places it on his or her tongue.
5. The subject, while still holding his or her nose, guesses the flavor of the candy. The experimenter records the guess in Table 18.4.
6. The subject releases his or her nose and guesses the flavor again. The experimenter records the guess and the actual flavor in Table 18.4.

Table 18.4 Test and Smell Experiment

Subject	Actual Flavor	Flavor While Holding Nose	Flavor After Releasing Nose
1			
2			
3			
4			
5			

Conclusions

- From your results, how would you say that smell affects the taste of LifeSavers candy?

  ____

- What do you conclude about the effect of smell on your sense of taste?

  ____

Laboratory Review 18

______ 1. What portion of the brain is largest in humans?

______ 2. What portion of the brain controls muscular coordination?

______ 3. What is the most inferior portion of the brain stem?

______ 4. What structures protect the spinal cord?

______ 5. Are motor neuron cell bodies located in the gray or white matter of the spinal cord?

______ 6. What type of neuron is found completely within the central nervous system?

______ 7. Which neuron's cell body is in the dorsal root ganglion?

______ 8. What part of the eye contains the sensory receptors for sight?

______ 9. Where on the retina is the blind spot located?

______ 10. What do you call the outer layer of the eye?

______ 11. What part of the ear contains the sensory receptors for hearing?

______ 12. Where in relation to the head is it most difficult to detect the location of a sound?

______ 13. In which portion of the ear are the malleus, incus, and stapes located?

______ 14. What layer of the skin contains sensory receptors?

______ 15. Are touch receptors distributed evenly or unevenly in the skin?

______ 16. What senses are dependent on chemoreceptors?

______ 17. The four taste sensations are sour, salty, bitter, and ______.

______ 18. What advantages are associated with the fact that the spinal cord and spinal nerves function below the level of consciousness?

______ 19. Identify the type of neuron responsible for transmitting nerve impulses from the spinal cord to an effector.

Thought Questions

20. Trace the path of light in the human eye through each structure or compartment—from the exterior to the retina. How do nerve impulses from the retina reach the brain?

(*Continued on reverse*)

21. Trace the path of sound waves in the human ear—from the tympanic membrane to the sensory receptors for hearing.

22. In a drag race, drivers must wait until the green light is illuminated before they can move their vehicle forward. Even the best drivers have a time delay between the illumination of the light and the forward movement of their vehicle (reaction time). Explain why this time delay exists based on the information presented in this exercise.

LABORATORY

19

Musculoskeletal System

Learning Outcomes

19.1 Anatomy of a Long Bone
- Locate and identify the portions of a long bone, and associate particular tissues with each portion.
- Using slides or other images, identify significant features of compact bone, spongy bone, and hyaline cartilage.

19.2 The Skeleton
- Using a preserved skeleton, model, or image, locate and identify the bones of the human skeleton, including the skull.

19.3 The Skeletal Muscles
- Using models, charts, or other images, locate and identify selected muscles.
- Illustrate the types of joint movements.
- Give examples of antagonistic pairs of muscles and differentiate the antagonistic actions involved.
- Distinguish between isometric and isotonic contractions.

19.4 Mechanism of Muscle Fiber Contraction
- Describe the structure of skeletal muscle.
- Describe an experiment that demonstrates the role of ATP and ions in the contraction of sarcomeres.

Introduction

The human skeletal system consists of the bones (206 in adults) and joints, along with the cartilage and ligaments that occur at the joints. The muscular system contains three types of muscles: smooth, cardiac, and skeletal. The term **musculoskeletal system** recognizes that contraction of skeletal muscles causes the bones to move. This lab examines the skeletal muscles.

In humans, the skeletal muscles are most often attached across a joint (Fig. 19.1). The biceps brachii muscle has two heads of origin (hence the name biceps) on different parts of the scapula. Notice the groove in the surface of the humerus in which one of the tendons of the biceps brachii muscle runs. Find the tendon of insertion of the biceps brachii muscle by feeling on the anterior surface of your elbow while contracting your biceps muscle. The triceps brachii muscle has three heads of origin—one on the scapula and two on the posterior humerus. Feel for the bone at your posterior elbow. This is the ulna, the site of insertion of the tendon for the triceps brachii muscle.

Muscles work in antagonistic pairs. For example, when the biceps brachii contracts, the bones of the forearm are pulled upward, and the triceps brachii relaxes; and when the triceps brachii contracts, the bones of the forearm are pulled downward, and the biceps brachii relaxes.

Figure 19.1 Muscular action. Muscles, such as these muscles of the arm (which have their origin on the scapula and their insertion on the bones of the forearm), cause bones to move.

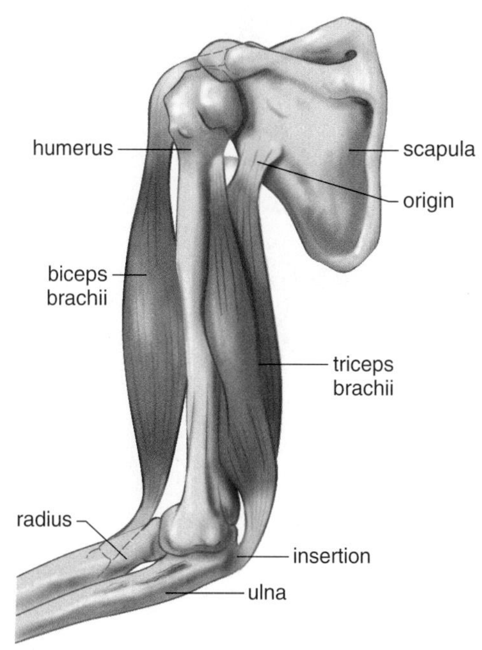

Figure 19.2 Anatomy of a bone from the macroscopic to the microscopic level.
A long bone is encased by the periosteum except at the ends, where it is covered by hyaline (articular) cartilage (see micrograph, *top left*). Spongy bone located in each epiphysis may contain red bone marrow. The medullary cavity contains yellow bone marrow and is bordered by compact bone, which is shown in the enlargement and micrograph (*right*). (hyaline cartilage: Magnification ×250; compact bone: Magnification ×100; spongy bone: Magnification ×4700)

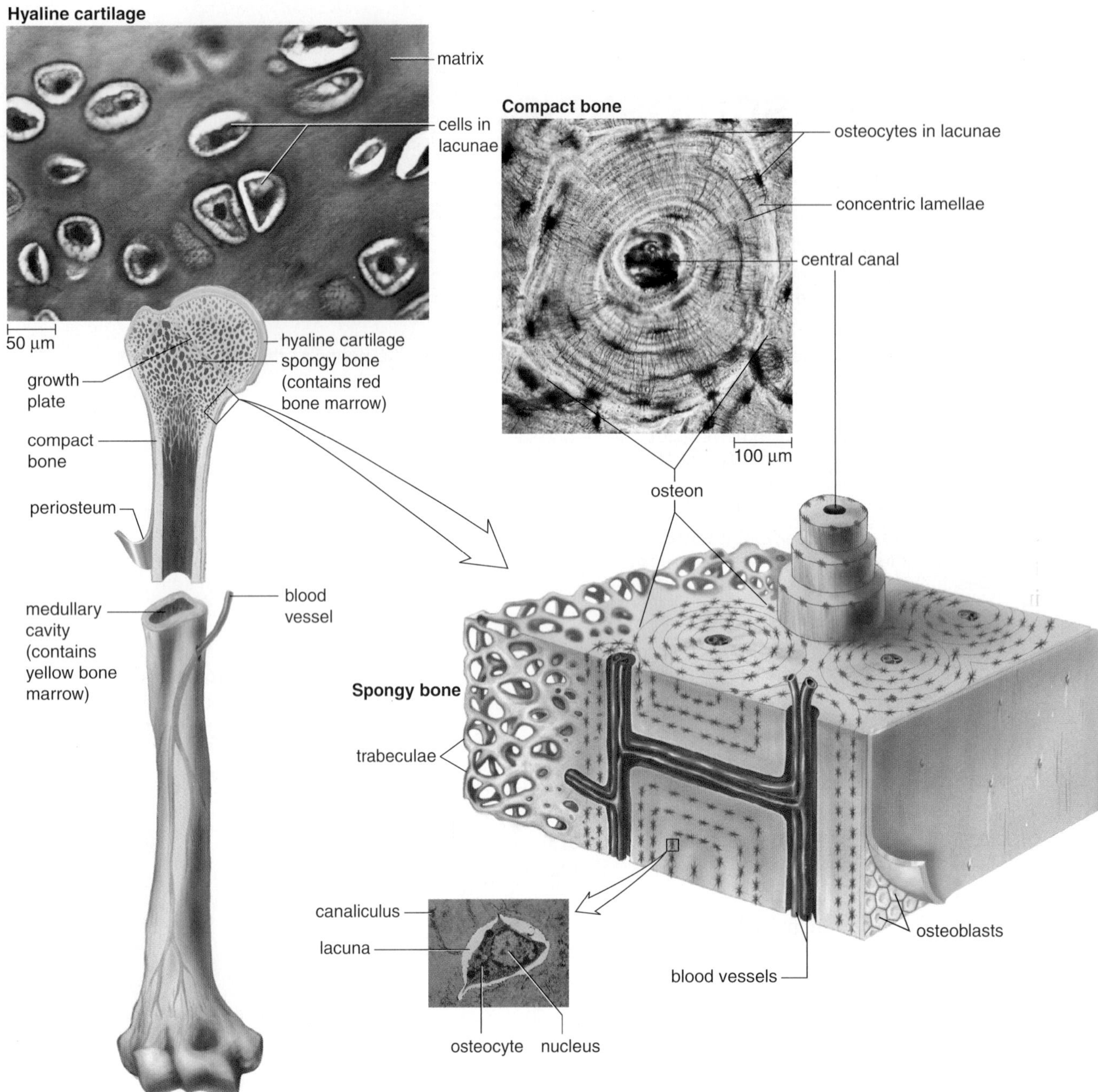

19.1 Anatomy of a Long Bone

Although the bones of the skeletal system vary considerably in shape as well as in size, a long bone, such as the human femur, illustrates the general principles of bone anatomy (Fig. 19.2).

Observation: Anatomy of a Long Bone

Examine the exterior and a longitudinal section of a long bone or a model of a long bone, and with the help of Figure 19.2, identify the following:

1. **Periosteum:** Tough, fibrous connective tissue covering that is continuous with the ligament and tendons that anchor bones. The periosteum contains blood vessels that enter the bone and service its cells.
2. **Epiphyses:** Expanded portions at each end of the bone (sing., *epiphysis*).
3. **Diaphysis:** Extended portion, or shaft, of a long bone that lies between the epiphyses.
4. **Hyaline (articular) cartilage:** Layer of cartilage where the bone articulates with (meets) another bone; decreases friction between bones during movement.
5. **Medullary cavity:** Cavity located in the diaphysis that stores yellow marrow, which contains a large amount of fat.

Observation: Tissues of a Long Bone

The medullary cavity is bounded at the sides by **compact bone** and at the ends by **spongy bone.** Beyond a thin shell of compact bone is the layer of articular cartilage. **Red marrow,** a specialized tissue that produces blood cells, occurs in the spongy bone of the skull, ribs, sternum, and vertebrae, and in the ends of the long bones.

1. Examine a prepared slide of compact bone, and with the help of Figure 19.2, identify:
   a. **Osteons:** Cylindrical structural units.
   b. **Lamellae:** Concentric rings of matrix.
   c. **Matrix:** Nonliving material maintained by osteocytes. Contains mineral salts (notably calcium salts) and protein.
   d. **Lacunae:** Cavities between the lamellae that contain osteocytes (bone cells).
   e. **Central canal:** Canal in the center of each osteon.
   f. **Canaliculi:** Tiny tubules that allow nutrients to pass between the osteocytes (sing., *canaliculus*).
2. Examine a prepared slide of spongy bone, and with the help of Figure 19.2, identify:
   a. **Trabeculae:** Bony bars and plates made of mineral salts and protein.
   b. **Lacunae:** Cavities scattered throughout the trabeculae that contain osteocytes.
   c. **Bone marrow:** Within large spaces separated by the trabeculae.
3. Examine a prepared slide of hyaline cartilage, and with the help of Figure 19.2, identify:
   a. **Lacunae:** Cavities in twos and threes scattered throughout the matrix, which contain chondrocytes (cells that maintain cartilage).
   b. **Matrix:** Material that is more flexible than bone because it consists primarily of protein.

Figure 19.3 Skull.
a. Lateral view. **b.** Inferior view. **c.** Facial view.

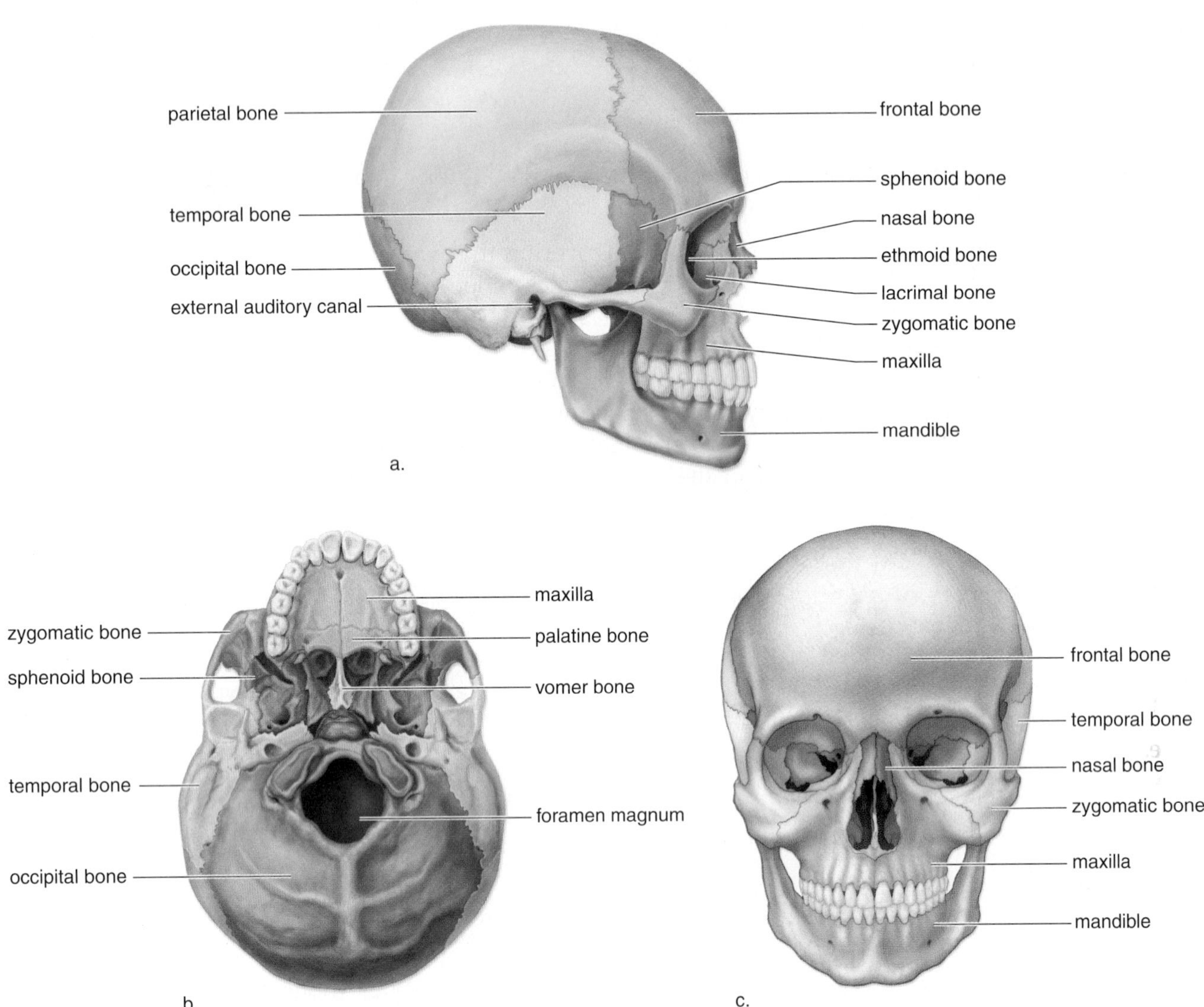

19.2 The Skeleton

The human skeleton is divided into axial and appendicular components. The **axial skeleton** is the main longitudinal portion, and includes the skull, the vertebral column, the sternum, and the ribs. The **appendicular skeleton** includes the bones of the appendages and their supportive pectoral and pelvic (shoulder and hip) girdles.

Observation: Axial Skeleton

Examine a human skeleton, and with the help of Figures 19.3 and 19.4, identify the **foramen magnum,** a large opening through which the spinal cord passes, and the following bones:

1. The **skull** is composed of many small bones fused together. Note the following in the cranium:
   a. **Frontal bone:** Forms forehead.
   b. **Parietal bones:** Extend to sides of skull.
   c. **Occipital bone:** Curves to form base of skull.
   d. **Temporal bones:** Located on sides of skull.
   e. **Sphenoid bone:** Helps form base and sides of skull, as well as part of the orbits.
2. Note the facial bones:
   a. **Mandible:** The lower jaw.
   b. **Maxillae:** The upper jaw and anterior portion of the hard palate.
   c. **Palatine bones:** Posterior portion of hard palate and floor of nasal cavity.
   d. **Zygomatic bones:** Cheekbones.
   e. **Nasal bones:** Bridge of nose.
3. The **vertebral column** provides support and houses the **spinal cord.** It is composed of many vertebrae separated from one another by intervertebral disks. The vertebral column customarily is divided into five series:
   a. Seven **cervical vertebrae** (forming the neck region)
   b. Twelve **thoracic vertebrae** (with which the ribs articulate)
   c. Five **lumbar vertebrae** (in the abdominal region)
   d. Five fused sacral vertebrae, called the **sacrum**
   e. Four fused caudal vertebrae forming the **coccyx** in humans
4. The twelve pairs of **ribs** and their associated muscles form a bony case that supports the thoracic cavity wall. The ribs connect posteriorly with the thoracic vertebrae, and some are also attached by cartilage directly or indirectly to the sternum. Those ribs without any anterior attachment are called **floating ribs.**

Figure 19.4 Human skeletal system.
a. Anterior view. **b.** Posterior view. The axial skeleton appears in blue, while the appendicular skeleton is shown in tan.

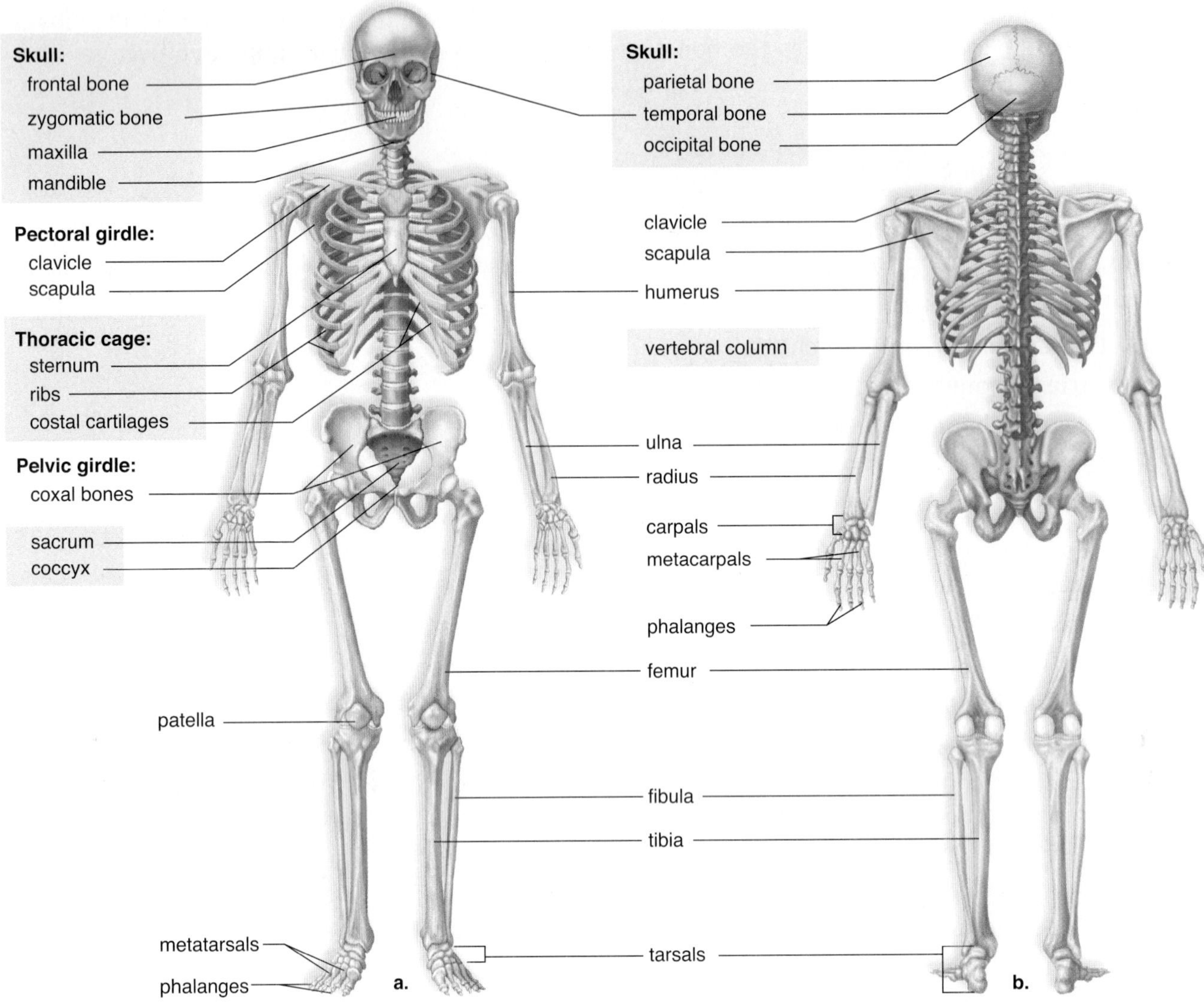

Observation: Appendicular Skeleton

Examine a human skeleton, and with the help of Figure 19.4, identify the following bones:

1. The **pectoral girdles,** which support the upper limbs, are composed of the **clavicle** (collarbone) and **scapula** (shoulder bone).
2. The upper limb (arm plus the forearm) is composed of the following:
   a. **Humerus:** The large long bone of the arm.
   b. **Radius:** The long bone of the forearm, with a pivot joint at the elbow that allows rotational motion.
   c. **Ulna:** The other long bone of the forearm, with a hinge joint at the elbow that allows motion in only one plane. Take hold of your elbow, and twist the forearm to show that the radius rotates over the ulna but the ulna doesn't move during this action.
   d. **Carpals:** A group of small bones forming the wrist.
   e. **Metacarpals:** Slender bones forming the palm.
   f. **Phalanges:** The bones of the fingers.
3. The **pelvic girdle** forms the basal support for the lower limbs and is composed of two **coxal** (hip) **bones.** The female pelvis is much broader and shallower than that of the male. The angle between the pubic bones looks like a U in females and a V in males.
4. The lower limb (the thigh plus the leg) is composed of a series of loosely articulated bones, including the following:
   a. **Femur:** The long bone of the thigh.
   b. **Patella:** Kneecap.
   c. **Tibia:** The larger of the two long bones of the leg. Feel for the bump on the inside of the ankle.
   d. **Fibula:** The smaller of the two long bones of the leg. Feel for the bump on the outside of the ankle.
   e. **Tarsals:** A group of small bones forming the ankle.
   f. **Metatarsals:** Slender anterior bones of the foot.
   g. **Phalanges:** The bones of the toes.

19.3 The Skeletal Muscles

This laboratory is concerned with skeletal muscles—those muscles that make up the bulk of the human body. With the help of Figure 19.5 and Tables 19.1 and 19.2, identify the major muscles of the body. Muscles are named for various characteristics, as shown in the following list:

1. **Size:** The gluteus *maximus* is the largest muscle, and it forms the buttocks.
2. **Shape:** The *deltoid* is shaped like a Greek letter delta, or triangle.
3. **Direction of fibers:** The *rectus* abdominis is a longitudinal muscle of the abdomen (*rectus* means straight).
4. **Location:** The *frontalis* overlies the frontal bone.
5. **Number of attachments:** The *biceps* brachii has two attachments, or origins.
6. **Action:** The extensor *digitorum* extends the fingers, or digits.

Figure 19.5 Human superficial skeletal muscles.
All the muscles of the human body are named in accordance with the structure and/or function of the underlying bone. The muscles highlighted here are those noted in the exercise entitled Observation: Antagonistic Pairs.

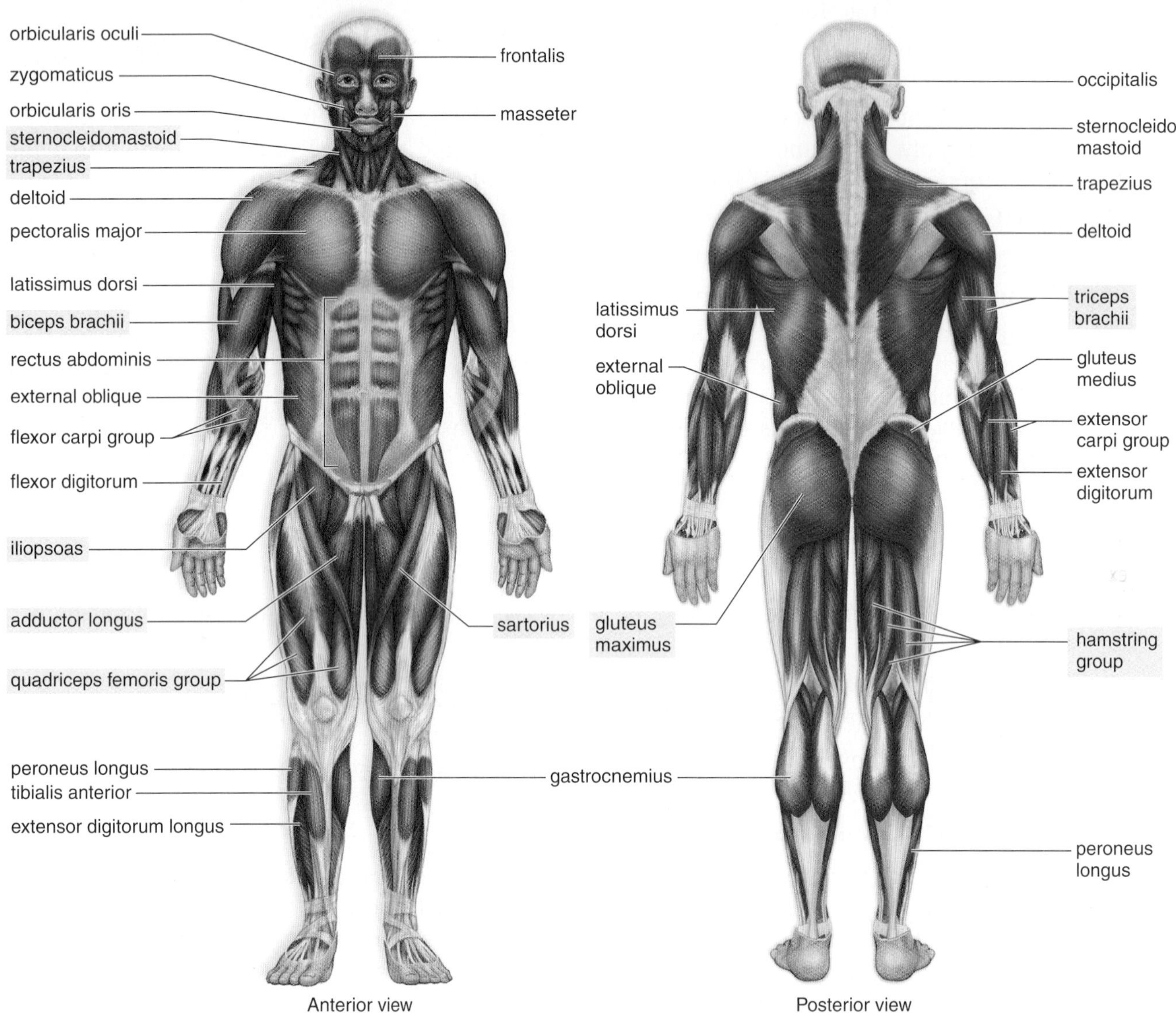

Table 19.1 Muscles (Anterior View)

Name	Action
Head and Neck	
Frontalis	Wrinkles forehead and lifts eyebrows
Orbicularis oculi	Closes eye (winking)
Zygomaticus	Raises corner of mouth (smiling)
Masseter	Closes jaw
Orbicularis oris	Closes and protrudes lips (kissing)
Upper Limb and Trunk	
External oblique	Compresses abdomen; rotates trunk
Rectus abdominis	Flexes spine
Pectoralis major	Flexes and adducts shoulder and arm ventrally (pulls arm across chest)
Deltoid	Abducts and raises arm at shoulder joint
Biceps brachii	Flexes forearm and supinates hand
Lower Limb	
Adductor longus	Adducts and flexes thigh
Iliopsoas	Flexes thigh at hip joint
Sartorius	Rotates thigh (sitting cross-legged)
Quadriceps femoris group	Extends leg
Peroneus longus	Everts foot
Tibialis anterior	Dorsiflexes and inverts foot
Flexor digitorum longus	Flexes toes
Extensor digitorum longus	Extends toes

Table 19.2 Muscles (Posterior View)

Name	Action
Head and Neck	
Occipitalis	Moves scalp backward
Sternocleidomastoid	Turns head to side; flexes neck and head
Trapezius	Extends head; raises and adducts shoulders dorsally (shrugging shoulders)
Upper Limb and Trunk	
Latissimus dorsi	Extends and adducts shoulder and arm dorsally (pulls arm across back)
Deltoid	Abducts and raises arm at shoulder joint
External oblique	Rotates trunk
Triceps brachii	Extends forearm
Flexor carpi group	Flexes hand
Extensor carpi group	Extends hand
Flexor digitorum	Flexes fingers
Extensor digitorum	Extends fingers
Buttocks and Lower Limb	
Gluteus medius	Abducts thigh
Gluteus maximus	Extends thigh (forms buttock)
Hamstring group	Flexes leg and extends thigh at hip joint
Gastrocnemius	Flexes leg and foot (tiptoeing)

Antagonistic Pairs

Skeletal muscles are attached to the skeleton, and their contraction causes the movement of bones at a joint. Because muscles shorten when they contract, they can only pull; they cannot push. Therefore, muscles work in **antagonistic pairs.** Usually, contraction of one member of the pair causes a bone to move in one direction, and contraction of the other member of the pair causes the same bone to move in an opposite direction.

Figure 19.6 demonstrates the following types of joint movements:

Flexion	Moving jointed body parts toward each other
Extension	Moving jointed body parts away from each other
Adduction	Moving a part toward a vertical plane running through the longitudinal midline of the body
Abduction	Moving a part away from a vertical plane running through the longitudinal midline of the body
Rotation	Moving a body part around its own axis; **circumduction** is moving a body part in a wide circle
Inversion	A movement of the foot in which the sole is turned inward
Eversion	A movement of the foot in which the sole is turned outward

These terms describe the action of the muscles listed in Tables 19.1 and 19.2.

Figure 19.6 Joint movements.
a. Flexion and extension. **b.** Adduction and abduction. **c.** Rotation and circumduction. **d.** Inversion and eversion. Circles indicate pivot points.

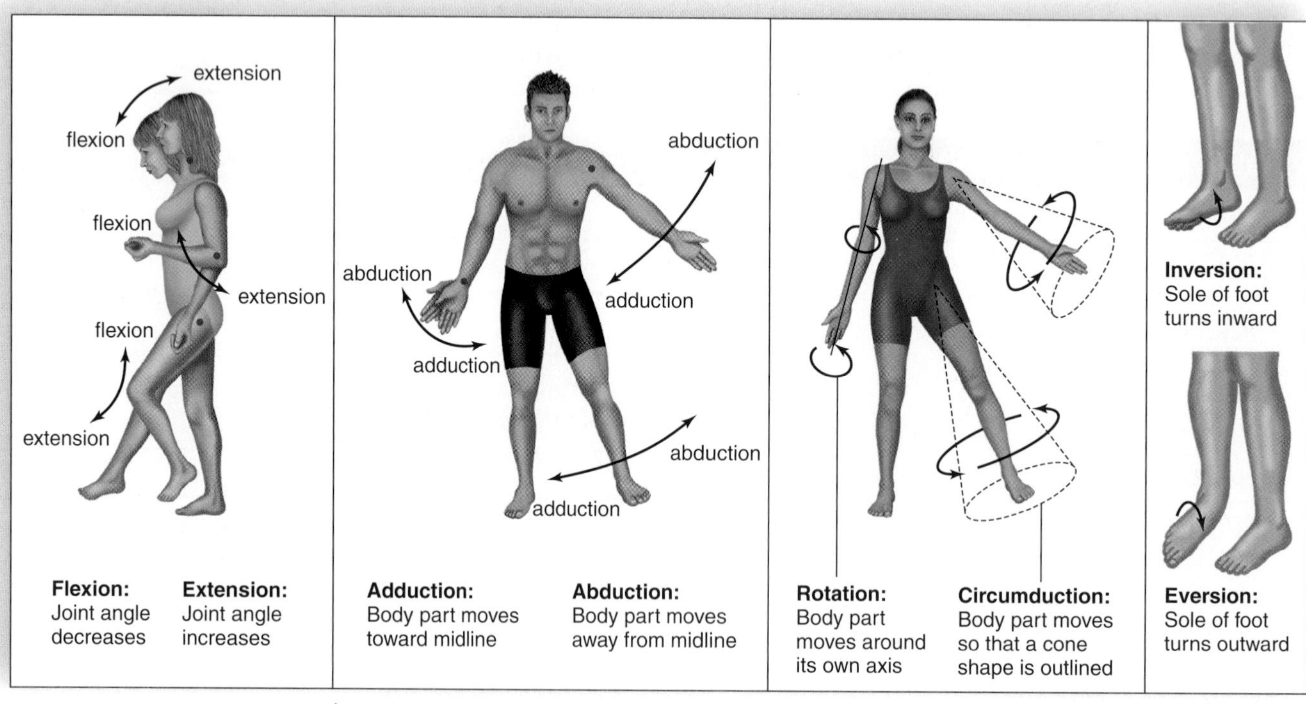

Observation: Antagonistic Pairs

Locate the following antagonistic pairs in Figure 19.5. In each case, state their opposing actions by inserting one of these functions: *flexes, extends, raises, lowers, adducts,* or *abducts.*

1. The biceps brachii ______________ the forearm.

   The triceps brachii ______________ the forearm.

2. The sternocleidomastoid ______________ the head.

   The trapezius ______________ the head.

3. The sartorius ______________________________ the thigh.

   The adductor longus ______________ the thigh.

4. The iliopsoas ______________ the thigh.

   The gluteus maximus ______________ the thigh.

5. The quadriceps femoris group ______________ the leg.

   The hamstring group ______________ the leg.

Isometric and Isotonic Contractions

A muscle contains many muscle fibers. When a muscle contracts, usually some fibers undergo isotonic contraction, and others undergo isometric contraction. When the tension of muscle fibers is sufficient to lift a load, many fibers change length as they lift the load. The muscle contraction is said to be **isotonic** (same tension). In contrast, when the tension of muscle fibers is used only to support rather than lift a load, the muscle contraction is said to be **isometric** (same length). The length of many fibers remains the same, but their tension still changes.

Experimental Procedure: Isometric and Isotonic Contractions

Note: The upper limb is composed of the arm plus the forearm.

Isotonic Contraction

1. Start with your left forearm resting on a table. Watch the anterior surface of your left arm while you slowly bend your elbow and bring your left forearm toward the arm. An isotonic contraction of the biceps brachii produces this movement.
2. If a muscle contraction produces movement, is this an isometric or isotonic contraction? ________

   ____

Isometric Contraction

1. Place the palm of your left hand underneath a tabletop. Push up against the table while you have your right hand cupped over the anterior surface of your left arm so that you can feel the muscle there undergo an isometric contraction.
2. Is the biceps brachii or the triceps brachii located on the anterior surface of the arm?

   ____

3. What change did you notice in the firmness of this muscle as it contracted? ______________

   ____

4. Did your hand or forearm move as you pushed up against the table? ______________
5. Given your answer to question 4, did this muscle's fibers shorten as you pushed up against the tabletop? ____

19.4 Mechanism of Muscle Fiber Contraction

A whole skeletal muscle is made up of many cells called **muscle fibers** (myofibrils) (Fig. 19.7). Muscle fibers are striated—that is, they have alternating light and dark bands. These striations can be observed in a light micrograph of muscle fibers in longitudinal section.

Electron microscopy has shown that striations are due to the placement of protein filaments of **myosin** and **actin.** During contraction, actin filaments move past myosin filaments, and units of the muscle, called **sarcomeres,** shorten. ATP serves as the immediate energy source for sarcomere contraction. Potassium (K^+) and magnesium (Mg^{2+}) ions are cofactors for the breakdown of ATP by myosin.

Observation: Skeletal Muscle

Examine a prepared slide of skeletal muscles, and identify long, multinucleated fibers arranged in a parallel fashion. Note that the muscle fibers are striated (Fig. 19.7). See also Section 11.3 for a photomicrograph of skeletal muscle.

Experimental Procedure: Muscle Fiber Contraction

1. Label two slides, slide 1 and slide 2. Mount a strand of glycerinated muscle fibers in a drop of *glycerol* on each slide. Place each slide on a millimeter ruler, and measure the length of the strand. Record these lengths in the first row in Table 19.3.
2. If there is more than a small drop of glycerol on the slides, soak up the excess on a piece of lens paper held at the edge of the glycerol farthest from the fiber strand.
3. To slide 1, add a few drops of a *salt solution* containing potassium (K^+) and magnesium (Mg^{2+}) ions, and measure any change in strand length. Record your results in Table 19.3.
4. To slide 2, add a few drops of *ATP solution,* and measure any change in strand length. Record your results in Table 19.3.

Table 19.3 Glycerinated Muscle Contraction

Solution	*Length (mm)*	
	Slide 1	**Slide 2**
Glycerol alone		
K^+/Mg^{2+} salt solution alone		—
ATP alone	—	
Both salt solution and ATP		

5. Now add *ATP solution* to slide 1. Measure any change in strand length, and record your results in Table 19.3.
6. To slide 2, add a few drops of the K^+/Mg^{2+} *salt solution,* and measure any change in strand length. Record your results in Table 19.3.
7. To demonstrate that you understand the requirements for contraction, state the function of each of the substances listed in Table 19.4.

Table 19.4 Summary of Muscle Fiber Contraction

Substance	**Function**
Myosin	
Actin	
K^+/Mg^{2+} salt solution	
ATP	

Figure 19.7 Microscopic structure of a skeletal muscle fiber.

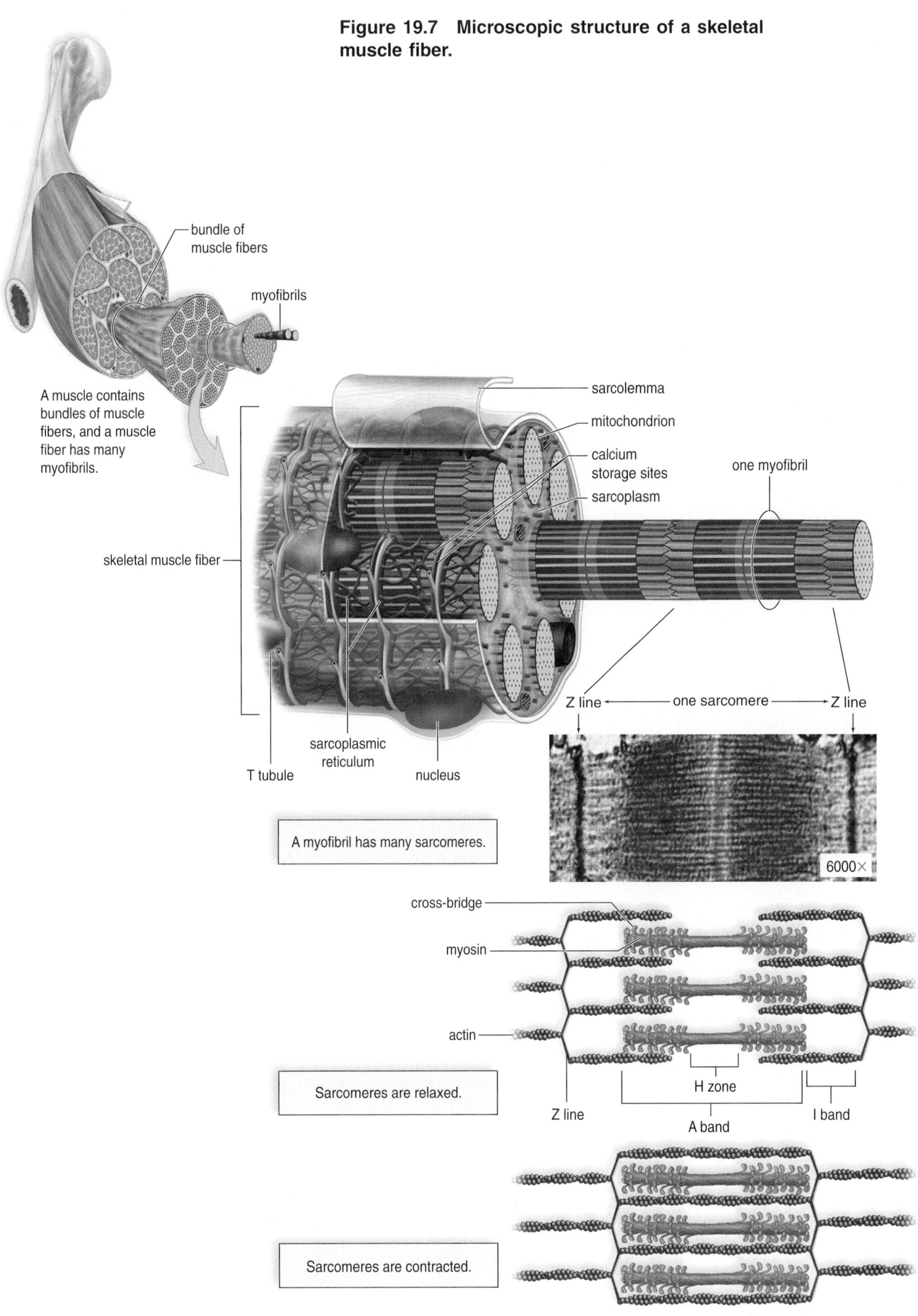

Laboratory Review 19

______ 1. Is compact bone located in the diaphysis or in the epiphyses?

______ 2. Does compact bone or spongy bone contain red bone marrow?

______ 3. What are bone cells called?

______ 4. What are the vertebrae in the neck region called?

______ 5. Name the strongest bone in the lower limb.

______ 6. What bones are part of a pectoral girdle?

______ 7. What type of joint movement occurs when a muscle moves a limb toward the midline of the body?

______ 8. What type of joint movement occurs when a muscle moves a body part around its own axis?

______ 9. Skeletal muscle is voluntary, and its appearance is ______ because of the placement of actin and myosin filaments.

______ 10. Glycerinated muscle requires the addition of what molecule to supply the energy for muscle contraction?

______ 11. Actin and myosin are what type of molecule?

______ 12. Does the quadriceps femoris group flex or extend the leg?

______ 13. Does the biceps brachii flex or extend the forearm?

______ 14. What muscle forms the buttocks?

______ 15. Name the muscle group that is antagonistic to the quadraceps femoris group.

Thought Questions

16. What bones protect the thoracic cavity?

17. When you see glycerinated muscle shorten, what is happening microscopically?

LABORATORY

20

Development

Learning Outcomes

20.1 Early Embryonic Stages
- Using slides, images, or models, identify and compare the morula, blastula, gastrula, and neurula in the sea star and frog.
- Describe the development of the neural tube in the frog.

20.2 Embryonic Germ Layers
- Name the three embryonic germ layers and the major organs that develop from each.
- Describe the significance of the process of induction during development.
- Compare the events of gastrulation in the sea star, the frog, the chick, and a human.

20.3 Chick Development
- Name and contrast the functions of the extraembryonic membranes in mammals and chicks.
- Identify the major organs in chick embryos, and describe, in general, how chick embryos develop.

20.4 Human Development
- Compare the development of humans with that of the other animals studied.
- Contrast the major events that occur in the preembryonic, embryonic, and fetal stages of human development, and illustrate key events in each stage using human models, photographs, or images.

Introduction

The early development of many species of animals is quite similar. This similarity is a unifying feature and is used to establish evolutionary relationships between species. The fertilized egg, or zygote, undergoes successive mitotic divisions collectively referred to as cleavage. The resultant ball of cells, the morula, resembles a mulberry or blackberry in appearance. Since there is little time between divisions for cells to grow in size, the amount of total cytoplasm in the morula remains approximately the same as that present in the original zygote. This results in an increase in the number of cells, but with each subsequent division, the cells grow smaller as the available amount of cytoplasm is divided among an ever-increasing number of cells. As cleavage continues, the cells in this solid ball undergo division in a manner that forms a layer of cells surrounding a hollow central cavity. This stage is known as a blastula. The fluid-filled cavity of the blastula is the blastocoel. Later, some of the surface cells fold inward, or invaginate, eventually forming a double-walled structure, the early gastrula. The outer layer is called the ectoderm and the inner layer is the endoderm. As the embryo continues to develop, a third layer will form between these layers, the mesoderm. With the three embryonic tissues, or germ layers, in place or developing, the embryo is now called a gastrula. Although not all animals will have a distinct gastrula stage, the formation of the germ layers is referred to as gastrulation. In particular, the presence of yolk (nutrient material) influences how the gastrula will develop.

In general, all later development can be associated with the three embryonic germ layers that give rise to different tissues and systems: (1) The ectoderm forms the nervous system and the skin plus its accessory structures (hair, nails, scales, feathers, etc.); (2) the endoderm forms the lining of the digestive system and respiratory system; and (3) the mesoderm gives rise to the cardiovascular, muscular, reproductive, and skeletal systems, and to connective tissue. The development of terrestrial vertebrate animals (reptiles, birds, and mammals) also includes the development of specialized support tissues that do not actually become part of the developing embryo. Since these tissues are not part of the makeup of the actual embryo, they are referred to as the extraembryonic membranes.

Development requires growth, differentiation, and morphogenesis.

Growth occurs when cells divide, get larger, and divide again.

Differentiation occurs when cells become specialized in structure and function. As is evidenced by the information in the preceding paragraphs, germ layers have a major role in determining the fate of development. For example, a muscle cell, derived from mesoderm, looks and acts quite differently than a nerve cell, which has its origins in ectoderm.

Morphogenesis occurs when body parts become shaped and patterned into a certain form. For example, your arm and leg are very different, even though they contain the same types of tissues.

20.1 Early Embryonic Stages

Successive division of the **zygote** (fertilized egg) results in 2-cell, 4-cell, 8-cell, and 16-cell stages, and finally in a many-celled stage called a **morula.** The morula becomes a **blastula,** and the blastula becomes a **gastrula.**

Sea Star Development

Echinoderms (e.g., sea stars, sea urchins) are useful for illustrating the stages of early development for multicellular animals. Sea stars develop in an aquatic environment and develop quickly into a larva that is capable of feeding itself. Explain why you would not expect a sea star's egg to be heavily laden with yolk. ______________________________

____

Observation: Sea Star Embryos

Examine whole-mount microscope slides of stained sea star embryos at the stages of development shown in Figure 20.1. Try to identify the following:

1. **Unfertilized egg:** Observe the large nucleus and the darkly staining nucleolus. The plasma membrane surrounds the cytoplasm, which contains a small amount of yolk. After fertilization, a fertilization membrane, which prevents the entrance of other sperm, can be seen outside the plasma membrane, and the distinct nucleus disappears.
2. **Cleavage:** Successive division of the embryo results in a 2-cell, 4-cell, 8-cell, 16-cell stages, and finally a many-celled stage. Cleavage occurs without an accompanying increase in size.
3. **Morula:** The morula is a ball of cells that is about the same size as the original zygote.

   Explain why the morula and zygote are about the same size. ______________________________

   ____
4. **Blastula:** The large number of embryonic cells of the morula arrange themselves into a blastula, a single-layered ball with a fluid-filled cavity, called the **blastocoel,** in the middle.
5. **Early gastrula:** The cells of the blastula fold inward to form a two-layered gastrula. The cavity produced by the infolded layer of cells is the **archenteron,** or primitive gut, which has an opening to the outside called the **blastopore.** The outer layer of cells is the **ectoderm,** and the inner layer is the **endoderm.**
6. **Late gastrula:** As development continues, two pouches (the coelomic sacs) form by outpocketing from the endoderm surrounding the gut. These pouches become part of the **coelom** (body cavity), and the walls of the lateral pouches become the third germ layer, the **mesoderm.**

Figure 20.1 Photographs of sea star developmental stages.
All animal embryos go through these same early developmental stages. (**a–f:** Magnification ×75.) Courtesy of Carolina Biological Supply Company, Burlington, N.C.

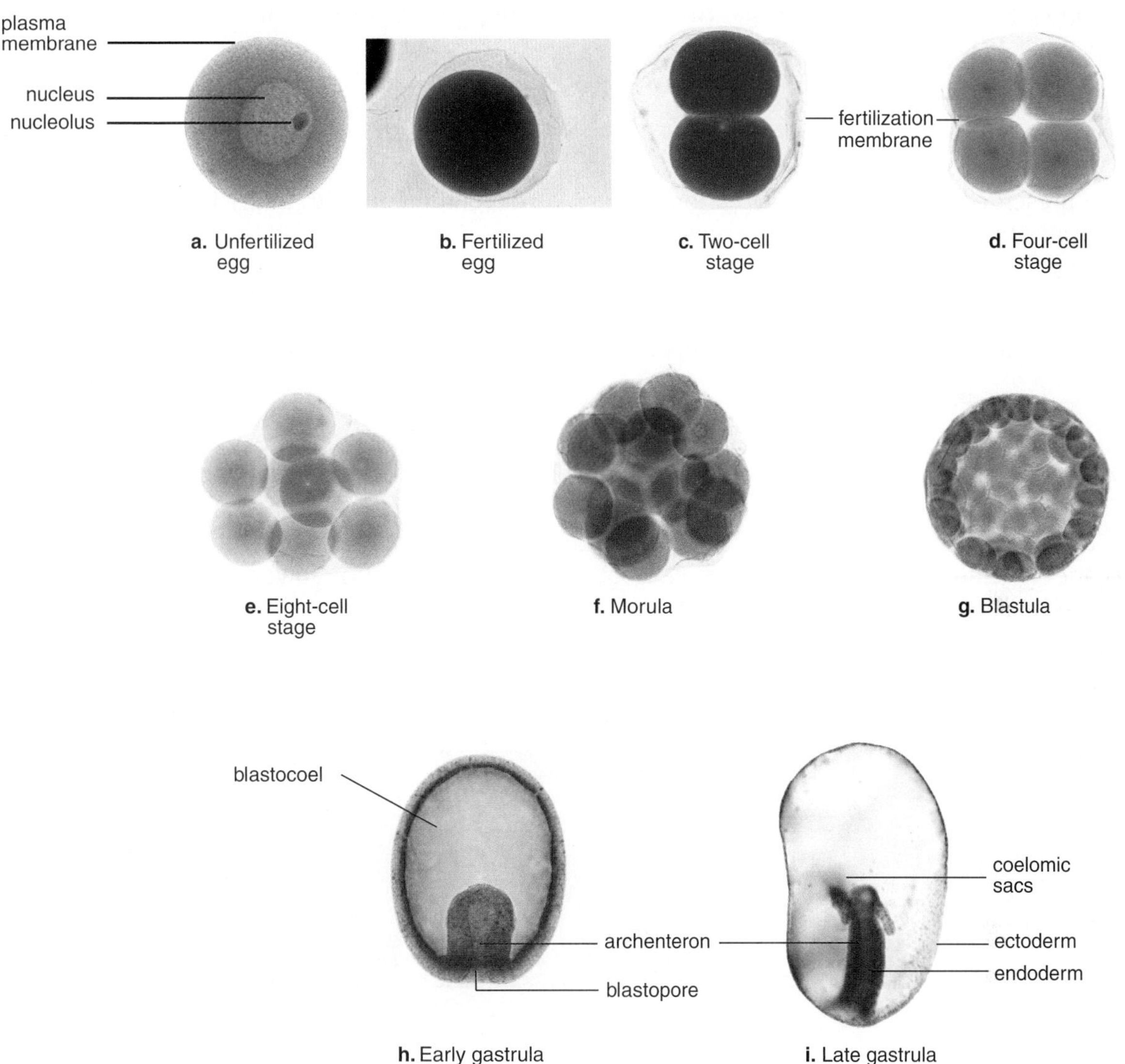

a. Unfertilized egg **b.** Fertilized egg **c.** Two-cell stage **d.** Four-cell stage **e.** Eight-cell stage **f.** Morula **g.** Blastula **h.** Early gastrula **i.** Late gastrula

Frog Development

Like sea star embryos, frog embryos develop in water, but the process takes a bit longer. Frog eggs contain more yolk than sea star eggs, making the developmental process for the frog appear different than what was seen for the sea star.

Observation: Preserved Frog Embryos

Examine preserved frog embryos at various stages of development (Fig. 20.2) in a petri dish, or use models, and identify the following stages:

1. **Fertilized egg:** Notice that the eggs are partially pigmented. The black side contains very little yolk (nutrient material) and is called the **animal pole.** The unpigmented, yolky side is called the **vegetal pole.** Did the sea star unfertilized egg have an animal pole and a vegetal pole?

   __________ Explain. ____

   ____

2. **Morula:** Cleavage begins at the animal pole (Fig. 20.3*a*). The first two cell divisions are polar. The third division is horizontal or equatorial between the poles. Cleavage continues until there is a morula. Notice that the cells of the animal pole are smaller and more numerous than those of the vegetal pole. The yolk-laden cells of the vegetal pole are slower to divide than those of the animal pole.

Figure 20.2 Photographs of frog developmental stages.
The presence of yolk alters the appearance of the usual stages somewhat. (**a:** Magnification ×25; **i:** Magnification ×22.5.)

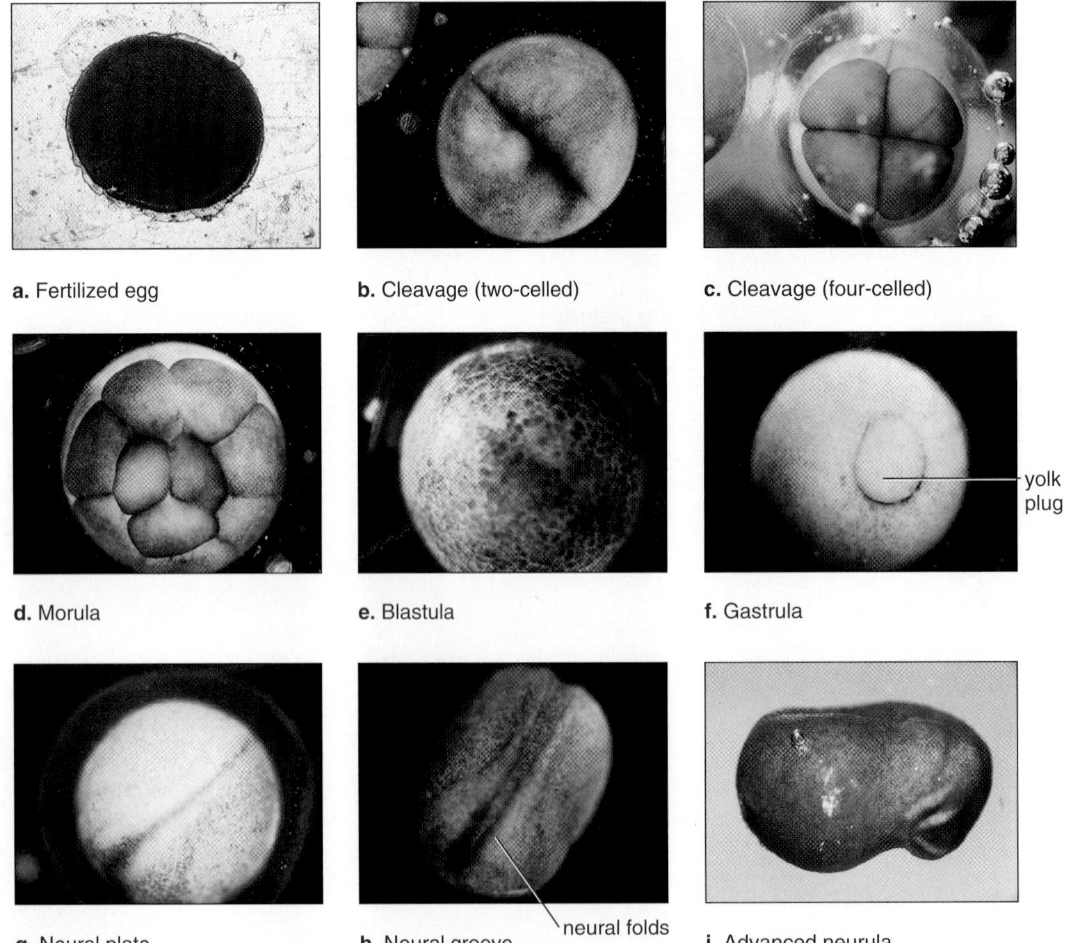

a. Fertilized egg **b.** Cleavage (two-celled) **c.** Cleavage (four-celled) **d.** Morula **e.** Blastula **f.** Gastrula **g.** Neural plate **h.** Neural groove **i.** Advanced neurula

3. **Blastula:** The blastula is a hollow ball of cells. However, the blastocoel (fluid-filled cavity) is found only at the animal pole. The yolk-laden cells of the vegetal pole do not divide rapidly and, therefore, do not help in the formation of the blastocoel. Compare the frog blastula with that of the sea star. ____

____

____

4. **Early gastrula:** Gastrulation is recognized by the presence of a crescentic slit. This is the location of the blastopore, where cells are invaginating (Fig. 20.3*b*). The blastopore later takes on a circular shape but is plugged by yolk cells that do not invaginate rapidly. In the frog, invagination is accomplished primarily by yolkless cells from the animal pole.
5. **Late gastrula:** The moderate amount of yolk also influences the formation of the mesoderm. This germ layer develops by invagination of cells at the lateral and ventral lips of the blastopore. Compare formation of the mesoderm in the frog with that in the sea star. ____________________

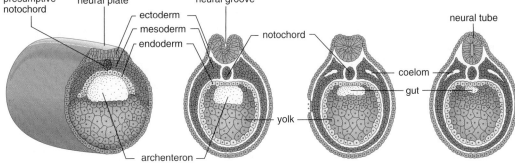

____

6. **Neurula:** During neurulation in the frog, two folds of ectoderm grow upward as the neural folds with a groove between them (Fig. 20.3*c*). The flat layer of ectoderm between them is the **neural plate.** The tube resulting from closure of the folds is the **neural tube,** which will become the spinal (nerve) cord and brain. An examination of the neurula in cross section shows that the nervous system develops directly above the **notochord,** a structure that arises from invaginated cells in the middorsal region.

Figure 20.3 Drawings of frog developmental stages.
Compare these series of drawings with the photographs in Figure 20.2.

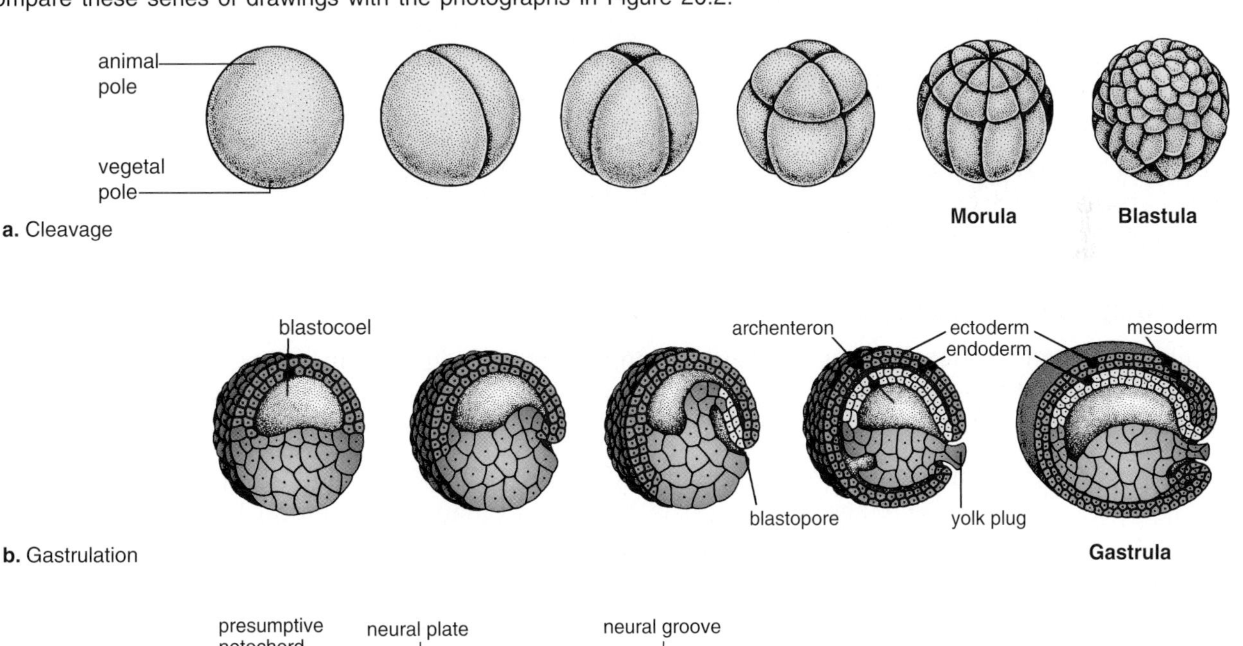

20.2 Embryonic Germ Layers

As mentioned previously, early development precedes the formation of three **embryonic germ layers.** In Table 20.1, list the three embryonic germ layers and the major organs that develop from each in the frog.

Induction

The notochord is said to induce the formation of the nervous system. Experiments have shown that, if contact with notochord tissue is prevented, no neural plate is formed. Even more dramatic are experiments in which presumptive (soon-to-be) notochord is transplanted under an area of ectoderm not in the dorsal midline. This ectoderm then is induced to differentiate into neural plate tissue, something it would not normally do.

Induction is believed to be one means by which development is usually orderly. The part of the embryo that induces the formation of an adjacent organ is said to be an **organizer** and is believed to carry out its function by releasing one or more chemical substances.

Table 20.1 Embryonic Germ Layer Organization

Germ Layers	Organs/Systems Associated with Germ Layer
1.	
2.	
3.	

20.3 Chick Development

Unlike sea stars and frogs, chicks develop on land, and there is no larval stage. Therefore, chick development is markedly different from that of sea stars and frogs because of three features: (1) The embryo is enclosed by extraembryonic membranes, (2) there is a large amount of yolk to sustain development, and (3) there is a hard outer shell.

Extraembryonic Membranes

The embryos of land vertebrates are surrounded and covered by membranes that do not become a part of the animal. These **extraembryonic membranes** are the **chorion, amnion, yolk sac,** and **allantois** (Fig. 20.4).

The chorion is the outermost membrane, and in chicks, it lies just below the porous shell, where it functions in gas exchange. In mammals, such as humans, the chorion forms the fetal portion of the placenta, through which nutrients, gases, and wastes are exchanged with the blood of the mother.

The amnion is a delicate membrane containing amniotic fluid that bathes the embryo. The embryo, suspended in this fluid medium, is thus protected from drying out and from mechanical shocks.

The yolk sac surrounds the yolk in bird and reptile eggs. In mammals, whose embryos generally contain very little yolk, the yolk sac is the first site of blood cell formation.

The allantois serves as a storage area for metabolic waste products in reptiles and birds. In placental mammals, waste storage and removal are solved by the mother's circulatory system, leaving the allantois to contribute blood vessels to the formation of the umbilical cord. Like the allantois, the yolk sac has its role redefined in mammal development, becoming incorporated into the formation of the umbilical cord, as can be seen in Figure 20.4*b*.

Observation: Raw Chick Egg

1. Carefully crack an unfertilized chick egg into a 5-inch finger bowl.
2. Examine the contents of the raw egg (Fig. 20.5). Note that some of the egg white **(albumen)** is denser and thicker than the remainder. These thicker masses are the **chalazae,** two spirally wound strands of albumen that result from the passage of the egg through the oviduct.
3. On the top of the yolk mass, note the germinal vesicle, a small, white spot containing the clear cytoplasm of the egg cell and the egg nucleus. Only the yolk with its germinal vesicle is the ovum. The various layers of albumen, including the chalazae, the shell membranes, and the shell, are all secreted by the oviduct as the ovum travels through it.

Figure 20.4 Extraembryonic membranes.
The chick and a human have the same extraembryonic membranes, but except for the amnion, they have different functions.

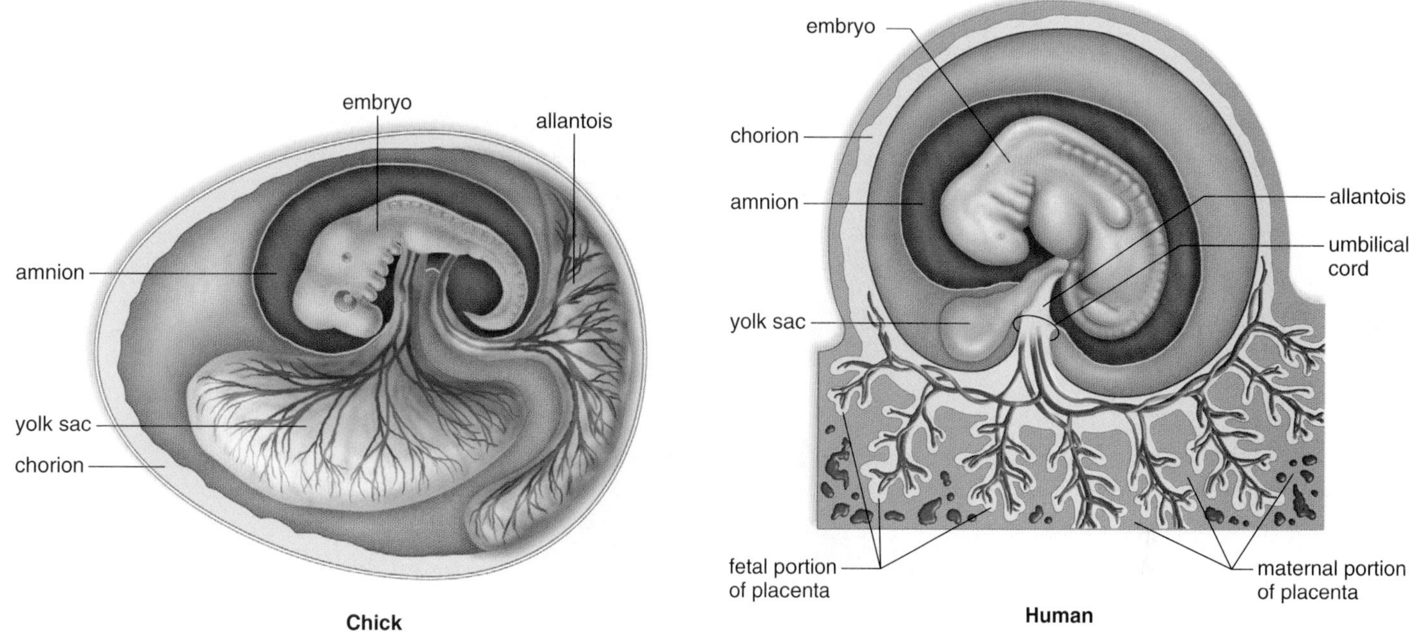

Figure 20.5 Unfertilized chick egg.
Chalazae are two spirally wound strands of albumen.

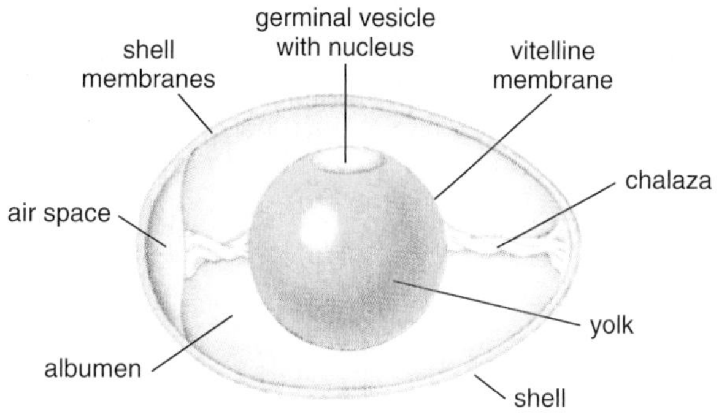

Observation: Twenty-Four-Hour Chick Embryo

1. Follow the standard procedure (see p. 271) for selecting and opening an egg containing a 24-hour chick embryo.
2. Neurulation is occurring anteriorly, and invagination is occurring posteriorly. Refer to Figure 20.6, and identify the following:
   a. **Embryo:** Lies atop the yolk.
   b. **Head fold:** The beginning of the embryo.
   c. **Nervous system:** In the process of developing and has a **neural tube** at the anterior end and **neural folds** toward the posterior end, where the neural tube is still in the process of forming.

Figure 20.6 Dorsal view of 24-hour chick embryo.
The head fold is the amnion that will eventually envelop the entire embryo.

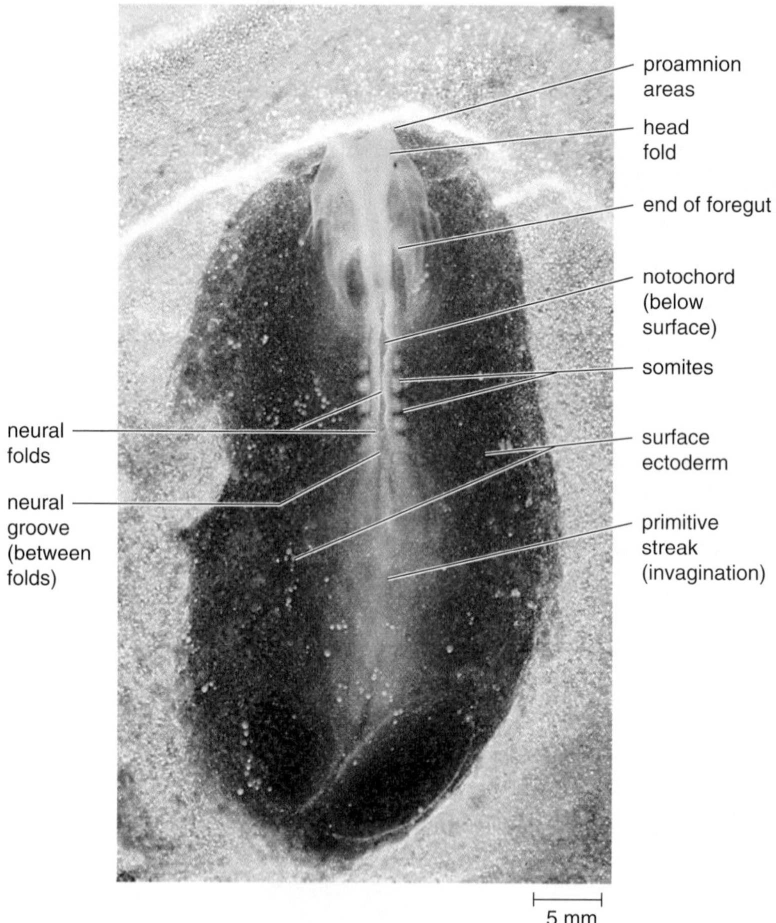

a. Photograph

d. **Primitive streak:** An elongated mass of cells. In the chick, the germ layers develop as flat sheets of cells. The mesodermal cells invaginate as the germ layers form and give rise to the early organs.

e. **Somites:** Blocks of developing muscle tissue that differentiate from mesoderm.

Figure 20.6 Twenty-four-hour chick embryo—*continued.*

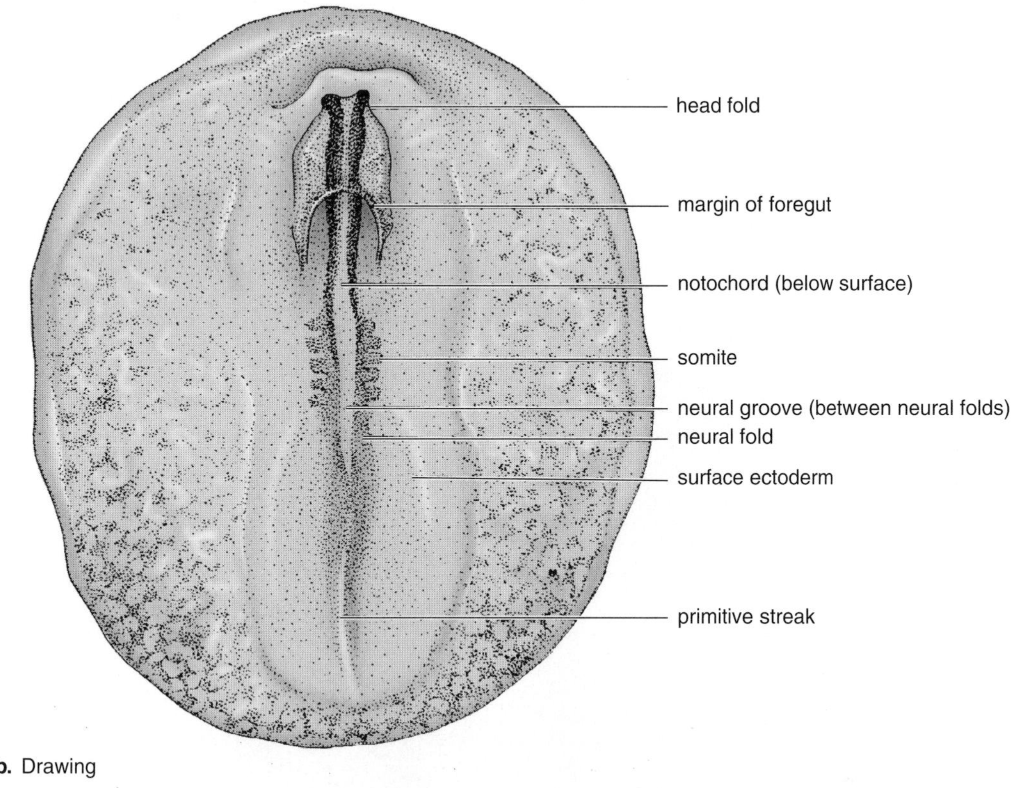

b. Drawing

Observing Live Chick Embryos

Use the following procedure for selecting and opening the eggs of live chick embryos:

1. Choose an egg of the proper age to remove from the incubator, and put a penciled X on the uppermost side. The embryo is just below the shell.
2. Add warmed chicken Ringer solution to a finger bowl until the bowl is about half full. (Chicken Ringer solution is an isotonic salt solution for chick tissue that maintains the living state.) The chicken Ringer solution should not cover the yolk of the egg.
3. On the edge of the dish, gently crack the egg on the side opposite the X.
4. With your thumbs placed over the X, hold the egg in the chicken Ringer solution while you pry it open from below and allow its contents to enter the solution. If you open the egg too slowly or too quickly, the shell may damage the delicate membranes surrounding the embryo.

Observation: Forty-Eight-Hour Chick Embryo

1. Follow the standard procedure (see p. 271) for selecting and opening an egg containing a 48-hour chick embryo.
2. The embryo has started to twist so that the head region is lying on its side. Refer to Figure 20.7*a* and *b*, and identify the following:
   a. **Shape of the embryo:** Embryo has started to bend. The head is now almost touching the heart.
   b. **Heart:** Has begun contracting and circulating blood. Can you make out the heart, and the aortic arches in the region below the head? Later, only one aortic arch will remain.

Figure 20.7 Forty-eight-hour chick embryo.
The most prominent organs are labeled.

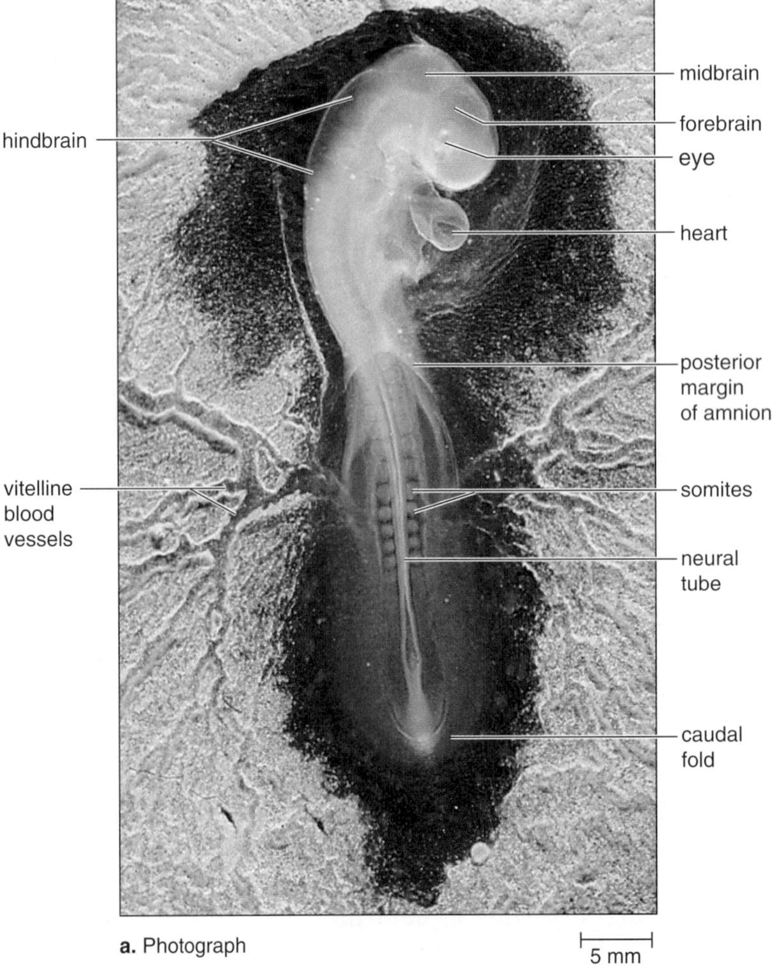

a. Photograph

c. **Vitelline arteries** and **veins:** Both extend over the yolk. The vitelline veins carry nutrients from the yolk to the embryo.

d. **Brain:** Has several distinct regions.

e. **Eye:** Has a developing lens.

f. **Margin (edge) of the amnion:** Can be seen above the vitelline arteries.

g. **Somites:** Now number 24 pairs.

h. **Caudal fold** of the amnion: The embryo will be completely enveloped when the head fold and caudal fold meet the margin of the amnion.

Figure 20.7 Forty-eight-hour chick embryo—*continued.*

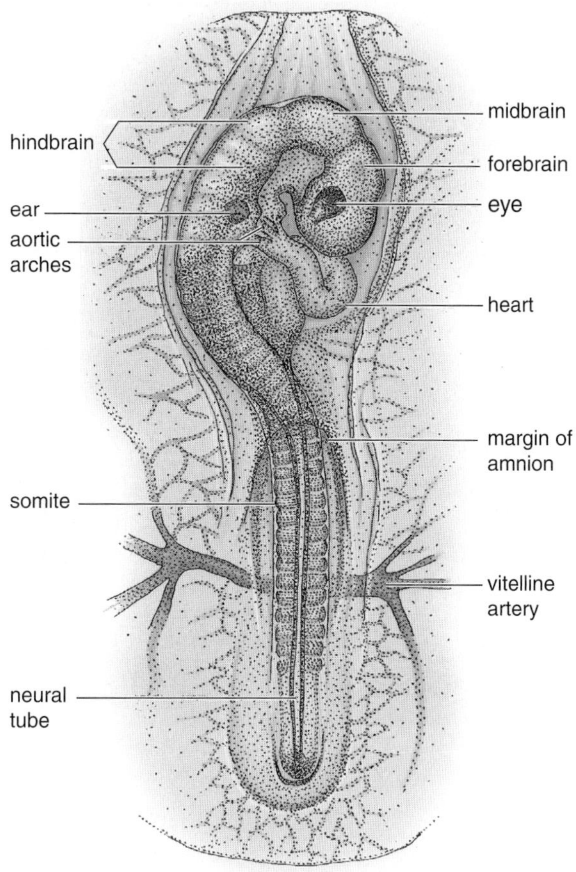

b. Drawing

Observation: Seventy-Two-Hour Chick Embryo

1. Follow the standard procedure (see p. 271) for selecting and opening an egg containing a 72-hour chick embryo.
2. The chick is now almost completely on its side, and the brain is even more flexed than before. Refer to Figure 20.8, and identify the following:
   a. **Brain:** Has a Ɔ shape and is very prominent.
   b. **Eye:** Has a distinctive lens.

Figure 20.8 Seventy-two-hour chick embryo.
Sense organs and limb buds have appeared.

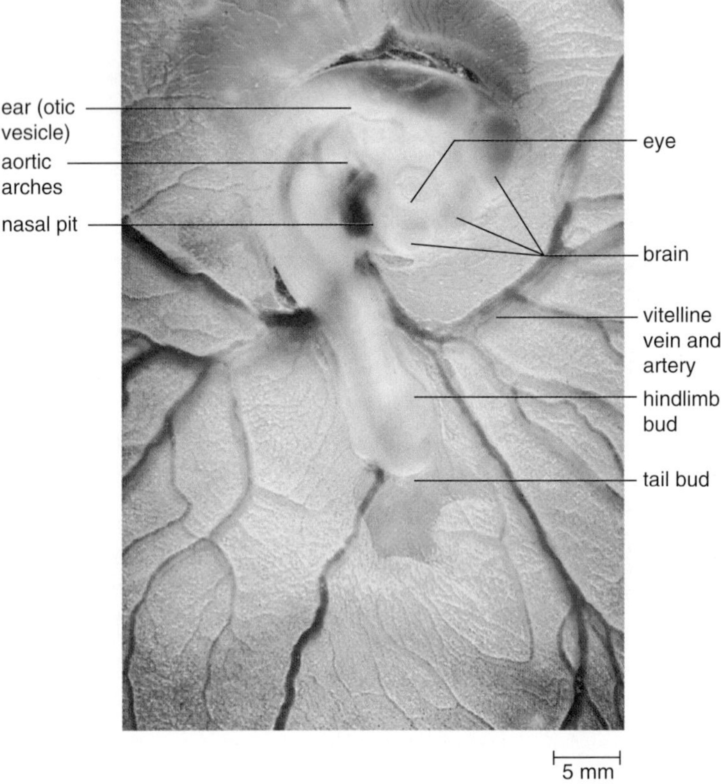

a. Photograph

c. **Ear (otic vesicle):** Is enlarged and more noticeable than before.

d. **Heart:** Has a distinctive ventricle and atrium and is actively pumping blood. There are pharyngeal pouches between the aortic arches.

e. **Vitelline blood vessels:** Extend over the yolk.

f. **Limb buds:** Become the wings and hind limbs. The posterior limb buds are easier to see at this point.

g. **Somites:** Now number 36 pairs.

h. **Tail bud:** Marks the end of the embryo.

Figure 20.8 Seventy-two-hour chick embryo—*continued.*

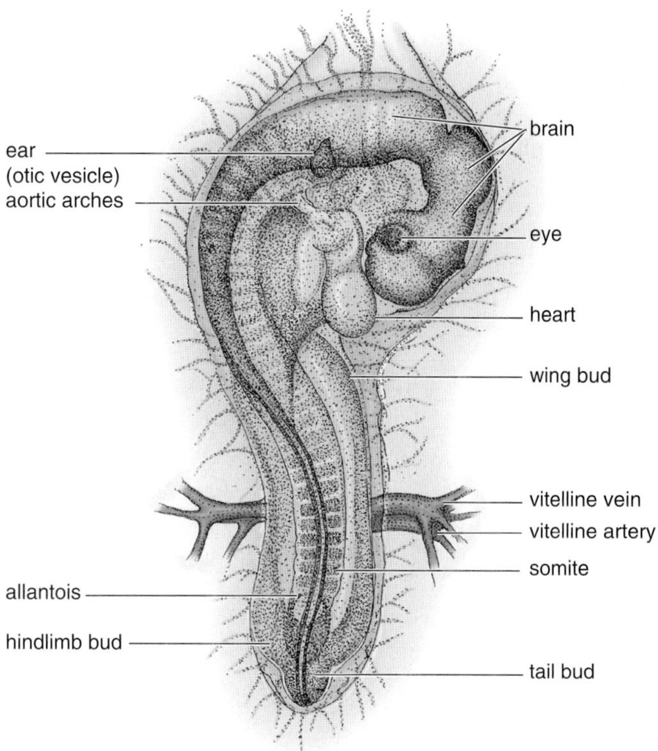

b. Drawing

Observation: Older Chick Embryos

As a chick embryo continues to grow, various organs differentiate further (Fig. 20.9). The neural tube closes along the entire length of the body and is now called the spinal cord. The allantois, an extraembryonic membrane, is seen as a sac extending from the ventral surface of the hindgut near the tail bud. The digestive system forms specialized regions, and there is both a mouth and an anus. The yolk sac, the extraembryonic membrane that encloses the yolk, is attached to the ventral wall, but when the yolk is used up, the ventral wall closes.

1. Follow the standard procedure (see p. 271) for selecting and opening an egg containing a 96-hour, or older, chick embryo.
2. How old is this embryo? ________ Describe what you see. ______________________________

____

____

Figure 20.9 Ninety-six-hour chick embryo.
Parts of the brain can now be seen. Courtesy of Carolina Biological Supply, Burlington, N.C.

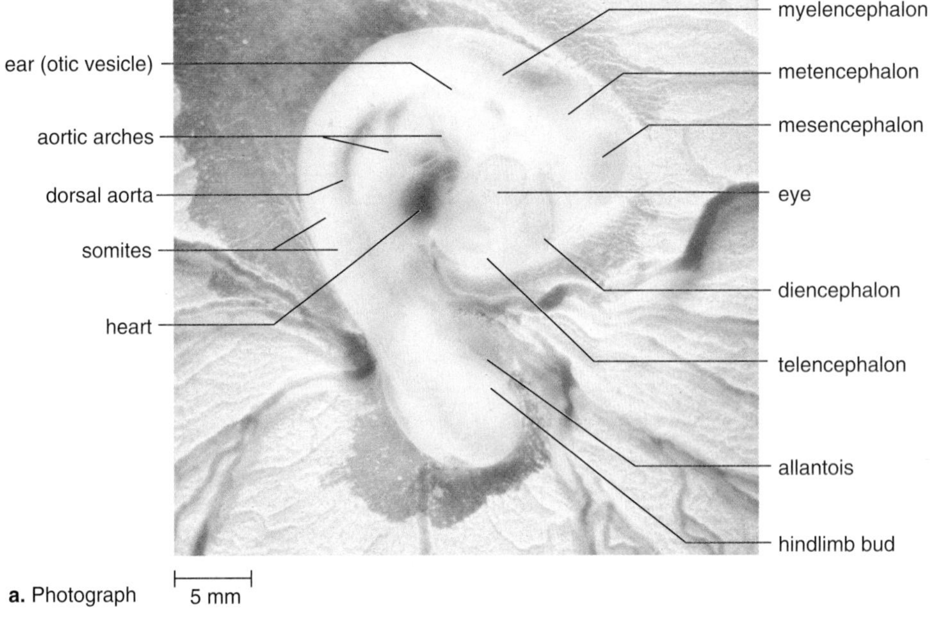

a. Photograph

Figure 20.9 **Ninety-six-hour chick embryo—*continued.***

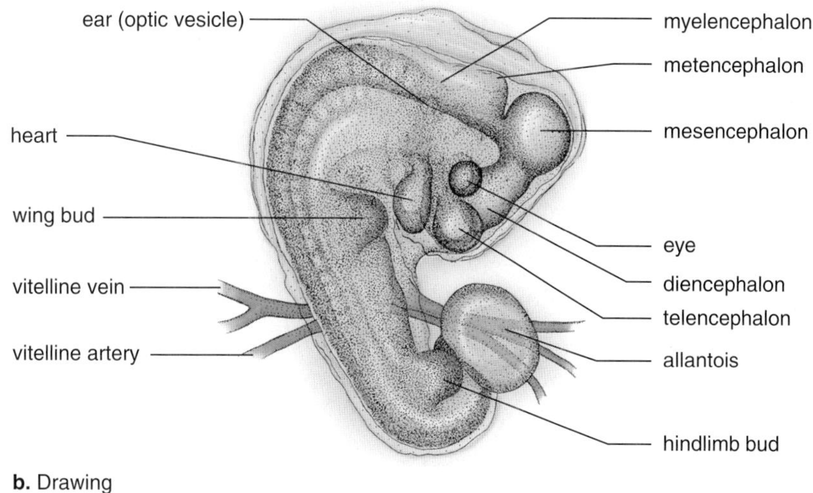

b. Drawing

Key:

Embryonic region	**Becomes**
Telencephalon	Cerebrum
Diencephalon	Thalamus, hypothalamus, posterior pituitary gland, pineal gland
Mesencephalon	Midbrain
Metencephalon	Cerebellum, pons
Myelencephalon	Medulla oblongata

20.4 Human Development

Human development is divided into preembryonic development, embryonic development, and fetal development. During **preembryonic development** (first two weeks after conception), the preembryo forms a ball of cells similar to a blastula, with an inner mass of cells that will become the embryo. As the preembryonic stage comes to an end, the germ layers begin to form and the extraembryonic membranes begin to develop. **Embryonic development** begins the third week after conception and continues through roughly the eighth to tenth week after conception. During this period, the germ layers and extraembryonic membranes complete their development. Once the germ layers have formed, they can initiate the formation of the internal organs (Fig. 20.10). At the end of the embryonic period, the embryo has a human appearance. During **fetal development** (a period that lasts for nearly seven months), the skeleton becomes ossified (bony), reproductive organs form, arms and legs fully develop, and the fetus enlarges in size and gains weight (Fig. 20.11).

Observation: Human Development

1. Study models of human development. Distinguish as many extraembryonic membranes as possible.
2. Try to determine the ages of your models by comparing them with Figures 20.10 and 20.11.

Figure 20.10 Human development.
Changes occurring during the third to the fifth week include the development of the extraembryonic membranes and the umbilical cord.

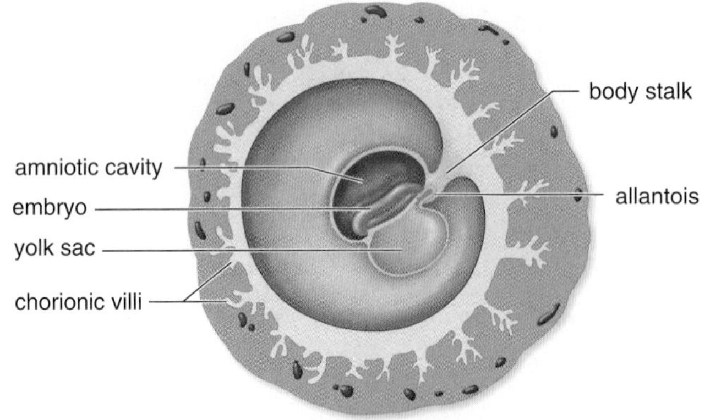

21 days

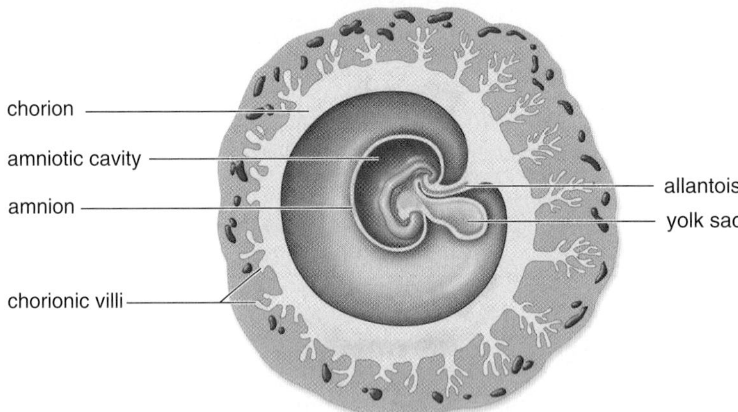

25 days

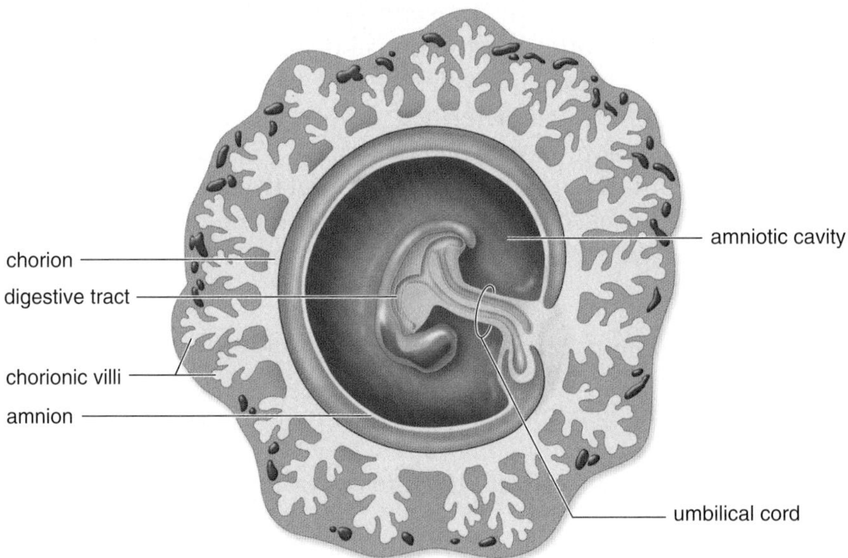

35+ days

Figure 20.11 Human development.
Changes occurring from the fifth week to the eighth month.

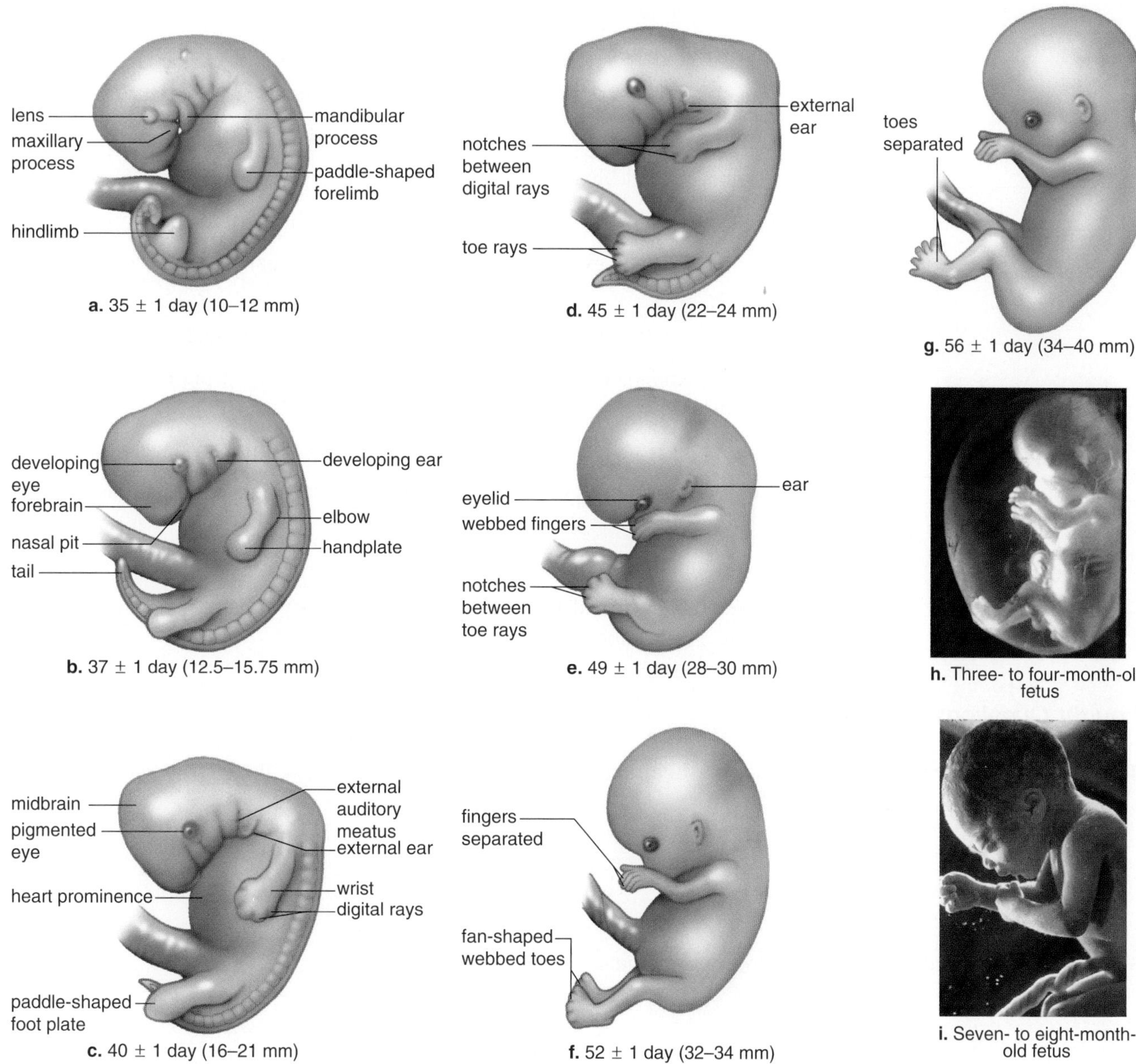

Laboratory Review 20

______________ 1. By what process does the embryo divide with no increase in size?

______________ 2. Name the stage of development when the embryo is a solid ball of cells.

______________ 3. Name the stage of development when the embryonic germ layers are forming.

______________ 4. The nervous system develops from which embryonic germ layer?

______________ 5. The intestinal tract develops from which embryonic germ layer?

______________ 6. What structure in an embryo induces formation of the nervous system?

______________ 7. Which group has four extraembryonic membranes—aquatic animals or land animals?

______________ 8. Which two organ systems appear before the others during chick development?

______________ 9. Is the human embryo or the fetus most likely to resemble a pig during development?

______________ 10. What term is used to refer to the last seven months of human development?

______________ 11. How does the morula stage of sea star development differ from the blastula stage?

______________ 12. What is the function of the amnion?

______________ 13. What structures are found in the 72-hour chick embryo that are not found in the 48-hour chick embryo?

Thought Questions

14. What factor causes a frog's morula, blastula, and gastrula to look different than those of a sea star?

15. Describe how induction may control development.

16. Why does the sea star embryo lack the yolk associated with the chicken embryo?

LABORATORY

21

Patterns of Inheritance

Learning Outcomes

21.1 One-Trait Crosses

- Explain how meiosis plays a central role in Mendel's law of segregation.
- Demonstrate the ability to differentiate the possible phenotypes and genotypes in individuals that either express a single trait or that do not.
- Describe the life cycle of *Drosophila* (fruit fly), and be able to discriminate these stages in a population of flies in a culture bottle.
- Use a Punnett square to predict the results of a monohybrid cross in both tobacco seedlings and *Drosophila.*

21.2 Human Inheritance

- Using data gathered by observing traits expressed by students in the laboratory, and their own observations regarding the expression of the traits in their parents and siblings, identify the possible student genotypes for the traits under consideration.
- Using Punnett squares, solve genetics problems involving autosomal dominant and autosomal recessive alleles.

21.3 Two-Trait Crosses

- Explain how meiosis plays a role in Mendel's law of independent assortment and illustrate how the two laws, segregation and independent assortment, are different.
- Demonstrate the ability to differentiate the possible phenotypes and genotypes in individuals that either express two separate traits, that express one but not the other, or that do not express either trait.
- Use a Punnett square to illustrate the results of dihybrid crosses in both corn plants and *Drosophila.*

21.4 X-Linked Crosses

- Demonstrate how the results of a cross in *Drosophila* for X-linked genes confirm what was predicted using a Punnett square cross.

21.5 Chi-Square Analysis

- Utilize the chi-square statistical test to help determine whether data do or do not support a hypothesis.

Introduction

Gregor Mendel, sometimes called the "father of genetics," formulated the basic laws of genetics examined in this laboratory. He determined that individuals have two alternate forms of a gene (two **alleles,** in modern terminology) for each trait in their body cells. Today, we know that alleles occur at specific locations on the chromosomes (Fig. 21.1). An individual can be homozygous dominant (two dominant alleles, *AA*), homozygous recessive (two recessive alleles, *aa*), or heterozygous (one dominant and one recessive allele, *Aa*). **Genotype** refers to an individual's genes, while **phenotype** refers to an individual's appearance. Homozygous dominant and heterozygous individuals show the dominant phenotype; homozygous recessive individuals show the recessive phenotype.

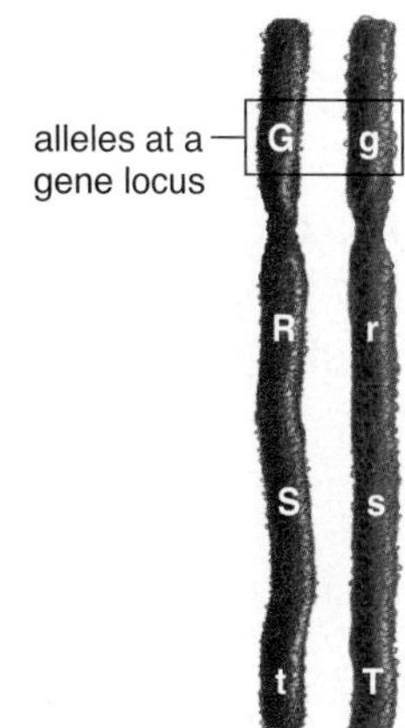

Figure 21.1 Gene locus. Each allelic pair, such as *Gg,* is located on a pair of chromosomes at a particular gene locus.

21.1 One-Trait Crosses

A single pair of alleles is involved in one-trait crosses. Mendel found that reproduction between two heterozygous individuals (*Aa*) resulted in both dominant and recessive phenotypes among the offspring. The expected phenotypic ratio among the offspring was 3:1. Three offspring had the dominant phenotype for every one that had the recessive phenotype.

Mendel realized that these results were obtainable only if the alleles of each parent separated during meiosis (otherwise, all offspring would inherit a dominant allele, and no offspring would be homozygous recessive). Therefore, Mendel formulated his first law of inheritance:

Law of Segregation

Each organism contains two alleles for each trait, and the alleles segregate during the formation of gametes. Each gamete then contains only one allele for each trait. When fertilization occurs, the new organism has two alleles for each trait, one from each parent.

Inheritance is a game of chance. Just as there is a 50% probability of heads or tails when tossing a coin, there is a 50% probability that a sperm or an egg will have an *A* or an *a* when the parent is *Aa*. Meiosis segregates the two alleles during the first of its two cell divisions. The chance of an equal number of heads or tails improves as the number of tosses increases. In the same way, the chance of an equal number of gametes with *A* and *a* improves as the number of cells going through meiosis and the resulting gametes increases. Therefore, the 3:1 ratio among offspring is more likely when a large number of sperm fertilize a large number of eggs.

Color of Tobacco Seedlings

In tobacco plants, a dominant allele (*C*) for chlorophyll gives the plants a green color, and a recessive allele (*c*) for chlorophyll causes a plant to appear white. If a tobacco plant is homozygous for the recessive allele (*c*), it cannot manufacture chlorophyll and thus appears white (Fig. 21.2).

Figure 21.2 Monohybrid cross.
These tobacco seedlings are growing on an agar plate. The white plants cannot manufacture chlorophyll.

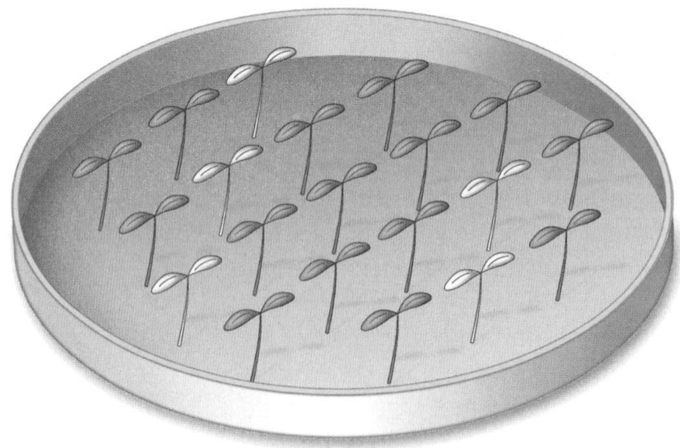

Experimental Procedure: Color of Tobacco Seedlings

1. Obtain a numbered agar plate on which tobacco seedlings are growing. They are the offspring of the cross *Cc* × *Cc*. Complete the Punnett square. What is the expected phenotypic ratio?

   ____

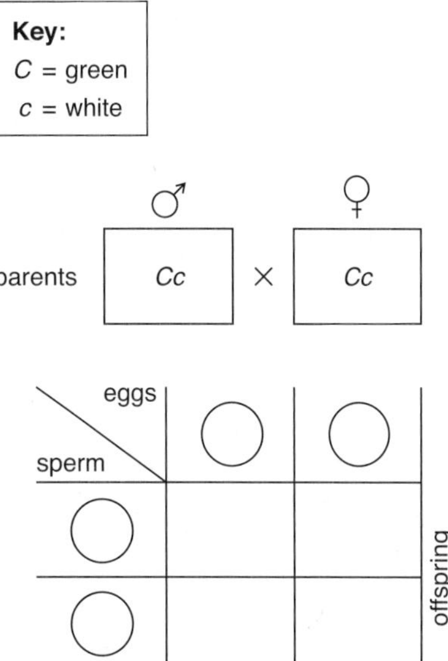

2. Using a binocular dissecting microscope, view the seedlings, and count the number that are green and the number that are white. Record the plate number and your results in Table 21.1.
3. Repeat steps 1 and 2 for two additional plates. Total the number that are green and the number that are white.
4. Complete Table 21.1 by recording the class data.

Table 21.1 Color of Tobacco Seedlings

	Number of Offspring	
	Green Color	**White Color**
Plate # ______		
Plate # ______		
Plate # ______		
Totals		
Class data		

Conclusions

- Calculate the actual phenotypic ratio you observed. ________ Do your results differ from the expected ratio? ________ Explain. ______________________________
  ____
- Do a chi-square test (see the end of this laboratory) to determine if the deviation from the expected results can be accounted for by chance alone. Chi-square test results: ____________
- Repeat these steps using the class data. Do your class data give a ratio that is closer to the expected ratio, and is the chi-square deviation insignificant? ________ Explain. ____________
  ____
- What results would you expect if the parent plants had the following genotypes?

  Cc × cc

Drosophila melanogaster

Both the adults and the larvae of the fruit fly *Drosophila melanogaster* feed on plant sugars and on the wild yeasts that grow on rotting fruit. The female flies first lay **eggs** on the same materials. After a day or two, the eggs hatch into small **larvae** that feed and grow for about eight days, depending on the temperature. During this period, they **molt** twice. Therefore, three larval periods of growth, called **instars,** occur between molts. When fully grown, the third-stage instar larvae cease feeding and **pupate.** During pupation, which lasts about four days, the larval tissues are reorganized to form those of the adult. The life cycle is summarized in Figure 21.3.

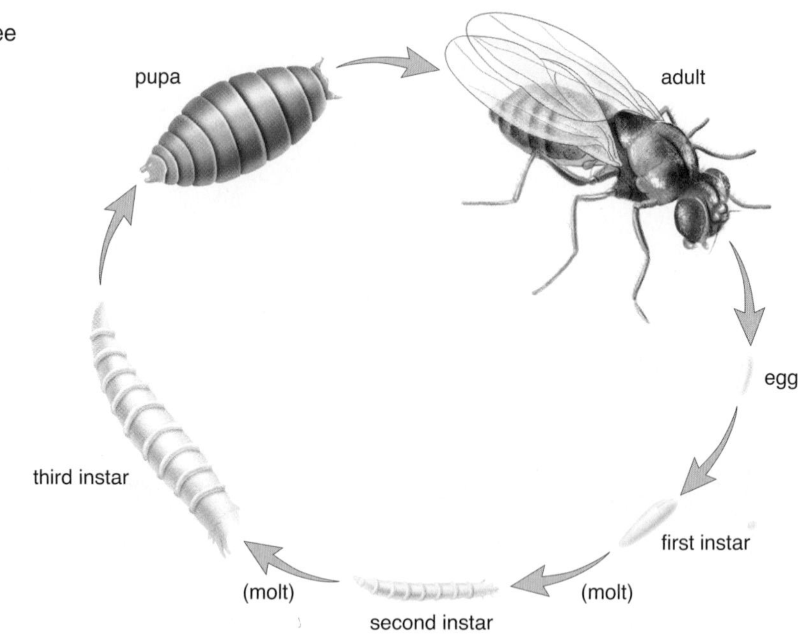

Figure 21.3 Life cycle of *Drosophila.* The adult female lays eggs that go through three larval stages before pupating. Metamorphosis, which occurs during pupation, produces the adult fly.

Observation: *Drosophila melanogaster*

Drosophila breeding experiments may run over a period of three weeks. Follow instructions carefully in order to obtain accurate results.

Culture

You will be provided with a stock culture of *Drosophila melanogaster* to examine. Answer the following questions:

1. Where in the culture vial are the adult flies? __________
2. The eggs? __________
3. The larvae? __________
4. The pupae? __________

Flies

You will be provided with slides, frozen flies, or live flies to examine. If the flies are alive, follow the directions on page 286 for using FlyNap™ to anesthetize them.

1. Put frozen or anesthetized flies on a white card, and use a camel-hair brush to move them around. Use a binocular dissecting microscope or a hand lens in order to see the flies clearly.
2. If you are looking at slides, you may use the scanning lens of your compound light microscope to view the flies.
3. Examine wild-type flies. Complete Table 21.2 for wild-type flies. Long wings extend beyond the body, and vestigial (short) wings do not extend beyond the body.
4. Examine mutant flies. Complete Table 21.2 for all the mutant flies you examined.

Table 21.2 Characteristics of Wild-Type and Mutant Flies

	Wild-Type	Ebony Body	Vestigial-Wing	Sepia-Eye	White-Eye
Wing length					
Color of eyes					
Color of body					

5. Use the following characteristics to distinguish male flies from female flies (Fig. 21.4):
   - The male is generally smaller.
   - The male has a more rounded abdomen than the female. The female has a pointed abdomen.

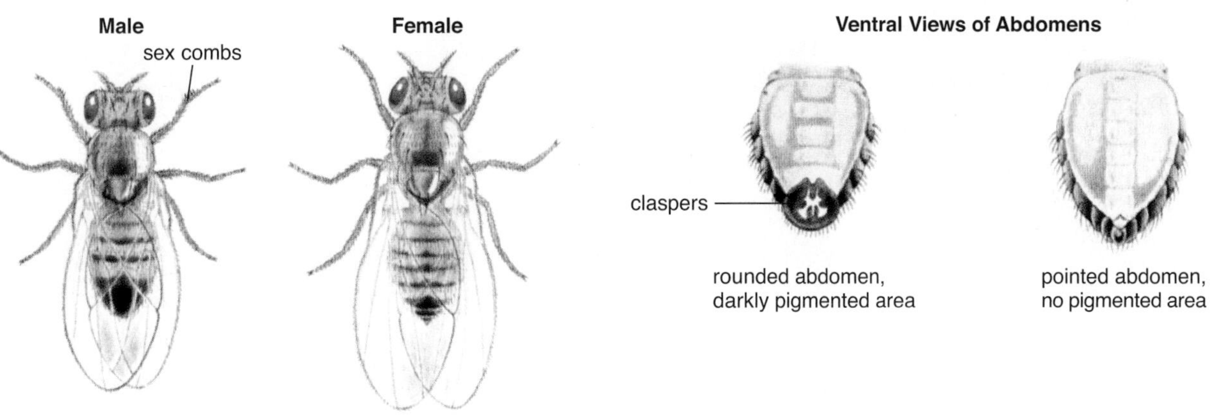

Figure 21.4 ***Drosophila* male versus *Drosophila* female.**

- The male has sex combs on the forelegs.
- Dorsally, the male is seen to have a black-tipped abdomen, whereas the female appears to have dark lines only at the tip.
- Ventrally, the abdomen of the male has a dark region at the tip due to the presence of claspers; this dark region is lacking in the female.

Anesthetizing Flies

The use of FlyNap™ allows flies to be anesthetized for at least 50 minutes without killing them.

1. Dip the absorbent end of a swab into the FlyNap™ bottle, as shown in Figure 21.5.
2. Tap the bottom of the culture vial on the tabletop to knock the flies to the bottom of the vial. (If the medium is not firm, you may wish to transfer the flies to an empty bottle before anesthetizing them. Ask your instructor for assistance, if this is the case.)
3. With one finger, push the plug slightly to one side. Remove the swab from the FlyNap™, and quickly stick the anesthetic end into the culture vial beside the plug so that the anesthetic tip is below the plug. Keep the culture vial upright with the swab in one place.
4. Remove the plug and the swab immediately after the flies are anesthetized (approximately 2 minutes in an empty vial, 4 minutes in a vial with medium), and spill the flies out onto a white file card. The length of time the flies remain anesthetized depends on the amount of FlyNap™ on the swab and on the number and age of the flies in the culture vial.
5. Transfer the anesthetized flies from the white file card onto the glass plate of a binocular dissecting microscope for examination, or use a hand lens.

Figure 21.5 FlyNap™.
Flies can be anesthetized for at least 50 minutes without being killed in the FlyNap™.

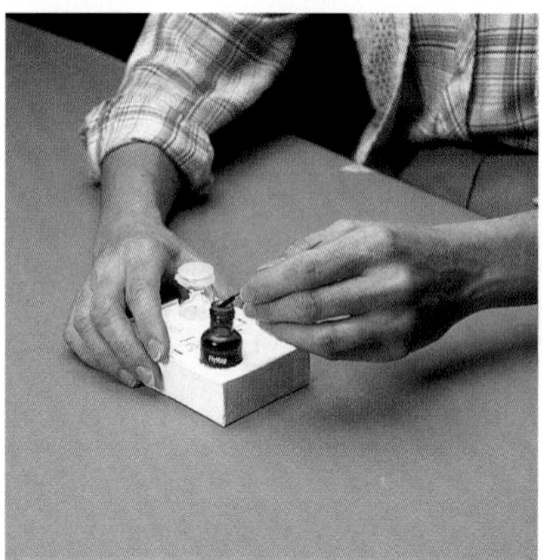

a. Moisten swab in FlyNap™.

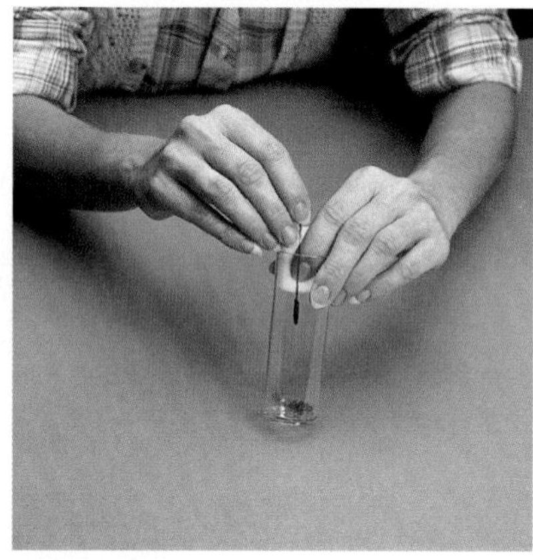

b. Insert swab into culture bottle.

c. Remove flies as shown.

Wing Length in *Drosophila*

In fruit flies, the allele for long wings (*L*) is dominant over the allele for vestigial (short) wings (*l*). You will be examining the results of the cross *Ll* × *Ll*. Complete this Punnett square:

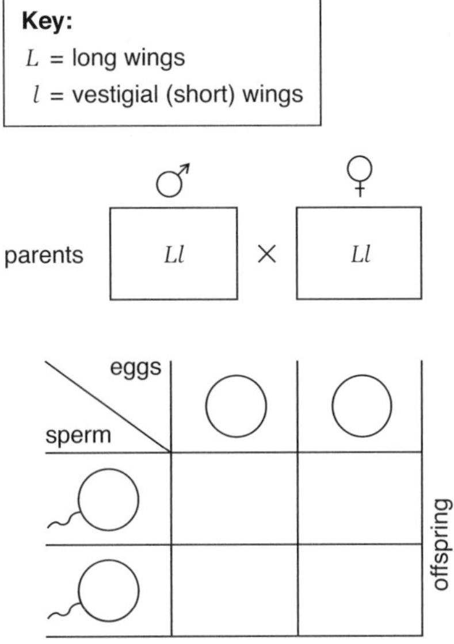

What is the expected phenotypic ratio among the offspring? ____________________

Experimental Procedure: Wing Length in Drosophila

The cross described here will take three weeks. The experiment can be done in one week, however, if your instructor provides you with a vial that already contains the results of the cross *Ll* × *Ll*. In this case, proceed directly to step 3.

1. **Week 1:** Place heterozygous flies in a prepared culture vial. Your instructor will show you how to use instant medium and dry yeast to prepare the vial. Label your culture. What is the phenotype of heterozygous flies? ____________________

   What is the genotype of heterozygous flies? ____________________
2. **Week 2:** Remove the heterozygous flies from the vial before their offspring pupate. Why is it necessary to remove these flies before you observe your results? ____________________
3. **Week 3:** Observe the results of the cross by counting the offspring. Follow the directions on page 286 for anesthetizing and removing flies. When counting, use the binocular dissecting microscope or a hand lens. Divide your flies into those with long wings and those with short wings. Record your results and the class results in Table 21.3.

Table 21.3 Wing Length in *Drosophila*

	Number of Offspring	
	Long Wings	**Vestigial Wings**
Your data		
Class data		

Conclusions

- Calculate the actual phenotypic ratio you observed. ________ Do your results differ from the expected ratio? ________ Explain. ________
- Do a chi-square test (see the end of this laboratory) to determine if the deviation from the expected results can be accounted for by chance alone. Chi-square test results: ________
- Repeat these steps using the class data. Do your class data give a ratio that is closer to the expected ratio, and is the chi-square deviation insignificant? ________ Explain. ________
- What is the genotype of a fly with short wings? ________

21.2 Human Inheritance

Human beings are subject to the same laws of genetics as are tobacco seedlings, corn, and fruit flies. Figure 21.6 illustrates both the phenotypes and associated genotypes for some traits that we will be considering.

Figure 21.6 Commonly inherited traits in human beings.
The alleles indicate which traits are dominant and which are recessive.

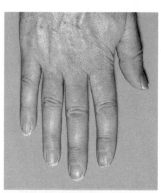

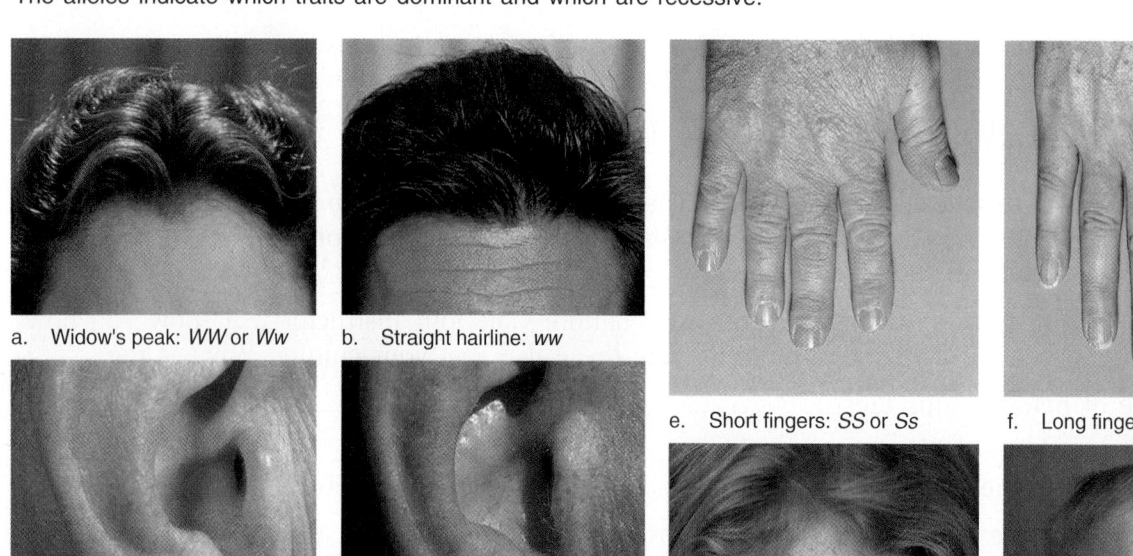

a. Widow's peak: *WW* or *Ww* b. Straight hairline: *ww*

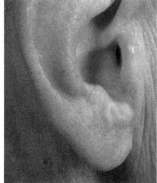

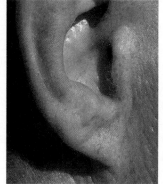

c. Unattached earlobes: *EE* or *Ee* d. Attached earlobes: *ee*

e. Short fingers: *SS* or *Ss* f. Long fingers: *ss*

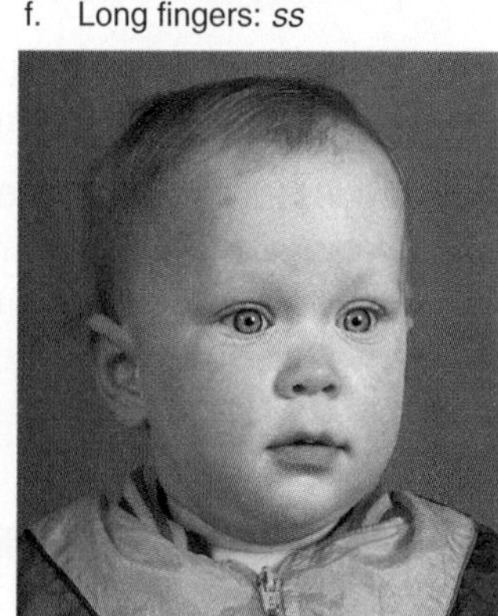

g. Freckles: *FF* or *Ff* h. No freckles: *ff*

Autosomal Dominant and Recessive Traits

The alleles for autosomal traits are carried on the non-sex chromosomes. If individuals are homozygous dominant (*AA*) or heterozygous (*Aa*), their phenotype is the dominant trait. If individuals are homozygous recessive (*aa*), their phenotype is the recessive trait.

Experimental Procedure: Human Traits

1. For this Experimental Procedure, you will need a lab partner to help you determine your phenotype for the traits listed in the first column of Table 21.4.
2. Determine your probable genotype. If you have the recessive phenotype, you know your genotype. If you have the dominant phenotype, you may be able to decide whether you are homozygous dominant or heterozygous by recalling the phenotype of your parents, siblings, or children. Circle your probable genotype in the second column of Table 21.4.
3. Your instructor will tally the class's phenotypes for each trait so that you can complete the third column of Table 21.4.
4. Complete Table 21.4 by calculating the percentage of the class with each trait. Are dominant phenotypes always the most common in a population? ____________ Explain. ____________

   ____

Table 21.4 Autosomal Human Traits

Trait: d = Dominant r = Recessive	Possible Genotypes	Number in Class	Percentage of Class with Trait
Hairline:			
Widow's peak (d)	*WW* or *Ww*	________	
Straight hairline (r)	*ww*		
Earlobes:			
Unattached (d)	*UU* or *Uu*	________	
Attached (r)	*uu*		
Skin pigmentation:			
Freckles (d)	*FF* or *Ff*	________	
No freckles (r)	*ff*		
Hair on back of hand:			
Present (d)	*HH* or *Hh*	________	
Absent (r)	*hh*		
Thumb hyperextension—"hitchhiker's thumb":			
Last segment cannot be bent backward (d)	*TT* or *Tt*	________	
Last segment can be bent back to 60° (r)	*tt*		
Bent little finger:			
Little finger bends toward ring finger (d)	*LL* or *Ll*	________	
Straight little finger (r)	*ll*		
Interlacing of fingers:			
Left thumb over right (d)	*II* or *Ii*	________	
Right thumb over left (r)	*ii*		

Genetics Problems

1. Nancy and the members of her immediate family have attached earlobes. Her maternal grandfather has unattached earlobes. What is the genotype of her maternal grandfather? ________ Nancy's maternal grandmother is no longer living. What could have been the genotype of her maternal grandmother? ____________________
2. Joe does not have a bent little finger, but his parents do. What is the expected phenotypic ratio among the parents' children? ____________________
3. Henry is adopted. He has hair on the back of his hand. Could both of his parents have had hair on the back of the hand? ________ Could both of his parents have had no hair on the back of the hand? ________ Explain. ____________________
____
4. Tarzan and Jane both have a widow's peak. Is it possible for them to have a child without one?
____

21.3 Two-Trait Crosses

Two-trait crosses involve two pairs of alleles. Mendel found that when two dihybrid individuals (*AaBb*) reproduce, the expected phenotypic ratio among the offspring is 9:3:3:1, representing four possible phenotypes. He realized that these results could be obtained only if the alleles of the parents separated independently of one another when the gametes were formed. From this, Mendel formulated his second law of inheritance:

> **Law of Independent Assortment**
> **Members of an allelic pair segregate (assort) independently of members of another allelic pair. Therefore, all possible combinations of alleles can occur in the gametes.**

20 mm

Figure 21.7 Dihybrid cross. Four types of kernels are seen on an ear of corn following a dihybrid cross: purple smooth, purple rough, yellow smooth, and yellow rough.

Color and Texture of Corn

In corn plants, the allele for purple kernel (*P*) is dominant over the allele for yellow kernel (*p*), and the allele for smooth kernel (*S*) is dominant over the allele for rough kernel (*s*) (Fig. 21.7).

Experimental Procedure: Color and Texture of Corn

1. Obtain an ear of corn from the supply table. You will be examining the results of the cross *PpSs* × *PpSs*. Complete this Punnett square:

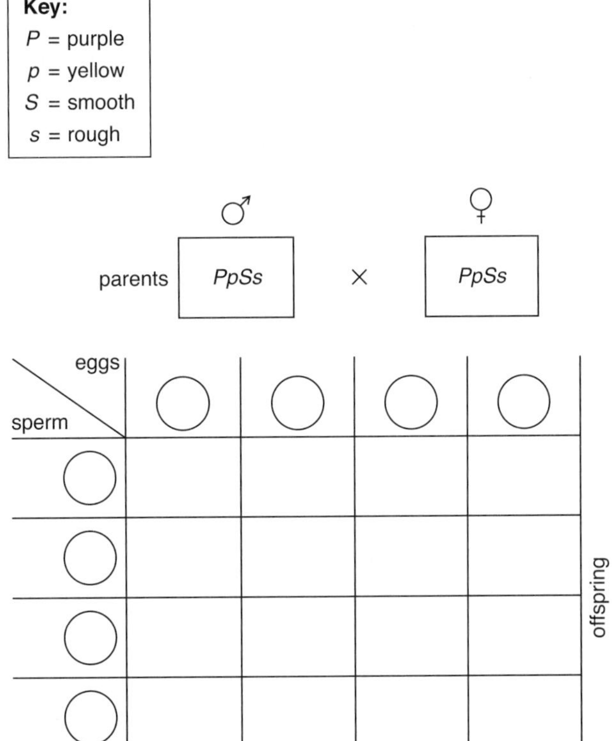

What is the expected phenotypic ratio among the offspring? ____________

2. Count the number of kernels of each possible phenotype listed in Table 21.5. Record the sample number and your results in Table 21.5. Use three samples, and total your results for all samples. Also record the class data.

Table 21.5 Color and Texture of Corn

	Number of Kernels			
	Purple Smooth	**Purple Rough**	**Yellow Smooth**	**Yellow Rough**
Sample # ______				
Sample # ______				
Sample # ______				
Totals				
Class data				

Conclusions

- From your data, which two traits seem dominant? ________ and ________________________
  Which two traits seem recessive? ________ and ________________________
- Calculate the actual phenotypic ratio you observed. ________ Do your results differ from the expected ratio? ________________________
- Use the chi-square test (see the end of this laboratory) to determine if the deviation from the expected results can be accounted for by chance alone. Chi-square test results: ____________
- Repeat these steps using the class data. Do your class data give a ratio that is closer to the expected ratio, and is the chi-square deviation insignificant? ________ Explain. ____________
  ____

Wing Length and Body Color in *Drosophila*

In *Drosophila,* long wings (L) are dominant over vestigial (short) wings (l), and gray body (G) is dominant over ebony (black) body (g). You will be examining the results of the cross $LlGg \times LlGg$. Complete this Punnett square:

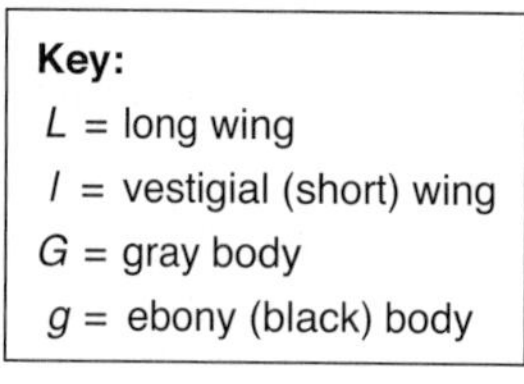

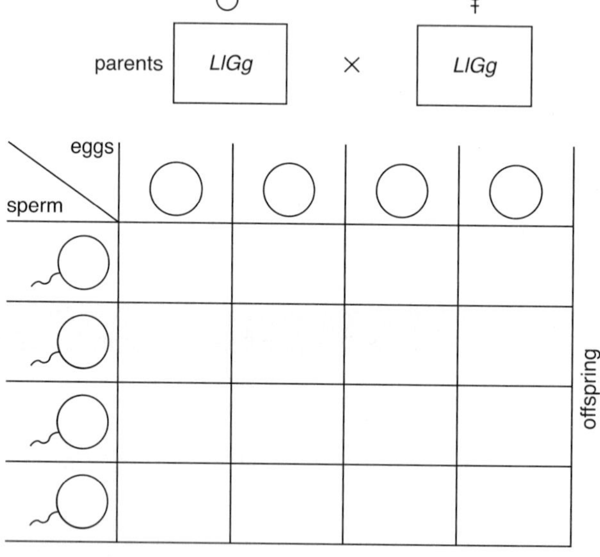

What are the expected phenotypic results of this cross? ________________________

Experimental Procedure: Wing Length and Body Color in Drosophila

The cross described here will take three weeks. The experiment can be done in one week if your instructor provides you with a vial that already contains the results of the cross *LlGg* × *LlGg*. In this case, proceed directly to step 3.

1. **Week 1:** Place heterozygous flies in a prepared culture vial. Your instructor will show you how to use instant medium and dry yeast to prepare the vial. Label your culture.

   What is the phenotype of heterozygous flies? ______________________

   What is the genotype of heterozygous flies? ______________________
2. **Week 2:** Remove the heterozygous flies from the vial before their offspring pupate. Why is it necessary to remove these flies before you observe your results? ______________________
3. **Week 3:** Observe the result of the cross by counting the offspring. Follow the standard directions (see p. 286) for anesthetizing and removing flies. Find one fly of each phenotype, and check with the instructor that you have identified them correctly before proceeding. When counting, use the binocular dissecting microscope or a hand lens. Divide your flies into the groups indicated in Table 21.6, and record your results. Also record the class data.

Table 21.6 Wing Length and Body Color in *Drosophila*

	Phenotypes			
	Long Wings, Gray Body	**Long Wings, Ebony Body**	**Vestigial Wings, Gray Body**	**Vestigial Wings, Ebony Body**
Number of offspring				
Class data				

Conclusions

- Calculate the actual phenotypic ratio you observed. ________ Do your results differ from the expected ratio? ________ Explain. ______________________

  ____
- Do a chi-square test (see the end of this laboratory) to determine if the deviation from the expected results can be accounted for by chance alone. Chi-square test results: ____________
- Repeat these steps using the class data. Do your class data give a ratio that is closer to the expected ratio, and is the chi-square deviation insignificant? ________ Explain. ____________

  ____
- What results would you expect if the following cross is made:

  LlGg × llgg

21.4 X-Linked Crosses

In animals, some alleles occur only on the X chromosome, and therefore, these alleles are said to be X-linked. Females inherit two X chromosomes, and males inherit an X and a Y. Males of this type can never be heterozygous for alleles carried on the X chromosome since they only have a single copy of the chromosome.

Red/White Eye Color in *Drosophila*

In fruit flies, red eyes (X^R) are dominant over white eyes (X^r). You will be examining the results of the cross $X^RX^r \times X^RY$. Complete this Punnett square:

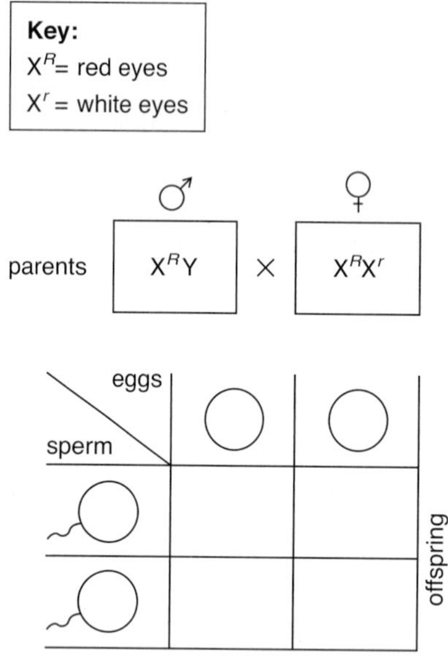

What are the expected phenotypic results of this cross?

Females: ______________________ Males: ______________________

Experimental Procedure: Red/White Eye Color in Drosophila

The cross described here will take three weeks. The experiment can be done in one week if your instructor provides you with a vial that already contains the results of the cross $X^RX^r \times X^RY$. In this case, proceed directly to step 3.

1. **Week 1:** Place the parental flies ($X^RX^r \times X^RY$) in a prepared culture vial. Your instructor will show you how to use instant medium and dry yeast to prepare the vial. Label your culture.

   What is the phenotype of the female and male flies you are using? ______________________

   What is the genotype of the female flies? ______________________

   What is the genotype of the male flies? ______________________
2. **Week 2:** Remove the parental flies from the vial before their offspring pupate.

   Why is it necessary to remove these flies before you observe your results? ______________________

   ____
3. **Week 3:** Observe the results of the cross by counting the offspring. Follow the standard directions (see p. 286) for anesthetizing and removing flies. When counting, use the binocular dissecting microscope or a hand lens. Divide your flies into the following groups: (1) red-eyed males, (2) red-eyed females, (3) white-eyed males, and (4) white-eyed females. Record your results in Table 21.7. Also record the class data.

Table 21.7 Red/White Eye Color in *Drosophila*

	Number of Offspring	
Your Data	**Red Eyes**	**White Eyes**
Males		
Females		
Class data Males		
Females		

Conclusions

- Calculate the actual phenotypic ratio you observed for males and females separately.

  Males: ____________________

  Females: ____________________
- Do your results differ from the expected ratio? ____________________
- Do a chi-square test (see the next section) to determine if the deviation from the expected results can be accounted for by chance alone. Chi-square test results: ____________________
- Repeat these steps using the class data. Do your class data give a ratio that is closer to the expected ratio, and is the chi-square deviation insignificant? ________ Explain. ____________________

  ____
- What results would you expect from the following cross?

  $X^RY \times X^rX^r$

21.5 Chi-Square Analysis

Your experimental results can be evaluated using the **chi-square** (χ^2) test. This is the statistical test most frequently used to determine whether data obtained experimentally provide a "good fit," or approximation, to the expected or theoretical data. Basically, the chi-square test can determine whether any deviations from the expected values are due to chance. Chance alone can cause the actual observed ratio to vary somewhat from the calculated ratio for a genetic cross. For example, a ratio of exactly 3:1 for a monohybrid cross is only rarely observed. Actual results will differ, but at some point, the difference is so great as to be unexpected. The chi-square test indicates this point. After this point, the original hypothesis (for example, that a monohybrid cross gives a 3:1 ratio) is not supported by the data.

The formula for the test is $\chi^2 = \Sigma(d^2/e)$

where χ^2 = chi-square

Σ = sum of

d = difference between expected and observed results (most often termed the *deviation*)

e = expected results

For example, in a one-trait cross involving fruit flies with long wings (dominant) and fruit flies with short wings (recessive), a 3:1 ratio is expected in the second generation (F_2). Therefore, if you count 160 flies, 40 are expected to have short wings, and 120 are expected to have long wings. But if you count 44 short-winged flies and 116 long-winged flies, then the value for the chi-square test would be as calculated in Table 21.8.

Table 21.8 Calculation of Chi-Square (Example)

Phenotype	Observed Number	Expected Results (e)	Difference (d)	d^2	Partial Chi-Square (d^2/e)
Short wings	44	40	4	16	16/40 = 0.400
Long wings	116	120	4	16	16/120 = 0.133
			Chi-square = $\chi^2 = \Sigma(d^2/e) = 0.400 + 0.133 = 0.533$		

Now look up this chi-square value (χ^2) in a table that indicates whether the probability (p) is that the differences noted are due only to chance in the form of random sampling error or whether the results should be explained on the basis of a different prediction (hypothesis). In Table 21.9, the notation C refers to the number of "classes," which in this laboratory would be determined by the number of phenotypic traits studied. The example involves two classes: short wings and long wings. However, as indicated in the table by $C - 1$, it is necessary to subtract 1 from the total number of classes. In the example, $C - 1 = 1$. Therefore, the χ^2 value (0.533) would fall in the first line of Table 21.9 between 0.455 and 1.074. These correspond with p values of 0.50 and 0.30, respectively. This means that, by random chance, this difference between the actual count and the expected count would occur between 30% and 50% of the time. In biology, it is generally accepted that a p value greater than 0.05 is acceptable and that a p value lower than 0.05 indicates that the results cannot be due to random sampling and therefore do not support (do not "fit") the original prediction (hypothesis). A chi-square analysis is used to *refute* (falsify) a hypothesis, not to *prove* it.

Table 21.9 Values of Chi-Square

	Hypothesis Is Supported							Hypothesis Is Not Supported		
	Differences Are Insignificant							**Differences Are Significant**		
p	0.99	0.95	0.80	0.50	0.30	0.20	0.10	0.05	0.02	0.01
$C - 1$										
1	0.00016	0.0039	0.064	0.455	1.074	1.642	2.706	3.841	5.412	6.635
2	0.0201	0.103	0.446	1.386	2.408	3.219	4.605	5.991	7.824	9.210
3	0.115	0.352	1.005	2.366	3.665	4.642	6.251	7.815	9.837	11.341
4	0.297	0.711	1.649	3.357	4.878	5.989	7.779	9.488	11.668	13.277
5	0.554	1.145	2.343	4.351	6.064	7.289	9.236	11.070	13.388	15.086

Source: Data from W.T. Keeton, et al., *Laboratory Guide for Biological Science,* 1968, p. 189.

Use Table 21.10 for performing a chi-square analysis of your results from a previous Experimental Procedure in this laboratory. If you performed a one-trait (monohybrid) cross, you will use only the first two lines. If you performed a two-trait (dihybrid) cross, you will use four lines.

χ^2 = ______________________

$C - 1$ = ______________________

p (from Table 21.9) = ______________________

Table 21.10 Calculation of Chi-Square

Phenotype	Observed Number	Expected Results (e)	Difference (d)	d^2	Partial Chi-Square (d^2/e)
					=
					=
					=
					=
			Chi-square = $\chi^2 = \Sigma(d^2/e)$ =		

Conclusions

- Do your results support your original prediction? ______________________
- If not, how can you account for this? ______________________

____

Laboratory Review 21

1. A cross gives a 3:1 phenotypic ratio. What are the genotypes of the parents?
2. According to Mendel's law of segregation, parents who both have the genotype *Aa* would produce what type of gametes?
3. What is the genotype of a man who has unattached earlobes but whose mother has attached earlobes?
4. When doing a genetic cross, why is it necessary to remove the parent flies before the pupae have hatched?
5. If you performed the *Drosophila* cross *LL* × *ll*, what phenotypic ratio would you expect among the offspring?
6. According to Mendel's law of independent assortment, how many different types of gametes would an *AaBb* parent have? An *AABb* parent?
7. What is the genotype of a homozygous long-winged fly that is heterozygous for gray body color?
8. What is the expected phenotype ratio among offspring if both parents are dihybrids?
9. Why do you expect class data to be closer to the expected ratio than your individual data?
10. What is the genotype of a white-eyed male fruit fly? A white-eyed female fruit fly?
11. Which gender can have white eyes if the female parent is homozygous dominant for red eyes and if the male parent has white eyes?
12. Why don't you expect to get a ratio that is exactly 3:1 for a monohybrid cross?
13. A p value less than 0.05 means what?
14. What is the phenotype of a tobacco plant with the genotype *Cc*?

Thought Questions

15. What are the results of the cross *AaBb* × *aaBb*?

16. You count 73 long-winged flies and 27 vestigial-winged flies from a cross between two heterozygous parents. How many flies would you have expected to have long wings?

17. Bob has attached earlobes and both of his parents have unattached earlobes. Sally, Bob's wife, has unattached earlobes. Sally's mother has attached earlobes and her father has unattached earlobes. Sally's brother has attached earlobes.
    a. Identify the genotypes of all involved.

    b. What is the probability that Bob and Sally's first child will have unattached earlobes?

LABORATORY

22

DNA Biology and Technology

Learning Outcomes

22.1 DNA Structure and Replication

- Explain how the structure of DNA facilitates replication.
- Explain why DNA replication is semiconservative and why this feature reduces the chances of errors during replication.

22.2 RNA Structure

- Compare DNA and RNA, discussing their similarities as well as what distinguishes one from the other.

22.3 DNA and Protein Synthesis

- Compare the events of transcription with those of translation during protein synthesis.
- Describe how DNA is able to store so much varied information.

22.4 Isolation of DNA

- Explain the importance of DNA technology in modern science.
- Describe how DNA can be isolated, and explain the procedure for testing DNA.

22.5 DNA Fingerprinting

- Explain what a DNA fingerprint is and how it is created.
- Demonstrate the ability to use the data generated by DNA fingerprinting.

Introduction

This laboratory pertains to molecular genetics and biotechnology. Molecular genetics is the study of the structure and function of **DNA (deoxyribonucleic acid),** the genetic material. Biotechnology is the manipulation of DNA for the benefit of human beings and other organisms. Significant advances in medicine, agriculture, and science in general can be attributed to the fields of molecular genetics and biotechnology.

First we will study the structure of DNA and see how that structure facilitates DNA replication in the nucleus of cells. DNA replicates prior to cell division; following cell division, each daughter cell has a complete copy of the genetic material. DNA replication is also needed in order to pass genetic material from one generation to the next. You may have an opportunity to use models to see how replication occurs.

Then we will study the structure of **RNA (ribonucleic acid)** and how it differs from that of DNA, before examining how DNA and RNA specify protein synthesis. The linear construction of DNA, in which nucleotide is linked to nucleotide, is paralleled by the linear construction of the primary structure of protein, in which amino acid is linked to amino acid. Essentially, we will see that the sequence of nucleotides in DNA codes for the sequence of amino acids in a protein. We will also review the role of three types of RNA in protein synthesis. DNA's code is passed to messenger RNA (mRNA), which moves to the ribosomes containing ribosomal RNA (rRNA). Transfer RNA (tRNA) brings the amino acids to the ribosomes.

Next you will isolate DNA, which will enable you to work with DNA firsthand. Finally, we will review the procedure for preparing a DNA fingerprint. If so instructed, you will actually construct a DNA fingerprint; otherwise, the exercises in the manual will allow you to see how it is done. DNA fingerprinting has many different practical applications. For example, it can help identify a criminal, match a child with its parents, and identify the victims of accidents.

22.1 DNA Structure and Replication

The structure of DNA lends itself to **replication,** the process that makes a copy of a DNA molecule. Accurate DNA replication is a necessary part of chromosome duplication, which precedes cell division. It also makes possible the passage of DNA from one generation to the next. It is this process that makes identical copies of DNA during the S stage of the cell cycle, allowing each new daughter cell to have exactly the same genetic information that was in the original cell.

DNA Structure

DNA is a polymer of nucleotide subunits (Fig. 22.1). Each nucleotide is composed of three molecules: deoxyribose (a 5-carbon sugar), a phosphate, and a nitrogen-containing base.

Figure 22.1 Overview of DNA structure.
Diagram of DNA double helix shows that the molecule resembles a twisted ladder. Sugar-phosphate backbones make up the sides of the ladder, and hydrogen-bonded bases make up the rungs of the ladder. Complementary base pairing dictates that A is bonded to T and G is bonded to C and vice versa. *Label the boxed nucleotide pair as directed in the next Observation.*

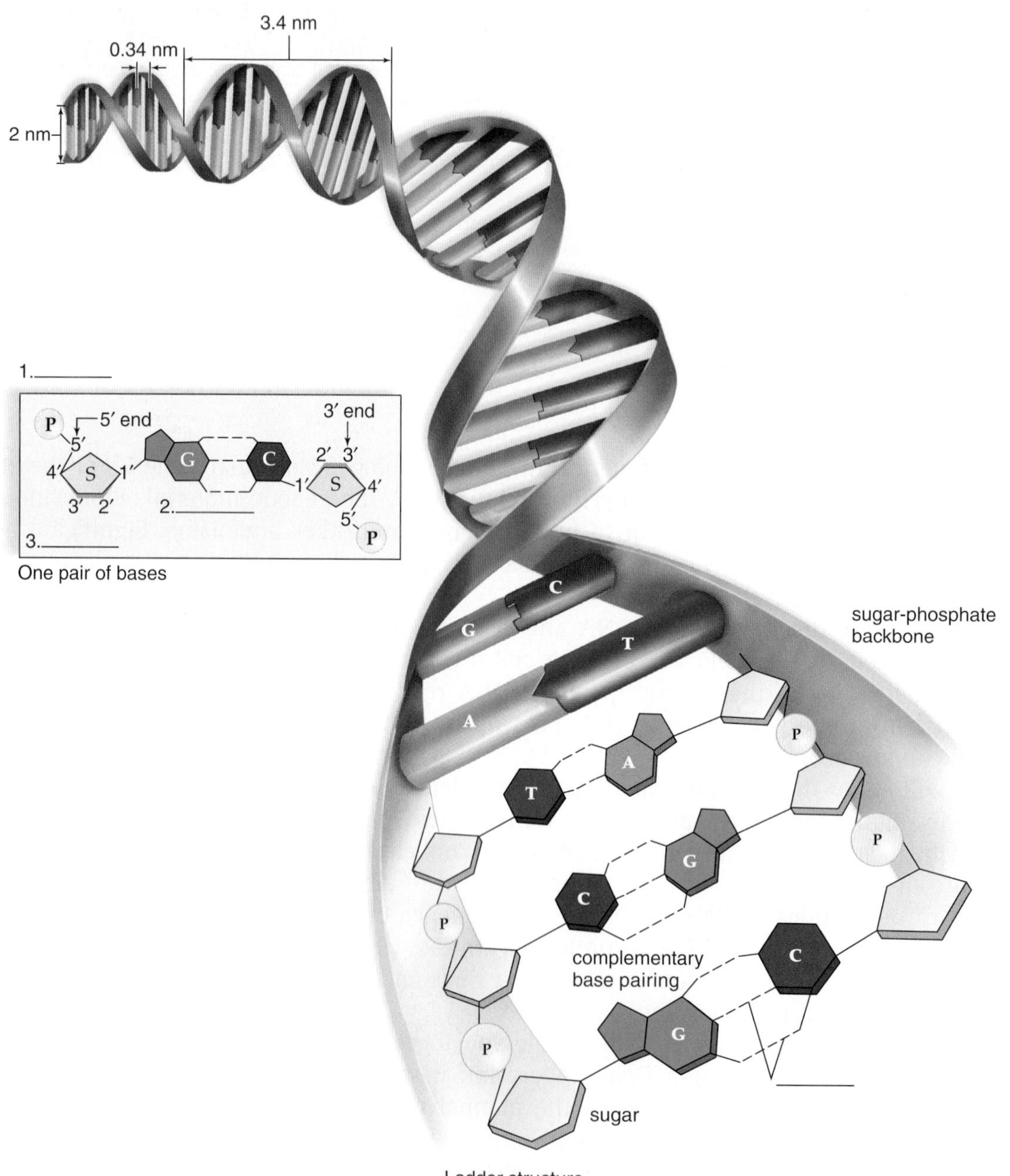

Observation: DNA Structure

1. A nucleotide pair is shown in Figure 22.1. If you are working with a kit, draw a representation of one of your nucleotides here. *Label phosphate, base, and deoxyribose in your drawing and in Figure 22.1.*

2. Notice the four types of bases: cytosine (C), thymine (T), adenine (A), and guanine (G). What is the color of the four types of bases in Figure 22.1? In your kit? Complete Table 22.1 by writing in the colors of the bases.

Table 22.1 Base Colors

	In Figure 22.1	In Your Kit
Cytosine		
Thymine		
Adenine		
Guanine		

3. Using Figure 22.1 as a guide, join several nucleotides together. Observe the entire DNA molecule. What types of molecules make up the backbone (uprights of ladder) of DNA (Fig. 22.1)?

   __________ and __________

   Notice that in the backbone, the phosphate of one nucleotide is bonded to a sugar of the next nucleotide by a covalent bond.
4. Using 22.1 as a guide, join the bases together with hydrogen bonds. Label a hydrogen bond in Figure 22.1. Dots are used to represent hydrogen bonds in Figure 22.1 because hydrogen bonds are (strong or weak)? ________
5. Notice in Figure 22.1 and in your model, that the base A is always paired with the base ________, and the base C is always paired with the base ________. This is called complementary base pairing.
6. In Figure 22.1, what molecules make up the rungs of the ladder? ______________________
7. Each half of the DNA molecule is a DNA strand. Why is DNA also called a double helix (Fig. 22.1)? ____

DNA Replication

During replication, the DNA molecule is duplicated so that there are two DNA molecules. We will see that complementary base pairing makes replication possible.

Observation: DNA Replication

1. Before replication begins, DNA is unzipped. Using Figure 22.2*a* as a guide, break apart your two DNA strands. What bonds are broken in order to unzip the DNA strands? ______________________
2. Using Figure 22.2*b* as a guide, attach new complementary nucleotides to each strand using complementary base pairing.
3. Show that you understand complementary base pairing by completing Table 22.2. You now have two DNA molecules (Fig. 22.2*c*). Are your molecules identical? ________
4. Because of complementary base pairing, each new double helix is composed of an ________ strand and a ________ strand. *Write old or new beside each strand in Figure 22.2a, b, and c, 1–10. Conservative* means to save something from the past. Why is DNA replication called semiconservative?

____

____

____

____

Figure 22.2 DNA replication.
Use of the ladder configuration better illustrates how replication takes place. **a.** The parental DNA molecule. **b.** The "old" strands of the parental DNA molecule have separated. New complementary nucleotides available in the cell are pairing with those of each old strand. **c.** Replication is complete.

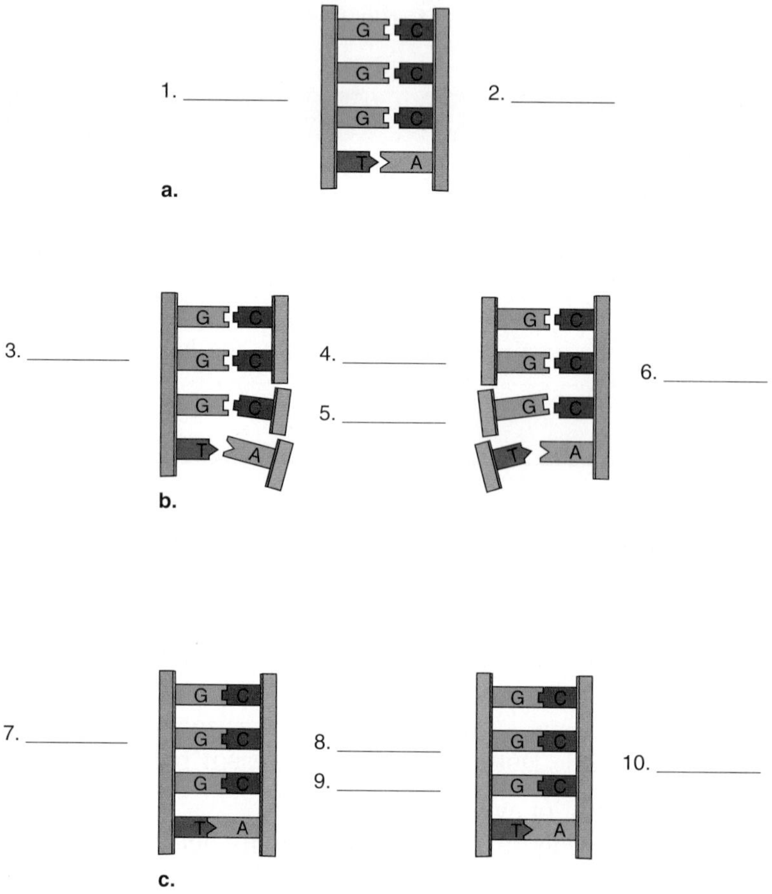

5. Genetic material has to be inherited from cell to cell and organism to organism. Consider that because of DNA replication, a chromosome is composed of two chromatids, and each chromatid is a complete DNA molecule. The chromatids separate during cell division so that each daughter cell receives a copy of each chromosome. Does replication provide a means for passing DNA from cell to cell and organism to organism?

____

Explain. ______________________________

____

____

Table 22.2 DNA Replication

Old strand	G	G	G	T	T	C	C	A	T	T	A	A	A	T	T	C	C	A	G	A	A	A	T	C	A	T	A
New strand																											

22.2 RNA Structure

Like DNA, RNA is a polymer of nucleotides (Fig. 22.3). In an RNA nucleotide, the sugar ribose is attached to a phosphate molecule and to a nitrogen-containing base, C, U, A, or G. Notice that in RNA, the base uracil replaces thymine as one of the pyrimidine bases. RNA is single stranded, whereas DNA is double stranded.

Observation: RNA Structure

1. If you are using a kit, draw a nucleotide for the construction of mRNA. *Label the ribose (the sugar in RNA), the phosphate, and the base in your drawing and in Figure 22.3.*

Figure 22.3 Overview of RNA structure.
RNA is a single strand of nucleotides. *Label the boxed nucleotide as directed in the next Observation.*

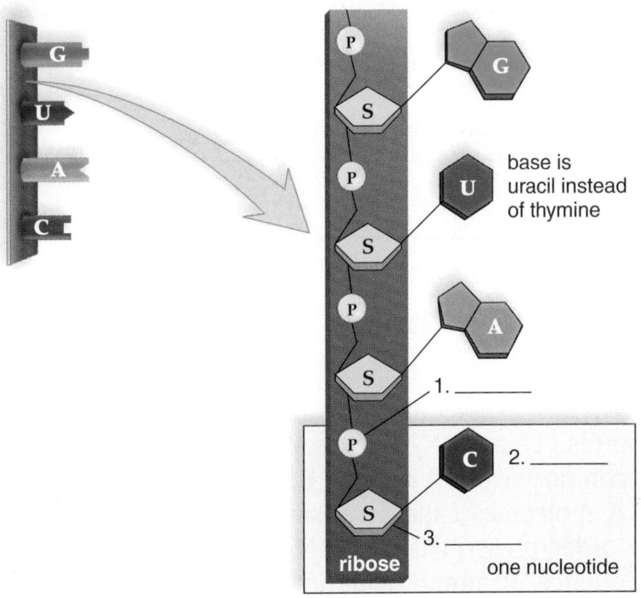

2. Complete Table 22.3 by writing in the colors of the bases in Figure 22.3 and in your kit.

Table 22.3 Base Colors

	In Figure 22.3	In Your Kit
Cytosine		
Uracil		
Adenine		
Guanine		

3. Notice that the base uracil substitutes for the base thymine in RNA. Complete Table 22.4 to show the several other ways RNA differs from DNA.

Table 22.4 DNA Structure Compared with RNA Structure

	DNA	RNA
Sugar	Deoxyribose	
Bases	Adenine, guanine, thymine, cytosine	
Strands	Double stranded with base pairing	
Helix	Yes	

22.3 DNA and Protein Synthesis

Protein synthesis requires the processes of transcription and translation. During **transcription,** which takes place in the nucleus, an RNA molecule called **messenger RNA (mRNA)** is made complementary to one of the DNA strands. This mRNA leaves the nucleus and goes to the ribosomes in the cytoplasm. Ribosomes are composed of **ribosomal RNA (rRNA)** and proteins in two subunits.

During **translation,** RNA molecules called **transfer RNA (tRNA)** bring amino acids to the ribosome, and they join in the order prescribed by mRNA.

In the end, the final sequence of amino acids in a protein is specified by DNA. This is the information that DNA, the genetic material, stores.

Transcription

During transcription, complementary RNA is made from a DNA template (Fig. 22.4). A portion of DNA unwinds and unzips at the point of attachment of RNA polymerase. A strand of mRNA is produced when complementary nucleotides join in the order dictated by the sequence of bases in DNA. Transcription occurs in the nucleus, and the mRNA passes out of the nucleus to enter the cytoplasm. *Label Figure 22.4.*

Figure 22.4 Messenger RNA (mRNA). Messenger RNA complementary to a section of DNA forms during transcription.

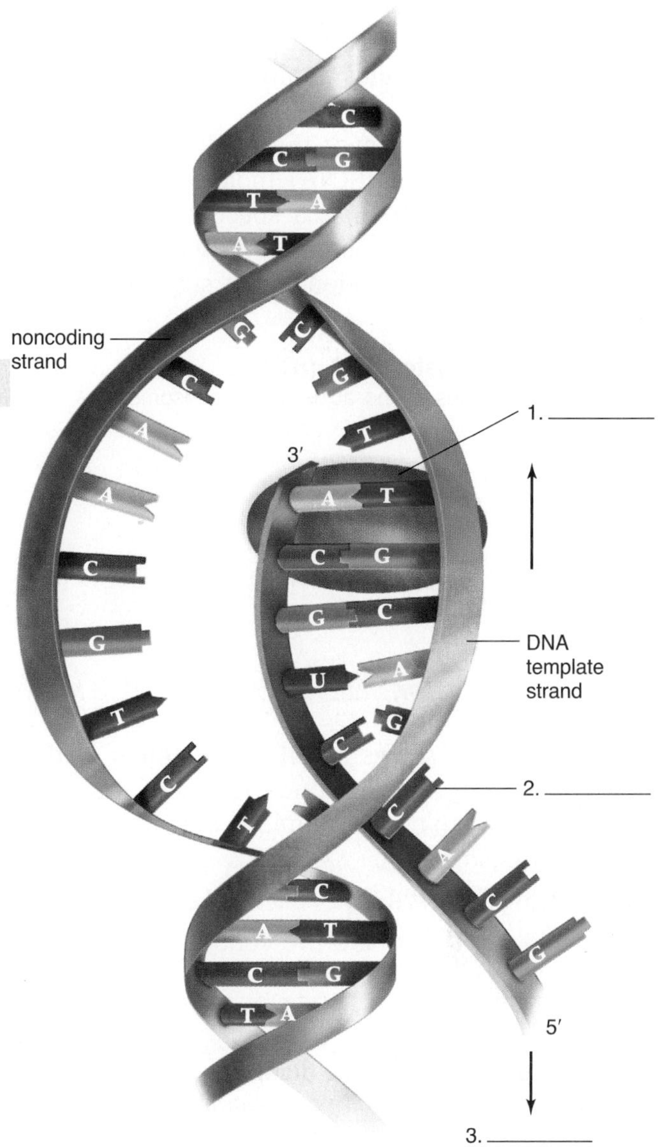

Observation: Transcription

1. If you are using a kit, unzip your DNA model so that only one strand remains. This strand is the **template strand,** the strand that is transcribed or copied into complementary base pairs of RNA.
2. Using Figure 22.4 as a guide, construct a messenger RNA (mRNA) molecule by first lining up RNA nucleotides that are complementary to the template strand of your DNA molecule. Join the nucleotides together to form mRNA. *Label mRNA in Figure 22.4.*
3. A portion of DNA has the sequence of bases shown in Table 22.5. *Complete Table 22.5 to show the sequence of bases in mRNA.*
4. If you are using a kit, unzip the mRNA transcript from the DNA. Locate the end of the strand that will move into the cytoplasm.

Table 22.5 Transcription

DNA	T A C A C G A G C A A C T A A C A T
mRNA	

Translation

DNA specifies the sequence of amino acids in a polypeptide because every three bases stands for an amino acid. Therefore, DNA is said to have a **triplet code.** The bases in mRNA are complementary to those in DNA, and therefore every three bases in mRNA (called a **codon**) stands for the same sequence of amino acids as does DNA. The correct sequence of amino acids in a polypeptide is the message that mRNA carries.

Messenger RNA leaves the nucleus and proceeds to the ribosomes, where protein synthesis occurs. Transfer RNA (tRNA) molecules are so named because they transfer amino acids to the ribosomes. Each tRNA has a specific amino acid at one end and a matching **anticodon** at the other end (Fig. 22.5). *Label Figure 22.5,* where the amino acid is represented as a colored ball, the tRNA is green, and the anticodon is the sequence of three bases.

Figure 22.5 Transfer RNA (tRNA).
Transfer RNA carries amino acids to the ribosomes.

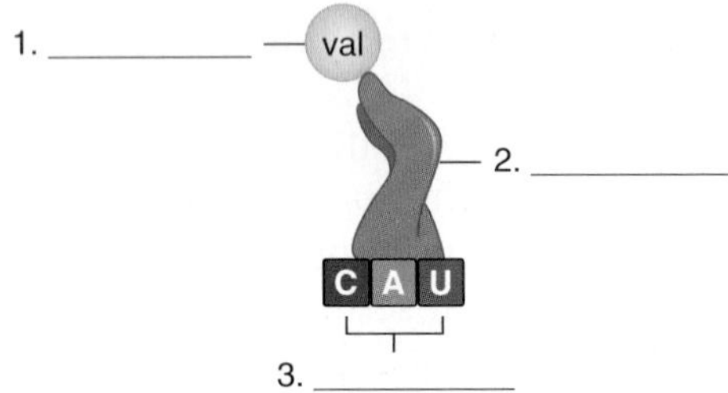

Observation: Translation

1. Figure 22.6 shows seven tRNA–amino acid complexes. Every amino acid has a name; in the figure, only the first three letters of the name are inside the ball. Using the mRNA sequence given in Table 22.6, number the tRNA–amino acid complexes in the order they will come to the ribosome.
2. If you are using a kit, arrange your tRNA–amino acid complexes in the proper order. Complete Table 22.6. Why are the codons and anticodons in groups of three? ____________________

____

Figure 22.6 Transfer RNA diversity.
Each type of tRNA carries only one particular amino acid, designated here by the first three letters of its name.

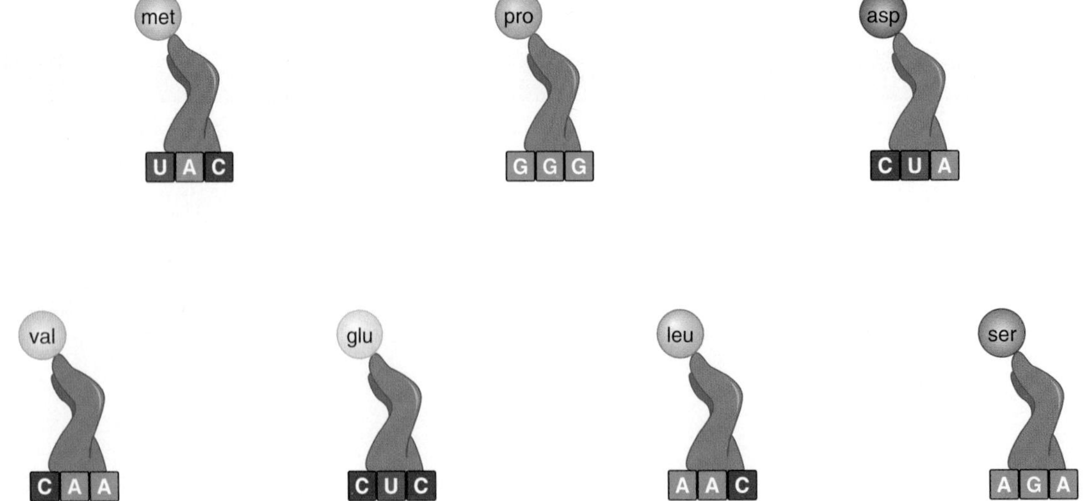

Table 22.6 Translation

mRNA codons	AUG	CCC	GAU	GUU	GAA	UUG	UCU
tRNA anticodons							
Amino acid*							

*Use three letters only. See Table 22.7 for the full names of these amino acids.

3. Figure 22.7 shows the manner in which the polypeptide grows. A ribosome has room for two tRNA complexes at a time. As the first tRNA leaves, it passes its amino acid or peptide to the second tRNA–amino acid complex. Then the ribosome moves forward, making room for the next tRNA–amino acid complex. This sequence of events occurs over and over again until the entire polypeptide is borne by the last tRNA to come to the ribosome. *In Figure 22.7, label the ribosome, the mRNA, and the peptide.*

Table 22.7 Names of Amino Acids

Abbreviation	Name
met	methionine
pro	proline
asp	aspartate
val	valine
glu	glutamine
leu	leucine
ser	serine

Figure 22.7 Protein synthesis.
1. A ribosome has room for two tRNA–amino acid complexes. 2. Before a tRNA leaves, an RNA passes its attached peptide to its neighboring tRNA–amino acid complex. 3. The ribosome moves forward, and the next tRNA–amino acid complex arrives.

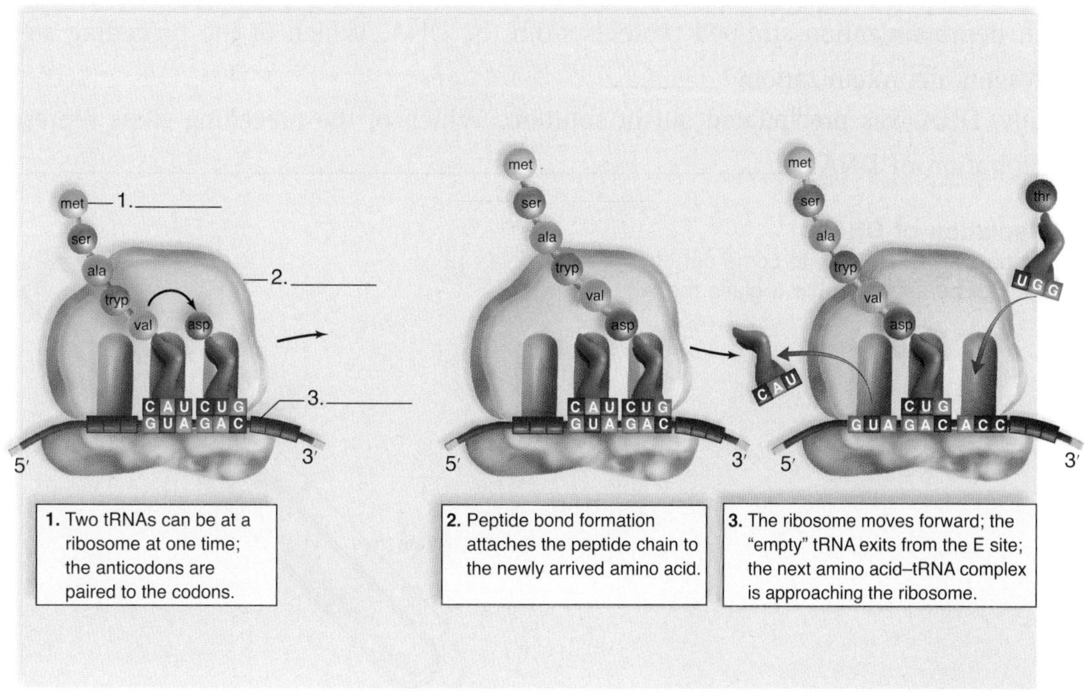
met
ser
ala
tryp
val
asp
thr
1. ________
2. ________
3. ________
C A U
C U G
G U A
G A C
U G G
A C C
5′
3′
1. Two tRNAs can be at a ribosome at one time; the anticodons are paired to the codons.
2. Peptide bond formation attaches the peptide chain to the newly arrived amino acid.
3. The ribosome moves forward; the "empty" tRNA exits from the E site; the next amino acid–tRNA complex is approaching the ribosome.

22.4 Isolation of DNA

In the following Experimental Procedure, you will isolate DNA from the cells of an organism using a modified procedure like that used worldwide in biotechnology laboratories. You will extract DNA from an onion filtrate that contains DNA in solution. To prepare the filtrate, your instructor homogenized an onion with a detergent. The detergent emulsifies and forms complexes with the lipids and proteins of the plasma membrane, causing them to precipitate out of solution. Cell contents, including DNA, become suspended in solution. The cellular mixture is then filtered to produce the filtrate that contains DNA.

The DNA molecule is easily degraded (broken down), so it is important to follow all instructions closely. Handle glassware carefully to prevent nucleases in your skin from contaminating the glassware.

Experimental Procedure: Isolating DNA

1. Obtain a large, clean test tube, and place it in an ice bath. Let stand for a few minutes to make sure the test tube is cold. Everything must be kept very cold.
2. Obtain approximately 4 ml of the *onion filtrate,* and add it to your test tube while keeping the tube in the ice bath.
3. Obtain and add 2 ml of cold *meat tenderizer solution* to the solution in the test tube, and mix the contents slightly with a stirring rod or Pasteur pipette. Let stand for 10 minutes.
4. Use a graduated cylinder or pipette to slowly add an equal volume (approximately 6 ml) of ice-cold *95% ethanol* along the inside of the test tube. Keep the tube in the ice bath, and tilt it to a 45° angle. You should see a distinct layer of ethanol over the white filtrate. Let the tube sit for 2 to 3 minutes.
5. Insert a glass rod or a Pasteur pipette into the tube until it reaches the bottom of the tube. *Gently* swirl the glass rod or pipette, always in the same direction. (You are not trying to mix the two layers; you are trying to wind the DNA onto the glass rod like cotton candy.) This process is called "spooling" the DNA. The stringy, slightly gelatinous material that attaches to the pipette is DNA (Fig. 22.8). If the DNA has been damaged, it will still precipitate, but as white flakes that cannot be collected on the glass rod.
6. Answer the following questions:
   a. This procedure requires homogenization. When did homogenization occur? ____________

   ____

   b. What was the purpose of homogenization? ____________
   Next, deproteinization stripped proteins from the DNA. Which of the preceding steps represents deproteinization? ________

   c. Finally, DNA was precipitated out of solution. Which of the preceding steps represents the precipitation of DNA? ____________

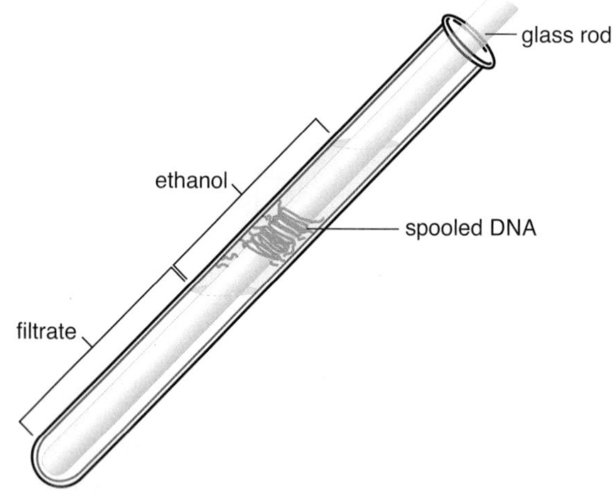

Figure 22.8 Isolation of DNA.
The addition of ethanol causes DNA to come out of solution so that it can be spooled onto a glass rod.

22.5 DNA Fingerprinting

A genome is all the genetic material in a set of chromosomes. The genome contains portions of DNA (i.e., genes) that code for the various proteins and specialized RNAs (tRNAs and rRNAs). Other DNA portions are sometimes called "junk" because they do not code for proteins or RNAs—these sections are repeats of the same short sequence of bases over and over again. Detectable differences in the noncoding portions of the genome comprise an individual's **DNA fingerprint.** DNA fingerprinting has many uses.

DNA fingerprinting is used by (1) police and courts to identify a person who has committed a crime; (2) genetic counselors to determine if an individual has or will develop a genetic disorder; (3) lawyers to determine relatedness between individuals; and (4) scientists to determine the identity of individuals based on minimal remains after death. DNA fingerprinting is also used by (1) conservation biologists to determine the genetic dissimilarity of males versus females for breeding purposes; (2) evolutionary biologists to construct evolutionary trees; and (3) taxonomists to distinguish species.

DNA fingerprinting requires three steps (Fig. 22.9):

1. Fragmentation of usually a selected portion of the genome. The DNA is digested with restriction enzymes, and the result is different-sized fragments unique to the individual.
2. Gel electrophoresis. The DNA fragments are separated according to their length, and the result is a DNA pattern unique to the individual.

Figure 22.9 DNA fingerprinting.
During DNA fingerprinting, DNA samples are digested to fragments. The fragments are separated by gel electrophoresis, and then the resulting fragment length pattern is observed. In this example, the sample in well I could be from the crime scene, making the DNA in well II from the criminal; or the DNA in well I could be from a parent, making the donor of the sample in well II his child; or sample I could be from a body part, making sample II the deceased individual. In the last case, sample II would have been taken from some object known to belong to the individual in question.

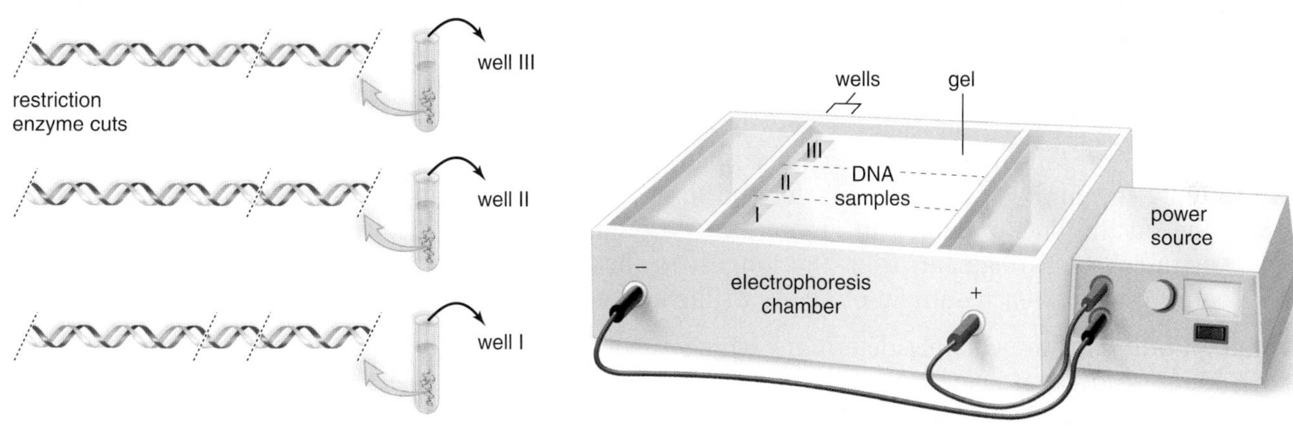

1. Digest DNA samples with restriction enzymes.

2. Apply samples to gel, and perform electrophoresis.

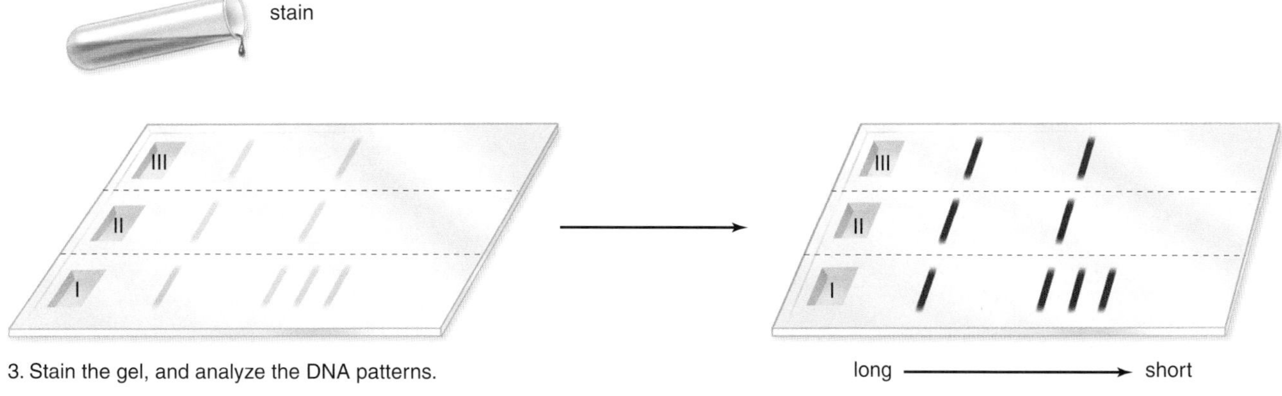

3. Stain the gel, and analyze the DNA patterns.

3. Analysis of the DNA pattern. The pattern is revealed by using radioactive probes or by staining. If you are dealing with just a small portion of the genome, staining of the gel is sufficient to reveal the fragment pattern.

Fragmentation of DNA

Fragmentation of DNA is carried out by restriction enzymes. Restriction enzymes are so named because they restrict the growth of viruses. These enzymes occur naturally in bacteria where they cut up any viral DNA that enters the cell. There are many different types of restriction enzymes, and each one has a preferred location for cutting DNA. For example, a restriction enzyme called *Eco*RI cuts a DNA double helix like this wherever it finds this sequence of bases:

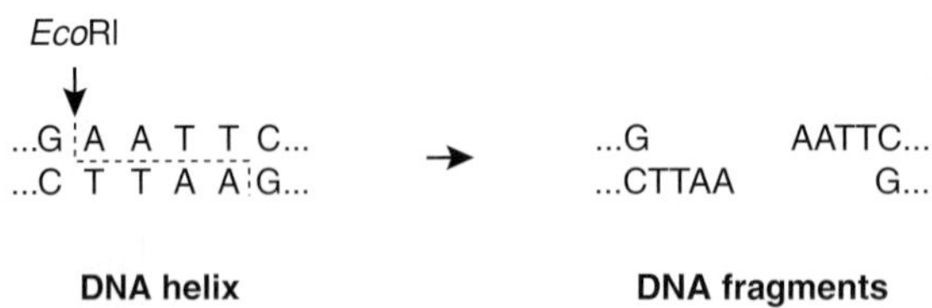

Experimental Procedure: Fragmentation of DNA

1. If you are working with a kit, you will be given a sample of "crime scene DNA." A sample can come from blood, semen, hair follicles, or almost any tissue left behind by the perpetrator of a crime. You will also be given DNA samples for several suspects.
2. Number as many test tubes as you need, and add a DNA sample and restriction enzyme solution to each tube. Mix the contents, and incubate the tubes at 37°C for at least 45 minutes.
3. Figure 22.10 gives examples of fragmentation using the enzyme *Eco* RI. Following digestion, the crime scene (CS) DNA has four cuts. The length of a DNA fragment is the number of *base pairs* in that fragment. The number of base pairs in each of the CS fragments has been recorded for you. (Note that the restriction enzyme breaks apart certain bases, and therefore they can no longer be counted as *base pairs.*)
4. Supply the number of base pairs for the three suspect (S) samples. Write the number of base pairs on the lines provided in Figure 22.10. Complete Table 22.8 by recording the base-pair length for each fragment, from the longest to the shortest. Use only the appropriate number of columns for each sample of DNA. Crime scene DNA has been done for you.

Although you can immediately determine which suspect is the guilty party, remember that the experimenter does not actually know the fragment lengths until gel electrophoresis is done. Why not? Because at this point, the fragments for each sample are mixed together in the test tube.

Figure 22.10 Fragmentation of DNA samples by the restriction enzyme *Eco* RI.
Complete as directed by the text.

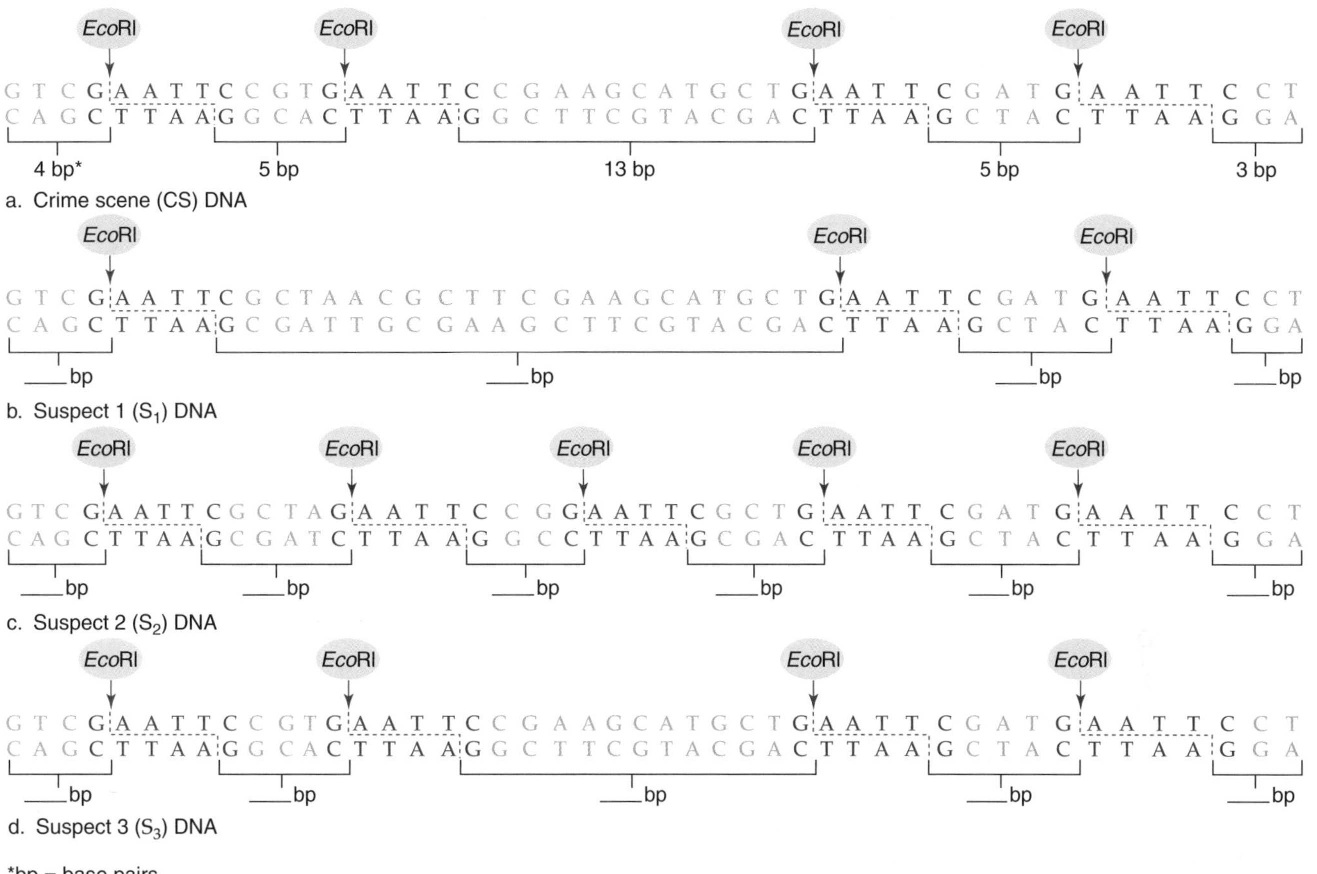

a. Crime scene (CS) DNA

b. Suspect 1 (S_1) DNA

c. Suspect 2 (S_2) DNA

d. Suspect 3 (S_3) DNA

*bp = base pairs

Table 22.8 Fragmentation of DNA

	Fragment	Fragment	Fragment	Fragment	Fragment	Fragment
Crime scene (CS) DNA	4 bp	5 bp	13 bp	5 bp	3 bp	None
Suspect 1						
Suspect 2						
Suspect 3						

Gel Electrophoresis

The two most widely used techniques for separating molecules in biotechnology are chromatography and gel electrophoresis. Chromatography separates molecules on the basis of their solubility and size. **Gel electrophoresis** separates molecules on the basis of their charge and size.

During gel electrophoresis, charged molecules migrate across a span of gel (gelatinous slab) because they are placed in a powerful electrical field. In the present experiment, the fragment mixture for each DNA sample is placed in a small depression in the gel called a well. The gel is placed in a powerful electrical field. The electricity causes the DNA fragments, which are negatively charged, to move through the gel to the positive pole. The gel acts as a sieve to retard the passage of fragments according to their length. In the end, the longest fragment in each sample is closest to the negative pole, and the shortest fragment is closest to the positive pole.

Almost all DNA gel electrophoresis is carried out using horizontal gel slabs. First, the gel is poured onto a glass plate, and the wells are formed. After the samples are added to the wells, the gel and the glass plate are put into an electrophoresis chamber, and buffer is added. The fragments begin to migrate after the electrical current is turned on. With staining, the fragments appear as a series of bands spread from one end of the gel to the other.

Experimental Procedure: Gel Electrophoresis

> **Caution: Gel electrophoresis** Students should wear personal protective equipment: safety goggles and smocks or aprons while loading gels and during electrophoresis and protective gloves while staining.

1. If you are working with a kit, use Figure 22.11 as a guide to prepare your gel, and fill the wells with your digested DNA samples. Place the gel in the electrophoresis chamber so that the wells are closest to the negative electrode. Add buffer to just cover the wells. Close the lid, connect the electrical leads, and turn on the power supply.
2. When electrophoresis is over, turn off the power and remove the lid from the electrophoresis chamber. Carefully slide the gel into a staining tray. Add stain and let the gel stain overnight.
3. Although gels run from left to right, it will be easier to study the DNA pattern by turning the gel so that the wells are positioned at the top of the gel. Predict the pattern for crime scene (CS) DNA by writing in the base-pair length above the four fragments shown in Figure 22.12. *Remember that the longest fragment is closest to the well, and fragments of the same base-pair length migrate the same distance. When two fragments of the same length migrate together, the gel band is slightly wider.*

Figure 22.11 Equipment and procedure for gel electrophoresis.

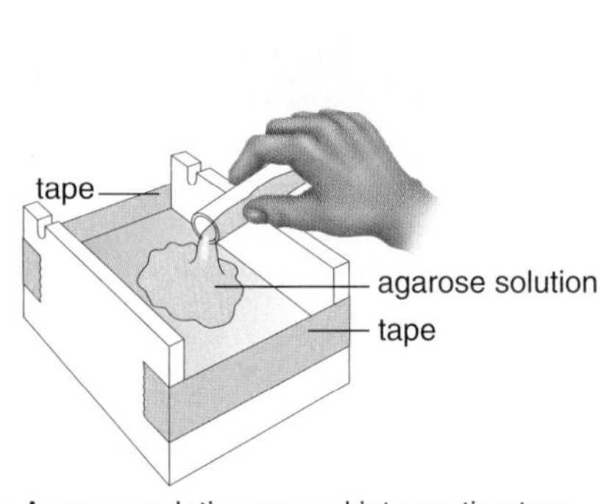

a. Agarose solution poured into casting tray

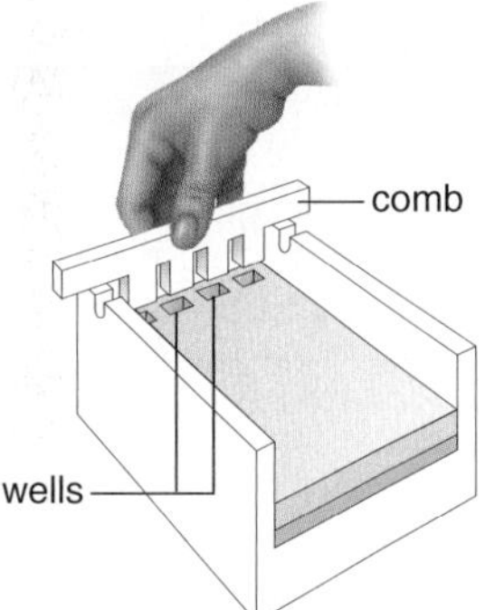

b. Comb that forms wells for samples

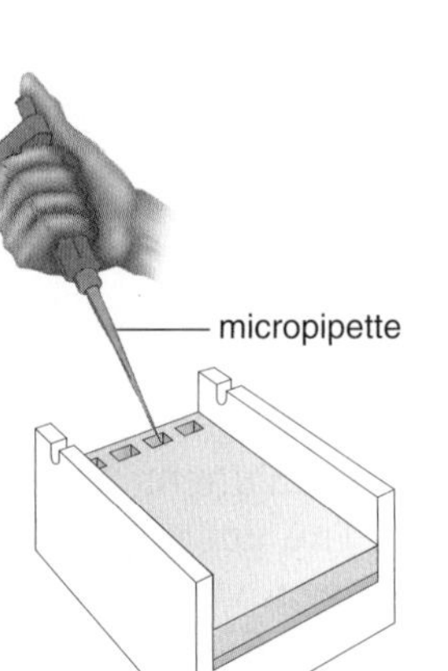

c. Wells that can be loaded with samples

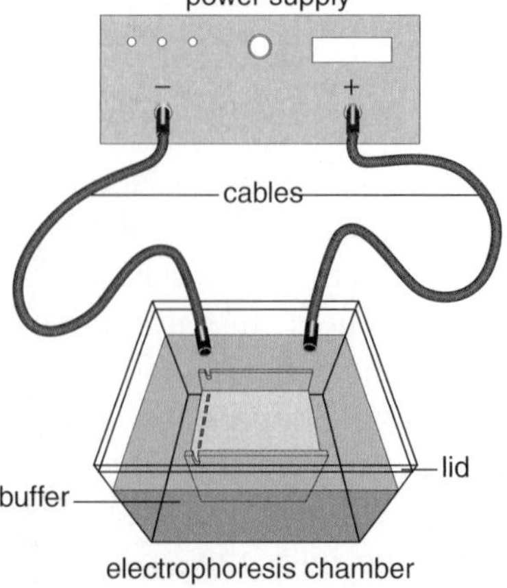

d. Electrophoresis chamber and power supply

Figure 22.12 DNA fingerprint for the crime scene DNA sample.

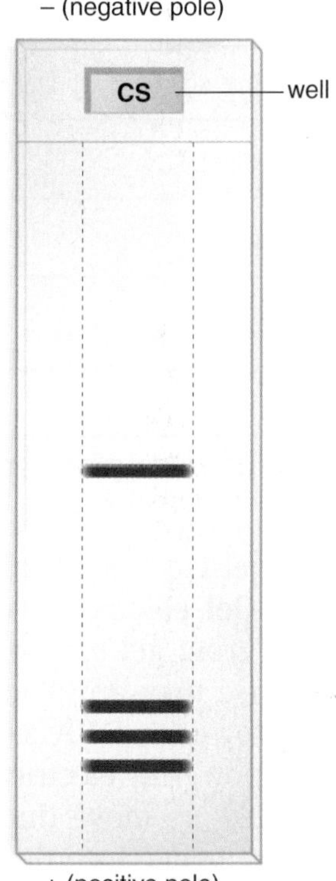

Analyzing the DNA Pattern

1. If you are working with a kit, pour off the stain, and destain with water for 15 minutes.
2. Figure 22.13 shows the fragment patterns for the DNA samples in Figure 22.11. Using the data from Table 22.8, record the base-pair length for each fragment just above each fragment. *Remember that the longest fragment is closest to the well, and fragments of the same base-pair length migrate the same distance. When two fragments of the same length migrate together, the gel band is slightly wider.*
3. Which of the suspects' DNA samples matches the DNA sample collected at the crime scene? ______________

Figure 22.13 DNA fingerprints.
These are the DNA fingerprints for the crime scene DNA sample and the three suspects' samples. In this case, comparing DNA fragments patterns allows you to determine who committed the crime. In other instances, DNA fragment patterns allow you to determine who is the parent of a child, whether a person has a genetic disorder, or the identity of the remains following death.

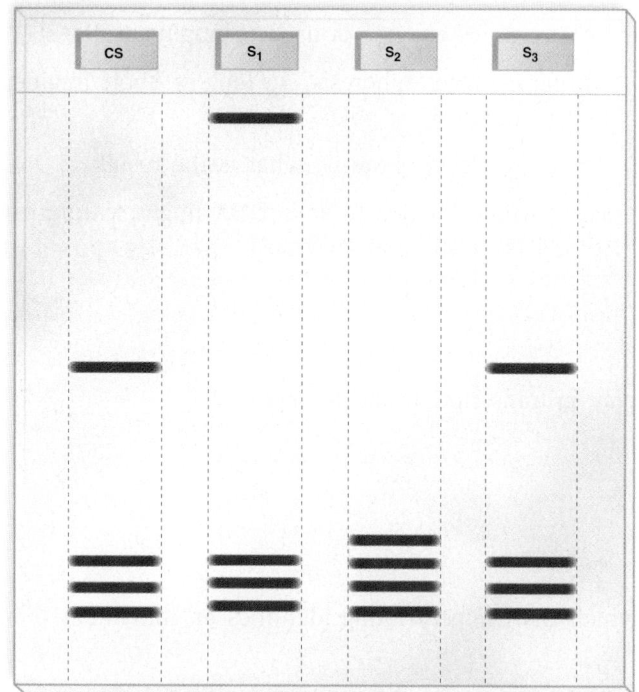

Laboratory Review 22

_______________ 1. The DNA structure resembles a twisted ladder. What molecules make up the sides of the ladder?

_______________ 2. What makes up the rungs of the ladder?

_______________ 3. Do the two DNA double helices following DNA replication have the same, or a different, composition?

_______________ 4. If DNA has 20% of adenine bases, what would be the percentage of thymine?

_______________ 5. If the codons are AUG, CGC, and UAC, what are the anticodons?

_______________ 6. Where does protein synthesis take place?

_______________ 7. During transcription, what type of RNA is formed?

_______________ 8. In what part of the cell does translation occur?

_______________ 9. During translation, what type of RNA carries amino acids to the ribosomes?

_______________ 10. What type of enzyme is used to fragment DNA?

_______________ 11. During gel electrophoresis, do long or short fragments travel more quickly toward the positive pole?

_______________ 12. Following electrophoresis, what is the resulting DNA pattern called?

_______________ 13. What would happen to your DNA fingerprinting results if you had forgotten to add the restriction enzyme?

Thought Questions

14. What role does mRNA play in transcription and translation?

15. Explain the manner in which DNA fingerprinting identifies an individual.

16. Below is a sequence of bases associated with the template DNA strand:

TAC CCC GAG CTT

a. Identify the sequence of bases in the mRNA resulting from the transcription of the above DNA sequence.

b. Identify the sequence of bases in the tRNA anticodon that will bind with the first codon on the mRNA identified above.

LABORATORY

23

Genetic Counseling

Learning Outcomes

23.1 Chromosomal Inheritance

- Explain how a karyotype is prepared and the role it can play in evaluating some genetic disorders.
- Describe how nondisjunction occurs and cite several examples of genetic disorders that can occur as a result of numerical sex chromosome abnormalities.

23.2 Genetic Disorders: The Present

- Evaluate a pedigree and determine if any pattern of autosomal dominant, autosomal recessive, or X-linked recessive inheritance exists.
- Solve genetics problems involving autosomal dominant traits, autosomal recessive traits, or X-linked recessive traits.
- Identify appropriate methods of genetic testing for specific genetic disorders and describe what options parents might have should they have a child with a given genetic disorder.

23.3 Genetic Disorders: The Future

- Relate specific DNA base sequence irregularities to genetic disorders and speculate on how molecular biology might one day be able to correct the genetic defects that result in a genetic disease.

Introduction

Chromosomal inheritance has a marked effect on the general anatomy and physiology of the individual. If by chance the individual has an abnormality in number or structure, a syndrome results. A **syndrome** is a group of symptoms that appear together and tend to indicate the presence of a particular disorder.

Genes, the units of heredity located on the chromosomes, determine whether an individual has a genetic disorder. Individuals do not have an autosomal recessive disorder unless they have inherited two mutated alleles. For an autosomal dominant disorder, one mutated allele is sufficient. Most sex-linked disorders are carried on the X chromosome, and males receive only one X chromosome; therefore, they will display an X-linked recessive disorder even if they receive only a single mutated allele.

We now understand that a mutated gene has an altered DNA base sequence. For example, in cystic fibrosis (Fig. 23.1), DNA codes for a sequence of amino acids in a particular membrane protein; this protein is abnormal and usually has a missing amino acid. This results in an altered shape, which entraps the chloride ions inside cells. Water entering these cells leaves behind a very thick mucus, which clogs the bronchi and bronchioles, resulting in respiratory infections. Other organs in the body are also affected.

Figure 23.1 Cause of cystic fibrosis.

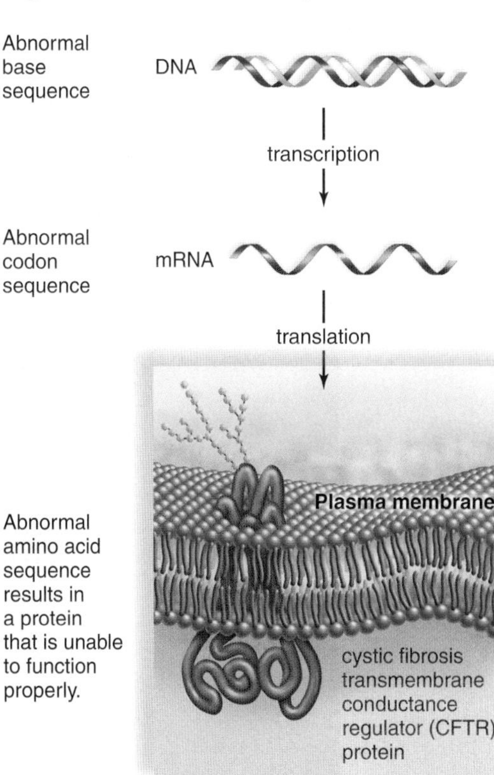

23.1 Chromosome Inheritance

To view an individual's chromosome inheritance, cells can be treated and photographed just prior to division. Before birth, cells can be obtained by amniocentesis, a procedure in which a physician uses a long needle to withdraw a portion of the amniotic fluid containing fetal cells. In chorionic villi sampling, cells are removed from the chorion. In any case, the fetal cells are cultured, and then a karyotype of the chromosomes is prepared. Karyotypes can also be done using white blood cells from an adult (Fig. 23.2). Notice that a karyotype displays homologues. They are called homologues

Figure 23.2 Human karyotype preparation.

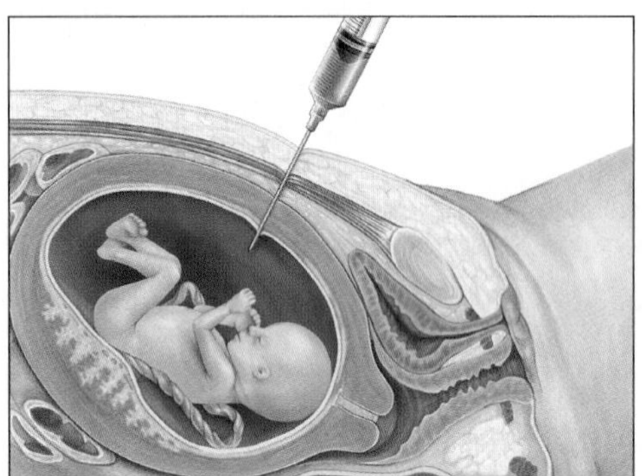

a. During amniocentesis, a long needle is used to withdraw amniotic fluid containing fetal cells.

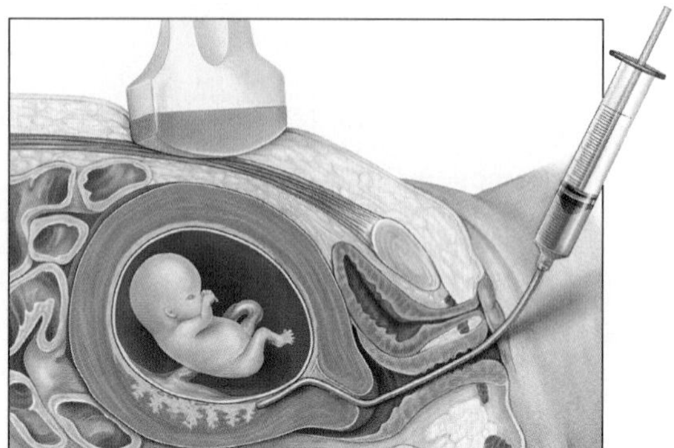

b. During chorionic villi sampling, a suction tube is used to remove cells from the chorion, where the placenta will develop.

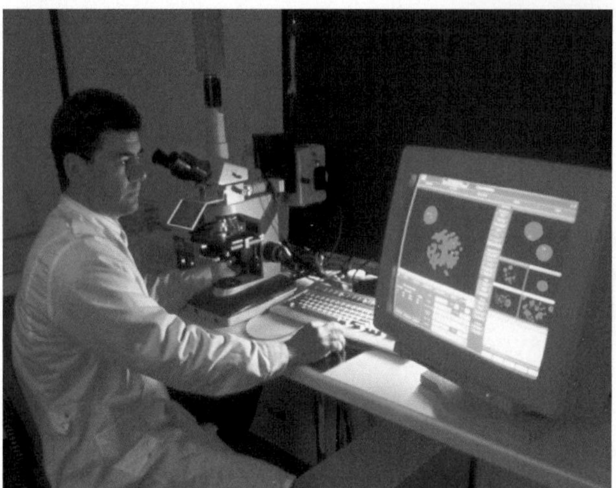

c. Cells are microscopically examined and photographed. A computer is used to arrange chromosomes by pairs.

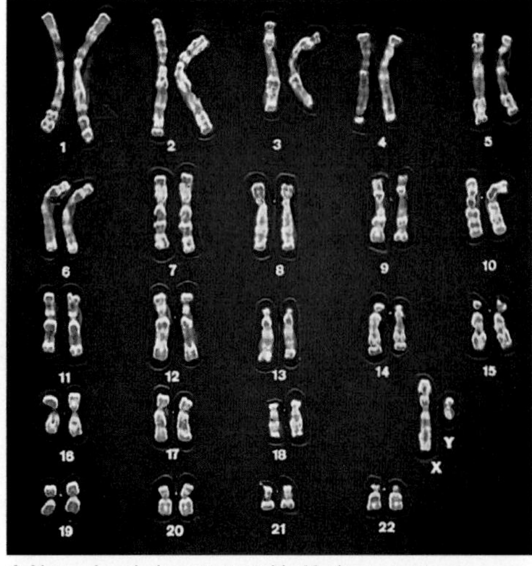

d. Normal male karyotype with 46 chromosomes.

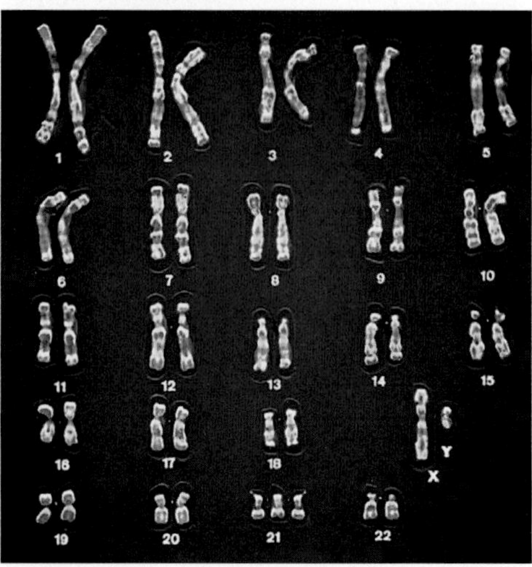

Down syndrome karyotype with an extra chromosome 21.

because they have the same size and shape. In the karyotype of males, it is possible to see that the **X chromosome** is the larger and the **Y chromosome** is the smaller of the sex chromosomes. Females have two X chromosomes.

Numerical Sex Chromosome Abnormalities

Gamete formation in humans involves meiosis, the type of cell division that reduces the chromosome number by one-half because the homologues separate during meiosis I. When homologues fail to separate during meiosis I, called nondisjunction, gametes with too few (n − 1) or too many (n + 1) chromosomes result. Nondisjunction can also occur during meiosis II if the chromatids fail to separate and the daughter chromosomes go into the same daughter cell.

Figure 23.3*a* shows nondisjunction of the X chromosomes in humans during oogenesis. Three viable abnormal chromosomal types can occur: Turner syndrome (XO), poly-X syndrome (XXX), and Klinefelter syndrome (XXY). Figure 23.3*b* shows the results of nondisjunction of XY chromosomes during spermatogenesis. In addition to the three abnormal chromosomal types already mentioned, Jacobs syndrome (XYY) is possible.

Figure 23.3 Nondisjunction of sex chromosomes.
a. Nondisjunction during oogenesis produces the types of eggs shown. Fertilization with normal sperm results in the syndromes noted. (Nonviable means existence is not possible.) **b.** Nondisjunction during meiosis I of spermatogenesis results in the types of sperm shown. Fusion with normal eggs results in the syndromes noted. Nondisjunction during meiosis II of spermatogenesis results in the types of sperm shown. Fusion with normal eggs results in the syndromes shown.

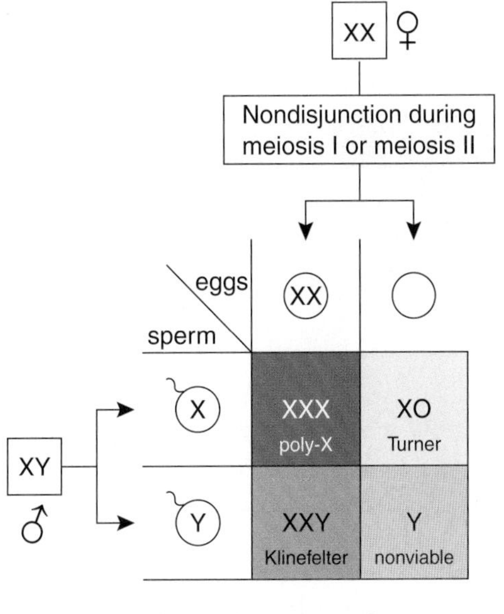

a. Nondisjunction during oogenesis

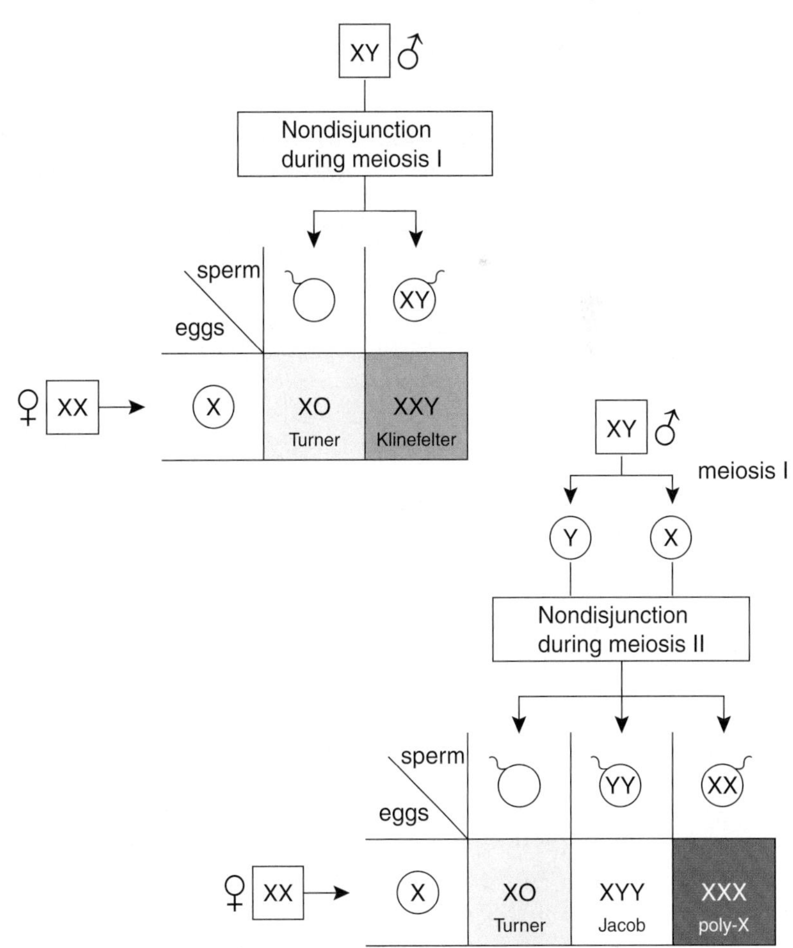

b. Nondisjunction during spermatogenesis

A female with **Turner syndrome** (XO) has only one sex chromosome, an X chromosome; the O signifies the absence of the second sex chromosome. Because the ovaries never become functional, these females do not undergo puberty or menstruation, and their breasts do not develop. Generally, females with Turner syndrome have a short build, folds of skin on the back of the neck, difficulty recognizing various spatial patterns, and normal intelligence. With hormone supplements, they can lead fairly normal lives.

When an egg having two X chromosomes is fertilized by an X-bearing sperm, an individual with **poly-X syndrome** results. The body cells have three X chromosomes, and therefore 47 chromosomes. It might be supposed that poly-X females are especially feminine, but this is not the case. Although they tend to have learning disabilities, poly-X females have no apparent physical abnormalities, and many are fertile and have children with a normal chromosome count.

When an egg having two X chromosomes is fertilized by a Y-bearing sperm, a male with **Klinefelter syndrome** results. This individual is male in general appearance, but the testes are underdeveloped, and the breasts may be enlarged. The limbs of XXY males tend to be longer than average, muscular development is poor, body hair is sparse, and many XXY males have learning disabilities.

Jacobs syndrome can be due to nondisjunction during meiosis II of spermatogenesis. These males are usually taller than average, suffer from persistent acne, and tend to have speech and reading problems. At one time, it was suggested that XYY males were likely to be criminally aggressive, but the incidence of such behavior has been shown to be no greater than that among normal XY males.

Complete Table 23.1 to show how a physician would recognize each of these syndromes from a karyotype.

*Experimental Procedure: Nondisjunction**

Table 23.1 Numerical Sex Chromosome Abnormalities

Syndrome	Karyotype
Turner	
Poly-X	
Klinefelter	
Jacobs	

Building the Chromosomes

1. Obtain the following materials: 36 red pop beads, 26 blue (or green) pop beads, and eight magnetic centromeres.
2. Build three duplicated sex chromosomes (2 Xs and 1 Y) as follows:

   Two red X chromosomes: Each chromatid will have nine red pop beads. Place the centromeres so that two beads are above each centromere and seven beads are below each centromere. Bring the centromeres together.

   One blue X chromosome: Each chromatid will have nine blue pop beads. Place the centromere so that two beads are above each centromere and seven beads are below each centromere. Bring the centromeres together.

   Y chromosome: Each chromatid will have four blue pop beads. Place the centromeres so that two beads are above each centromere and two beads are below each centromere. Bring the centromeres together.

*Exercise courtesy of Victoria Finnerty, Sue Jinks-Robertson, and Gregg Orloff of Emory University, Atlanta, GA.

Simulating Normal Oogenesis

Construct a primary oocyte by placing one blue and one red X chromosome together in the middle of your work area. (The blue X chromosome came from the father, and the red X chromosome came from the mother.) Have the chromosomes go through meiosis I and meiosis II. What is the chromosome constitution of each of the four meiotic products? Each "egg" has ________ (number) ________ (type) chromosome(s).

Simulating Normal Spermatogenesis

Construct a primary spermatocyte by placing a red X and a blue Y chromosome together in the middle of your work area. (The red X chromosome came from the mother, and the blue Y chromosome came from the father.) Have the chromosomes go through meiosis I and meiosis II. What is the chromosome constitution of each of the four meiotic products? Two sperm have one ________ chromosome, and two sperm have one ________ chromosome.

Simulating Fertilization

In the Punnett square provided here, fill in the products of fertilization using the *type* of gamete that resulted from normal oogenesis and the *types* of gametes that resulted from normal spermatogenesis. (Disregard the color of the chromosomes.)

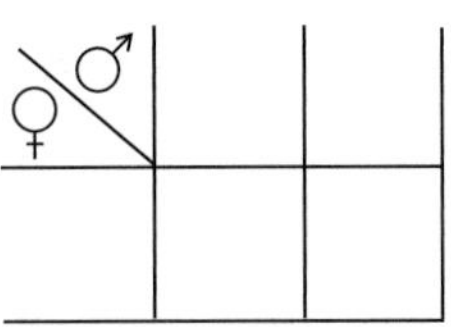

Simulating Nondisjunction During Meiosis I

1. Construct a primary oocyte and a primary spermatocyte as before. Assume that nondisjunction occurs during meiosis I, but the chromatids separate at the centromere during meiosis II. What is the chromosome constitution of each of the four meiotic products:

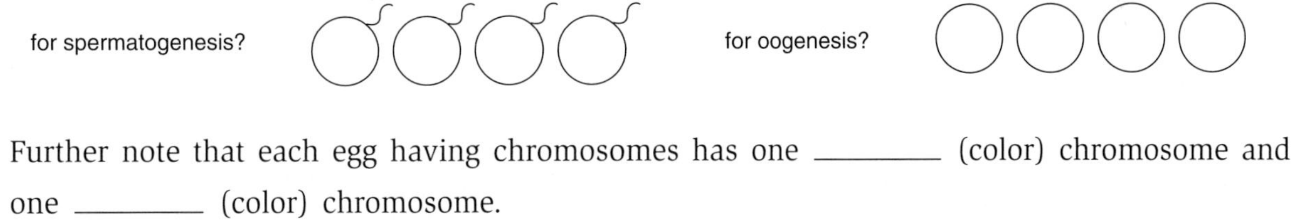

   Further note that each egg having chromosomes has one ________ (color) chromosome and one ________ (color) chromosome.

2. In the space provided here, draw a Punnett square, and fill in the products of fertilization using (a) normal sperm × *types* of abnormal eggs as in Figure 23.3*a* and (b) normal egg × *types* of abnormal sperm, as in Figure 23.3*b*. (Disregard the color of the chromosomes.)

a. **b.**

Conclusions

- What syndromes are the result of (a)? ______
- Are all offspring viable (capable of living)? ______ Explain. ______

______

- What syndromes are the result of (b)? ______
- Are all offspring viable? ______ Explain. ______

______

Simulating Nondisjunction During Meiosis II

1. Construct a primary oocyte and a primary spermatocyte as before. Assume that meiosis I is normal, but the chromatids of the chromosomes fail to separate during meiosis II. What is the chromosome constitution of each of the four meiotic products:

for spermatogenesis?

for oogenesis?

2. In the space provided here, draw a Punnett square, and fill in the products of fertilization using (a) normal sperm × *types* of abnormal eggs, as in Figure 23.3*a*, and (b) normal egg × *types* of abnormal sperm, as in Figure 23.3*b*.

a.

b.

Conclusions

- What syndromes are the result of (a)? ______
- Are all offspring viable? ______ Explain. ______

______

- What syndromes are the result of (b)? ______
- Are all offspring viable? ______ Explain. ______

______

23.2 Genetic Disorders: The Present

Today, genetic counselors can decide the pattern of inheritance by studying pedigrees.

Pedigrees

A **pedigree** shows the inheritance of a genetic disorder within a family and can help determine whether any particular individual has an allele for that disorder. Then a Punnett square can be done to determine the chances of a couple producing an affected child.

In a pedigree, Roman numerals indicate the generation, and Arabic numerals indicate particular individuals in that generation. The symbols used to indicate normal and affected males and females, reproductive partners, and siblings are shown in Figure 23.4.

Figure 23.4 Pedigree symbols.

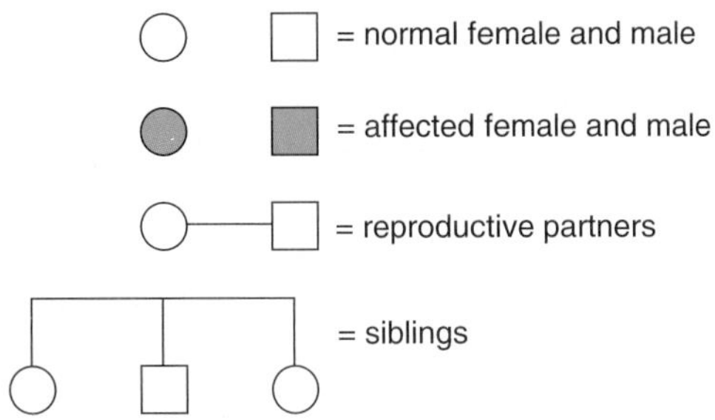

Pedigree Analyses

For each of the following pedigrees, determine how a genetic disorder is passed. Is the inheritance pattern autosomal dominant, autosomal recessive, or X-linked recessive? Also, decide the genotype of particular individuals in the pedigree. Remember that the *genotype* indicates the dominant and recessive alleles present and the *phenotype* is the actual physical appearance of the trait in the individual. A pedigree indicates the phenotype, and you can reason out the genotype.

1. Study the following pedigree:

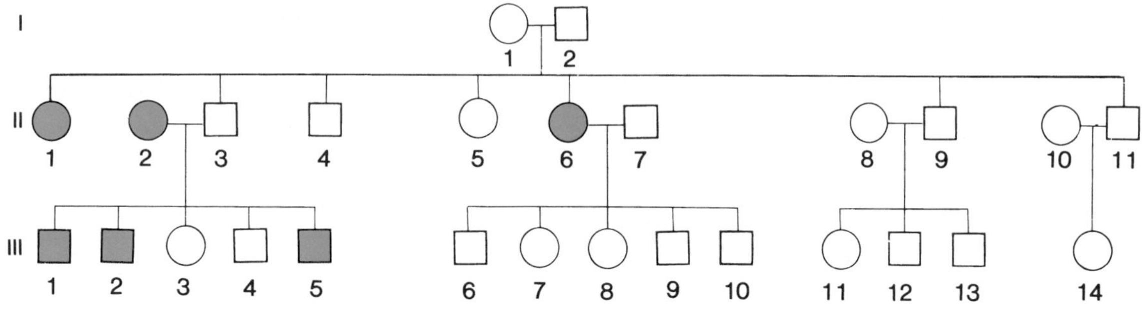

a. What is the inheritance pattern for this genetic disorder? ______________________

b. What is the genotype of the following individuals? Use *A* for the dominant allele and *a* for the recessive allele.

Generation I, individual 1: ______________________

Generation II, individual 1: ______________________

Generation III, individual 8: ______________________

2. Study the following pedigree:

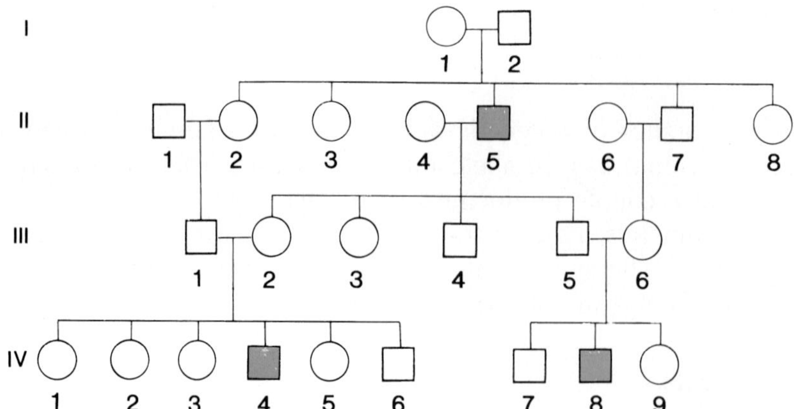

a. What is the inheritance pattern for this genetic disorder? ____________

b. What is the genotype of the following individuals?

Generation I, individual 1: ____________

Generation II, individual 8: ____________

Generation III, individual 1: ____________

3. Study the following pedigree:

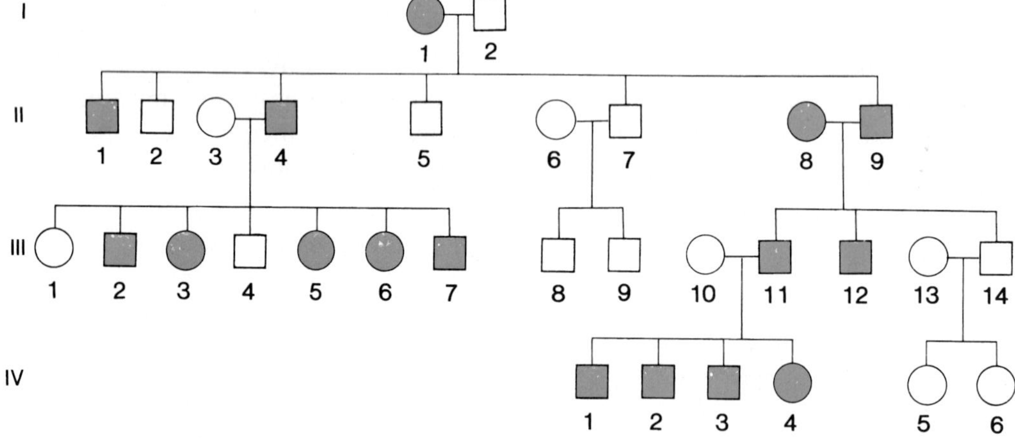

a. What is the inheritance pattern for this genetic disorder? ____________

b. What is the genotype of the following individuals?

Generation I, individual 1: ____________

Generation II, individual 7: ____________

Generation III, individual 4: ____________

Generation III, individual 11: ____________

Inheritance of a Genetic Disorder

After a counselor has decided the inheritance pattern of genetic disorder, a couple can be advised about the various chances of having a child with a disorder.

Autosomal Dominant and Autosomal Recessive

With regard to disorders on the autosomal chromosomes, counselors are familiar with the crosses shown in Figures 23.5 and 23.6.

Figure 23.5 Heterozygous-by-heterozygous cross.

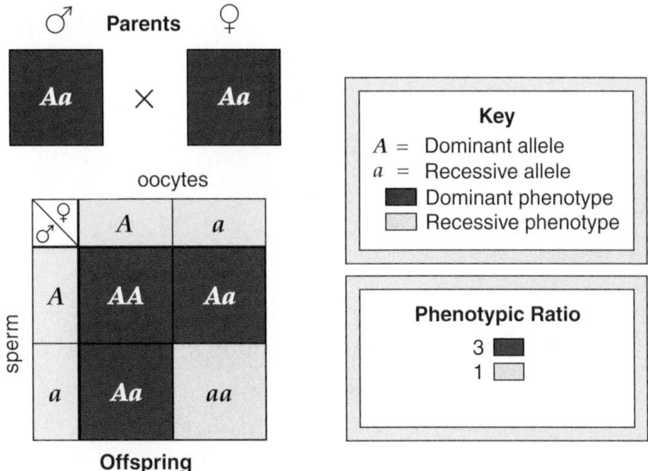

Figure 23.6 Heterozygous-by-homozygous recessive cross.

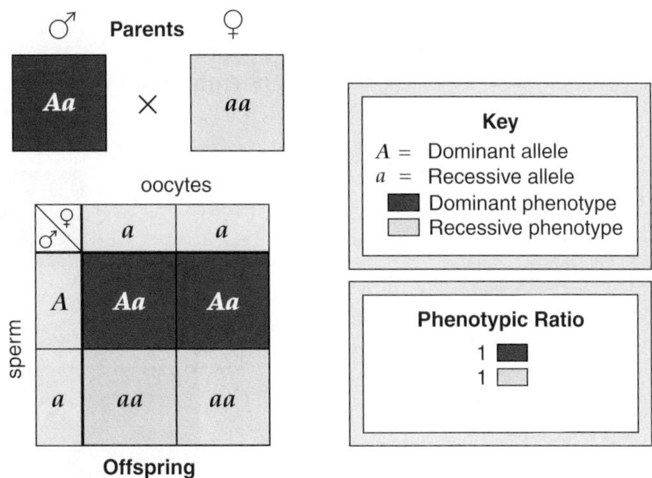

1. With reference to Figure 23.5, if a genetic disorder is recessive and the parents are heterozygous, what are the chances that an offspring will have the disorder? ____________
2. With reference to Figure 23.5, if a genetic disorder is dominant and the parents are heterozygous, what are the chances that an offspring will have the disorder? ____________
3. With reference to Figure 23.6, if the parents are heterozygous by homozygous recessive, and the genetic disorder is recessive, what are the chances that the offspring will have the disorder?

   ____

4. With reference to Figure 23.6, if the parents are heterozygous by homozygous recessive, and the genetic disorder is dominant, what are the chances that an offspring will have the disorder?

   ____

Genetics Problems

1. Neurofibromatosis is a dominant disorder. If a woman is heterozygous and reproduces with a homozygous normal man, what are the chances that a child will have neurofibromatosis?
   ____
2. Cystic fibrosis is a recessive disorder. A man and a woman are both carriers for cystic fibrosis. What are the chances a child will have cystic fibrosis? ______________
3. Huntington disease is a dominant disorder. Mary is 25 years old and as yet has no signs of Huntington disease. However, her mother does have Huntington disease, but her father is free of the disorder. What are the chances that Mary will develop Huntington disease?
   ____
4. Phenylketonuria (PKU) is a recessive disorder. Mr. and Mrs. Smith appear to be normal, but they have a child with PKU. What are the genotypes of Mr. and Mrs. Smith?
   ____
5. Tay-Sachs is an autosomal recessive disorder. Is it possible for two individuals who do not have Tay-Sachs to have a child with the disorder? Explain. ______________

X-Linked Recessive

Counselors are familiar with the crosses shown in Figures 23.7 and 23.8. Notice that in these crosses the possible genotypes and phenotypes are as follows:

Females	**Males**
X^BX^B = normal vision	X^BY = normal vision
X^BX^b = normal vision (carrier)	X^bY = color blindness
X^bX^b = color blindness	

Figure 23.7 Cross of a normal-visioned man and a carrier woman.

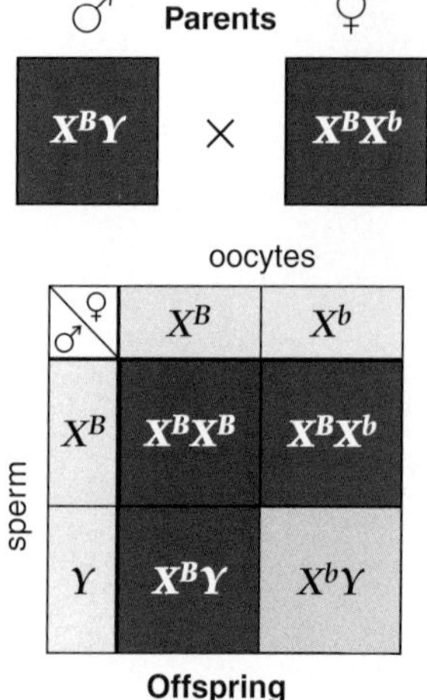

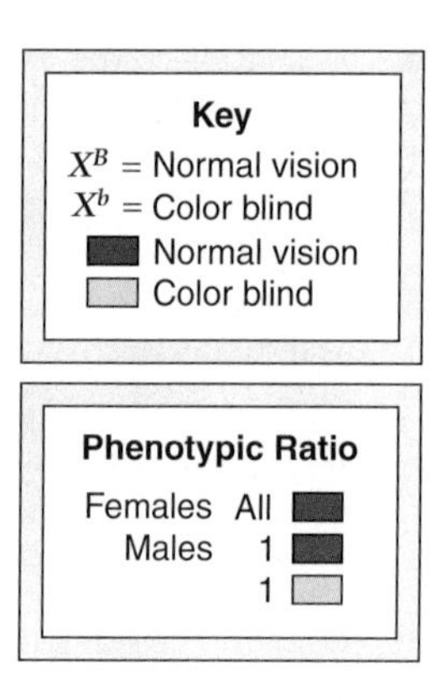

Figure 23.8 Cross of a color blind man and a normal woman.

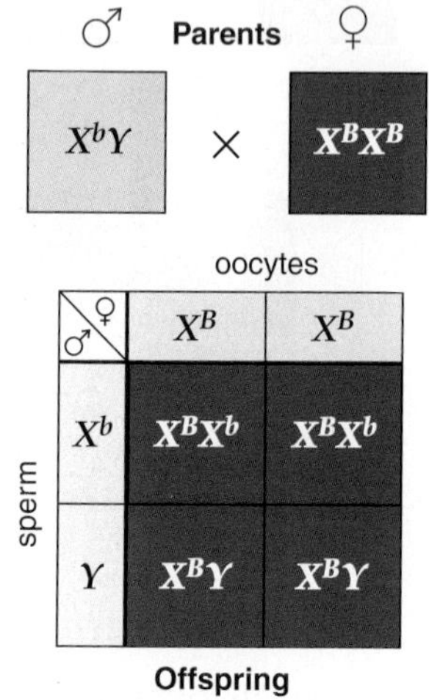

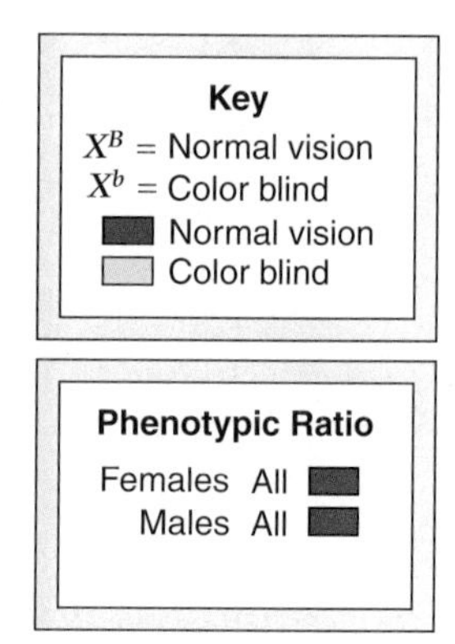

1. With reference to Figure 23.7, if the mother is a carrier and the father has normal vision, what are the chances that a daughter will be color blind? ____________

A daughter will be a carrier? ____________ A son will be color blind? ____________

2. With reference to Figure 23.8, if the mother has normal vision and the father is color blind, what are the chances that a daughter will be color blind? ____________

A daughter will be a carrier? ____________ A son will be color blind? ____________

Experimental Procedure: X-Linked Traits

1. Your instructor will provide you with a color blindness chart. Have your lab partner present the chart to you. Write down the words or symbols you see, but do not allow your partner to see what you write, and do not discuss what you see. This is important because color-blind people see something different than do people who are not color blind.
2. Now test your lab partner as he or she has tested you.
3. Are you color blind? ________ If so, what is your genotype? ____________
4. If you are a female and are not color blind, you can judge whether you are homozygous or heterozygous by knowing if any member of your family is color blind. If your father is color blind, what is your genotype? ________ If your mother is color blind, what is your genotype? ________ If you know of no one in your family who is color blind, what is your probable genotype? ________

Genetics Problems

1. A woman with normal color vision, whose father was color blind, marries a man with normal color vision. What do you expect to see among their offspring? ____________

   ____

   What would you expect to see if it was the normal-visioned man's father who was color blind?

   ____

2. John's father is color blind but his mother is not color blind. Is John necessarily color blind? ____________ Why? ____________ Could he be color blind? ____________ Why? ____________
3. A person with Turner syndrome has hemophilia. Her mother does not have hemophilia, but her father does. In which parent did nondisjunction occur, considering that the single X came from the father? ________ Is it possible to tell if nondisjunction occurred during meiosis I or meiosis II? ________ Explain. ____________

   ____

   (*Hint:* Use Figure 23.3 to help solve this problem.)

Testing for Genetic Disorders

Tests are now available for a large number of genetic disorders (Table 23.2). For example, chromosomal tests are available for cystic fibrosis, neurofibromatosis, and Huntington disease. Blood tests can identify carriers of thalassemia and sickle-cell disease. By measuring enzyme levels in blood, tears, or skin cells, carriers of enzyme defects can also be identified for certain inborn metabolic errors, such as Tay-Sachs disease. From this information and the knowledge of the pattern of inheritance, a counselor can sometimes predict the chances of a couple having a child with the disorder.

If a woman is already pregnant, chorionic villi sampling can be done early, and amniocentesis can be done later in the pregnancy. These procedures, which were illustrated in Figure 23.2, allow the testing of embryonic and fetal cells, respectively, to determine if the unborn child has a genetic

Table 23.2 Tests and Treatments for Some Human Genetic Disorders

Name	Description	Chromosome	Incidence Among Newborns in the U.S.	Status
Autosomal Recessive Disorders				
Cystic fibrosis	Mucus in the lungs and digestive tract is thick and viscous, making breathing and digestion difficult.	7	One in 2,500 Caucasians	Allele located; chromosome test now available;* treatment being investigated
Tay-Sachs disease	Neurological impairment and psychomotor difficulties develop early, followed by blindness and uncontrollable seizures; death usually occurs before age 5.	15	One in 3,600 Jews of eastern European descent	Biochemical test now available*
Phenylketonuria	The inability to metabolize phenylananine; if a special diet is not begun, mental retardation develops.	12	One in 5,000 Caucasians	Biochemical test now available; treatment available*
Sickle-cell disease	Sickle-shaped red blood cells causing poor circulation, anemia, and internal hemorrhaging.	11	One in 100 African Americans	Chromosome test now available*
Autosomal Dominant Disorders				
Neurofibromatosis	Benign tumors occur under the skin or deeper.	17	One in 3,000	Allele located; chromosome test now available*
Huntington disease	Minor disturbances in balance and coordination develop in middle age and progress toward severe neurological disturbances, leading to death.	4	One in 20,000	Allele located; chromosome test now available*
X-Linked Recessive Disorders				
Hemophilia A	Propensity for bleeding, often internally, due to the lack of a blood clotting factor	X	One in 15,000 male births	Treatment available
Duchenne muscular dystrophy	Muscle weakness develops early and progressively intensifies until death occurs, usually before age 20.	X	One in 5,000 male births	Allele located; biochemical tests of muscle tissue available; treatment being investigated

*Prenatal testing is done.

disorder. If so, treatment may be available even before birth, or parents may decide whether or not to end the pregnancy.

If a woman is not yet pregnant, how might a couple who test positive for an abnormal allele ensure that a child born to them will be normal? In most instances, it is possible to perform in vitro fertilization and then preimplantation genetic diagnosis (PGD) in order to determine the genotype of the embryo with regard to particular genetic disorders. During preimplantation analysis, one of the cells of an eight-celled embryo is removed, and its genotype is determined (Fig. 23.9). The other seven cells will continue to develop normally. Only embryos that test normal are implanted in the uterus of the female.

By 2004, nearly a decade after PGD was developed, nearly 1000 children have been born worldwide following PGD screening. In the future, PGD might be coupled with gene therapy so that any embryo would be suitable for implantation.

Figure 23.9 Preimplantation genetic diagnosis.
One cell from an eight-celled embryo can be tested for abnormal alleles; if all tested alleles are normal, the seven-celled preembryo can complete normal development.

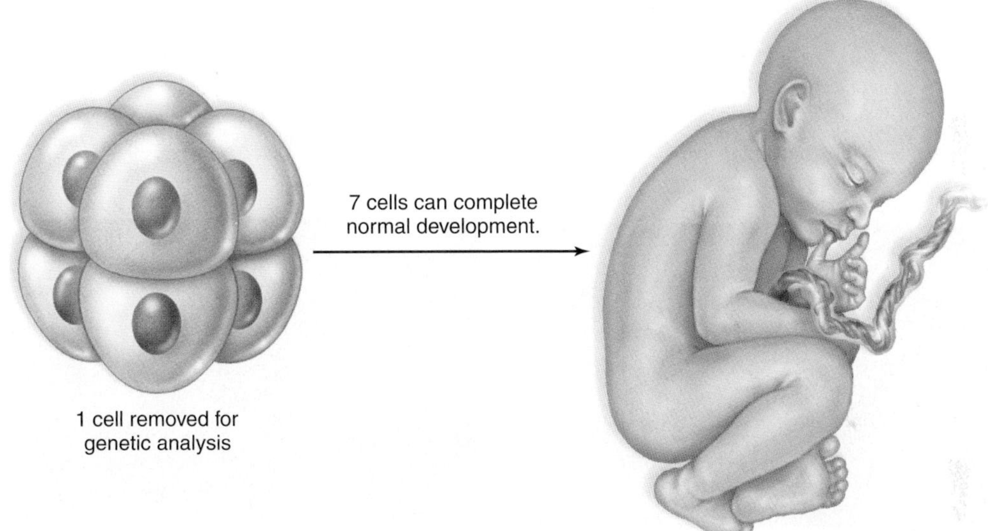
7 cells can complete normal development.
1 cell removed for genetic analysis

23.3 Genetic Disorders: The Future

The base sequence of the DNA in all the chromosomes is an organism's genome. Now that the Human Genome Project is finished, we know the normal order of all the 3.6 billion nucleotide bases in the human genome. Someday it will be possible to sequence anyone's genome within a relatively short period of time, and thereby determine what particular alterations in base sequence signify that he or she has a disorder or will have one in the future. In this laboratory, you will study how an alteration in base sequence causes a person to have sickle-cell disease.

In persons with sickle-cell disease, the red blood cells aren't biconcave disks, as are normal red blood cells—they are sickle-shaped. Sickle-shaped cells can't pass along narrow capillary passageways. They clog the vessels and break down, causing the person to suffer from poor circulation, anemia, and poor resistance to infection. Internal hemorrhaging leads to further complications, such as jaundice, episodic pain in the abdomen and joints, and damage to internal organs.

Sickle-shaped red blood cells are caused by an abnormal hemoglobin (Hb^S). Individuals with the Hb^AHb^A genotype are normal; those with the Hb^SHb^S genotype have sickle-cell disease and those with the Hb^AHb^S have sickle-cell trait. Persons with sickle-cell trait do not usually have sickle-shaped cells unless they experience dehydration or mild oxygen deprivation.

Genomic Sequence for Sickle-Cell Disease

Examine Figure 23.10*a* and *b*, which show the DNA base sequence, the mRNA codons, and the amino acid sequence for a portion of the gene for Hb^A and the same portion for Hb^S.

Figure 23.10 Sickle-cell disease.
Sickle-cell disease occurs when *(a)* the DNA base sequence in one location has changed from CTC *(b)* to CAC in both alleles for Hb^A.

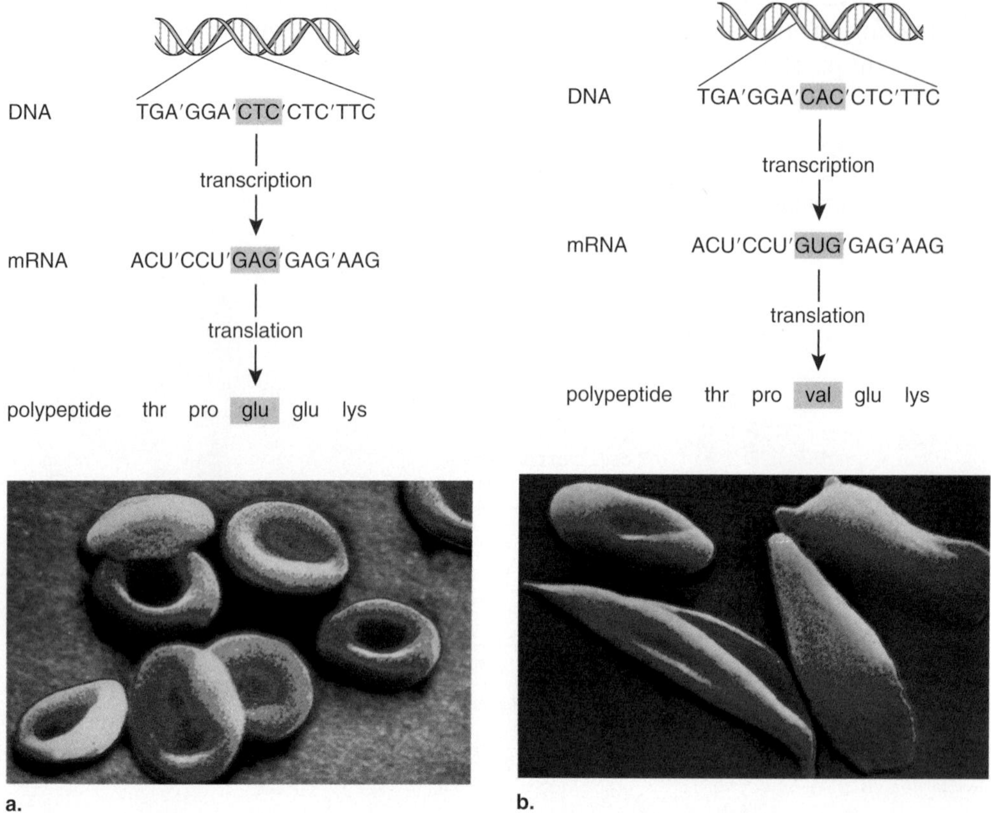

1. In what one base does Hb^A differ from Hb^S? Hb^A ______ Hb^S ______
2. What are the codons that contain this base? Hb^A ______ Hb^S ______
3. What is the amino acid difference? Hb^A ______ Hb^S ______

This amino acid difference causes the polypeptide chain in sickle-cell hemoglobin to pile up as firm rods that push against the plasma membrane and deform the red blood cell into a sickle shape:

$-CH_2-CH_2-C(=O)O^-$ glutamate (polar *R* group)	$-CH_2-CH(CH_3)_2$ valine (nonpolar *R* group)

Gel Electrophoresis

If so instructed, you will carry out gel electrophoresis as directed by a kit and described in Laboratory 22, Figure 22.11. Gel electrophoresis allows us to detect the molecule difference between hemoglobin in a person with sickle-cell disease, a normal person, and a person with sickle-cell trait (Fig. 23.11). If you are not doing the gel electrophoresis, continue with Analyzing the Electrophoresed Gel below.

Experimental Procedure: Gel Electrophoresis

Obtain three samples of hemoglobin provided by your kit. They are labeled sample A, B, and C.

> **Caution: Gel electrophoresis** Students should wear personal protective equipment: safety goggles and smocks or aprons while loading gels and during electrophoresis and protective gloves while staining.

Then, as directed by your kit, carry out electrophoresis of these samples.

Analyzing the Electrophoresed Gel

1. Sickle-cell hemoglobin (Hb^S) migrates slower toward the positive pole than normal hemoglobin (Hb^A) because the amino acid valine has no polar *R* groups, whereas the amino acid glutamate does have a polar *R* group.
2. In Figure 23.11, which lane contains only Hb^S, signifying that the individual is Hb^SHb^S? ______
3. Which lane contains only Hb^A, signifying that the individual is Hb^AHb^A? ______
4. Which lane contains both Hb^S and Hb^A, signifying that the individual is Hb^AHb^S? ______

Conclusions

- You are a genetic counselor of the future. A young couple seeks your advice because sickle-cell disease occurs among the family members of each. You order DNA base sequencing to be done. The results come back that at one of the loci for normal hemoglobin, each has the abnormal sequence CAC instead of CTC. The other locus is normal. What are the chances that this couple will have a child with sickle-cell disease? ______________________________
- How is it possible for this couple to ensure that only normal embryos come to term?

____

Figure 23.11 Gel electrophoresis of hemoglobins.

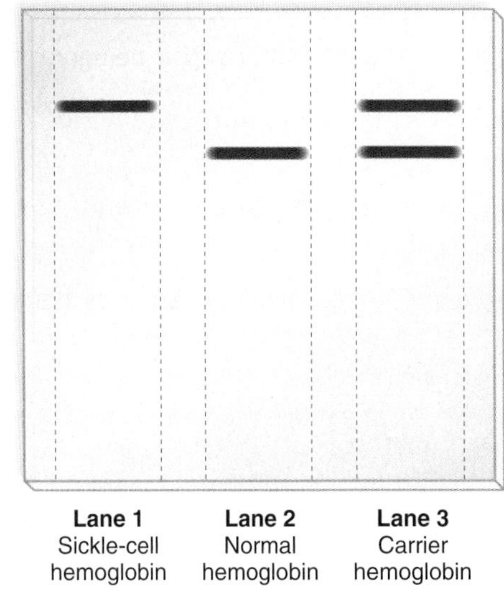

Laboratory Review 23

_______________ 1. What term refers to paired chromosomes arranged by size and shape?

_______________ 2. What pair of chromosomes is not homologous in a normal male karyotype?

_______________ 3. What syndrome is inherited when an egg carrying two X chromosomes is fertilized by a sperm carrying one Y chromosome?

_______________ 4. What abnormal meiotic event leads to the syndrome described in question 3? In which parent?

_______________ 5. What two types of sperm result if nondisjunction of sex chromosomes occurs during meiosis I of spermatogenesis?

_______________ 6. Name a common autosomal trisomy.

_______________ 7. What does a genetic counselor construct to show the inheritance pattern of a genetic disorder within a family?

_______________ 8. If an individual exhibits the dominant trait, what two genotypes are possible? (Use *A* for the dominant allele and *a* for the recessive allele.)

_______________ 9. What is the genotype of a man who has sickle-cell trait?

_______________ 10. The alleles of which parent, regardless of the phenotype, determine color blindness in a son?

_______________ 11. If only males are affected in a pedigree, what is the likely inheritance pattern for the trait?

_______________ 12. If the parents are not affected and a child is affected, what is the inheritance pattern?

_______________ 13. What is the genotype of a female with hemophilia?

Thought Questions

14. An egg contains two X chromosomes that carry the same alleles. Did nondisjunction take place during meiosis I or meiosis II? Explain.

15. What inheritance pattern in a pedigree would allow you to decide that a trait is X-linked?

16. Why are X-linked disorders such as hemophilia generally more common in males?

LABORATORY

24

Evidences of Evolution

Learning Outcomes

24.1 Fossil Record

- Describe several types of fossils and explain how fossils help to establish the sequence of earth's history.
- Explain how scientists use fossils to establish relationships between different forms of life.

24.2 Comparative Anatomy

- Explain how comparative anatomy gives evidence of common descent.
- Compare the human skeleton with the chimpanzee skeleton, and illustrate how the differences between the two can be related to each individual's unique way of life.

24.3 Biochemical Evidence

- Explain how biochemistry aids the study of the evolutionary relationships among organisms.
- Explain how biochemical evidence adds additional support to the concept of common descent.

Introduction

Evolution is the process by which life has changed through time. A **species** is a group of organisms that share similar characteristics and common genes, and are capable of interbreeding and producing fertile offspring. A **population** is all the members of a species living in a particular area and is likely to be affected as a group by changes in the environment. When environmental conditions do change, they may allow individuals with particular genetic variations to capture more resources. Should such changes occur, these more competitive individuals tend to survive and have more offspring than the unchanged or less competitive members. Therefore, each successive generation will include more members with the particular variation that made the bearer more competitive. Eventually, most members of a population and the species will have the same adaptations to their environment. Should a population that has undergone this type of genetic change become isolated from other populations of the same species by either physical or behavioral barriers, over a period of time, the population may become so genetically distinct that it can no longer interbreed and produce viable offspring. It is in this way that new species are able to emerge from an existing species.

Adaptations to various ways of life explain why life is so diverse. However, evolution, which has been ongoing since the origin of life, is also an explanation for the unity of life. All organisms share the same characteristics of life because they can trace their ancestry to the first cell or cells. Many different lines of evidence support this hypothesis of common descent, and the more varied the evidence supporting the hypothesis, the more certain the hypothesis becomes. As you have studied biology, you have seen a number of these unifying features. DNA is the molecule of life in living organisms. Multicellular organisms use the same basic chemical processes to sustain life. Cell division, both mitosis and meiosis, is practiced by essentially all living things. Even at the smallest level, cells use the same small group of 20 amino acids to make their own unique proteins. While the organisms themselves may appear to be different, there is much that unifies life.

In this Laboratory, you will study three types of data that support the hypothesis of common descent: (1) the fossil record, (2) comparative anatomy (embryological and adult), and (3) biochemical comparison. Fossils are the remains or evidence of some organism that lived long ago. In some cases, fossils are nothing more than casts left by something as simple as a leaf that was covered by sand. At the other extreme are the massive skeletons of fossil animals trapped alive in quicksand or tar pits, their bones mineralized and essentially turned into stone. In the history of life as detected by fossils, prokaryotic cells preceded eukaryotic cells, and among animals, invertebrates preceded vertebrates. And, importantly, the age of the deposits surrounding the fossils reflect the same orderly progression; the older, more primitive forms of life are found in the oldest rock formations, while the more advanced fossils are found in younger rock. Humans began evolving about 5 million years ago, but modern humans do not appear in the fossil record until about 100,000 years ago. A comparative study of the anatomy of modern groups of organisms has shown that each group has homologous structures. For example, all vertebrate animals have essentially the same type of skeleton. Homologous structures signify relatedness through evolution. Almost all living organisms use the same basic biochemical molecules, including DNA, ATP, and many identical or nearly identical enzymes. Proteins and DNA are analyzed to determine how closely related the two groups are.

24.1 Fossil Record

The **geological timescale** (Table 24.1) pertains to the history of the earth from its formation 4 to 4.5 billion years ago to the present. The ages of rocks can be measured in years by analyzing naturally occurring radioactive elements found in minute quantities in certain rocks and minerals. Before this method was discovered, geologists depended on a relative dating system; that is, they reasoned that any given stratum (layer of sediment) is younger than the stratum of the earth's crust just beneath it.

Because certain fossils are associated with particular strata, geologists are able to relate various strata around the world. *Fossils* are the remains and traces of past life or any other direct evidences of past life. By the 1860s, fossil-containing rocks in western Europe had been divided into three great eras: Paleozoic (ancient life), Mesozoic (middle life), and Cenozoic (recent life). Each era was divided into periods, and periods, in turn, were divided into epochs. The first period of the Paleozoic is called the Cambrian, and the time before the Cambrian period is called the Precambrian. The oldest fossils that have been found date from the Precambrian. For reasons that are still being explored, the fossil record improves dramatically starting with the Cambrian period.

Notice in Table 24.1 that divisions of the geological time scale generally tend to become increasingly shorter: For example, the Cenozoic era is much shorter than the Paleozoic era, and the Neogene period is much shorter than the Paleogene period. Notice also that time is measured as "millions of years ago," and therefore, larger numbers represent an earlier time than smaller numbers.

Use Table 24.1 to answer the following questions:

1. During the ____________ era and the ____________ period, the first vascular plants appeared. How many million years ago was this? ____________________
2. During the ____________ era and the ____________ period, angiosperm (flowering plant) diversity occurred. How many million years ago was this? ____________________

Use Figure 24.1 to answer the following questions:

3. During what period did the trilobites appear? ____________________
   When did they become extinct? ____________________
4. During what period did the insects first appear? ____________________
   How many million years ago was this? ____________________
5. When did multicellular organisms first appear in the fossil record? ____________________
6. What evolved first, the prokaryotic or the eukaryotic cell? ____________________

Use Figures 24.1 and 24.2 to answer the following questions:

7. During what period were the cycads most abundant? ____________________
   What types of animals were also prevalent at this time? ____________________
8. During what period did the angiosperms evolve? ____________________
   How do you know they are the most prominent plants today? ____________________
   ____
9. When did vertebrates first appear in the fossil record? ____________________

Table 24.1 The Geological Time Scale: Major Divisions of Geological Time with Some of the Major Evolutionary Events of Each Geological Period

Era	Period	Epoch	Millions of Years Ago	Plant Life	Animal Life
		Holocene	0–0.01	Destruction of tropical rain forests by humans accelerates extinctions.	AGE OF HUMAN CIVILIZATION
Cenozoic* (from the present to 66.4 million years ago)				*Significant Mammalian Extinction*	
	Neogene	Pleistocene	0.01–2	Herbaceous plants spread and diversify.	Modern humans appear.
		Pliocene	2–6	Herbaceous angiosperms flourish.	First hominids appear.
		Miocene	6–24	Grasslands spread as forests contract.	Apelike mammals and grazing mammals flourish; insects flourish.
		Oligocene	24–37	Many modern families of flowering plants evolve.	Browsing mammals and monkeylike primates appear.
	Paleogene	Eocene	37–58	Subtropical forests with heavy rainfall thrive.	All modern orders of mammals are represented.
		Paleocene	58–66	Angiosperms diversify.	Primitive primates, herbivores, carnivores, and insectivores appear.
Mesozoic (from 66.4 to 245 million years ago)				*Mass Extinction of Dinosaurs and Most Reptiles*	
	Cretaceous		66–144	Flowering plants spread; coniferous trees decline.	Placental mammals appear; modern insect groups appear.
	Jurassic		144–208	Cycads and other gymnosperms flourish.	Dinosaurs flourish; birds appear.
				Mass Extinction Affecting All Life Forms	
	Triassic		208–245	Cycads and ginkgoes appear; forests of gymnosperms and ferns dominate.	First mammals appear; first dinosaurs appear; corals and molluscs dominate seas.
				Mass Extinction Affecting All Life Forms	
	Permian		245–286	Conifers appear.	Reptiles diversify; amphibians decline.
	Carboniferous		286–360	Age of great coal-forming forests: club mosses, horsetails, and ferns flourish.	Amphibians diversify; first reptiles appear; first great radiation of insects.
Paleozoic (from 245 to 570 million years ago)				*Mass Extinction Affecting All Life Forms*	
	Devonian		360–408	First seed ferns appear.	Jawed fishes diversify and dominate the seas; first insects and first amphibians appear.
	Silurian		408–438	Low-lying vascular plants appear on land.	First jawed fishes appear.
				Mass Extinction Affecting All Life Forms	
	Ordovician		438–505	Marine algae flourish.	Invertebrates spread and diversify; jawless fishes, first vertebrates appear.
	Cambrian		505–570	Marine algae flourish.	Invertebrates with skeletons are dominant.
Precambrian time (from 570 to 4,600 million years ago)			700	Multicellular organisms appear.	
			2,100	First complex (eukaryotic) cells appear.	
			3,100–3,500	First prokaryotic cells in stromatolites appear.	
			4,500	Earth forms.	

*Many authorities divide the Cenozoic era into the Tertiary period (contains Paleocene, Eocene, Oligocene, Miocene, and Pliocene) and the Quaternary period (contains Pleistocene and Holocene).

Figure 24.1 Geological history of selected animals.
The relative abundance of each group during any particular time period is indicated by the width of the band.

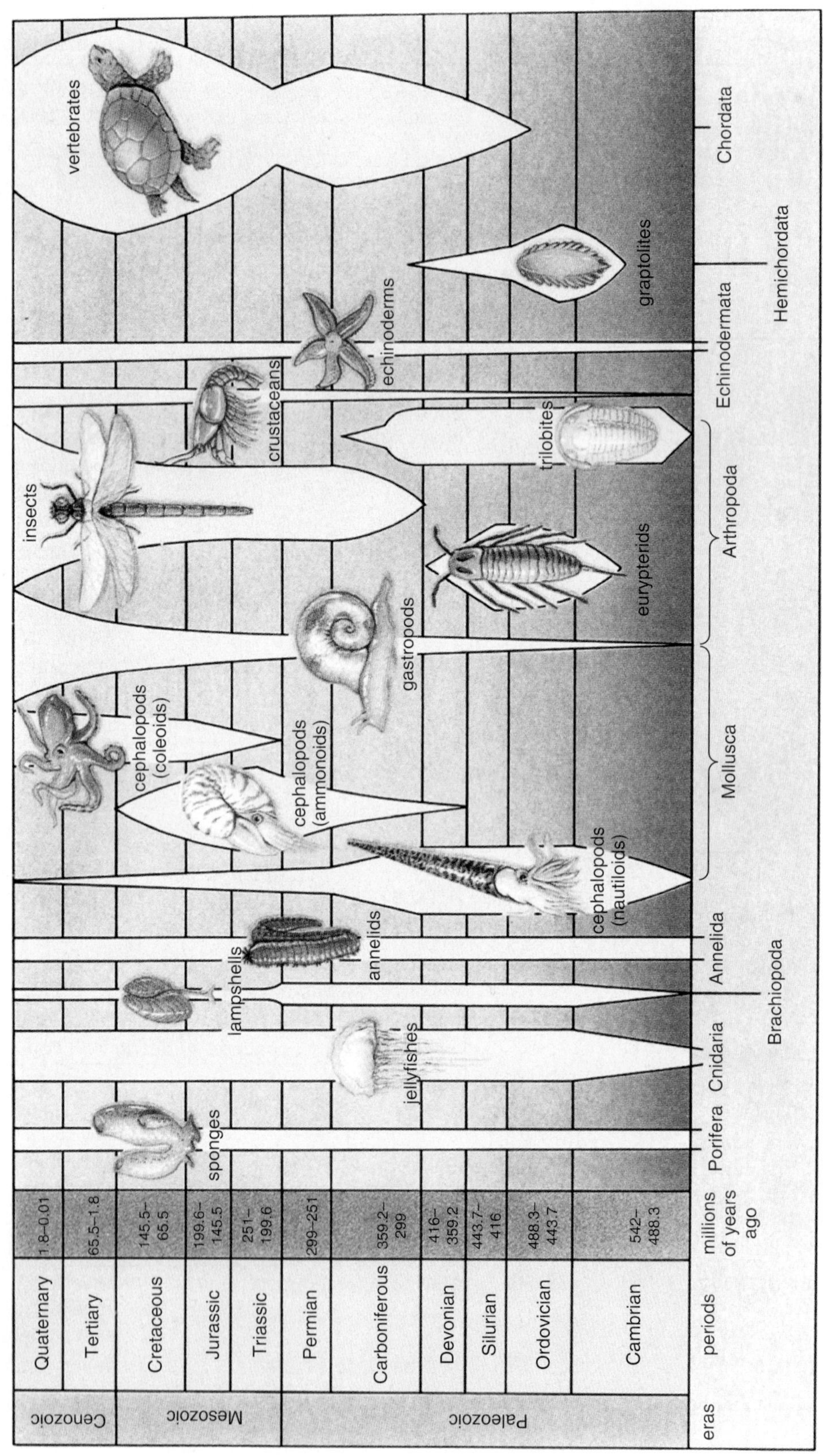

Figure 24.2 Geological history of selected algae, fungi, and plants.
The relative abundance of each group during any particular time period is indicated by the width of the band.

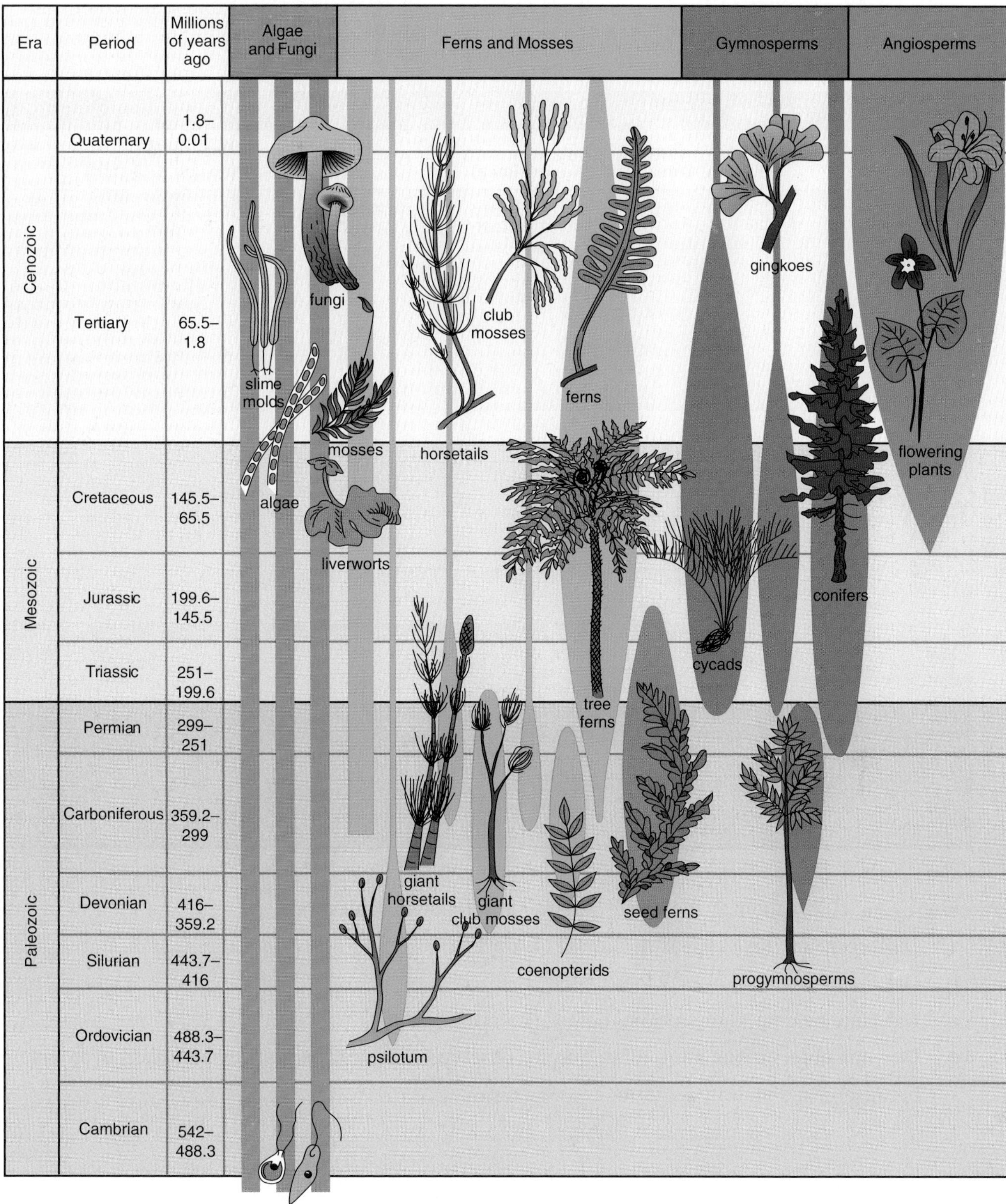

Observation: Fossils

1. Obtain a box of selected fossils from each geological era (Cenozoic, Mesozoic, Paleozoic). Fill in Table 24.2 from the identification key or the fossil labels.

Table 24.2 Fossils

Era	Period	Type of Fossil (Phylum, Class, or Common Name)	Description (May Include a Sketch)
Cenozoic			
Mesozoic			
Paleozoic			

2. From your observation of the fossils, answer the following questions:
   a. Did organisms first appear in the sea or on land? __________
   b. Did invertebrates evolve before vertebrates? __________
   c. Did cone-bearing plants evolve before flowering plants? __________
   d. Do your observations show that, despite observed changes, fossils can be linked over time because of a similarity in form? __________

24.2 Comparative Anatomy

In the study of evolutionary relationships, organisms or parts of organisms are said to be **homologous** if they exhibit similar basic structures and embryonic origins. If these organisms or parts of organisms are similar in function only, they are said to be **analogous.** Only homologous structures indicate an evolutionary relationship and are used to classify organisms.

Comparison of All Adult Vertebrate Forelimbs

The limbs of all vertebrates are homologous structures. Homologous structures share a basic pattern, although there may be specific differences. The similarity of homologous structures is explainable by descent from a common ancestor.

Observation: Vertebrate Forelimbs

1. The central diagram in Figure 24.3 represents the forelimb bones of the ancestral vertebrate. The basic components are the humerus (h), ulna (u), radius (r), carpals (c), metacarpals (m), and phalanges (p) in the five digits.
2. Carefully compare and label in Figure 24.3 the corresponding forelimb bones of the frog, the lizard, the bird, the bat, the cat, and the human. In particular, note the specific modifications that have occurred in some of the bones to meet the demands of a particular way of life.
3. Fill in Table 24.3 to indicate which bones in each specimen appear to most resemble the ancestral condition and which differ most from the ancestral condition.

Table 24.3 Comparison of Vertebrate Forelimbs

Animal	Bones That Resemble Common Ancestor	Bones That Differ from Common Ancestor
Frog		
Lizard		
Bird		
Bat		
Cat		
Human		

4. Relate the change in bone structure to mode of locomotion in two examples.

   Example 1: ______________________________

   Example 2: ______________________________

Figure 24.3 Vertebrate forelimbs.
Because all vertebrates evolved from a common ancestor, their forelimbs share homologous structures.

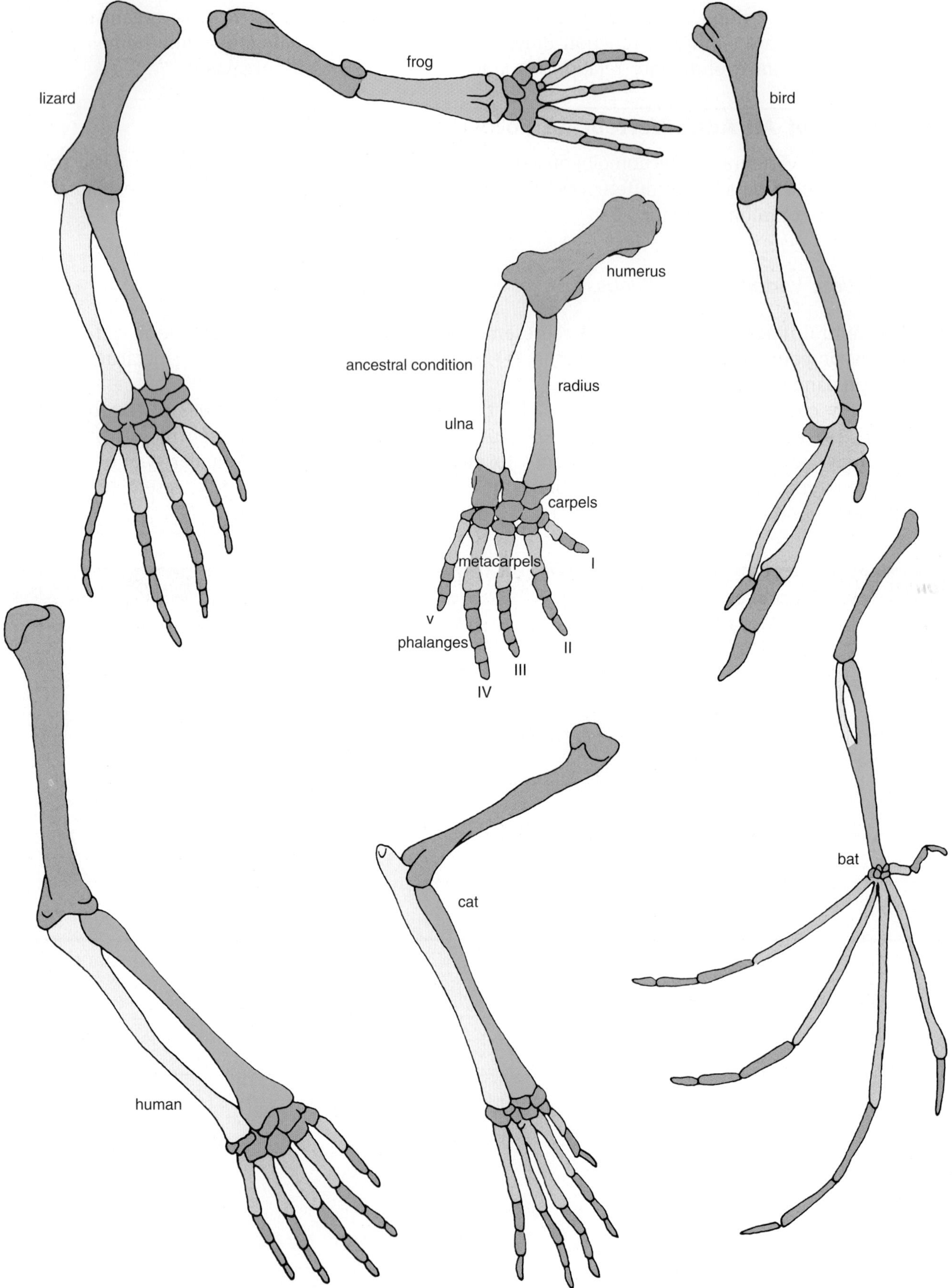

Comparison of Chimpanzee and Human Skeletons

Chimpanzees and humans are closely related, as is apparent from the comparison of the classifications of humans and chimpanzees in Table 24.4. Are chimpanzees and humans both primates?

________ At what category does the classification of chimpanzees and the classification of humans first differ? ____________________________________

Table 24.4 Comparison of Human and Chimpanzee Classifications

Classification	Chimpanzees	Humans
Domain	Eukarya	Eukarya
Kingdom	Animalia	Animalia
Phylum	Chordata	Chordata
Class	Mammalia	Mammalia
Order	Primate	Primate
Family	Pongidae	Hominidae
Genus	*Pan*	*Homo*
Species	*troglodytes*	*sapiens*

Observation: Chimpanzee and Human Skeletons

Comparison of Skeletons

Chimpanzees are arboreal and climb in trees. While on the ground, they tend to knuckle-walk, with their hands bent. Humans are terrestrial and walk erect.

Examine chimpanzee and human skeletons (Fig. 24.4), and answer the following questions:

1. **Head and torso:** Where are the head and trunk with relation to the hips and legs—thrust forward over the hips and legs or balanced over the hips and legs? *Record your answer in Table 24.5.*
2. **Spine:** Which animal has a long and curved lumbar region, and which has a short and stiff lumbar region (top set of red arrows)? *Record your answer in Table 24.5.*

   How does this contribute to an erect posture in humans? ____________________________

   ____
3. **Pelvis:** Chimps sway when they walk because lifting one leg throws them off balance. Which animal has a narrow and long pelvis, and which has a broad and short pelvis? *Record your answer in Table 24.5.*
4. **Femur:** In humans, the femur better supports the trunk. In which animal is the femur angled between articulations with the pelvic girdle and the knee (bottom set of red arrows)? In which animal is the femur straight with no angle? *Record your answer in Table 24.5.*
5. **Knee joint:** In humans, the knee joint is modified to support the body's weight. In which animal is the femur larger at the bottom and the tibia larger at the top? *Record your answer in Table 24.5.*
6. **Foot:** In humans, the foot is adapted for walking long distances and running with less chance of injury. In which animal is the big toe opposable? Which foot has an arch? *Record your answers in Table 24.5.*

Figure 24.4 Adult skeletons.
a. Human skeleton compared to an **(b)** ape skeleton. See the text for an explanation of the red arrows. (© 2001 Time Inc. reprinted by permission.)

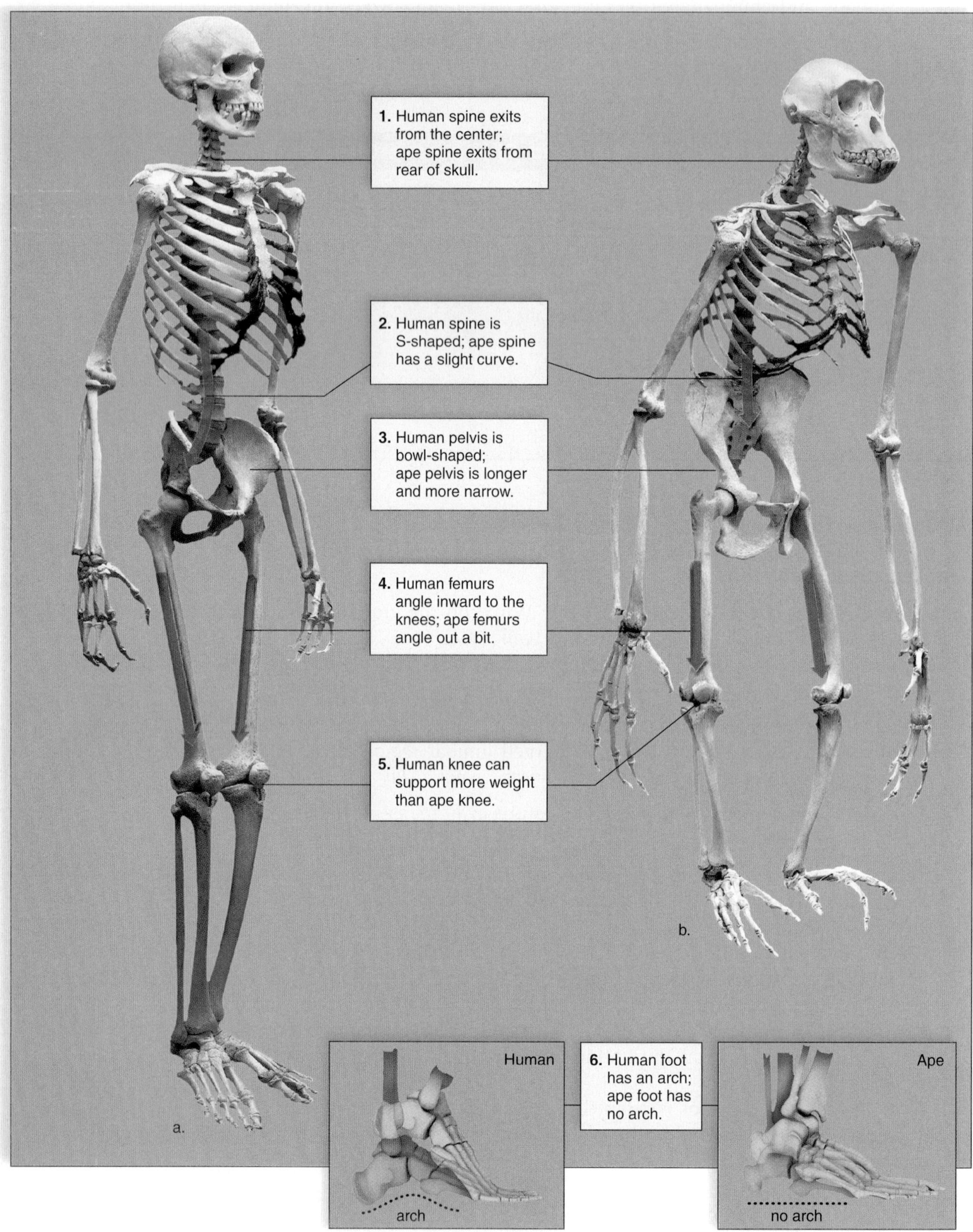
1. Human spine exits from the center; ape spine exits from rear of skull.
2. Human spine is S-shaped; ape spine has a slight curve.
3. Human pelvis is bowl-shaped; ape pelvis is longer and more narrow.
4. Human femurs angle inward to the knees; ape femurs angle out a bit.
5. Human knee can support more weight than ape knee.
6. Human foot has an arch; ape foot has no arch.
Human
arch
Ape
no arch
a.
b.

Table 24.5 Comparison of Chimpanzee and Human Skeletons		
Skeletal Part	**Chimpanzee**	**Human**
Head and torso		
Spine		
Pelvis		
Femur		
Knee joint		
Foot:		
Opposable toe		
Arch		

Figure 24.5 Chimpanzee skull compared with a human skull.
Although the general shapes of these skulls are different, they have similarities.

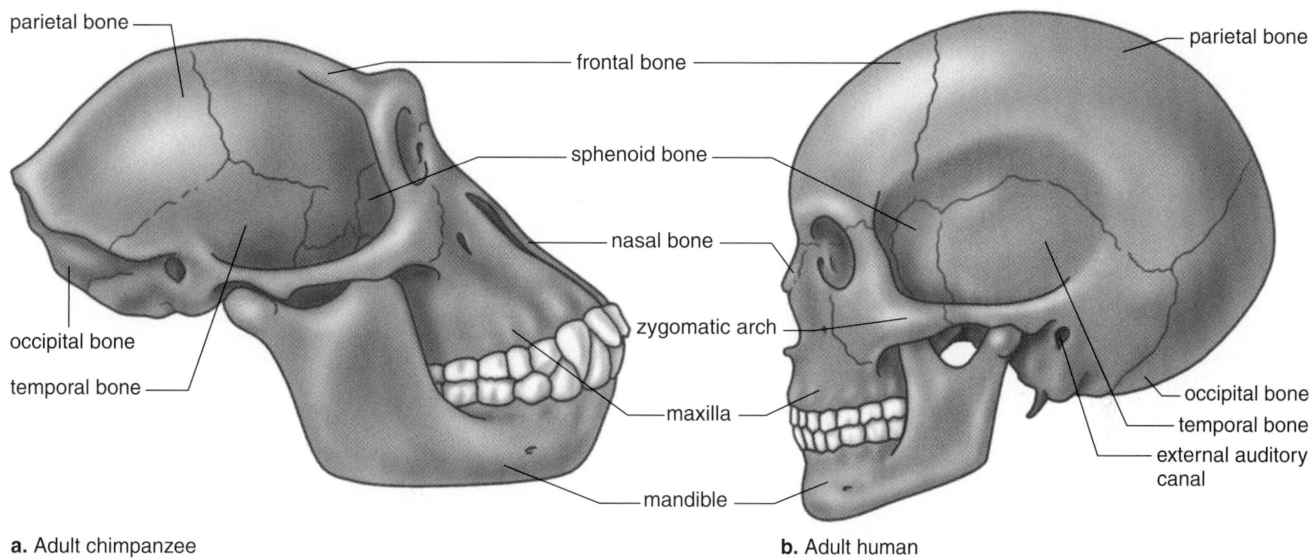

a. Adult chimpanzee

b. Adult human

7. Examine the position and shape of the parietal bones in both the chimpanzee and human skulls (Fig. 24.5). How does the chimpanzee skull differ from the human skull in this respect? ________________
8. Compare the shape and position of the occipital bones in the chimpanzee and human skulls. ________________
9. Compare the slope of the frontal bones of the chimpanzee and human skulls. How are they different? ________________
10. For which skull is the supraorbital ridge (the region of frontal bone just above the eye socket) thicker? ________________

11. What is the position of the mouth and chin in relation to the profile for each skull? ________

____

What effect has the evolutionary change in the positions of these bones had on the shape of the face? ____

12. Examine the teeth in the adult chimpanzee and adult human skulls. Are the shapes and types of teeth similar in both? ____________ Diet can account for many of the observed differences. Humans are omnivorous. A diet rich in meat does not require strong grinding teeth or well-developed facial muscles. Chimpanzees are vegetarians, and a vegetarian diet requires strong facial muscles that attach to bony projections.

Comparison of Vertebrate Embryos

The anatomy shared by vertebrates extends to their embryological development. For example, as embryos, they all have a post-anal tail, somites (segmented blocks of mesoderm lying on either side of the notochord), and paired pharyngeal pouches and bordering gill arches. In aquatic animals, these pouches and arches become functional gills (Fig. 24.6). In humans, the first pair of pouches becomes the cavity of the middle ear and auditory tube, the second pair becomes the tonsils, and the third and fourth pairs become the thymus and parathyroid glands.

Figure 24.6 Vertebrate embryos.
During early developmental stages, vertebrate embryos have certain characteristics in common.

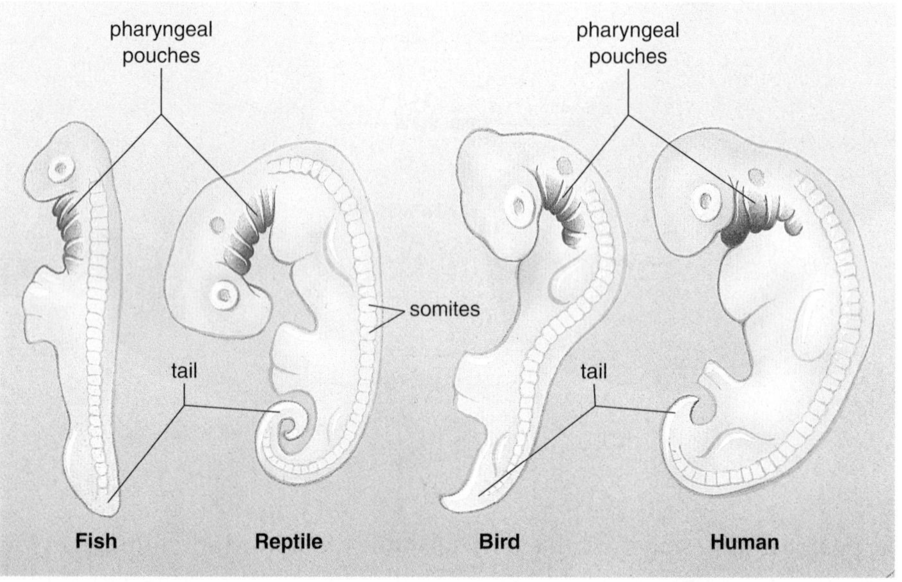

Observation: Chick and Pig Embryos

1. Obtain prepared slides of vertebrate embryos at comparable stages of development. Observe each of the embryos using a binocular dissecting microscope.
2. List five similarities of the embryos:

   a. ____

   b. ____

   c. ____

d. ______________________________

e. ______________________________

3. Why do the embryos resemble one another so closely? ______________________________

____

24.3 Biochemical Evidence

Almost all living organisms use the same basic biochemical molecules, including DNA, ATP, and many identical and nearly identical enzymes. In addition, living organisms utilize the same DNA triplet code and the same 20 amino acids in their proteins. There is no obvious functional reason these elements need to be so similar. Therefore, their similarity is best explained by descent from a common ancestor.

Antigen-Antibody Reactions

The immune system makes **antibodies** (proteins) that react with foreign proteins, termed **antigens.** The antigen-antibody reaction is specific: An antibody will react only with its particular antigen. In the laboratory, this reaction can be observed when a precipitate, a substance separated from the solution, appears (Fig. 24.7).

Biochemists have used the antigen-antibody reaction to determine the degree of relatedness between animals. In one technique, human serum (containing human proteins) is injected into the bloodstream of a rabbit, and the rabbit makes antibodies against the human serum. Some of the rabbit's blood is then drawn off, and the sensitized serum that contains the antibodies is separated from it. This sensitized rabbit serum will react strongly (determined by the amount of precipitate) against a new sample of human blood serum. The rabbit serum will also react against serum from other animals. *The more closely related an animal is to humans, the more the precipitate forms.*

Figure 24.7 Antigen-antibody reaction.
When antigens react with antibodies, a complex forms that appears as a precipitate.

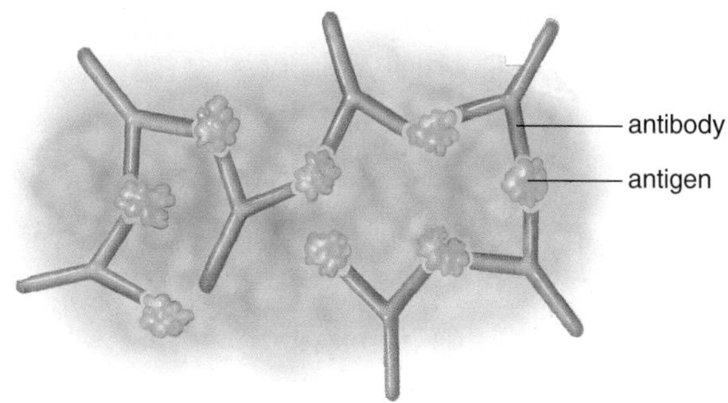

Experimental Procedure: Protein Similarities

1. Obtain a chemplate (a clear glass tray with wells), one bottle of synthetic *human blood serum,* one bottle of synthetic *rabbit blood serum,* and five bottles (I–V) of *blood serum test solution.*
2. Put two drops of synthetic rabbit blood serum in each of the six wells in the chemplate. Label the wells 1–6.
3. Add 2 drops of synthetic human blood serum to each well, and stir with the plastic stirring rod that was attached to the chemplate. The rabbit serum has now been "sensitized" to human serum. (This simulates the production of antibodies in the rabbit's bloodstream in response to the human blood proteins.)
4. Rinse the stirrer. (The large cavity of the chemplate may be filled with water to facilitate rinsing.)
5. Add 4 drops of *blood serum test solution III* (tests for human blood proteins) to well 6.

   Describe what you see. ______________________________

   ____

   This well will serve as the basis by which to compare all the other samples of test blood serum.
6. Now add 4 drops of *blood serum test solution I* to well 1. Stir and observe. Rinse the stirrer. Do the same for each of the remaining *blood serum test solutions* (*II–V*)—adding II to well 2, III to well 3, etc. Be sure to rinse the stirrer after each use.
7. At the end of 10 and 20 minutes, record the amount of precipitate in each of the six wells in Figure 24.8. Well 6 is recorded as having ++++ amount of precipitate after both 10 and 20 minutes. Compare the other wells with this well (+ = trace amount; 0 = none). Holding the plate slightly above your head at arm's length and looking at the underside toward an overhead light source will allow you to more clearly determine the amount of precipitate.

Figure 24.8 Biochemical evidence for evolution.
The greater the amount of precipitate, the more closely related an animal is to humans.

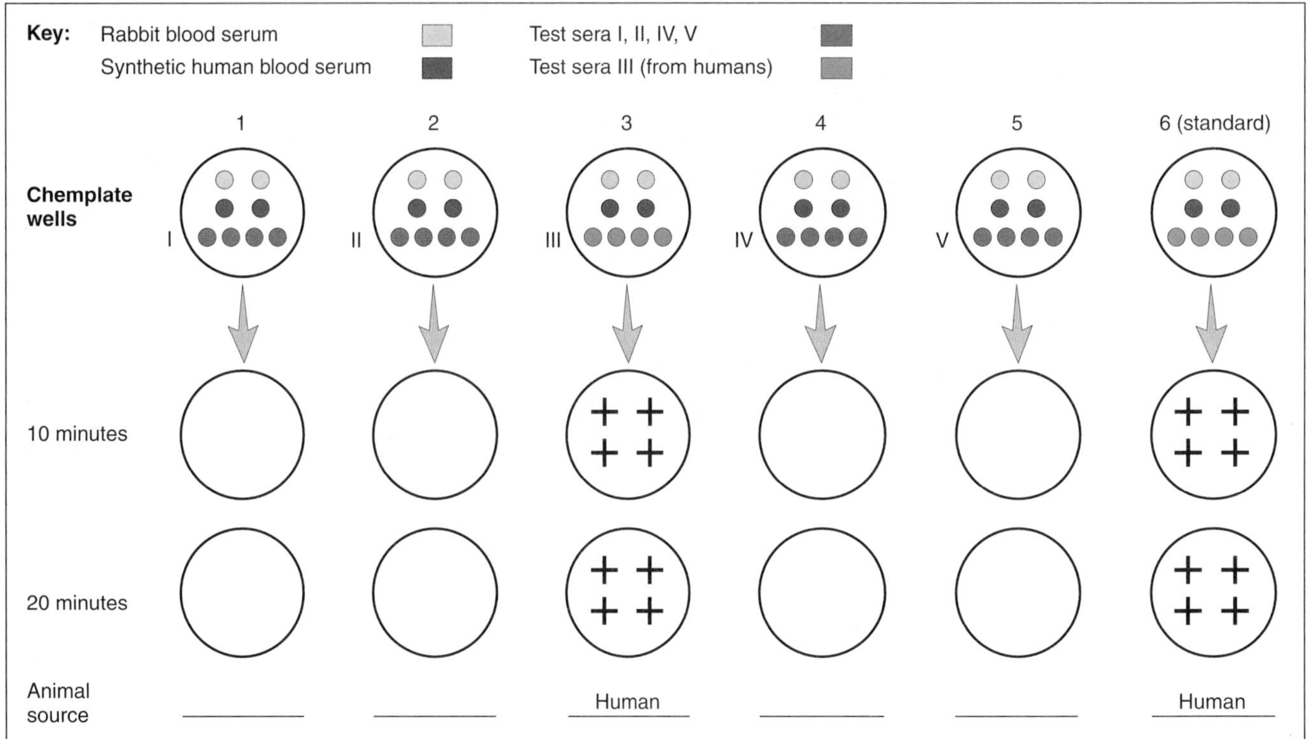

Conclusions

- The last row in Figure 24.8 tells you that the test serum in well 3 is from a human. How do your test results confirm this? ______________________________
- Aside from humans, the test sera (supposedly) came from a pig, a monkey, an orangutan, and a chimpanzee, stated in the correct order, from least related to most related, to humans.
- Judging by the amount of precipitate, complete the last row in Figure 24.8 by indicating which serum you believe came from which animal. On what do you base your conclusions?

____

____

Laboratory Review 24

______________ 1. State the simplest definition of evolution.

______________ 2. What does the geological timescale encompass—tens of millions or hundreds of millions of years?

______________ 3. What is the best explanation for the fact that fossils do not resemble their modern-day representatives?

______________ 4. Fossils are the __________ of past life.

______________ 5. All vertebrates go through similar embryological stages. What does this suggest?

______________ 6. Which skeleton—chimpanzee or human—has a narrow and long pelvis?

______________ 7. Which has thicker supraorbital ridges, the chimpanzee skull or the human skull?

______________ 8. Which term—homologous or analogous—means that components are similar in structure?

______________ 9. During development, all vertebrates have ________, even though only fish have gill slits as adults.

______________ 10. In this laboratory, what type of biochemical reaction was used to determine relatedness?

______________ 11. What does it indicate if antibodies to the serum of one species react strongly against the serum of another species?

______________ 12. All apes and humans are members of what order?

______________ 13. What four characteristics do all vertebrates share in common during embryological development?

Thought Questions

14. If a characteristic is found in bacteria, fungi, pine trees, snakes, and humans, when did it most likely evolve? Why?

(*Continued* on reverse)

15. What do mutations have to do with amino acid changes in a protein? How do such changes help determine relatedness of organisims?

16. Homologous structures arise from common ancestry; analogous structures do not. Given this information, how might you explain the existence of analogous structures?

Laboratory Notes

LABORATORY

25

Microbiology

Learning Outcomes

25.1 Bacteria
- Explain how the structure of a bacterium contributes to its ability to survive a host's defense mechanisms and cause disease.
- Using slides or prepared images, demonstrate the ability to distinguish the basic bacterial cell shapes.
- Explain the value of Gram staining and illustrate the correct protocol for using this staining technique in a laboratory setting.
- Identify and state the importance of cyanobacteria.

25.2 Protists
- Distinguish several differences between green and golden-brown algae.
- Contrast the types of locomotion employed by the protozoans observed in this laboratory activity.
- Describe the plasmodium of a plasmodial slime mold.

25.3 Fungi
- Compare and contrast the life cycles of zygospore, sac, and club fungi.
- Using either slides or images, identify and describe the reproductive structures of zygospore, sac, and club fungi.

Introduction

To simplify scientific study, similar organisms are grouped, or classified, into large categories; these larger groups are then subdivided on the basis of distinctive features that set the smaller subdivisions apart. Large groups with general similarities are grouped in categories referred to as **domains.** There are three domains of life: the **Bacteria,** the **Archaea,** and the **Eukarya** (Figure 25.1). The domains Bacteria and Archaea consist of single-celled prokaryotes. These unicellular organisms lack membrane-bound organelles, including a nucleus. The Archaea are represented by extremely primitive prokaryotes, some of which are capable of living in the most extreme environments on earth, habitats that are characterized by either high temperatures or high salinity. The Eukarya, eukaryotic organisms whose cells are more complex due to the presence of membrane-bound organelles, include the Protista, the Fungi, plants, and animals. In today's laboratory, we are studying representative examples of bacteria, protists, and fungi, at least some of which are entirely microscopic. **Bacteria** are prokaryotes in the **domain Bacteria,** while **protists** and **fungi** are eukaryotes in the **domain Eukarya.** This means that bacteria, placed in a different domain from the other two, are distantly related to them. In this laboratory, we will study bacteria chiefly as pathogenic, or disease-causing, organisms. While the disease-causing bacteria may have the greatest familiarity to the public at large, there are many other forms of bacteria that do not cause diseases and actually play vital roles in the environment. The protists are a diverse group because they include the photosynthetic algae, the often motile protozoans, and the fungus-like

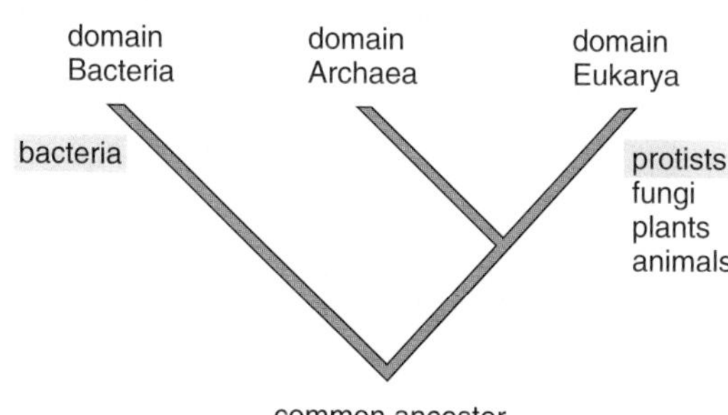

Figure 25.1 The domains.
The evolutionary relationship between the three domains are shown in this diagram. Today's laboratory will study the bacteria in domain Bacteria and the protists, one of the kingdoms in domain Eukarya.

slime molds. Unlike the other two groups, fungi are usually multicellular, reproduce by producing wind-blown spores, and along with bacteria, are most often saprotrophic. That means that they release digestive enzymes into the environment and absorb the products of digestion across the plasma membrane.

	Domain	Cell Structure	Nutrition
Bacteria	Bacteria	Unicellular	Most heterotrophic
Protists			
Protozoans	Eukarya	Unicellular	Heterotrophic
Algae	Eukarya	Unicellular, colonial, filamentous, multicellular	Photosynthetic
Slime molds	Eukarya	Unicellular/multinucleated plasmodium	Heterotrophic
Fungi	Eukarya	Multicellular hyphae	Heterotrophic

25.1 Bacteria

In this laboratory, you will first relate the general structure of a bacterium to its ability to cause disease. The specific shape, growth habit, and staining characteristics of bacteria are often used to identify them. Therefore, you will observe a variety of bacteria using the microscope. You will also perform one of the most important and discriminating tests, the Gram-staining protocol. Aside from their medical importance, bacteria are essential in ecosystems because, along with fungi, they are decomposers that break down dead organic remains, and thereby return inorganic nutrients to plants and other soil-borne organisms.

Pathogenic Bacteria

Pathogenic bacteria are infectious agents that cause disease. Infectious bacteria are able to invade and multiply within a host. Some also produce **toxins,** chemical agents that can be deadly to a host. Antibiotic therapy is often an effective treatment against a bacterial infection, but there is growing concern among medical professionals about strains of bacteria that have become resistant to common antibiotic therapies.

We will explore how it is possible to relate the structure of a bacterium to its ability to be invasive and avoid destruction by the immune system. We will also consider what morphological and physiological attributes allow bacteria to be resistant to antibiotics and to pass the necessary genes on to other bacteria.

Observation: Structure of a Bacterium

1. Study the structure of a generalized bacterium in Figure 25.2, and if available, examine a model or view a CD-Rom of a bacterium.
2. Identify:
   a. **capsule:** A covering composed of polysaccharide and/or protein that has a thick, gummy consistency. Capsules often allow bacteria to stick to surfaces such as teeth. They also prevent phagocytic white blood cells from engulfing and destroying them.
   b. **pilus:** Elongated, hollow appendage used to transfer DNA from one cell to another during a process called conjugation. Genes that allow bacteria to be resistant to antibiotics can be passed in this manner.
   c. **flagellum:** A specialized appendage that allows a bacterium to be motile.

d. **cell wall:** A semirigid covering that contains peptidoglycan, a mesh of carbohydrate and protein.

e. **plasma membrane:** A phospholipid bilayer that contains various proteins and regulates the entrance and exit of molecules into and out of the cell. Resistance to antibiotics can be due to plasma membrane alterations that do not allow the drug to bind to the membrane or cross the membrane, or to a plasma membrane that increases the elimination of the drug from the bacteria.

f. **ribosomes:** Particles composed of protein and RNA that function at the site of protein synthesis. Some bacteria possess antibiotic-inactivating enzymes that make them resistant to antibiotics.

g. **nucleoid:** The location of the bacterial chromosome.

Conclusions

- Which portions of a bacterial cell aid the ability of a bacterium to cause infections?

  ____

- Which portions of a bacterial cell aid the ability of a bacterium to be resistant to antibiotics? ____

- Which portions of a bacterial cell aid the motility of a bacterium? ______________________

  ____

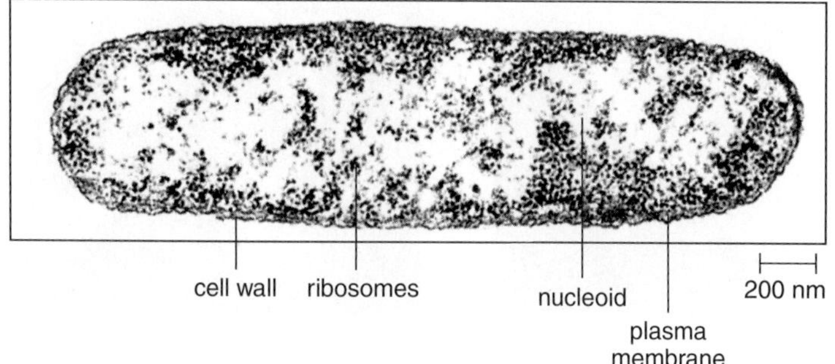

a. Electron micrograph

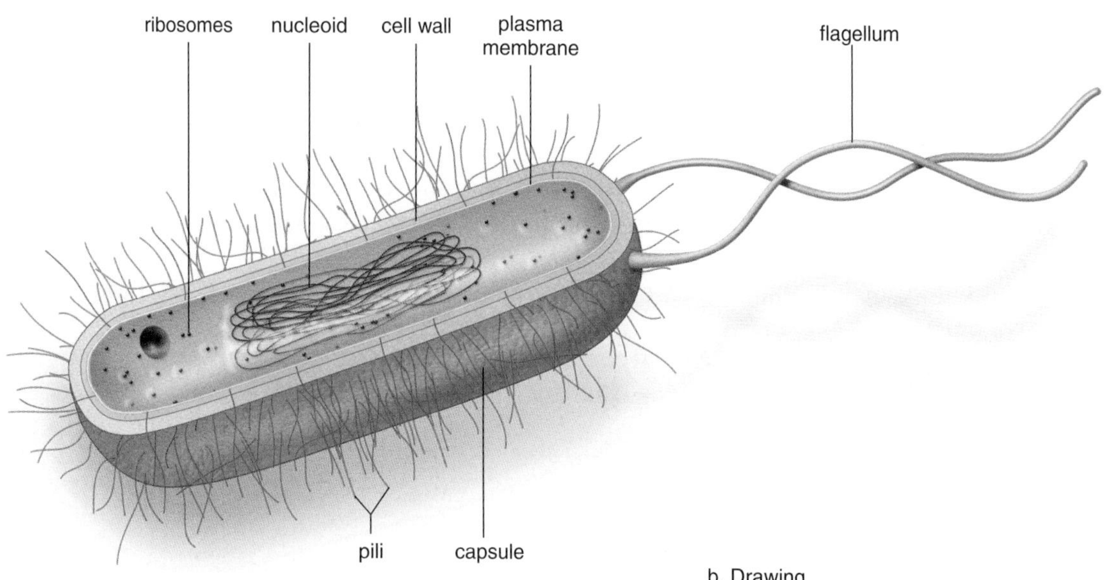

b. Drawing

Figure 25.2 Generalized structure of a bacterium.

Identification of Bacteria by Morphology

The bacteria featured in Figure 25.3 are all single-celled; therefore, they are unicellular. Shape and arrangement help identify bacteria. Most bacteria have one of three shapes. Bacteria that exhibit a spiral shape are called spirilla, (sing., **spirillum**) (Fig. 25.3*a*). Bacteria that are rod-shaped are called bacilli (sing., **bacillus**) (Fig. 25.3*b*). Bacteria that are spherical are called cocci (sing., **coccus**) (Fig. 25.3*c*).

Bacteria have a variety of shapes and arrangements in addition to the three mentioned. The cocci and bacilli can appear in a variety of sizes: large or small spheres, long or short rods. There are some intermediate forms called coccobacilli. Curved rods are called *vibrios.* Some bacteria exist as chains, packets, or clusters of cells. *Diplococcus* means that a coccus forms a chain consisting of only two bacteria. A coccus that forms a longer chain is called a *streptococcus*. A chain of bacilli is called a *streptobacillus.* Packets of four cells are called a tetrad, while packets of eight cells are called a sarcina. More common are bacteria that form clusters. When the cluster is made of cocci, the organism is a *staphylococcus.*

Observation: Cell Morphologies

1. View the microscope slides of bacteria that are on display. What magnification is required to view bacteria? ______________________
2. Using Figure 25.3 as a guide, identify the three different shapes of bacteria.

Figure 25.3 Typical shapes of common bacteria.
a. Spirillum, a spiral-shaped bacterium. **b.** Bacilli, rod-shaped bacteria. **c.** Cocci, round bacteria.

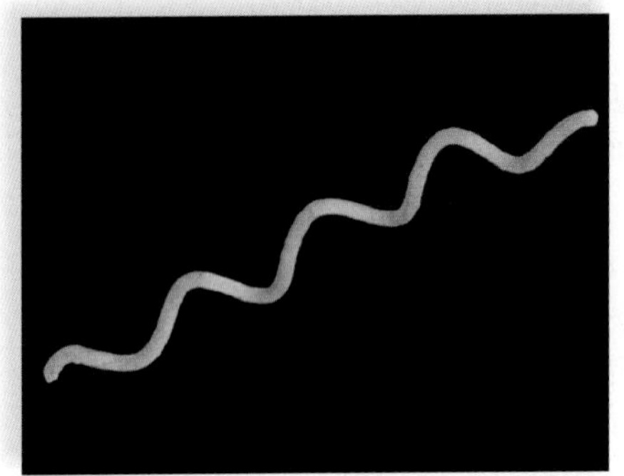

a. Spirillum: *Spirillum volutans* SEM 3,520×

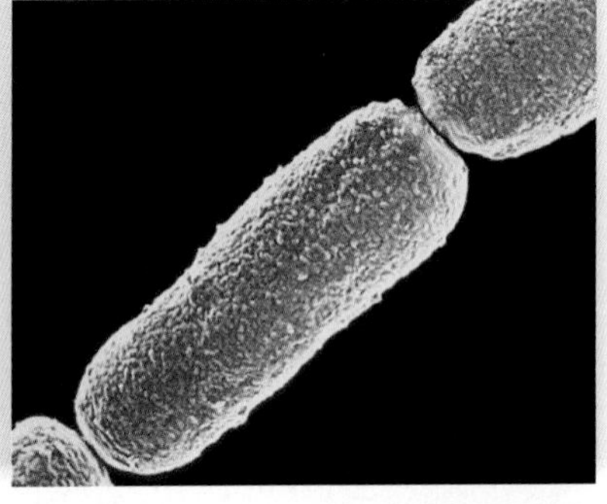

b. Bacilli: *Bacillus anthracis* SEM 35,000×

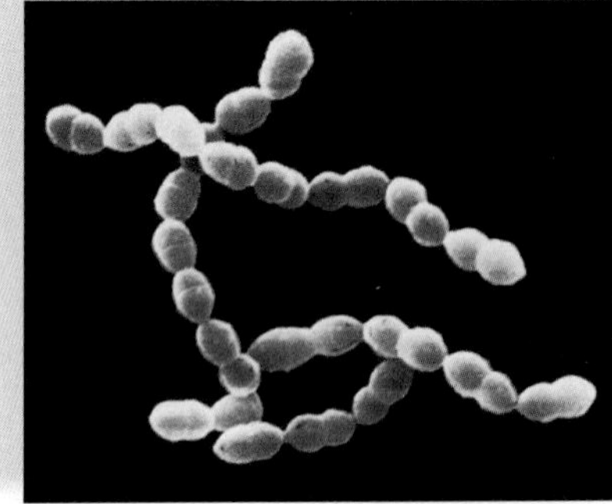

c. Cocci: *Streptococcus thermophilus* SEM 6,250×

Observation: Colonies

Agar Plates

Agar is a semisolid medium used to grow bacteria.

1. View agar plates that have been inoculated with bacteria and then incubated. Notice the "colonies" of bacteria growing on the plates. Each colony contains cells that are all descended from one original cell.
2. Compare the colonies' color, surface, and margin, and note your observations in Table 25.1.

Table 25.1 Agar Staining

Plate Number	Description of Colonies

Identification of Bacteria by Gram Stain

Most bacterial cells are protected by a cell wall that contains a unique molecule called peptidoglycan. Bacteria are commonly differentiated by using the **Gram stain** procedure, which distinguishes bacteria that have a thick layer of peptidoglycan (gram-positive) from those that have a thin layer of peptidoglycan (gram-negative). Gram-positive bacteria retain a crystal violet-iodine complex and stain blue-purple, whereas gram-negative bacteria decolorize and counterstain red-pink with safranin (Figure 25.4).

Based upon the information above, would you predict that an antibiotic that controls bacteria by preventing the formation of cell walls would work better on gram-positive or gram-negative bacteria? Why? ______________________________

Experimental Procedure: Gram Stain

1. Use one designated square of a slide that has six squares.
2. With a sterile cotton swab, obtain a sample from around your teeth or inside your nose.
3. Carefully roll the swab across your allotted square. Body samples must be spread out thinly and evenly on the slide.
4. Allow the smear to air-dry.
5. Fix the smear by flooding the slide with *absolute methanol* for 1 minute. Allow the smear to dry before staining.
6. Flood the smear with *Gram Crystal Violet,* and wait for 1 to 2 minutes.
7. Gently rinse off the crystal violet with cold tap water.
8. Flood the smear with *Gram Iodine,* and allow it to react for 1 minute.
9. Gently rinse off the iodine with cold tap water.
10. Gently rinse the smear with *Gram Decolorizer* until the solution rinses colorlessly from the slide (approximately 20 to 30 seconds).
11. Immediately rinse the smear with cold tap water.
12. Flood the smear with *Gram Safranin,* and allow it to stain for 15 to 30 seconds.
13. Gently rinse off the safranin with cold tap water.
14. Blot off excess water with a paper towel, and allow the smear to air-dry.
15. Examine microscopically. (This will require the use of the oil immersion lens.)

Proper means of disposal of swabs and other contact material will be provided in your laboratory. Those wastes containing human tissue or fluid residue should not be disposed of in regular classroom trash receptacles.

Conclusion

The Gram stain is one way to distinguish bacteria from each other. Are these bacteria gram-positive or gram-negative? __________

Figure 25.4 Generalized structure of a bacterium.
a. Gram-positive cells have a thick layer of peptidoglycan. **b.** Gram-negative cells have a very thin layer of peptidoglycan. **c.** This difference causes Gram-positive cells to stain purple and Gram-negative cells to stain reddish-pink.

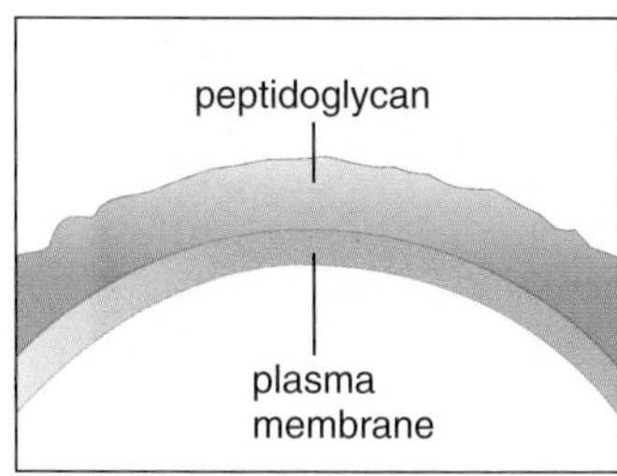

a. Gram-positive cell

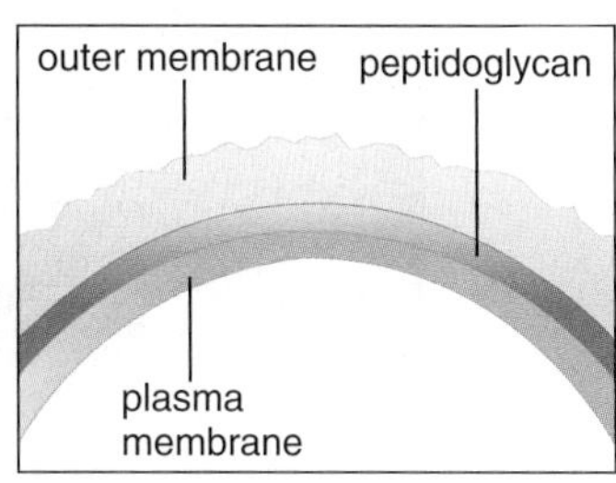

b. Gram-negative cell

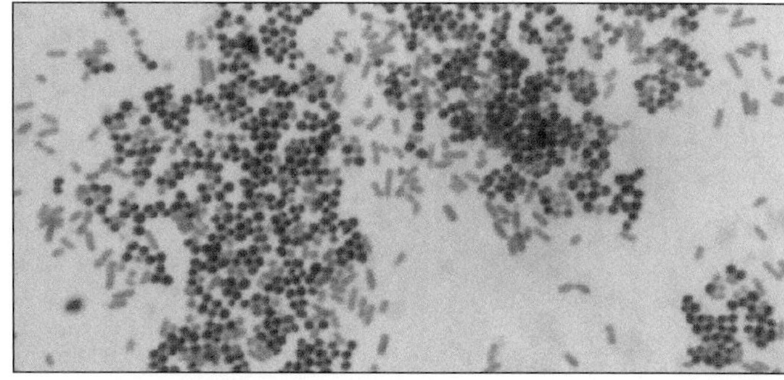

c. Micrograph of Gram stained bacteria

Cyanobacteria

Some bacteria are photosynthesizers that use solar energy to produce their own food. **Cyanobacteria** are believed to have arisen some 3.7 billion years ago and are thought to have been the first organisms to release oxygen into the atmosphere. Their importance as a source of oxygen, even today, should not be underemphasized. At one time, cyanobacteria were identified as blue-green algae, but today we classify them as a type of bacterium. You have to keep in mind that they have isolated thylakoids where photosynthesis occurs, but unlike plant cells, these thylakoids are not enclosed within chloroplasts.

Observation: Cyanobacteria

Oscillatoria

1. Prepare a wet mount of an *Oscillatoria* culture, if available, or examine a prepared slide, using high power (45×) or oil immersion (if available). This is a filamentous cyanobacterium with individual cells lined up like links in a chain or pennies stacked one on top of the other (Fig. 25.5*a*).
2. *Oscillatoria* takes its name from the characteristic oscillations that you may be able to see if your sample is alive. If you have a living culture, are oscillations visible? ________

Anabaena

1. Prepare a wet mount of an *Anabaena* culture, if available, or examine a prepared slide, using high power (45×) or oil immersion (if available). This is also a filamentous cyanobacterium, although its individual cells are barrel-shaped (Fig. 25.5*b*).
2. Note the thin nature of this strand. If you have a living culture, what is its color? ____________ (Prepared slides are artificially stained and may not resemble the color of the living cyanobacterium.)
3. What differences do you observe when comparing *Oscillatoria* and *Anabaena*? What similarities exist? ____

Figure 25.5 ***Oscillatoria*** **(a) and** ***Anabaena*** **(b).**

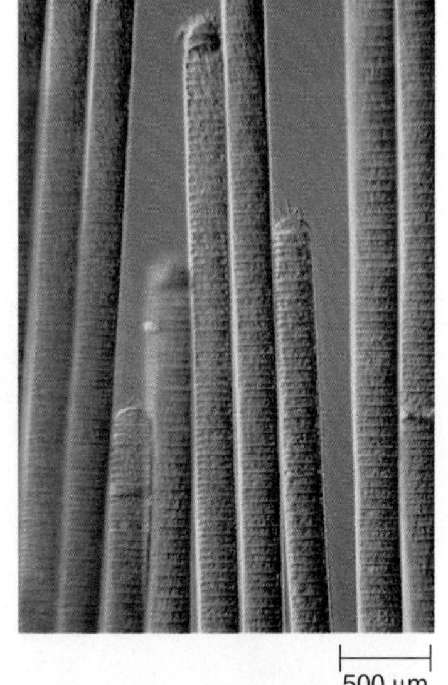

a.

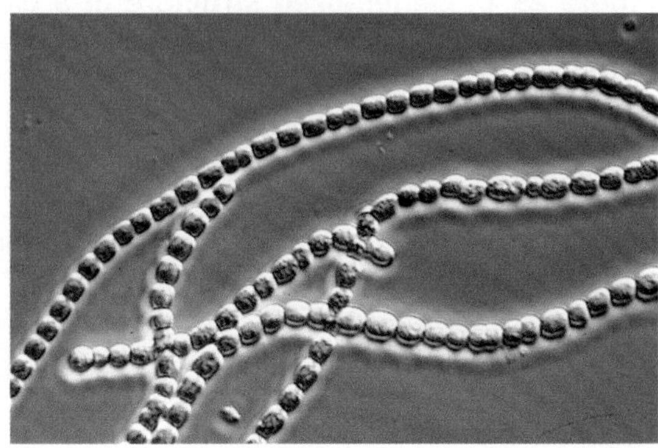

b.

25.2 Protists

Protists (kingdom Protista) are all eukaryotes, even though they may be unicellular. The protists include the algae, which photosynthesize in the same manner as plants; the protozoans, which are heterotrophic by ingestion; and the slime molds, which creep along the forest floor as a plasmodium.

Algae

The **algae,** whether green algae, red algae, brown algae, or golden-brown algae, all photosynthesize, as do plants. Why aren't they considered plants? Because they never protect the zygote and other reproductive structures the way plants do. Aside from releasing oxygen into the environment, algae play an important role in aquatic ecosystems—both freshwater and marine—because they are *producers.* They are called producers because they produce food for themselves and all members of an ecosystem.

In today's laboratory, we will focus on filamentous and colonial forms of green algae and observe slides of golden-brown algae, also known as diatoms. While this lab will not feature red algae and brown algae, they are typically large or macroscopic organisms that play important roles in the environment. Kelp and other common brown algae are included along with some green and red algae in what we call seaweed. Kelp is an important part of the marine ecosystem and is a source of important additives for the food and cosmetics industries. Red algae also play important roles in marine environments, particularly around coral reefs. In addition to their roles in ecosystems, the red algae also have economic value. The agar that was used to make the bacterial plates used in Section 25.1 is a product obtained from red algae.

Observation: A Sampling of Algae

To exemplify algae, you will examine a filamentous form (*Spirogyra*), a colonial form (*Volvox*), and a unicellular form (diatoms). The seaweeds seen along the coasts are also algae.

1. Obtain and examine a slide of *Spirogyra* (Fig. 25.6). The most prominent feature of the cells is the spiral, ribbonlike chloroplast. The nucleus is in the center of the cell, anchored by cytoplasmic strands. Your slide may show **conjugation,** a sexual means of reproduction illustrated in Figure 25.6*b*. If it does not, obtain a slide that does show this process. Conjugation tubes form between two adjacent filaments, and the contents of one set of cells enter the other set. As the nuclei fuse, a zygote is formed. The zygote overwinters: and in the spring, meiosis and, subsequently, germination occur.

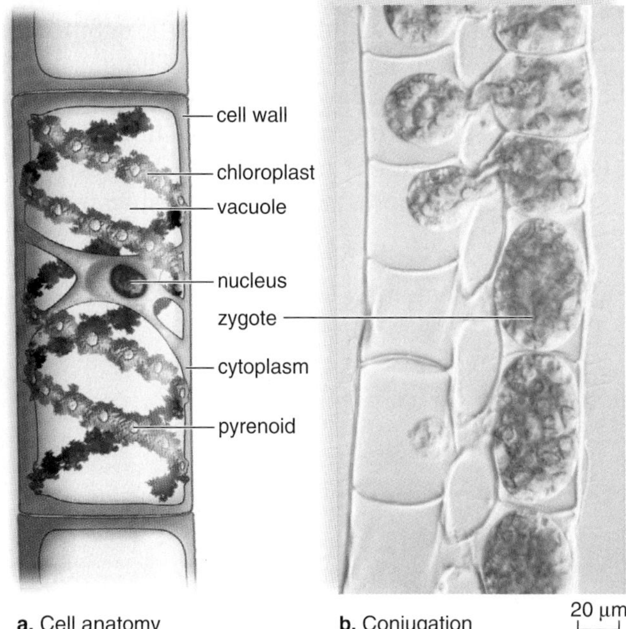

Figure 25.6 ***Spirogyra.***
a. *Spirogyra* is a filamentous green alga, in which each cell has a ribbonlike chloroplast. **b.** During conjugation, the cell contents of one filament enter the cells of another filament. Zygote formation follows.

2. Obtain and examine a slide of *Volvox* (Fig. 25.7). *Volvox* is a green algal colony, a group of individual cells that aggregate together forming a ball or mass of cells. It is motile (capable of locomotion) because the thousands of cells that make up the colony have flagella. These cells are connected by delicate cytoplasmic extensions.

   Volvox is capable of both asexual and sexual reproduction. Certain cells of the adult colony can divide to produce **daughter colonies** (Fig. 25.7) that reside for a time within the parental colony. A daughter colony escapes the parental colony by releasing an enzyme that dissolves away a portion of the matrix of the parental colony. During sexual reproduction, some colonies of *Volvox* have cells that produce sperm, and others have cells that produce eggs.

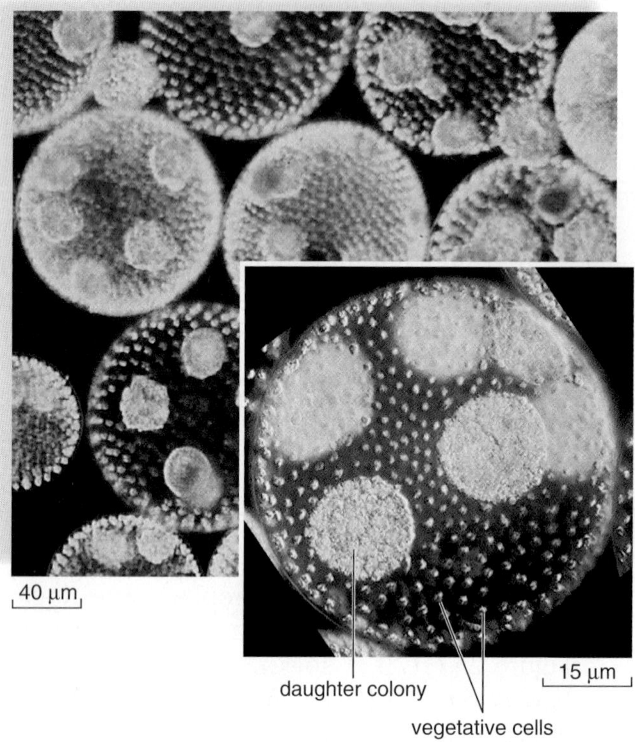

Figure 25.7 ***Volvox.***
Volvox is a colonial green alga. The adult *Volvox* colony often contains daughter colonies, asexually produced by special cells.

3. Obtain and examine a slide of diatoms (Fig. 25.8). Diatoms possess a yellow-brown pigment in addition to chlorophyll. The diatom cell wall is in two sections, with the larger one fitting over the smaller as a lid fits over a box. Since the cell wall is impregnated with silica, diatoms are said to "live in glass houses." The glass cell walls of diatoms do not decompose, so they accumulate in thick layers that are subsequently mined as diatomaceous earth and used in filters and as a natural insecticide. Diatoms, being photosynthetic and extremely abundant, are important food sources for the small heterotrophs (organisms that must acquire food from external sources) in both marine and freshwater environments.

Figure 25.8 Diatoms.
Diatoms, photosynthetic protists of the oceans.

Protozoans

The term **protozoan** refers to unicellular eukaryotes and is often restricted to unicellular, heterotrophic eukaryotes that ingest food by forming **food vacuoles.** Other vacuoles, such as **contractile vacuoles** that rid the cell of excess water, are also typical. Usually protozoans have some form of locomotion; some use **pseudopodia,** some move by **cilia,** and some use **flagella** (Fig. 25.9). Sporozoans—such as *Plasmodium vivax,* which causes a common form of malaria—are immobile.

Observation: Protozoans

1. You may already have had the opportunity to observe *Euglena* in laboratory 2. However, your instructor may want you to observe these organisms again.
2. If a video, CD-Rom, or film is available, watch it, and note the various forms of protozoans.
3. Prepare wet mounts or examine prepared slides of protozoans as directed by your instructor.
4. Note the means of locomotion for each protozoan in Figure 25.9.

Figure 25.9 Protozoan diversity.
Protozoans are motile by the means illustrated: **(a)** *Amoeba;* **(b)** *Paramecium;* **(c)** *Trypanosoma* in host; **(d)** *Trypanosoma* structure.

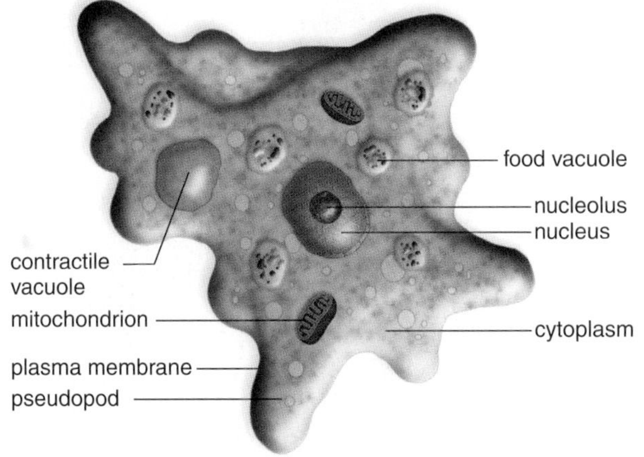

a. *Amoeba* moves by pseudopods

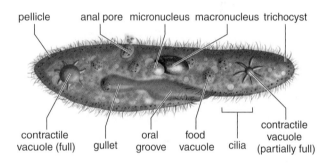

b. *Paramecium* moves by cilia

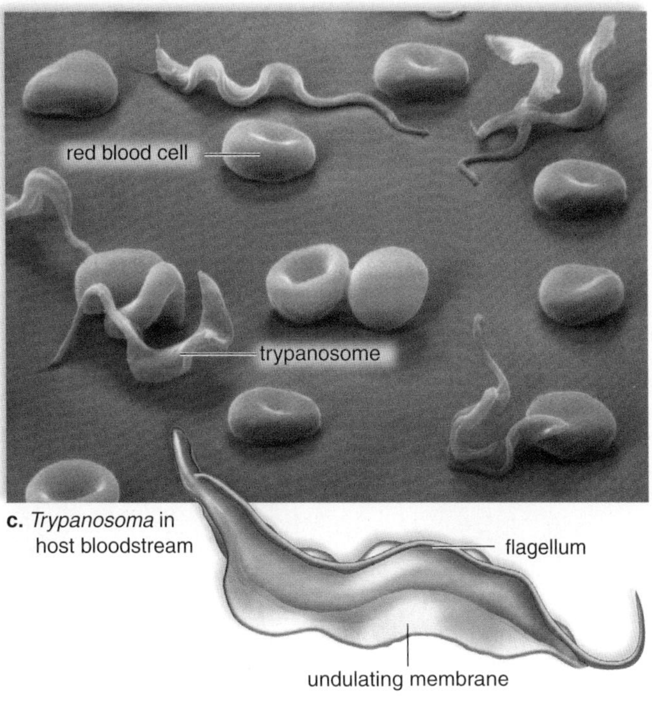

c. *Trypanosoma* in host bloodstream

d. *Trypanosoma* moves by flagella

Pond Water

Pond water typically contains various examples of protozoans and algae, the protists studied in this laboratory.

Observation: Pond Water

1. Prepare a wet mount of a sample of pond water. Be sure to select some of the sediment on the bottom and a few strands of filamentous algae.
2. Identify the organisms you see by consulting Figure 25.10. Those with chloroplasts are algae, and those without chloroplasts are protozoans. *Difflugia* and *Arcella* are amoebas. No line is provided for *Asplanchna, Philodina, Keratella,* and *Chaetonotus* because these are actually very small animals.

Figure 25.10 Microorganisms found in pond water. Drawings are not actual sizes of the organisms.

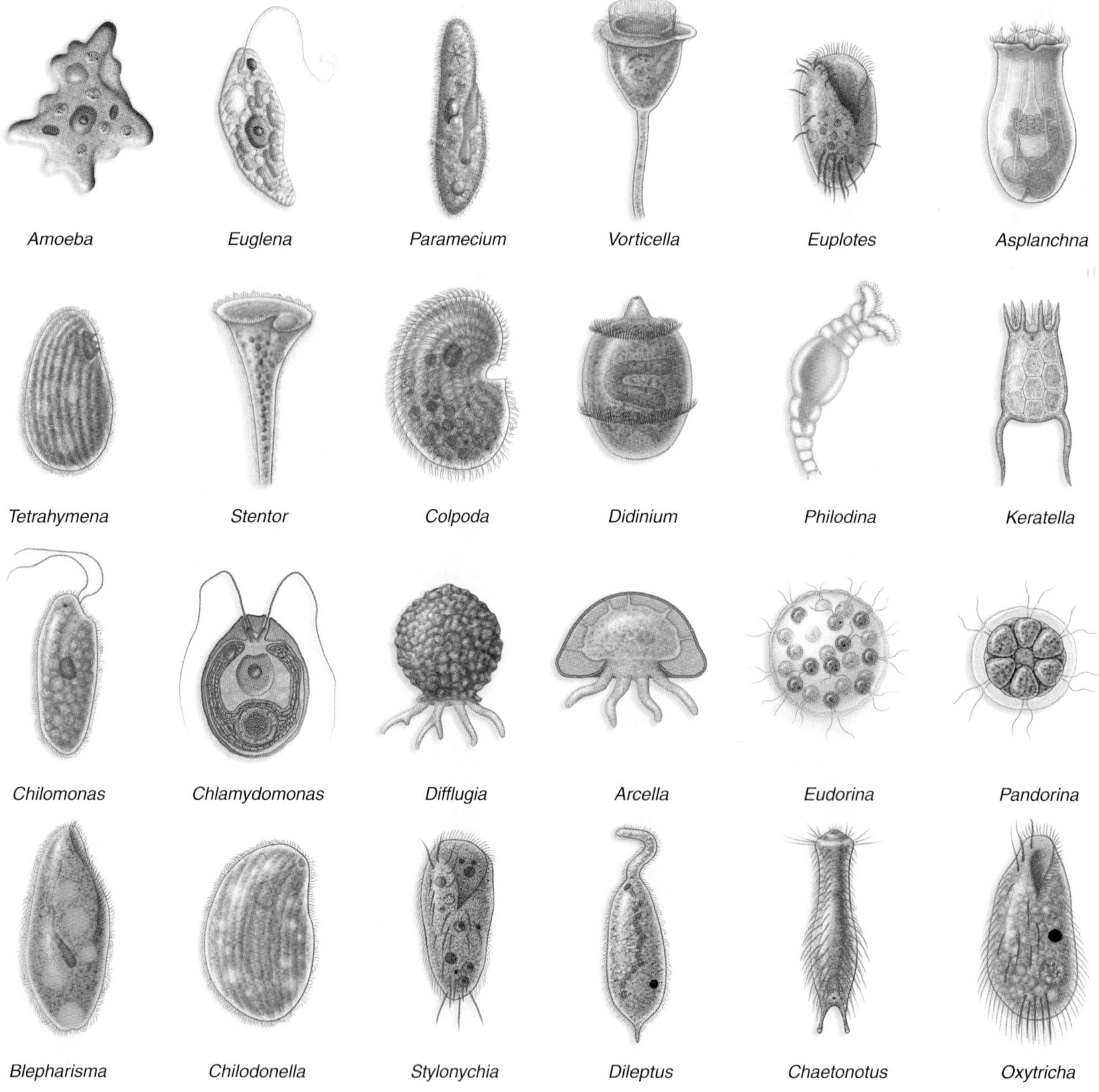

Slime Molds

Slime molds are called the funguslike protists. They are saprotrophic like fungi, and they do form spores during the haploid stage of their life cycle.

There are two types of slime molds: cellular slime molds and plasmodial slime molds. **Cellular slime molds** usually exist as individual amoeboid cells, which aggregate on occasion to form a pseudoplasmodium. **Plasmodial slime molds** usually exist as a **plasmodium,** a fan-shaped, multi-nucleated mass of cytoplasm. The plasmodium of a plasmodial slime mold creeps along, phagocytizing decaying plant material in a forest or an agricultural field. During times unfavorable for growth, such as a drought, the plasmodium develops many sporangia. A **sporangium** is a reproductive structure that produces spores by meiosis. In some plasmodial slime molds, the spores become flagellated cells, and in others, they are amoeboid. In any case, they fuse to form a zygote that develops into a plasmodium (Fig. 25.11).

Observation: Plasmodial Slime Mold

1. Obtain a plate of *Physarum* growing on agar. Examine the plate carefully under the dissecting microscope.
2. Describe what you see. ______________
____
____
____

Figure 25.11 Plasmodial slime mold.
The diploid adult forms sporangia during sexual reproduction, when conditions are unfavorable to growth. Haploid spores germinate, releasing haploid amoeboid or flagellated cells that fuse.

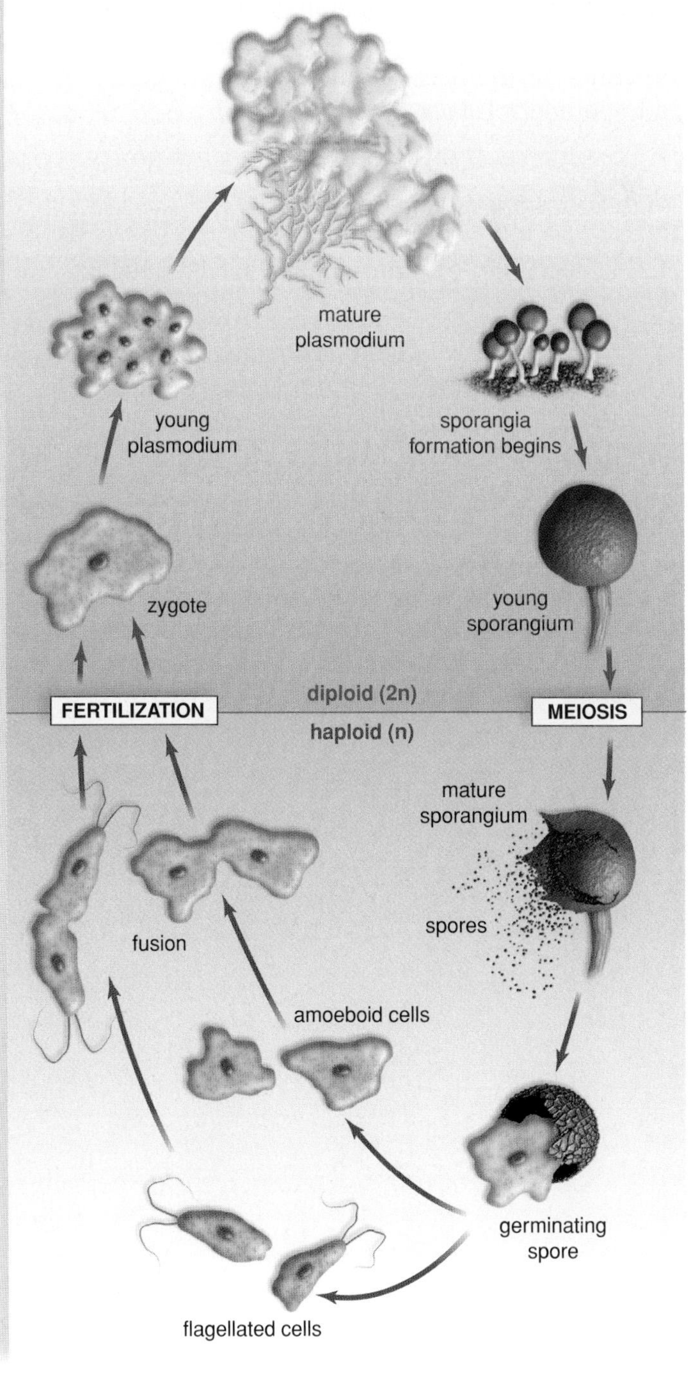

25.3 Fungi

Fungi (Kingdom Fungi) are saprotrophic in the same manner as bacteria. Both fungi and bacteria are often referred to as "organisms of decay" because they break down dead organic matter and release inorganic nutrients for plants. The Fungi include organisms that exhibit quite distinct lifestyles. Using modern molecular biology techniques, scientists are now sorting out the true relationships of this large group of organisms. A fungal body, called a **mycelium,** is composed of many strands called **hyphae.** Sometimes, the nuclei within a hypha are separated by walls, and sometimes they are not.

Fungi produce windblown **spores** (small, haploid bodies with a protective covering) when they reproduce sexually or asexually.

Black Bread Mold

In keeping with its name, black bread mold grows on bread and any other type of bakery goods. Notice in Figure 25.12, sporangia at the tips of aerial hyphae produce spores in both the asexual and sexual life cycles. As can be seen in Figure 25.12, a **zygospore** is diploid (2n); otherwise, all structures in the asexual and sexual life cycles of bread mold are haploid (n). Identify the hyphae and sporangia (black dots).

Observation: Black Bread Mold

1. If available, examine bread that has become moldy. Do you recognize black bread mold on the bread? ____________
2. Obtain a petri dish that contains living black bread mold. Observe with a dissecting microscope. *Identify the mycelium and a sporangium.*
3. View a prepared slide of *Rhizopus,* using both a dissecting microscope and the low-power setting of a light microscope. The absence of cross walls in the hyphae is an identifying feature of zygospore fungi.
4. Compare the micrograph on the left (Figure 25.13) with your slide of *Rhizopus* and identify and label the structures in Figure 25.13 that are visible during asexual reproduction: sporangiophore, rhizoid, and sporangium. Also compare your slide to the micrograph on the right and identify and label structures seen during sexual reproduction: stolon, gametangia, and zygospore.

Figure 25.12 Black bread mold, *Rhizopus stolonifer.*
Windblown spores are produced during both asexual and sexual reproduction.

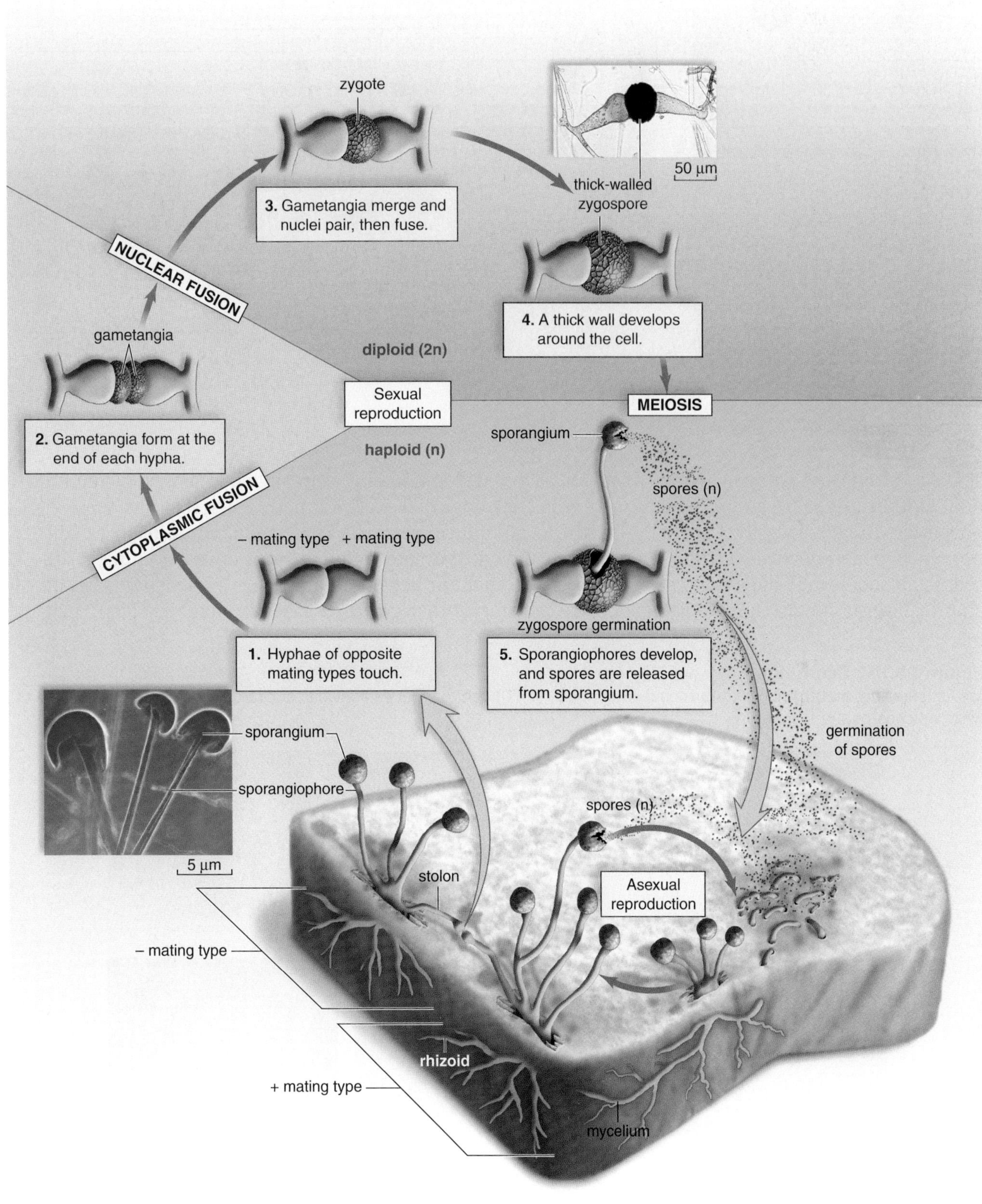

Figure 25.13 Microscope slides of black bread mold.
a. Asexual life cycle **b.** Sexual life cycle

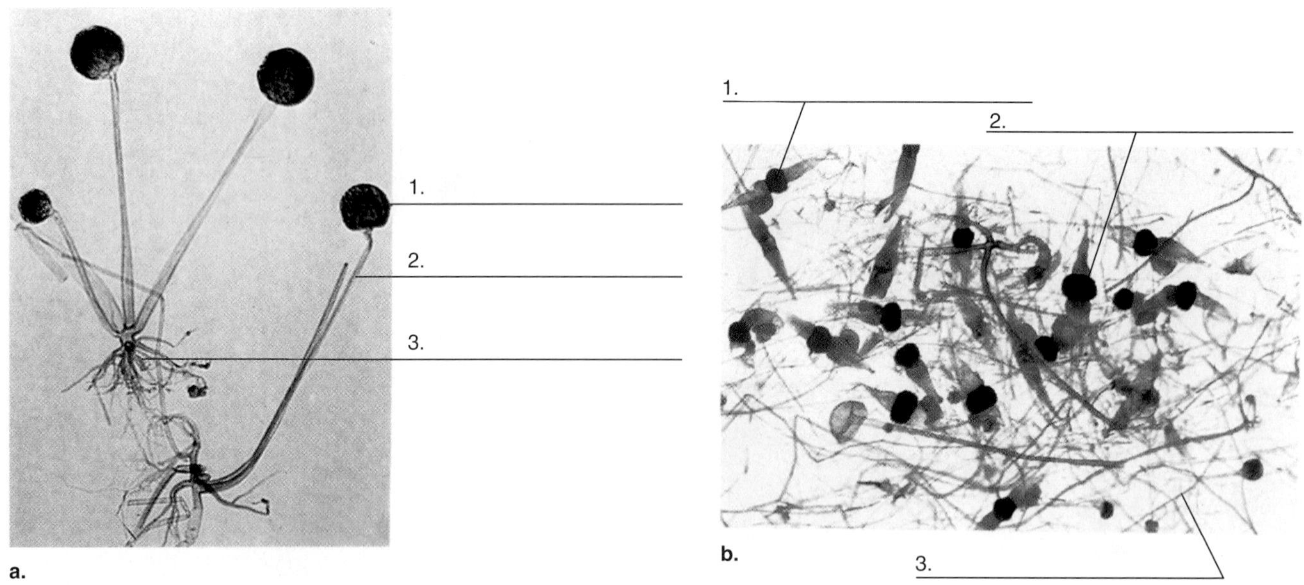

Sac Fungi

The **sac fungi** are distinguished by a sac, or ascus, which contains cells involved in sexual reproduction. Within this large group of fungi is a group of fungi that were formerly referred to as the "imperfect fungi" since they are only known to reproduce asexually by forming **conidiospores** on upright hyphae known as conidiophores. Even though they are different from other sac fungi in the sense that no sexual stage has yet been observed, modern genetic analysis has established a relationship to the sac fungi. This group includes several forms that cause diseases in humans (Fig. 25.14).

Figure 25.14 Sac fungi.
(a) Athlete's foot and **(b)** ringworm are caused by tinea. **(c)** Thrush, or oral candidiasis, is characterized by the formation of white patches on the tongue.

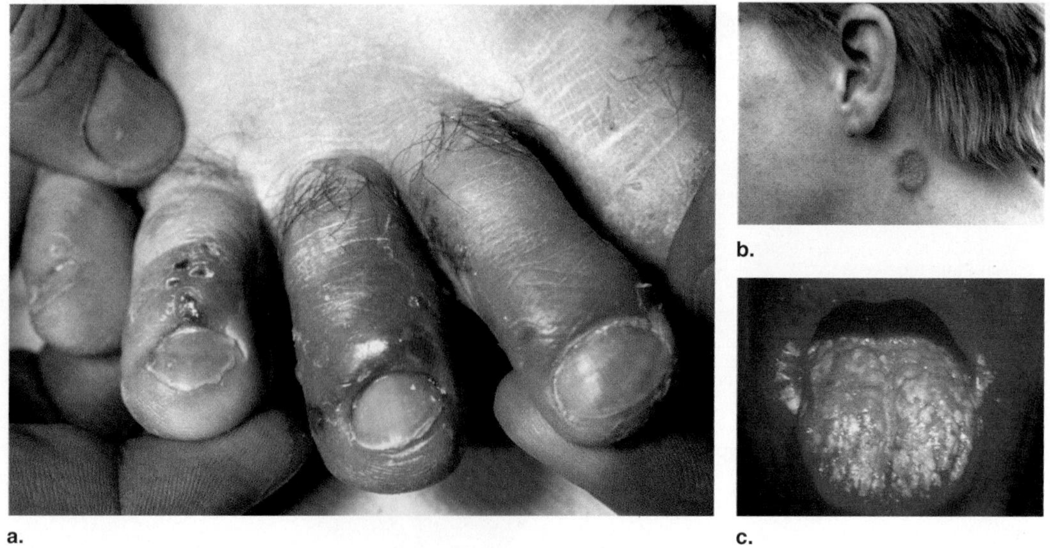

Observation: Aspergillus and Penicillium

1. Obtain a prepared slide of *Penicillium.* Locate the periphery of the mass under low power, and then switch to high power. You should now be able to see conidiophores.
2. View a prepared slide of *Aspergillus*, and observe how the conidiophore arrangement differs from that of *Penicillium.*
3. Label the conidiophore and conidiospores in the following diagram of *Aspergillus:*

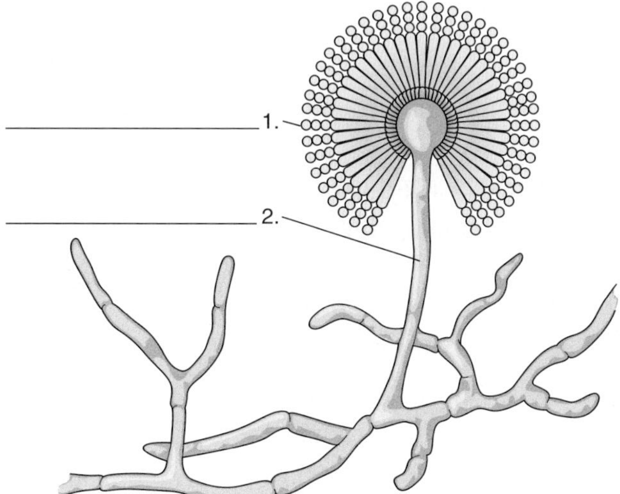

Club Fungi

Club fungi are just as familiar as black bread mold to most laypeople because they include the mushrooms. A gill mushroom consists of a stalk and a terminal cap with gills on the underside (Fig. 25.15). The cap, called a **basidiocarp,** is a fruiting body that arises following the union of + and − hyphae. The gills bear basidia, club-shaped structures, where nuclei fuse and meiosis occurs during spore production. The spores are called **basidiospores.**

Observation: Mushrooms

1. Obtain an edible mushroom—for example, *Agaricus*—and identify as many of the following structures as possible:
   a. **Stalk:** The upright portion that supports the cap.
   b. **Annulus:** A membrane surrounding the stalk where the immature (button-shaped) mushroom was attached.
   c. **Cap:** The umbrella-shaped basidiocarp of the mushroom.
   d. **Gills:** On the underside of the cap, radiating lamellae on which the basidia are located.
   e. **Basidia:** On the gills, club-shaped structures where basidiospores are produced.
   f. **Basidiospores:** Spores produced by basidia.
2. View a prepared slide of a cross section of *Coprinus*. Using all three microscope objectives, look for the gills, basidia, and basidiospores.
3. Can you see individual hyphae in the gills? ________________
4. Are the basidiospores inside or outside of the basidia? ________________
5. What type of nuclear division does the zygote undergo to produce the basidiospores? ________________
6. Can you suggest a reason for some of the basidia having fewer than four basidiospores? ________________
7. What happens to the basidiospores after they are released? ________________

Figure 25.15 Club fungi.
Life cycle of a mushroom. Sexual reproduction is the norm. After hyphae from two opposite mating types fuse, the dikaryotic mycelium is long-lasting. On the gills of basidiocarp, nuclear fusion results in a diploid nucleus within each basidium. Meiosis and production of basidiospores follow. Germination of a spore results in a haploid mycelium.

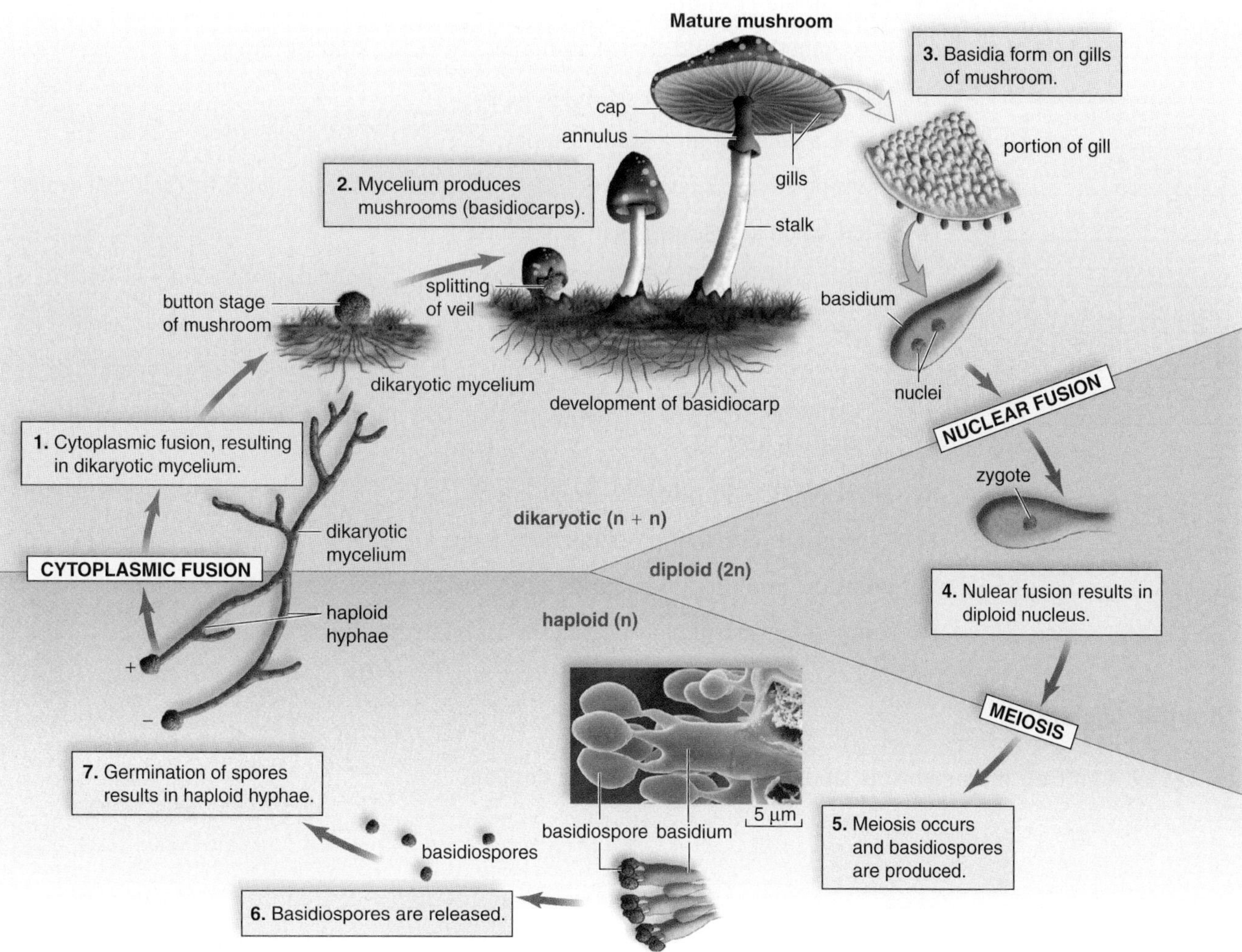

Laboratory Review 25

1. What role do bacteria and fungi play in ecosystems?
2. What type of semisolid medium is used to grow bacteria?
3. What is the scientific name for spherical bacteria?
4. It is sometimes said that diatoms live in what kind of "houses"?
5. What type of nutrition do algae have?
6. Name a colonial alga studied today.
7. Gram-positive bacteria have a thick layer of what substance in their cell walls?
8. What color are Gram-negative bacteria following Gram staining?
9. Once called the blue-green algae, cyanobacteria are now classified as what?
10. What do you call the projection that allows amoeboids to move and feed?
11. What do you call the multinucleate stage of the plasmodial slime mold?
12. The stalk and cap of a mushroom that rise above the substratum are termed what?
13. What type of nutrition do fungi have?
14. Explain the designation "imperfect fungi."
15. What do fungi produce during both sexual and asexual reproduction?
16. Name three morphological shapes associated with bacterial cells.

Thought Questions

17. Why aren't all the organisms studied today in the same domain?

18. Fungi always reproduce by producing haploid spores during both asexual and sexual reproduction. In general, how does sexual reproduction differ from asexual reproduction among fungi?

19. What advantage does conjugation provide over traditional modes of asexual reproduction?

20. What differences exist between prokaryotic and eukaryotic cells that allow antibiotics to destroy one cell type and not the other?

LABORATORY

26

Seedless Plants

Learning Outcomes

26.1 Evolution and Diversity of Plants
- Describe the four main events in the evolution of plants.
- Associate each of these events with a major group of plants.
- Describe the plant life cycle and the concept of the dominant generation.
- Contrast the role of meiosis in the plant and animal life cycles.
- Explain why you would expect the sporophyte generation to be dominant in a plant adapted to a land existence.

26.2 Nonvascular Plants
- Describe the two groups of nonvascular plants studied, the mosses and liverworts.
- Describe the life cycle of a moss, including the appearance of its two generations.
- Describe, in general, how nonvascular plants are adapted to living and reproducing on land.

26.3 Seedless Vascular Plants
- Describe the seedless vascular plants alive today.
- Describe the life cycle of the fern and contrast the appearances of its two generations.
- Describe, in general, how ferns are adapted to reproducing on land.

Introduction

Plants (domain Eukarya, kingdom Plantae) are multicellular, photosynthetic organisms adapted to living on land. They dominate much of the terrestrial landscape and produce organic food for themselves and all other terrestrial heterotrophic organisms. Organisms such as plants that have the ability to synthesize their own organic molecules from inorganic molecules are referred to as autotrophs and it is their ability to produce food that makes them essential to all other organisms in food chains. (Herbivores feed directly on plants, while carnivores feed on animals that have eaten plants.) Through the process of photosynthesis, plants and other photosynthetic organisms contribute to the production of the oxygen needed by heterotrophs, those organisms that cannot synthesize their own organic molecules from inorganic raw materials. In addition, industrial society depends on the solar energy stored by plants many millions of years ago, which has now taken the form of coal. When coal or wood is burned, plants purify the air by taking up the carbon dioxide and other pollutants that are given off.

Plants were among the first organisms to live on land. Photosynthetic organisms had to precede heterotrophic organisms on land. Why? ____

____

Plants require structural support to oppose the force of gravity and to lift their leaves up toward the sun. The first terrestrial plants that evolved lacked the necessary support tissues that would allow them to hold themselves above the ground. As a consequence, those plants without **vascular tissues** have remained small, seldom having growth that extends for more than a few centimeters above the ground. Plants with vascular tissue are able to provide support for more complex growth and transport water to and nutrients from the leaves. Much of a plant's body is covered by a waxy **cuticle** that prevents water loss. Unlike algae, all plants have tissue layers that keep the embryo from drying out. Features that help plants live on land (as opposed to living in water) are the primary focus of this Laboratory.

In this laboratory, we will examine the nonvascular plants and the seedless vascular plants. Laboratory 27 focuses on seed plants (the gymnosperms and the angiosperms).

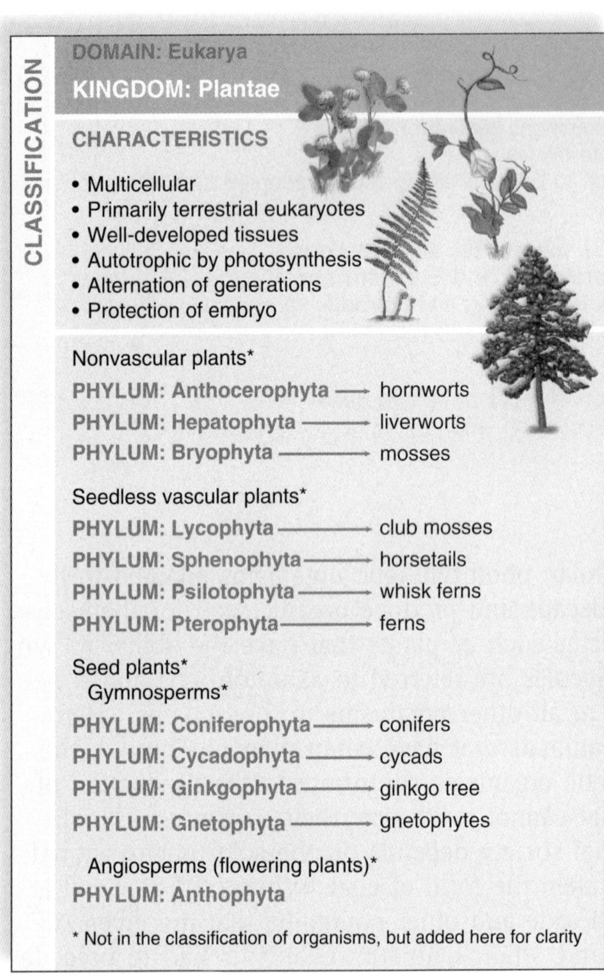

CLASSIFICATION

DOMAIN: Eukarya

KINGDOM: Plantae

CHARACTERISTICS

- Multicellular
- Primarily terrestrial eukaryotes
- Well-developed tissues
- Autotrophic by photosynthesis
- Alternation of generations
- Protection of embryo

Nonvascular plants*

PHYLUM: Anthocerophyta → hornworts

PHYLUM: Hepatophyta → liverworts

PHYLUM: Bryophyta → mosses

Seedless vascular plants*

PHYLUM: Lycophyta → club mosses

PHYLUM: Sphenophyta → horsetails

PHYLUM: Psilotophyta → whisk ferns

PHYLUM: Pterophyta → ferns

Seed plants*

Gymnosperms*

PHYLUM: Coniferophyta → conifers

PHYLUM: Cycadophyta → cycads

PHYLUM: Ginkgophyta → ginkgo tree

PHYLUM: Gnetophyta → gnetophytes

Angiosperms (flowering plants)*

PHYLUM: Anthophyta

* Not in the classification of organisms, but added here for clarity

Flowers

Angio-
sperms

Seeds

Gymno-
sperms

Seed plants

Vascular plants

Vascular tissue

Embryo protection

Common
ancestor

Seedless
vascular
plants

Nonvascular
plants
(bryotphytes)

Nonvascular plants

Green algae

CAMBRIAN

ORDOVICIAN

SILURIAN

DEVONIAN

CARBONIFEROUS

PERMIAN

TRIASSIC

JURASSIC

CRETACEOUS

Periods

PALEOZOIC

MESOZOIC

Eras

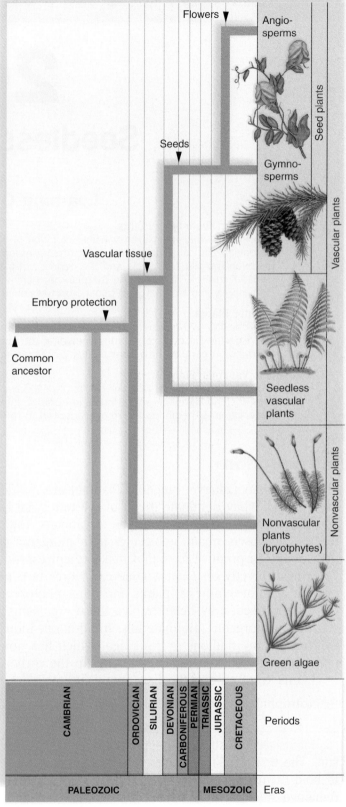

Figure 26.1 Evolutionary history of plants.
The evolution of plants contains four significant innovations. Protection of a multicellular embryo was seen in the first plants to live on land. The evolution of vascular tissue was another important adaptation to land living. The evolution of the seed increased the chance of survival for the next generation. The evolution of the flower fostered the use of animals as pollinators and the use of fruits to aid in the dispersal of seeds.

26.1 Evolution and Diversity of Plants

Figure 26.1 shows the evolutionary history and classification of plants. Four events mark the evolution of plants and their adaptation to the terrestrial environment: **Nonvascular plants** protect the zygote from drying out; evolution of vascular tissue is seen in the **seedless vascular plants;** the seed first appears in the gymnosperm; and only angiosperms have flowers.

Plants have a two-generation life cycle, called alternation of generations, that involves sporic meiosis:

1. The multicellular **sporophyte** (diploid) **generation** produces haploid spores by meiosis. Spores develop into a haploid generation.
2. The multicellular **gametophyte** (haploid) **generation** produces **gametes** (eggs and sperm) by mitosis. The gametes then unite to form a diploid zygote.

Alternation of Generations

In plants, the two generations are dissimilar, and one dominates the other. The dominant generation is larger and exists for a longer period of time. In nonvascular plants, the gametophyte generation is dominant, and the sporophyte generation is dependent on the gametophyte generation. Neither generation has vascular tissue. Windblown spores disperse the species. In seedless vascular plants, the sporophyte generation, which has vascular tissue, is dominant. Windblown spores develop into a separate gametophyte generation, which exists independent of the sporophyte generation. In seed plants (Laboratory 27), the sporophyte generation is dominant and the gametophyte generation is dependent. In these plants, seeds disperse the species.

Figure 26.2 contrasts the plant life cycle **(alternation of generations)** with the animal life cycle **(diploid).**

1. When does meiosis occur in the animal life cycle? ______________________
2. When does meiosis occur in the plant life cycle? ______________________
3. Which generation produces gametes in plants? ______________________
4. The sporophyte is the only generation to have transport (vascular) tissue. Which generation (sporophyte or gametophyte) is better adapted to a land environment? ______________________
   Explain. ______________________
5. Both plant and animal life cycles contain haploid gametes. How is the diploid condition restored?
   ____

Figure 26.2 Plant and animal life cycles.
a. The alternation of generations life cycle is typical of plants. **b.** The diploid life cycle is typical of animals.

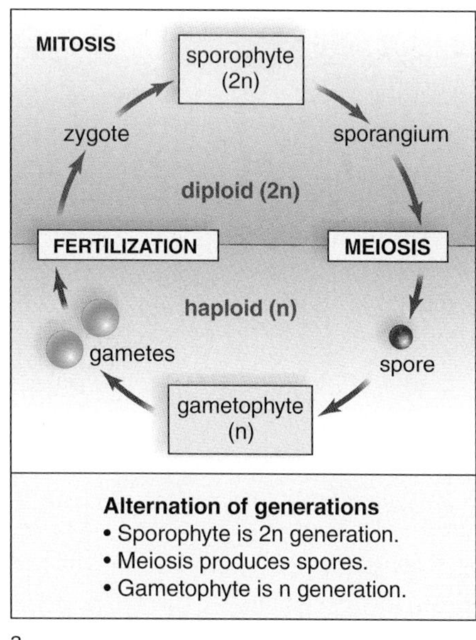

a.

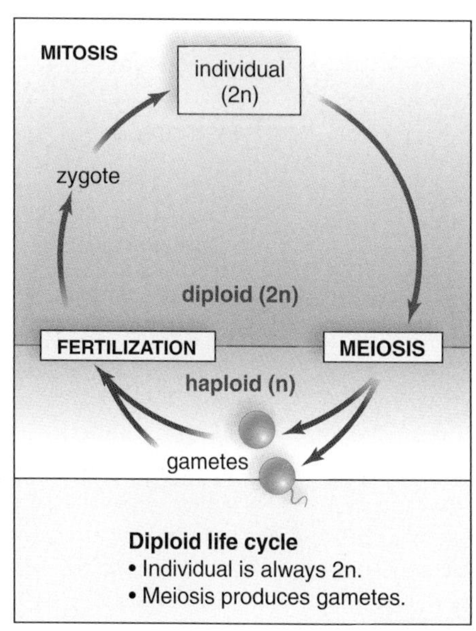

b.

26.2 Nonvascular Plants

The nonvascular plants include the **mosses, liverworts,** and **hornworts.** Lacking the necessary supportive tissues needed to grow upright, these plants tend to grow close to the ground. The gametophyte is dominant in all nonvascular plants. The gametophyte produces eggs within archegonia and flagellated sperm in **antheridia.** The sperm swim to the egg, and the embryo develops within the **archegonium.** The nonvascular sporophyte grows out of the archegonium. The sporophyte is dependent on the female gametophyte (Fig. 26.3).

Figure 26.3 Moss life cycle.
In mosses, the haploid generation (gametophyte) is dominant.

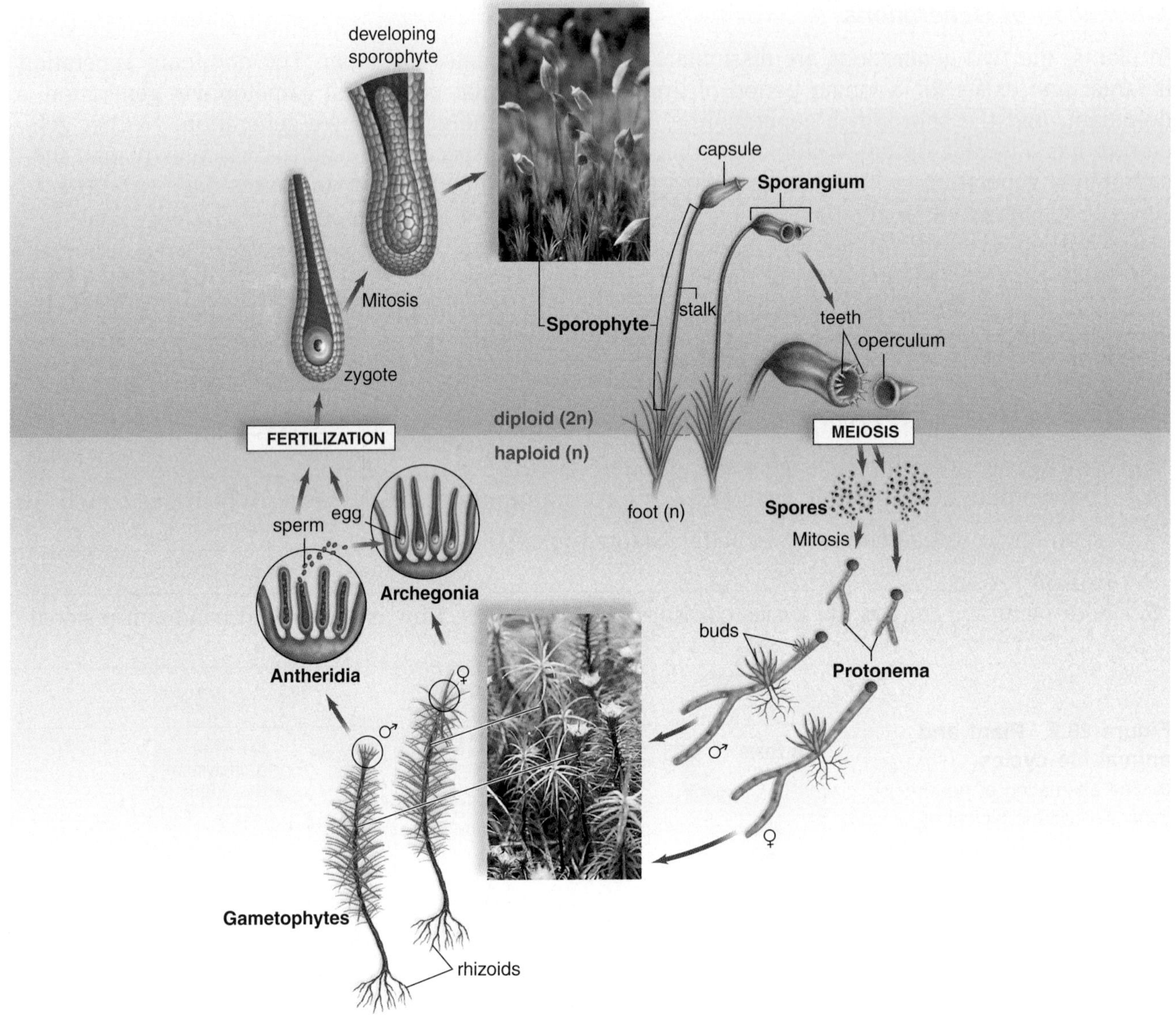

1. Put a check mark beside the phrases that describe nonvascular plants:

I	II
_______ No vascular tissue to transport water	_______ Vascular tissue to transport water
_______ Flagellated sperm that swim to egg	_______ Sperm protected from drying out
_______ Dominant gametophyte	_______ Dominant sporophyte

2. Which listing of features (I or II) would you expect to find in a plant fully adapted to a land environment? _______ Explain. _______________
3. In nonvascular plants, windblown spores disperse the species, and some species forcefully expel their spores. Are windblown spores an adaptation to reproduction on land? _______ Explain. _______________

Observation: Moss Gametophyte

Living Specimen or Plastomount

Obtain a living moss gametophyte or a plastomount of this generation. Describe its appearance. _______________

The leafy green shoots of a moss are said to lack true roots, stems, and leaves because, by definition, roots, stems, and leaves are structures that contain vascular tissue.

Microscope Slide

1. Study a slide of the top of a moss shoot that contains antheridia, the reproductive structures where sperm are produced (Fig. 26.4). What is the chromosome number (choose 2n or n) of the sperm (see Fig. 26.3)? _______ Are the surrounding cells haploid or diploid? _______
2. Study a slide of the top of a moss shoot that contains archegonia, the reproductive structures where eggs are produced (Fig. 26.5). What is the chromosome number of the egg? _______ Are the surrounding cells haploid or diploid? _______ When sperm swim from the antheridia to the archegonia, a zygote results. The zygote develops into the sporophyte. Is the sporophyte haploid or diploid? _______________

Figure 26.4 Moss antheridia.
Flagellated sperm are produced in antheridia.

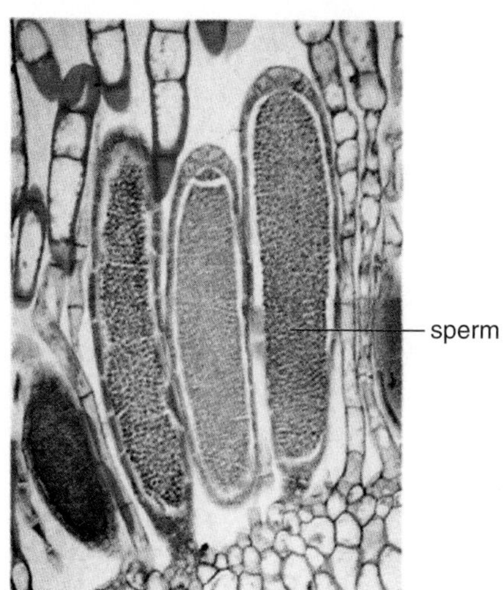

Figure 26.5 Moss archegonia.
Eggs are produced in archegonia.

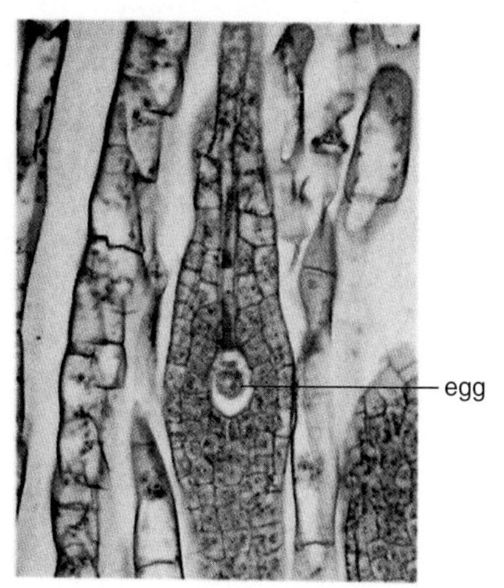

Observation: Moss Sporophyte

Living Sporophyte

1. Examine the living sporophyte of a moss in a minimarsh, or obtain a plastomount of a shoot with the sporophyte attached. Identify the capsule **(sporangium)** where spores are produced and released through a lid (the **operculum**).
2. *Bracket and label the gametophyte and sporophyte in Figure 26.6a.* Place an *n* beside the gametophyte and a *2n* beside the sporophyte.

Figure 26.6 Moss sporophyte.
a. The moss sporophyte is dependent on the female gametophyte. **b.** The sporophyte produces spores by meiosis.

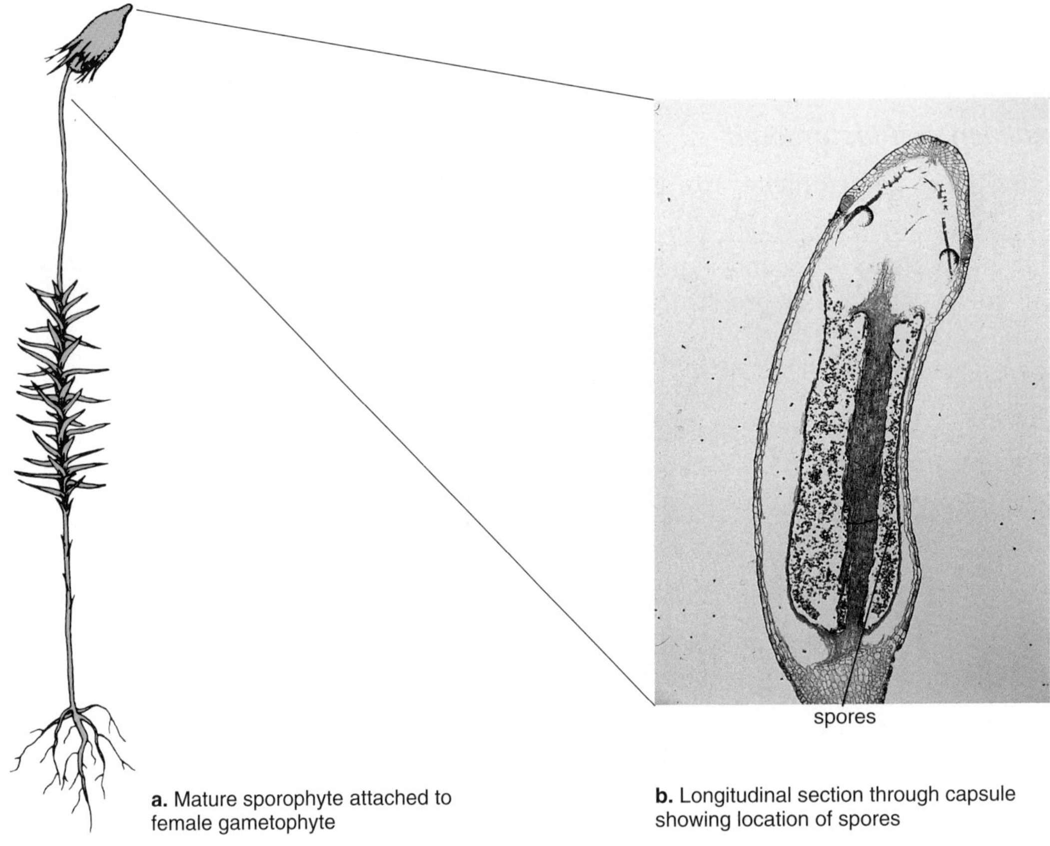

a. Mature sporophyte attached to female gametophyte

b. Longitudinal section through capsule showing location of spores

Microscope Slide

Examine a slide of a longitudinal section through a moss sporophyte (Fig. 26.6*b*). Identify the stalk and the sporangium, where spores are being produced. By what process are the spores being produced? ____________________

When spores germinate, what generation begins to develop? ____________________

Why is it proper to say that spores disperse the offspring? ____________________

____

Observation: Liverworts

Obtain a living sample of the liverwort *Marchantia* (Fig. 26.7), and examine it under the binocular dissecting microscope. What generation is this sample? ________ Identify the following:

1. The gametophyte thallus, or plant body, consists of lobes. Each lobe is about a centimeter or so in length; the upper surface is smooth, and the lower surface bears numerous **rhizoids** (rootlike hairs).
2. **Gemma cups** on the upper surface of the thallus. These contain groups of cells called **gemmae** that can asexually start a new plant.
3. Disk-headed stalks, which bear antheridia, where flagellated sperm are produced.
4. Umbrella-headed stalks, which bear archegonia, where eggs are produced. Following fertilization, tiny sporophytes arise from the archegonia.

Figure 26.7 Liverwort, *Marchantia*.
a. Gemmae can detach and start a new plant. **b.** Antheridia are present in disk-shaped structures, and **(c)** archegonia are present in umbrella-shaped structures.

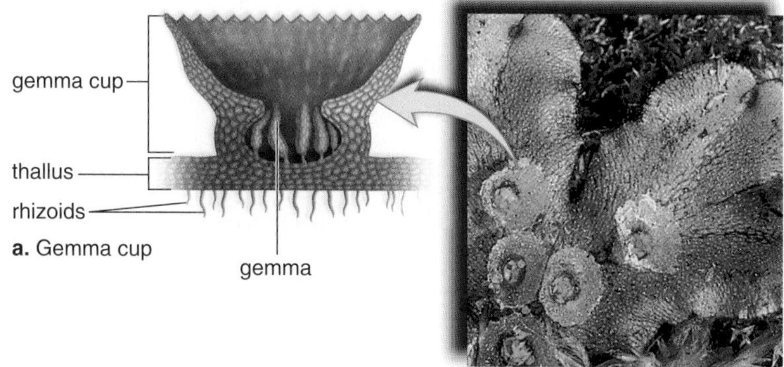

a. Gemma cup

Thallus with gemmae cups

b. Male gametophytes bear antheridia

c. Female gametophytes bear archegonia

26.3 Seedless Vascular Plants

Seedless vascular plants include **whisk ferns, club mosses, horsetails,** and **ferns.** These plants were prevalent and quite large during the Carboniferous period, when conditions were swampy. At that time, they formed the coal deposits still used today.

The dominant sporophyte has adaptations for living on land. It has vascular tissue and produces windblown spores. The spores develop into a separate gametophyte generation that is very small (less than 1 centimeter). The gametophyte generation lacks vascular tissue and produces flagellated sperm.

1. Place a check mark beside the phrases that describe seedless vascular plants:

I	II
________ Independent gametophyte	________ Gametophyte that is dependent on sporophyte, which has vascular tissue
________ Flagellated sperm	________ Sperm protected from drying out

2. Which listing (I or II) would you expect to find in a plant fully adapted to a land environment?

   ________ Explain. ____

   ____

   Are seedless vascular plants fully adapted to living on land? ________________________

Observation: Whisk Ferns

Psilotum (division Psilotophyta) is representative of whisk ferns, named for their resemblance to whisk brooms.

1. Examine a preserved specimen of *Psilotum,* and note that it has no leaves. The underground stem, called a **rhizome,** gives off upright, aerial stems with a dichotomous branching pattern, where bulbous sporangia are located (Fig. 26.8).

   What generation are you examining? ________________________

2. Label *the aerial stem, the rhizome, and the sporangia* in Figure 26.8*b*.

Figure 26.8 Whisk fern, *Psilotum.*
The whisk fern has no roots or leaves—the branches carry on photosynthesis. The sporangia are yellow. Label **(b)** per the text directions in the Observation on this page.

Observation: Club Mosses

In club mosses (division Lycopodophyta), a branching rhizome (horizontal underground stem) sends up aerial stems less than 30 cm tall.

1. Examine a living or preserved specimen of *Lycopodium* (Fig. 26.9*a*).
2. Note the shape and size of the leaves and the method of branching of the stems.
3. Note the terminal clusters of leaves, called **strobili,** that are club-shaped and bear sporangia.
4. *Label the aerial stem, the rhizome, the branches, the leaves and the strobili in Figure 26.9a.*

Observation: Horsetails

In horsetails (division Equisetophyta), a rhizome produces aerial stems that stand about 1.3 meters.

1. Examine *Equisetum,* a horsetail, and note the minute, scalelike leaves (Fig. 26.9*b*).
2. Feel the stem. *Equisetum* contains a large amount of silica in its stem. For this reason, these plants are sometimes called scouring rushes and may be used by campers for scouring pots.
3. Strobili appear at the tips of the stems, or else special buff-colored stems bear the strobili. Sporangia are in the strobili.
4. *Label the rhizome, the nodes, the branches, the leaves, and the strobilus in Figure 26.9b.*

Figure 26.9 Club moss and horsetail.
a. In the club moss *Lycopodium, green* photosynthetic stems are covered by scalelike leaves, and spore-bearing leaves are clustered in strobili. **b.** In the horsetail *Equisetum,* whorls of branches or tiny leaves appear in the joints of the stem. The sporangia are borne in strobili.

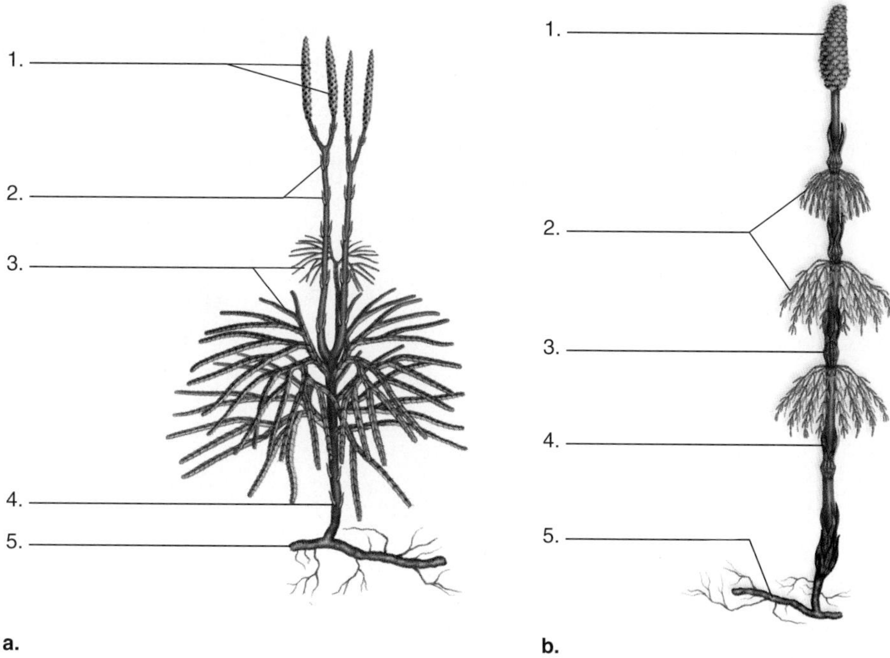

Ferns

Ferns are quite diverse and range in size from those that are low growing and resemble mosses to those that are as tall as trees. The rhizome grows horizontally, which allows ferns to spread without sexual reproduction (Fig. 26.10). The gametophyte (called a **prothallus**) is small (about 0.5 cm) and usually heart-shaped. The prothallus contains both archegonia and antheridia. Still, ferns are largely restricted to moist, shady habitats because sexual reproduction requires adequate moisture. Why?

Figure 26.10 Fern life cycle.
The rhizome and fronds are the sporophyte, and the gametophyte is the heart-shaped prothallus.

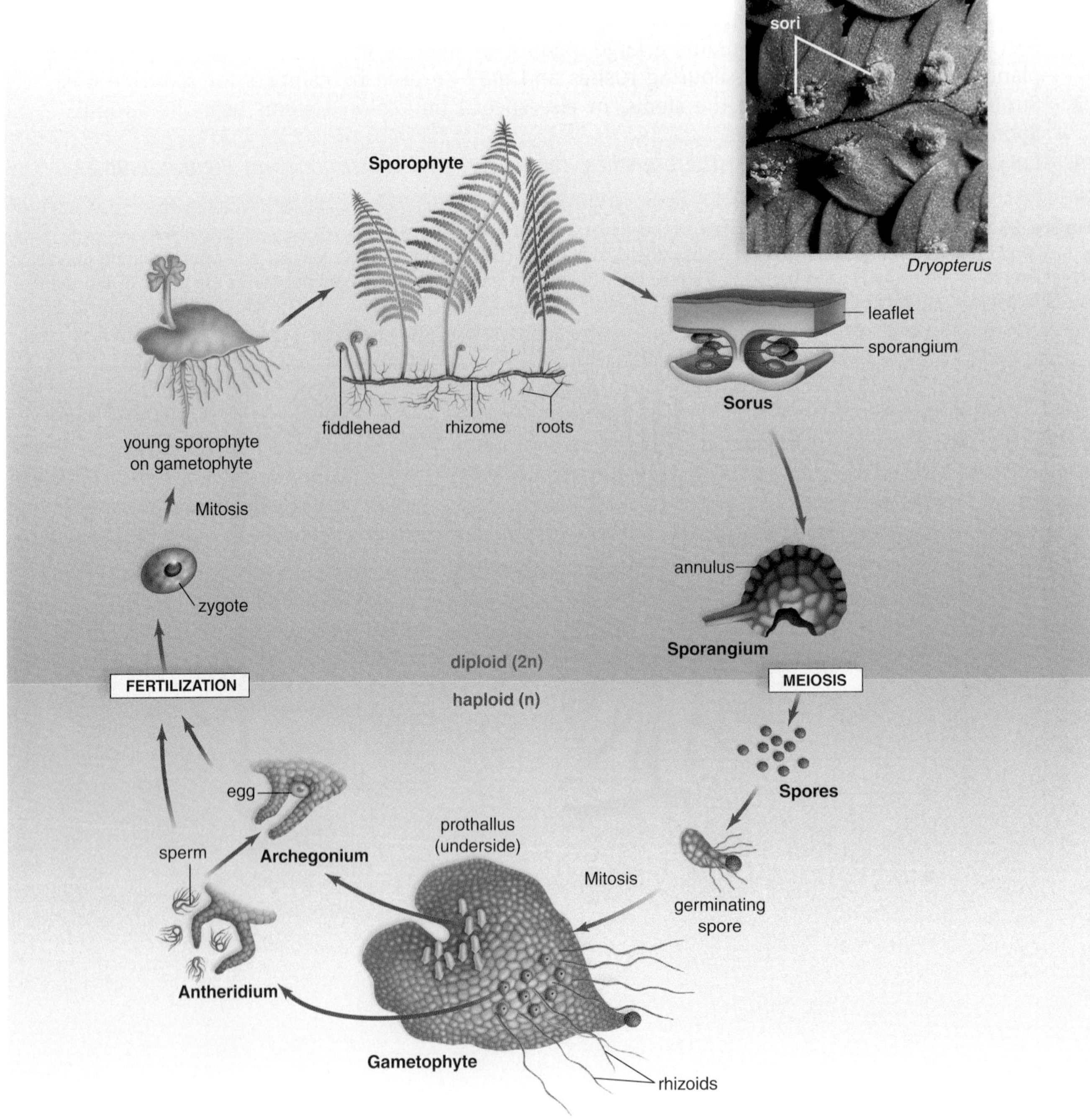

Table 26.1 Fern Diversity	
Type of Fern	**Description of Frond**

Fern spores are produced by meiosis in structures called sporangia, which in many species occur on the underside of large leaves called **fronds.** The spores are released when special cells of the **annulus** (a line or ring of thickened cells on the outside of the sporangium) dry out and the sporangium opens. How do ferns disperse the offspring? ______________________________
Observe the ferns on display, and then complete Table 26.1.

Observation: Fern Sporophyte

Study the life cycle of the fern (Fig. 26.10), and find the sporophyte generation. This large, complexly divided leaf is known as a frond. Fronds arise from an underground stem called a rhizome.

Living or Preserved Frond

Examine a living or preserved specimen of a frond, and on the underside, notice a brownish clump called a **sorus** (pl., *sori*), each a cluster of many sporangia (Fig. 26.11).

What is being produced in the sporangia? ______________________________

Given that this is the generation called the fern, what generation is dominant in ferns? ____________

Figure 26.11 Underside of frond leaflets.
Sori occur on the underside of frond leaflets.

Microscope Slide of Sorus

1. Examine a prepared slide of a cross section of a frond leaflet. Using Figure 26.12 as a guide, locate the fern leaf above and the sorus below.
2. Within the sorus, find the sporangia and spores. Look for an **indusium** (not present in all species), a shelflike structure that protects the sporangia until they are mature.

   Does this fern have an indusium? ________

Figure 26.12 Cross section of a frond leaflet.
A drawing of the internal anatomy of a sorus depicts many sporangia, where spores are produced.

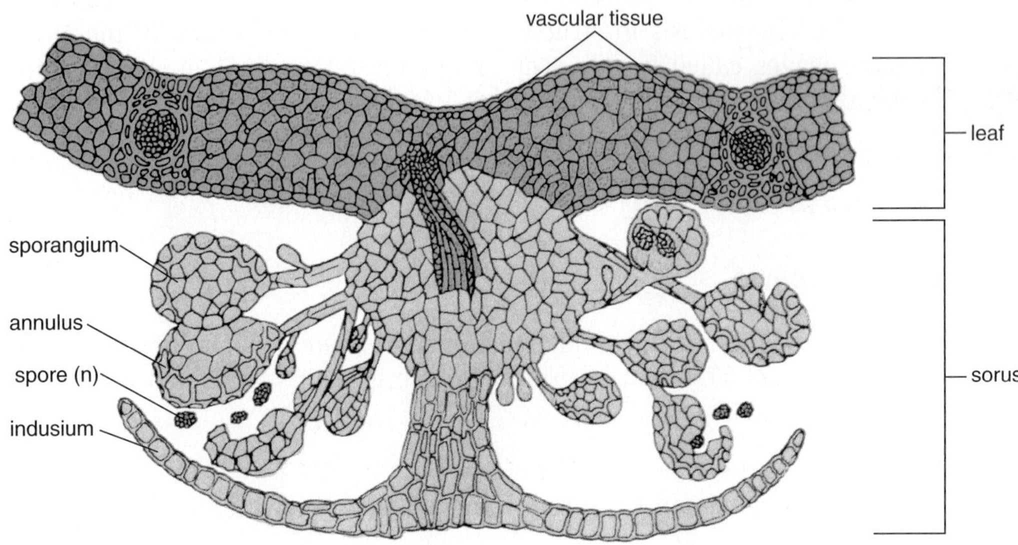

Observation: Fern Gametophyte

Plastomount

1. Examine a plastomount showing the fern life cycle.
2. Notice the prothallus, a small, heart-shaped structure.

   Can you find this structure in your fern minimarsh (if available)? ________ The prothallus is the gametophyte generation of the fern. Most persons do not realize that this structure exists as a part of the fern life cycle.

   What is the function of this structure? ____

   ____

Microscope Slide

1. Examine whole-mount slides of fern prothallium-archegonia and fern prothallium-antheridia (Fig. 26.13).
2. If you focus up and down very carefully on the archegonia, you may be able to see an egg inside. What is being produced inside the antheridia? ________ When sperm produced by the antheridia swim to the archegonia in a film of water, what results? ____________________ This structure develops into what generation? ____________________

Conclusions

- How are ferns dispersed from one area to another? ____________________
- Is either generation in the fern dependent for any length of time on the other generation? ____ Explain. ____________________

Figure 26.13 Fern prothallus.
The underside of the heart-shaped fern prothallus contains archegonia where eggs are produced and antheridia where flagellated sperm are produced.

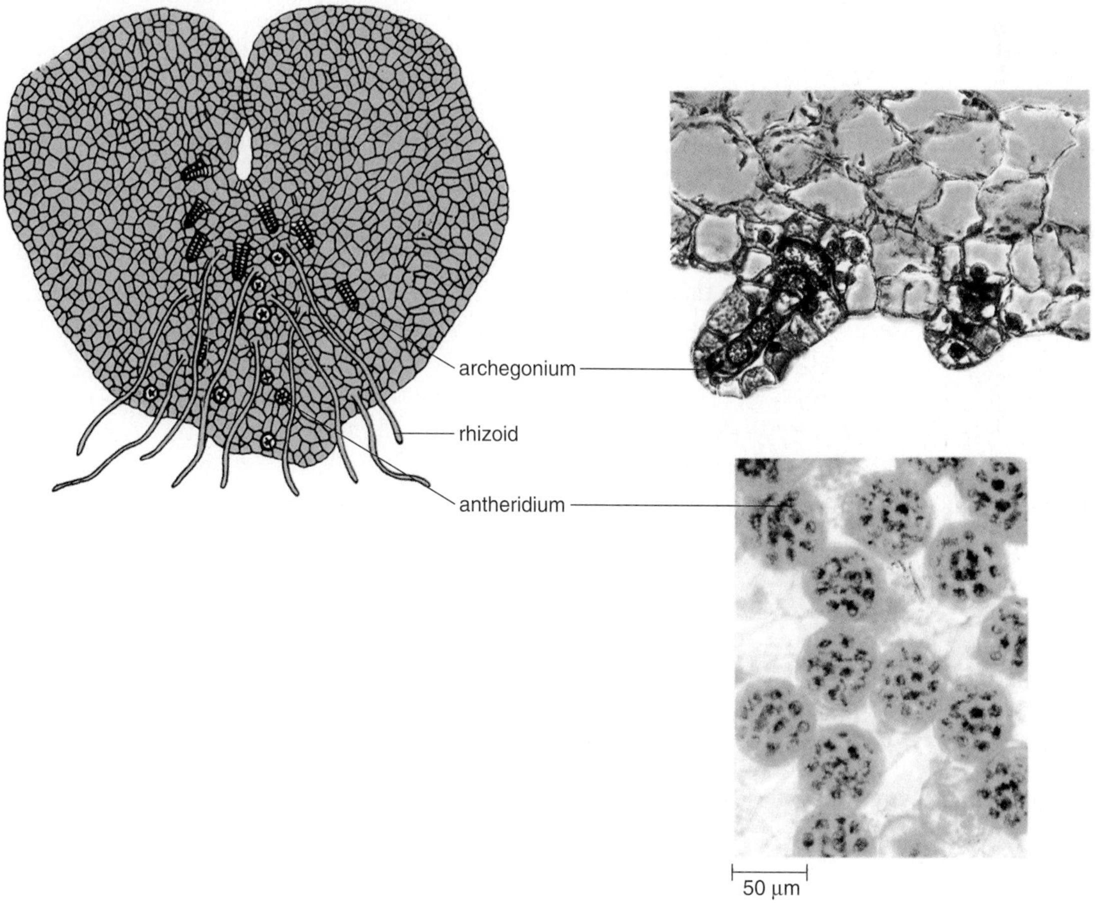

Laboratory Review 26

1. What type of life cycle do plants have?
2. Which generation in the plant life cycle produces gametes?
3. Which generation in the plant life cycle has vascular tissue?
4. What generation is dominant in mosses?
5. Leafy green shoot describes which generation of a moss?
6. What reproductive structure in mosses produces eggs?
7. What type of structure in mosses disperses the offspring?
8. What type of habitat do mosses need for sexual reproduction to occur?
9. What do ferns have that mosses lack?
10. What type of plant studied today has stems impregnated with silica?
11. Which generation is dominant in seedless vascular plants?
12. In whisk ferns, which part of the plant carries out photosynthesis?
13. A small, heart-shaped structure describes which generation in ferns?
14. Are sperm protected from drying out in the fern life cycle?
15. What do you call the clusters of sporangia often found on fern fronds?
16. What allows vascular plants to typically grow taller than nonvascular plants?

Thought Questions

17. Contrast the life cycle of plants to that of animals (e.g., human beings).

18. Aside from appearance, how is the gametophyte generation in ferns similar to that of mosses? How is it different?

19. During the alternation of generations life cycle typical of plants, spores are produced via meiosis within the sporangium. Why must these spores develop into a gametophyte by mitosis rather than meiosis?

LABORATORY

27

Seed Plants

Learning Outcomes

Introduction
- Describe the life cycle of seed plants, and illustrate how their life cycle is adapted to a land environment.

27.1 Gymnosperms
- City examples of each of the three divisions of gymnosperms.
- Describe the life cycle of a pine tree and all the structures involved.
- Distinguish between pollination and fertilization.
- Describe the structure of a pine cone, and contrast pollen cones with seed cones.

27.2 Angiosperms
- Describe the life cycle of a flowering plant and all the structures involved, including the male gametophyte and female gametophyte.
- Explain the significance of double fertilization in flowering plants.
- Describe the dispersal mechanisms of angiosperm seeds.
- Identify the parts of a flower, and contrast monocot and eudicot flowers.

Introduction

Sexual reproduction in seed plants differs from that in seedless plants in two important ways. In the first case, seed plants are not dependent upon water for gametes to achieve fertilization. The second difference involves the formation of a seed, a protective structure that protects the embryonic plant. These two features make seed plants well suited for a life on land. Both **gymnosperms** and **angiosperms** are groups of plants that disperse the species by means of seeds. Figure 27.1 depicts the general life cycle of seed plants.

Figure 27.1 Alternation of generations in flowering plants, a seed plant.

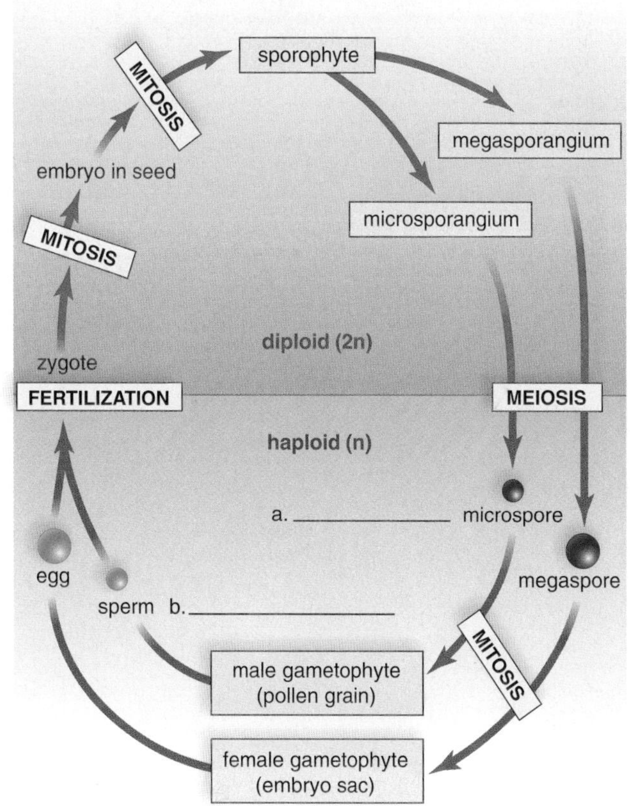

27.1 Gymnosperms

The term *gymnosperm* means "naked seed"—that is, the seeds are not enclosed by fruit as are the seeds of angiosperms. Three divisions of gymnosperms are especially familiar:

Phylum Cycadophyta: cycads
Phylum Ginkgophyta: ginkgoes
Phylum Pinophyta: conifers

Representative specimens of these plants may be available for you to examine. Compare them with Figure 27.2.

Conifers

The **conifers** (phylum Pinophyta) are by far the largest group of gymnosperms. Pines, hemlocks, and spruces are evergreen conifers because their leaves remain on the tree through all seasons. The cypress tree and larch (tamarack) are examples of conifers that are not evergreen.

Figure 27.2 Gymnosperms.
Three divisions of gymnosperms are well known: **(a)** cycads, **(b)** ginkgoes, and **(c)** conifers.

a. Cycad, *Encephalartos humlis.*

b. Ginkgo, *Ginkgo biloba.* Female maidenhair tree with seeds.

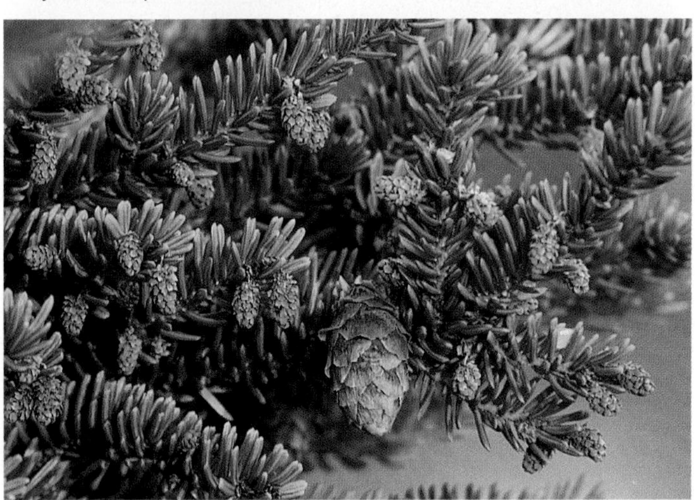

c. Conifer, *Picea.* Spruce with pollen cones and seed cones.

Figure 27.3 Pine life cycle.
The sporophyte is the tree. The male gametophytes are windblown pollen grains shed by pollen cones. The female gametophytes are retained within ovules on seed cones. The ovules develop into windblown seeds.

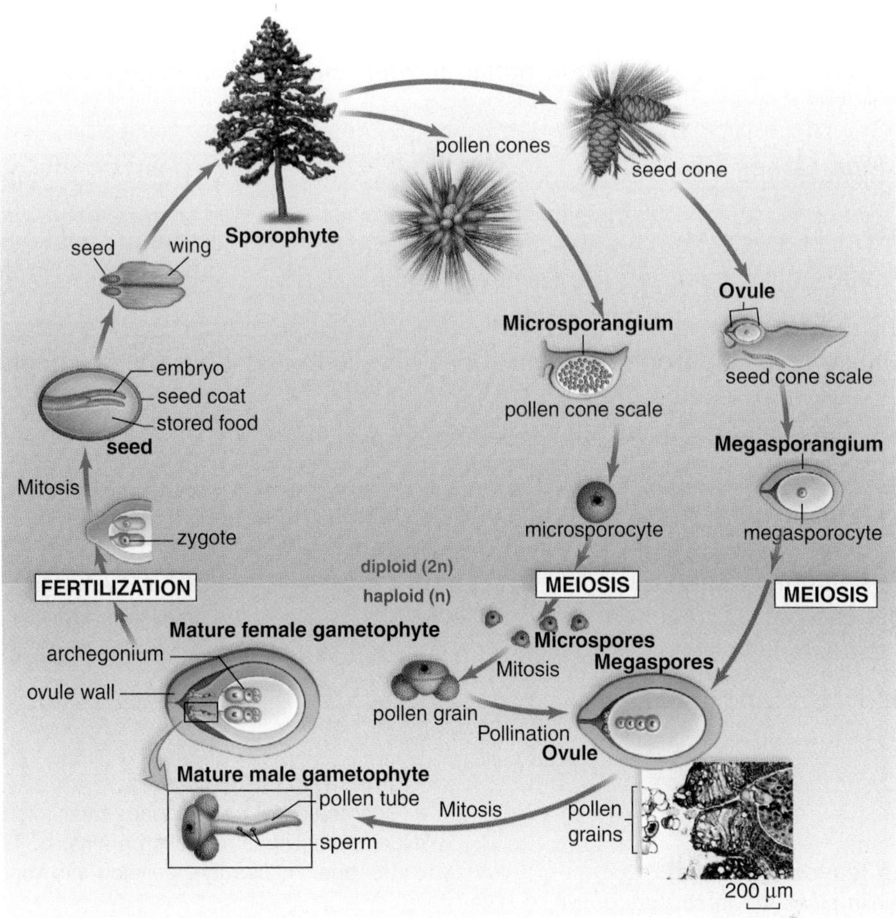

Pine Trees

The pine tree is the dominant sporophyte generation (Fig. 27.3). Vascular tissue extends from the roots, through the stem, to the leaves. Pine tree leaves are needlelike, leathery, and covered with a waxy, resinous cuticle. The **stomata,** openings in the leaves for gas exchange, are sunken. The structure of the leaf and the leaf's internal anatomy are adaptive to a drier climate.

In pine trees, the sporangia are located in cones. Pollen cone scales contain **microsporocytes** (microspore mother cells), which undergo meiosis to produce **microspores.** Microspores (n) become sperm-bearing **male gametophytes** (pollen grains). Seed cone scales bear ovules where **megasporangia** produce **megaspores.** A megaspore (n) becomes an egg-bearing **female gametophyte.**

Pollination occurs when pollen grains are windblown to the seed cones. After **fertilization,** the egg becomes a sporophyte (2n) embryo enclosed within the ovule, which develops a seed coat. The seeds are winged and are dispersed by the wind.

1. Which part of the pine life cycle represents the sporophyte? ________________
2. What two types of spores are produced by meiosis? Are these structures haploid or diploid? ________________
3. Which part of the pine life cycle represents the male gametophyte and female gametophyte (n) generations, and where is each generation located? ________________
4. Where does fertilization occur? ________________
5. What generation is located within the seed? ________________

Observation: Pine Leaf

Obtain a cluster of pine leaves (needles). A very short, woody stem is at its base. Each type of pine has a typical number of leaves in a cluster (Fig. 27.4). How many leaves are in the cluster you are examining? ________ What is the common name of your specimen? ______________________________

Observation: Pine Cones

Preserved Cones

1. Compare a pine pollen cone with a pine seed cone. ______________________________

   ____
2. Remove a single scale **(sporophyll)** from the pollen cone and from the seed cone, and observe each with a dissecting microscope.
3. Note the two microsporangia on the lower surface of each scale of the pollen cone (Fig. 27.5*a*). Why is this called a pollen cone? ______________________________
   Pollen cones remain on the tree for only one or two months.
4. Note two structures on the upper surface of a scale from a seed cone (Fig. 27.5*b*). Depending on the cone's maturity, these structures may still be ovules, or they may be the seeds that develop from the ovules after fertilization. These structures hold the megasporangia and, later, the female gametophyte generation. Seed cones remain on the tree for as long as three years. Why so long? ______________________________

Figure 27.4 Pine leaves (needles).
a. The needles of white pines are in clusters of five. **b.** The needles of pitch pines are in clusters of three. **c.** The needles of red pines are in clusters of two.

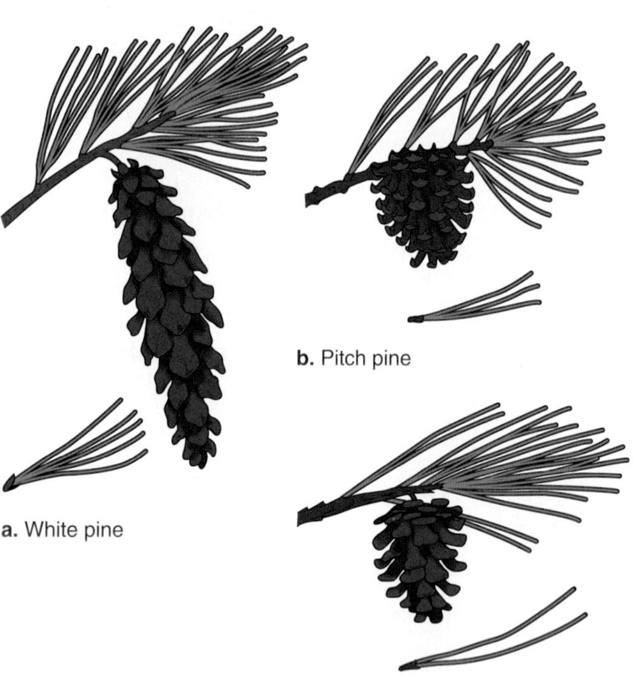

Figure 27.5 Pine cones.
a. The scales of pollen cones bear microsporangia where microspores become pollen grains. **b.** The scales of seed cones bear ovules that develop into winged seeds.

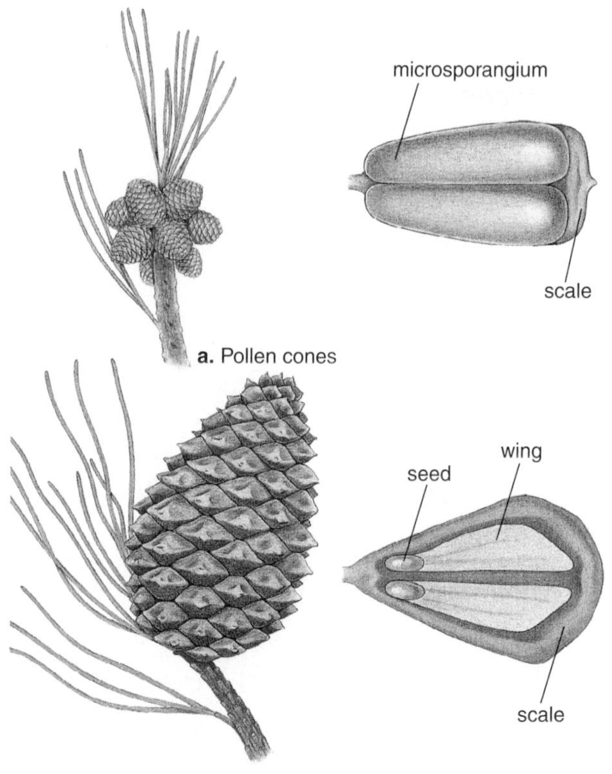

Figure 27.6 Pine pollen cone.
Pollen cones bear **(a)** microsporangia in which microspores develop into pollen grains. **b.** Enlargement of pollen grains.

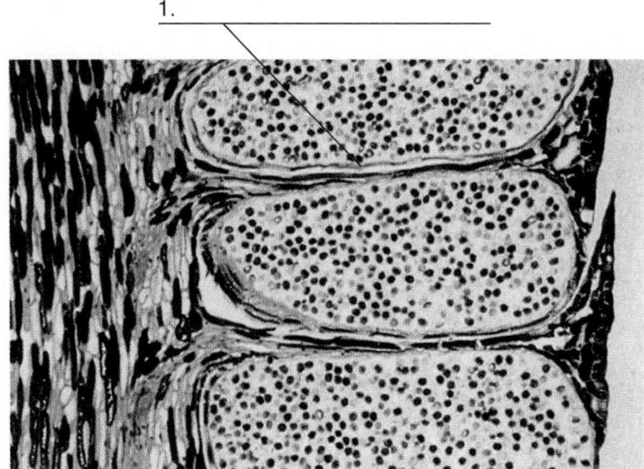

a. Longitudinal section through pine pollen cone, showing pollen grains within microsporangia.

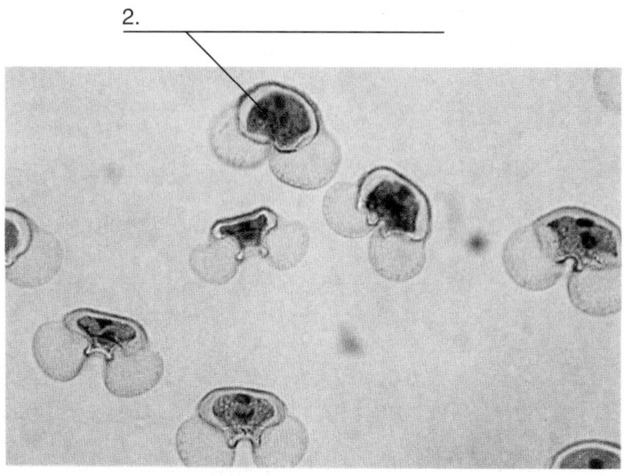

b. Enlargement of pollen grains.

Microscope Slides

1. Examine a prepared slide of a longitudinal section through a mature pine pollen cone. Label a microsporangium in Figure 27.6*a* and a winged pollen grain in Figure 27.6*b*. Describe a pine pollen grain. ______________________________

   ____

   One pollen grain cell will divide to become two nonflagellated sperm, one of which fertilizes the egg after pollination. The other cell forms the **pollen tube,** through which a sperm travels to the egg.
2. Examine a prepared slide of a longitudinal section through a mature pine seed cone. Seed cone scales bear ovules, each of which is composed of a megasporangium and its surrounding integument. The megasporangium contains a megasporocyte (megaspore mother cell), which undergoes meiosis to produce four megaspores, three of which disintegrate. This megaspore (n) becomes an egg-bearing female gametophyte. *Label the megasporangium and the megasporocyte in Figure 27.7. Also, label the pollen grains that you can see just outside the ovule.*

Figure 27.7 Seed cone.
Seed cones bear ovules, each of which at one time contains a megasporangium shown here in longitudinal section. Note pollen grains near the entrance.

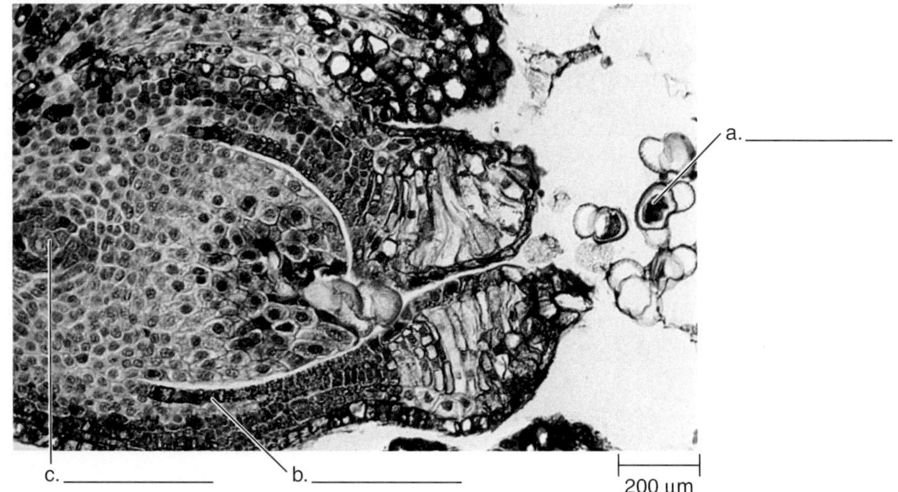

Observation: Pine Seeds

A seed contains an embryonic sporophyte and nutrient material. Examine some pine seeds, if they are available. In seed plants, when seeds are dispersed so are offspring. How are pine seeds dispersed? ______________________________

Some cones release the seeds only after a fire. The cones of a cedar tree are fleshy and are eaten by birds. Some seeds pass through the birds' digestive tract intact.

What does the word *gymnosperm* mean? ______________________________

____

Are the pine seeds covered by tissue donated by the original sporophyte? ______ Explain. ______________________________

____

Both seeds and pollen grains have wings associated with them. What does this suggest about their mechanism of dispersal? ______________________________

27.2 Angiosperms

Flowering plants (phylum Magnoliophyta) are the dominant plants today. They occur as trees, shrubs, vines, and garden plants. A typical leaf is composed of a thin **petiole** and an expanded **blade,** the part of the leaf that is typically some shade of green. Most angiosperms are said to have broad leaves because of the blade's shape.

The life cycle of a flowering plant (Fig. 27.8) is like that of a pine tree (see Fig. 27.3), except for these innovations:

- The often brightly colored flower contains the microsporangia and the megasporangia. Pollen may be windblown or carried by animals (e.g., insects). What additional methods do flowering plants use to accomplish pollination? ______________________________

  ____
- Flowering plants practice **double fertilization.** A pollen grain contains two sperm; one fertilizes the egg, and the other joins with the two polar nuclei to form **endosperm** (3n), which serves as food for the developing embryo.
- Flowering plants have seeds enclosed within fruits. Fruits protect the seeds and aid in seed dispersal. Sometimes, animals eat the fruits, and after the digestion process, the seeds are deposited far away from the parent plant. The term *angiosperm* means "covered seeds."

Laboratory 10 examined other differences in flowering plants.

1. The pollen grain contains nonflagellated sperm in seed plants. The sperm-bearing pollen grain is windblown or carried by animals to the vicinity of the egg. How have seed plants solved, through evolution, the problem of getting the gametes together? ______________________________

   ____
2. The pollen grain does not normally germinate until it is near the egg-bearing female gametophyte. How have seed plants solved, through evolution, the problem of protecting the gametes from drying out? ______________________________
3. The seed, which contains an embryonic sporophyte and stored food, is covered by a protective seed coat. How have seed plants solved, through evolution, the problem of protecting the embryo from drying out? ______________________________

   ____

Figure 27.8 Flowering plant life cycle.
The parts of the flower involved in reproduction are the anthers of stamens and the ovules in the ovary of a carpel.

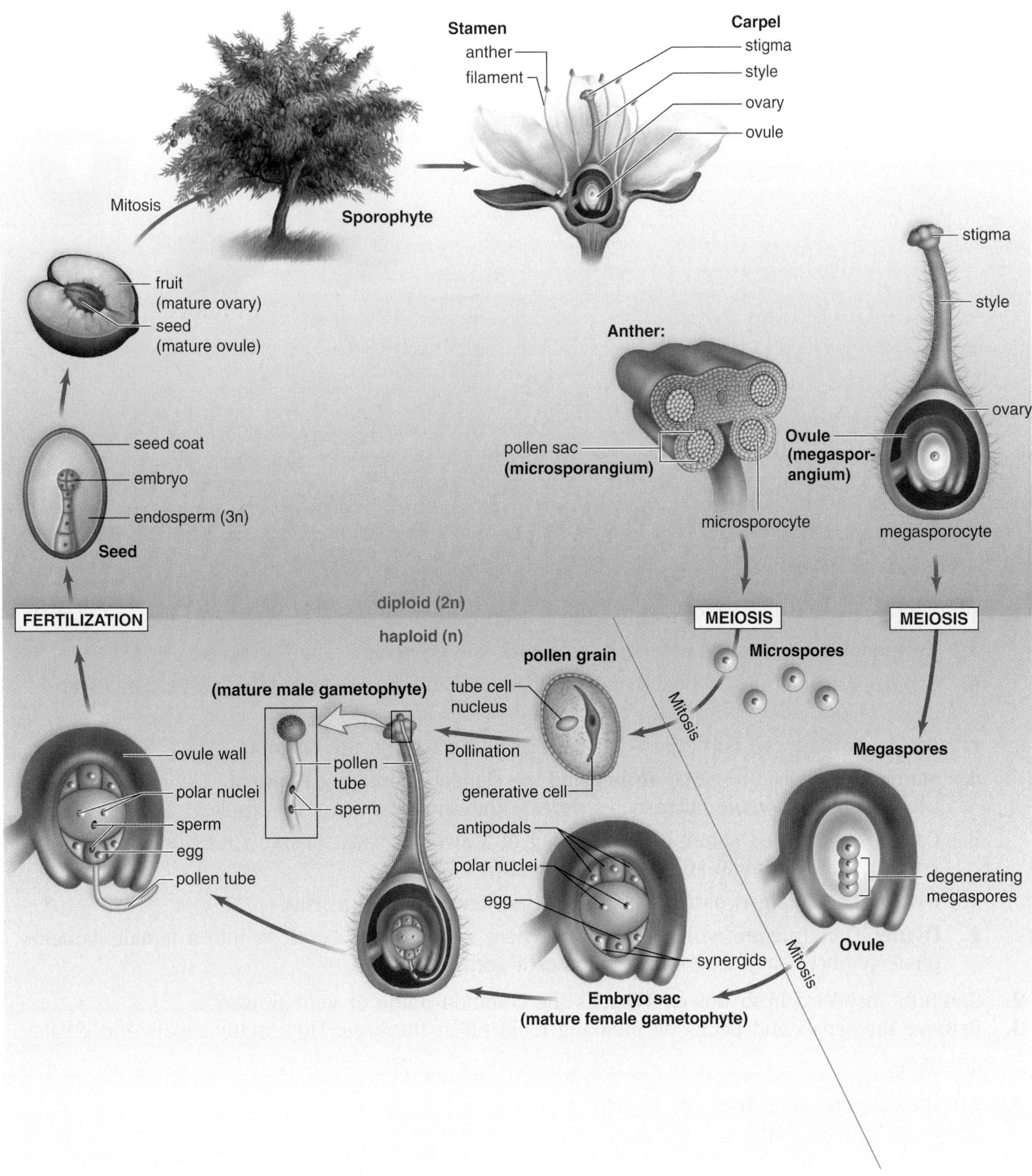

Figure 27.9 Generalized flower.
A flower has four main kinds of parts: sepals, petals, stamens, and a carpel. A stamen has an anther and filament. A carpel has a stigma, style, and ovary. An ovary contains ovules.

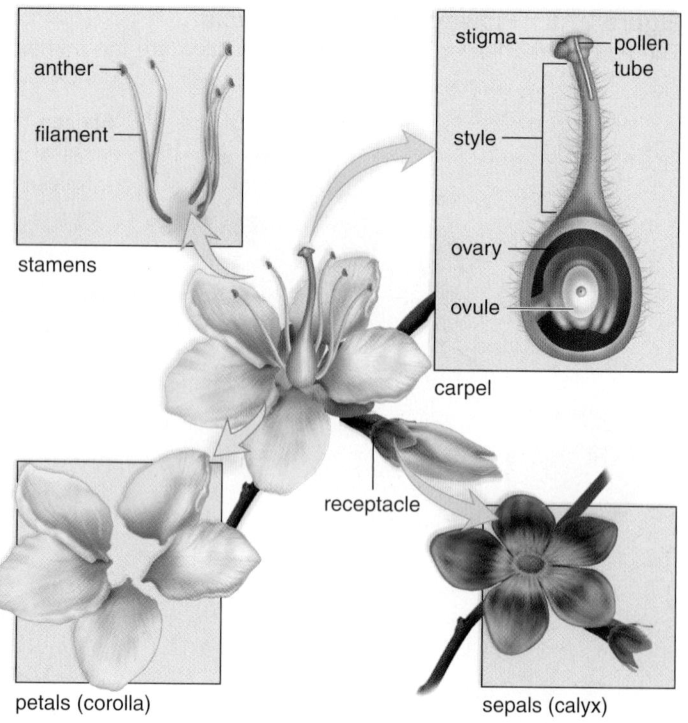

Observation: A Flower

1. With the help of Figure 27.9, identify the following structures on a model of a flower:
   a. **Receptacle:** The portion of a stalk to which the flower parts are attached.
   b. **Sepals:** An outermost whorl of modified leaves, collectively called the **calyx.** Sepals are green in most flowers. They protect a bud before it opens.
   c. **Petals:** Usually colored leaves that collectively constitute the **corolla.**
   d. **Stamen:** A swollen terminal **anther** and the slender supporting **filament.** The anther contains two pollen sacs, where microspores develop into male gametophytes (pollen grains).
   e. **Carpel:** A modified sporophyll consisting of a swollen basal ovary; a long, slender **style** (stalk); and a terminal **stigma** (sticky knob).
   f. **Ovary:** The enlarged part of the carpel that develops into a fruit.
   g. **Ovule:** The structure within the ovary where a megaspore develops into a female gametophyte (embryo sac). The ovule becomes a seed.
2. Carefully inspect a fresh flower. What is the common name of your flower? ______________
3. Remove the sepals and petals by breaking them off at the base. How many sepals and petals are there? ______________________________
4. Are the stamens taller than the carpel? ______________________________
5. Remove a stamen, and touch the anther to a drop of water on a slide. If nothing comes off in the water, crush the anther a little to release some of its contents. Place a coverslip on the drop, and observe with low- and high-power magnification. What are you observing? ______________
6. Remove the carpel by cutting it free just below the base. Make a series of thin cross sections through the ovary. Note that the ovary is hollow and that you can see nearly spherical bodies inside. What are these bodies? ______________________________

The Male Gametophyte

In flowering plants, **pollination** is the transfer of the pollen grain from the anther to the stigma, where the pollen grain germinates. The pollen grain's tube cell gives rise to the pollen tube. As it grows, the pollen tube passes through the stigma and grows through the style and into the ovary. Two sperm cells produced by division of the pollen grain's generative cell migrate through the pollen tube into the embryo sac.

Observation: Pollen Grain Slide

1. A pollen grain released from the anther has two cells. The larger of the two cells is the **tube cell,** and the smaller is the **generative cell.** Examine a prepared slide of pollen grains. Identify the tube cell and the generative cell. Sketch your observation here.

2. Examine a prepared slide of pollen grains with pollen tubes. You should be able to see the tube nucleus and two sperm cells. Why is the mature pollen grain called the male gametophyte? ______________________________
____

**Experimental Procedure: Pollen Grains*

Inoculate an agar-coated microscope slide with pollen grains. Invert the slide, and place it onto a pair of wooden supports in a covered petri dish. Leave the inverted slide in the covered petri dish for one hour. Then remove the slide, and examine it with the compound microscope. Have any of the pollen grains germinated? ______________________________

If so, describe. ______________________________
____

*Experimental Procedure from Richard Carter and Wayne R. Faircloth, *General Laboratory Studies,* Lab 17.2, 1991. Used with permission of Kendall/Hunt Publishing Company.

The Female Gametophyte

In the ovule, a megaspore undergoes three mitotic divisions to produce a seven-celled (eight-nuclei structure) called an **embryo sac.** One of these cells is an egg cell. The largest cell contains two **polar nuclei.** The embryo sac is the female gametophyte.

Why is the embryo sac called the female gametophyte? ______________________________

____

Observation: Embryo Sac Slide

1. Examine the demonstration slide of the mature embryo sac of *Lilium.* Identify the egg noted in Figure 27.10.
2. When the pollen tube delivers sperm to the embryo sac, double fertilization occurs. Describe double fertilization. ______________________________

   ____

   Following fertilization, the ovules develop into seeds, and the ovary develops into the fruit.

Figure 27.10 Embryo sac in a lily ovule.
An embryo sac is the female gametophyte of flowering plants. It contains seven cells, one of which is the egg. Not all cells are visible in this micrograph.

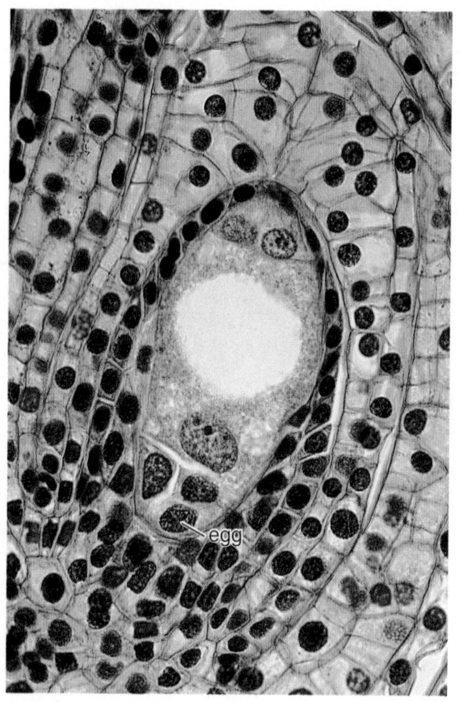

Observation: Two Fruits

Label the flower remnants, the fruit, and the seeds in the drawings of a pea pod and apple provided. The flesh of an apple comes from the enlarged receptacle that grows up and around the ovary, while the ovary largely consists of the core.

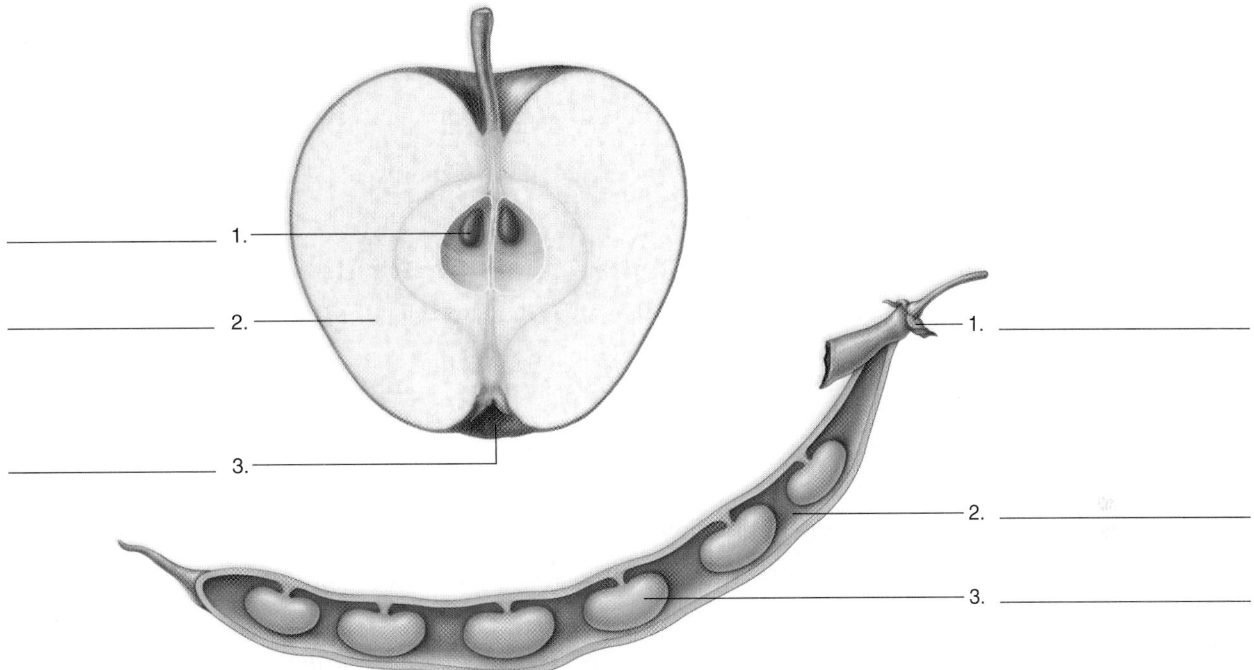

Monocots Versus Eudicots

The two classes of flowering plants—**monocots** (class Liliopsida) and **eudicots** (class Magnoliopsida)—are distinguished on the basis of the characteristics shown in Figure 27.11. A **cotyledon** is a seed leaf; it provides nutrient molecules to the growing embryo either by storing or absorbing nutrients from the endosperm. Monocots take their name from having one cotyledon in the seed, and eudicots take their name from having two cotyledons in the seed.

1. Reexamine the flower model from Observation: A Flower on p. 386, and on the basis of the number of flower parts, decide if the model represents a monocot or a eudicot. ______
______
2. Reexamine the fresh flower from Observation: A Flower on p. 386. Is this flower a monocot or a eudicot? ______

Figure 27.11 Monocots versus eudicots.

Monocots	Eudicots
Flower parts in threes and multiples of three	Flower parts in fours or fives and their multiples
a.	b.

Laboratory Review 27

_______________ 1. What structure transports sperm to the ovule in seed plants?

_______________ 2. Which group of plants practices double fertilization?

_______________ 3. Gymnosperms have naked seeds because they are not enclosed by _______________.

_______________ 4. What type of cell division produces microspores and megaspores in seed plants?

_______________ 5. In a conifer, is fruit present or absent?

_______________ 6. The pollen grain replaces what structure in the life cycle of seedless plants?

_______________ 7. One pine pollen grain cell divides to become two _______________, and the other cell forms the _______________.

_______________ 8. The carpel of a flower consists of what three structures?

_______________ 9. Name a type of gymnosperm that is not a conifer.

_______________ 10. On what structure would you be able to find the male gametophyte of a pine tree?

_______________ 11. In what structure would you be able to find the female gametophyte of a pine tree?

_______________ 12. Name the part of a flower that has a filament topped by the anther.

_______________ 13. What are the two classes of flowering plants? How do they differ from one another?

_______________ 14. How many sperm are in the pollen grain of a flowering plant?

Thought Questions

15. A pine tree, unlike a fern, is able to reproduce sexually in a dry environment. Explain.

16. What is the difference between pollination and fertilization?

17. Why are most pollen cones located at the tips of branches and not near the center of the tree?

18. How does the angiosperm life cycle differ from the gymnosperm life cycle? In what ways are these two lifes cycles similar to one another?

LABORATORY

28

Introduction to Invertebrates

Learning Outcomes

28.1 Classification of Animals

- Discuss the anatomical criteria that serve as a basis for the classification of animals.
- Categorize the phyla in the evolutionary tree according to level of organization, symmetry, body plan, germ layers, type of coelom, and presence of segmentation.

28.2 Phylum Porifera (Sponges)

- Identify a sponge, and describe the anatomy and way of life of sponges.

28.3 Phylum Cnidaria (Cnidarians)

- Describe six anatomical features of cnidarians; identify these features on a slide or image of a hydra.
- Generalize the life cycle of cnidarians, and relate this life cycle to *Hydra* and *Obelia.*

28.4 Phylum Platyhelminthes (Flatworms)

- Describe seven anatomical features of flatworms; identify these features on a slide, an image, or a preserved specimen of a planarian.
- Contrast a hydra with a planarian according to significant features.

28.5 Phylum Nematoda (Roundworms)

- Describe four anatomical features of roundworms; identify these features on preserved specimens or slides of *Ascaris* as an example of this phylum.
- Compare roundworms with other types of animals studied today using common anatomical criteria.

Introduction

This is the first of three laboratories concerning the animal kingdom. Animals, which are classified in the table on page 393, are all multicellular and have varying degrees of motility. They are heterotrophic and digest their food in a digestive cavity. The animals in the phyla studied today (Porifera, Cnidaria, Platyhelminthes, and Nematoda) are all invertebrates. *Invertebrates* lack the endoskeleton of bone or cartilage found in *vertebrates.*

An **evolutionary tree** shows how various groups of organisms may be related to one another through the process of evolution. The tree shown in Figure 28.1 shows how the phyla of animals we will be studying are possibly related. Note that symmetry, number of germ layers, presence of a coelom, and development are all criteria used to determine relationships among animal phyla.

Symmetry refers to the organization of body structures relative to an axis of the body. Radial symmetry exists when structures circle an axis, whereas bilateral symmetry exists when the organism is divided in half by the axis. As this laboratory exercise will demonstrate, most animals develop all of their structures using either two or three developmental tissues or germ layers. The number of germ layers involved in development is an important criterion in animal classification. Classification is also based on the presence and type of body cavity, a fluid-filled space between the outer body wall and the internal digestive tract. Complex animals have a true coelom, a cavity completely lined by mesoderm. Segmentation, indicated by the repetition of body parts, is an advanced characteristic that leads to specialization as segments differentiate for specific purposes.

Figure 28.1 Evolution of animals.
All animals are believed to be descended from protists; however, the sponges may have evolved separately from the rest of the animals.

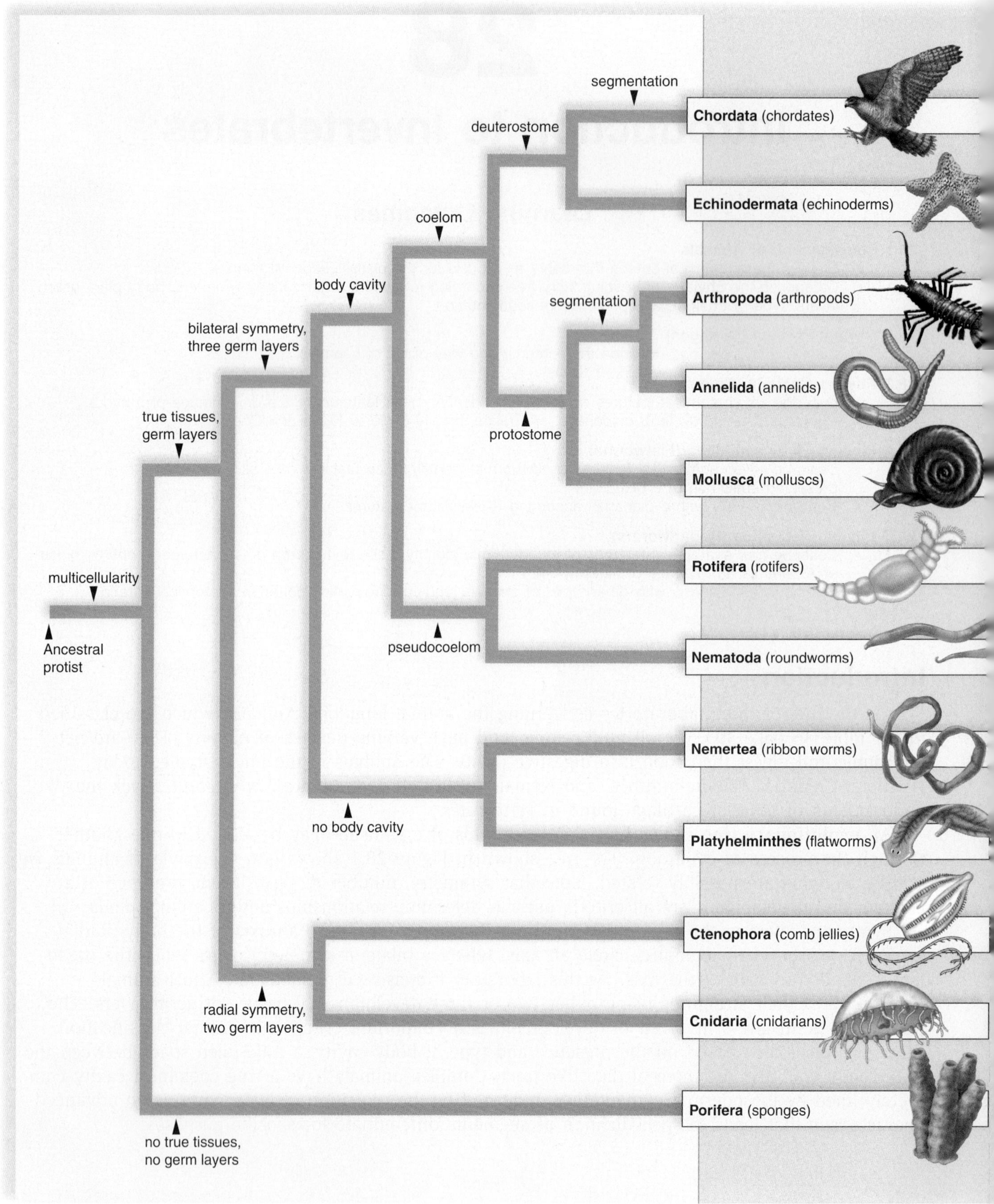

28.1 Classification of Animals

The classification of animals is based on the anatomical features given in Table 28.1, and each phylum will be examined according to these features. The phyla we will study in this laboratory differ anatomically; however, none are coelomates or segmented animals.

Table 28.1 Anatomical Criteria for Animal Classification

Level of Organization	
Cellular	Cells not organized into tissues
Tissues	Cells organized into tissues
Organs	Tissues organized into organs
Organ systems	Organs organized into systems
Germ Layers	
None	
Two	Two germ layers: the ectoderm (outside) and the endoderm (inside)
Three	Three germ layers: the ectoderm, the endoderm, and the mesoderm (middle); presence of mesoderm allows for development of various internal organs
Symmetry	
None	
Radial	Any longitudinal cut through the midpoint yields equal halves; this design allows animals to reach out in all directions
Bilateral	Only one longitudinal cut through the midpoint yields equal halves; such animals often have well-developed head regions (cephalization)
Body Plan	
None	
Sac plan	Mouth used for intake of nutrient molecules and exit of waste molecules
Tube-within-a-tube	Separate openings (mouth and anus) for food intake and waste exit; allows for specialization of parts along the digestive canal
Coelom	
Acoelomate	Have no coelom (fluid-filled extracellular spaces)
Pseudocoelomate	Have false coelom; coelom incompletely lined with mesoderm
Coelomate	Have true coelom, a fluid-filled body cavity completely lined with mesoderm
Segmentation	
Nonsegmented	No segmentation (repeating parts)
Segmented	A repeating series of parts from anterior to posterior; allows for specialization of some sections for different functions

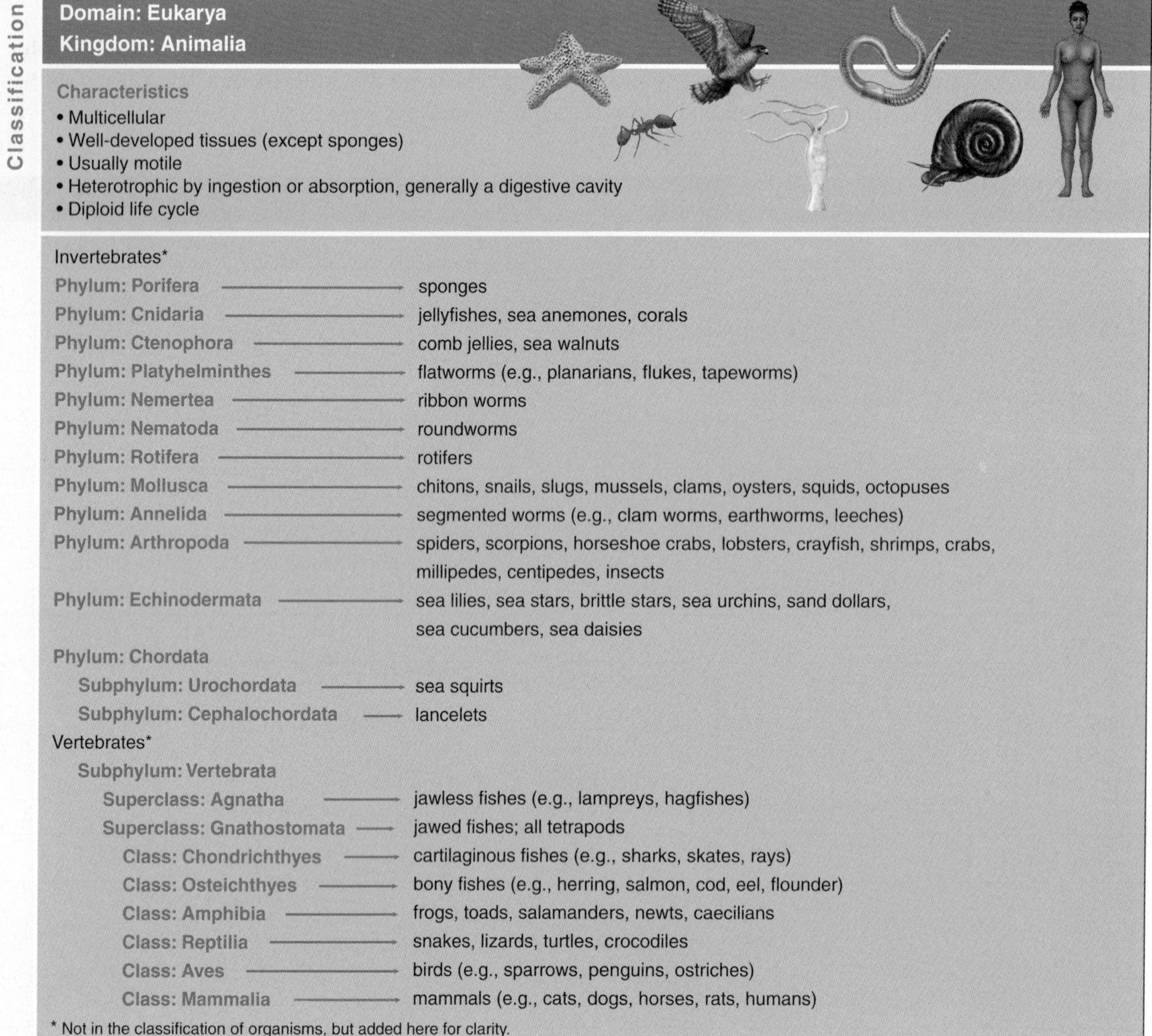

Classification

Domain: Eukarya
Kingdom: Animalia

Characteristics
- Multicellular
- Well-developed tissues (except sponges)
- Usually motile
- Heterotrophic by ingestion or absorption, generally a digestive cavity
- Diploid life cycle

Invertebrates*	
Phylum: Porifera	sponges
Phylum: Cnidaria	jellyfishes, sea anemones, corals
Phylum: Ctenophora	comb jellies, sea walnuts
Phylum: Platyhelminthes	flatworms (e.g., planarians, flukes, tapeworms)
Phylum: Nemertea	ribbon worms
Phylum: Nematoda	roundworms
Phylum: Rotifera	rotifers
Phylum: Mollusca	chitons, snails, slugs, mussels, clams, oysters, squids, octopuses
Phylum: Annelida	segmented worms (e.g., clam worms, earthworms, leeches)
Phylum: Arthropoda	spiders, scorpions, horseshoe crabs, lobsters, crayfish, shrimps, crabs, millipedes, centipedes, insects
Phylum: Echinodermata	sea lilies, sea stars, brittle stars, sea urchins, sand dollars, sea cucumbers, sea daisies
Phylum: Chordata	
Subphylum: Urochordata	sea squirts
Subphylum: Cephalochordata	lancelets
Vertebrates*	
Subphylum: Vertebrata	
Superclass: Agnatha	jawless fishes (e.g., lampreys, hagfishes)
Superclass: Gnathostomata	jawed fishes; all tetrapods
Class: Chondrichthyes	cartilaginous fishes (e.g., sharks, skates, rays)
Class: Osteichthyes	bony fishes (e.g., herring, salmon, cod, eel, flounder)
Class: Amphibia	frogs, toads, salamanders, newts, caecilians
Class: Reptilia	snakes, lizards, turtles, crocodiles
Class: Aves	birds (e.g., sparrows, penguins, ostriches)
Class: Mammalia	mammals (e.g., cats, dogs, horses, rats, humans)

* Not in the classification of organisms, but added here for clarity.

28.2 Phylum Porifera (Sponges)

Sponges live in water, mostly marine, attached to rocks, shells, and other solid objects. An individual sponge is typically shaped like a tube, cup, or barrel. Sponges grow singly or in colonies whose overall appearances vary widely (Fig. 28.2).

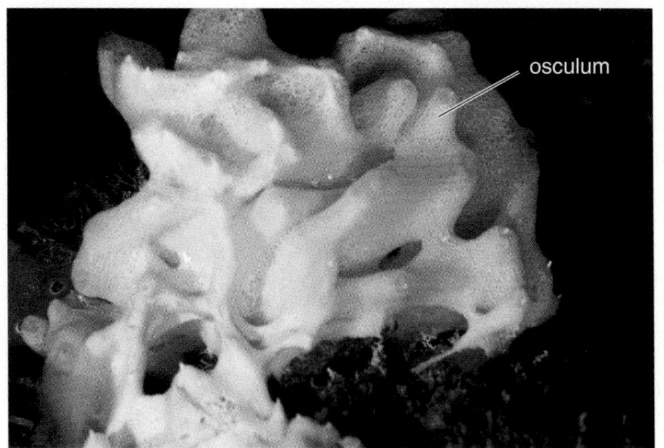

a. Class Calcarea: calcareous sponge, *Clathrina canariensis*

b. Class Demospongiae: bath sponge, *Xestospongia testudinaria*

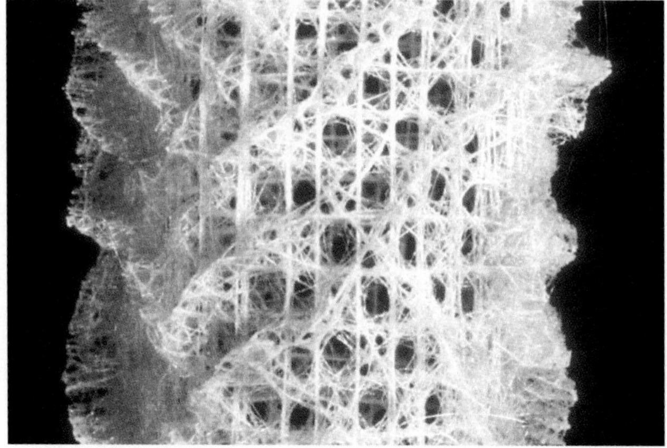

c. Class Hexactinellida: glass sponge, *Euplectella aspergillum*

Figure 28.2 Diversity of sponges.
a. Sponges in class Calcarea have spicules of calcium carbonate. **b.** In class Demospongiae, the skeleton is varied. Some members have a skeleton of spongin, some have glassy spicules, and some have a combination skeleton. **c.** Sponges in class Hexactinellida have glassy spicules.

Observation: Diversity of Sponges

Examine preserved, representative sponges and complete Table 28.2. Sponges are classified according to the type of **spicule** (little spike). Some have no spicules, and their skeleton is composed of a fibrous protein called spongin. These are the type of sponges that are sometimes still sold as *bath sponges*. Sponges that have sharp spicules made of calcium carbonate are called calcareous (chalk) sponges, and those whose spicules are made of silica are called glass sponges.

Table 28.2 Diversity of Sponges

Common Name of Specimen	Skeleton Made Of
1	
2	
3	
4	

Anatomy of Sponges

Sponges consist of loosely organized cells and have no well-defined tissues. They are asymmetrical or radially symmetrical and **sessile** (immotile). Sponges do contain specialized cells and therefore are considered to be a colony. They can reproduce asexually by budding or fragmentation, but they also reproduce sexually by producing eggs and sperm. Most sponges have the following three types of cells, as shown in Figure 28.3:

1. **Epidermal cells:** Flat cells that cover the outer surface and contain contractile fibers. Some surround pores that allow water to enter the sponge interior.
2. **Collar cells (choanocytes):** Flagellated cells lining the interior. Collar cells create water currents and filter suspended food particles from the water.
3. **Amoeboid cells:** Cells embedded in a noncellular matrix. Amoeboid cells digest and distribute food from collar cells, secrete spongin or spicules, and can form all other cells (totipotent).

Observation: Anatomy of Sponges

Preserved Sponge

1. Examine a preserved sponge. Note the main excurrent opening **(osculum)** and the multiple incurrent pores. Water is constantly flowing in through the pores and out the osculum. Label the arrows in the left-hand drawing of Figure 28.3 to indicate the flow of water. Use the labels *water out* and *water in through pores.*

Figure 28.3 Anatomy of a sponge.
The wall of a sponge has three types of cells: epidermal cells, amoebocytes (amoeboid cells), and choanocytes (collar cells).

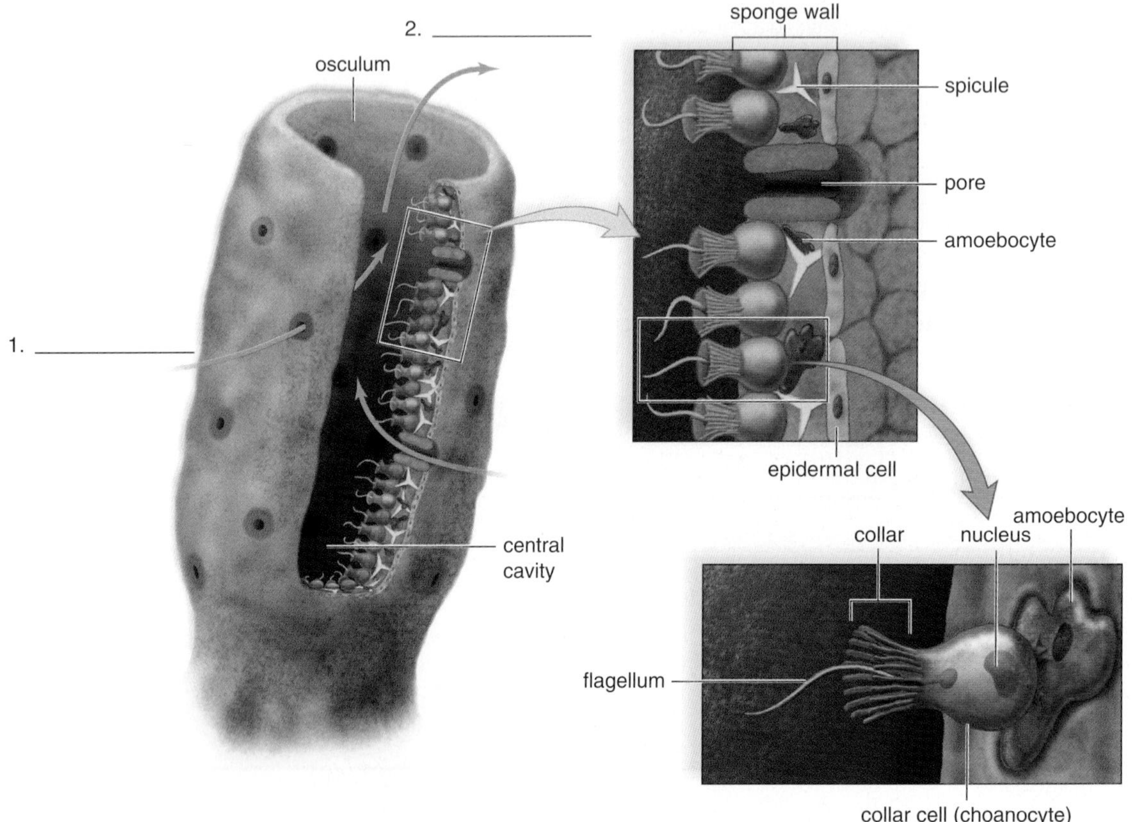

2. Examine a sponge specimen cut in half. Note the central cavity and the sponge wall. The wall is convoluted in some sponges, and the pores line small canals. Does this particular sponge have pore-lined canals? ___

3. You may be able to see spicules, fine projections over the body and especially encircling the osculum. Does this sponge have spicules? ___

Prepared Slides

1. Examine a prepared slide of *Grantia.*
   a. Find the collar cells that line the interior. A sponge is a sessile filter feeder. Collar cells phagocytize (engulf) tiny bits of food that come through the pores along with the water flowing through the sponge. They then digest the food in food vacuoles. Amoeboid cells pass some of the nutrient molecules to epidermal cells. Explain the expression *sessile filter feeder.* ___

   ___

   b. Find the epidermal cells. What is the shape of these cells? ___

   ___

   c. Are any amoeboid cells visible? ___ Where are they located? ___

2. Examine a prepared slide of sponge spicules. What do you see? ___

   ___

   Draw a sketch of a spicule.

Conclusions

- Among other functions, all animals have to acquire food, distribute nutrient molecules, carry on gas exchange, and excrete wastes. How do sponges, which have a cellular level of organization, carry on these functions? ___

  ___

- Sponges are the "have nots" of the animal kingdom. Explain this designation by referring to Table 28.1. Sponges do not have ___

  ___

- What features associated with sponges allow them to be classified in the animal kingdom?

  ___

28.3 Phylum Cnidaria (Cnidarians)

Cnidarians consist of a large number of mainly marine animals—for example, corals, sea anemones, and jellyfishes (Fig. 28.4). Stony corals have a calcium carbonate exoskeleton that contributes greatly to the building of coral reefs, areas of biological abundance in shallow tropical seas. Sea anemones, the "flowers" of the ocean, are also found around coral reefs. Jellyfishes are a part of the **zooplankton,** suspended animals that serve as food for larger animals in the ocean. The small and almost translucent hydras are one of the few freshwater species of cnidarians. They are found attached to underwater plants or rocks in most lakes and ponds.

Figure 28.4 Cnidarian diversity.
a. The life cycle of a cnidarian. Some cnidarians have both a polyp stage and a medusa stage; in others, one stage may be dominant or absent altogether. **b.** The anemone, sometimes called the flower of the sea, is a solitary polyp. **c.** Corals are colonial polyps residing in a calcium carbonate or proteinaceous skeleton. **d.** Portuguese man-of-war is a colony of modified polyps and medusae. **e.** True jellyfish undergo the complete life cycle; this is the medusa stage. The polyp is small.

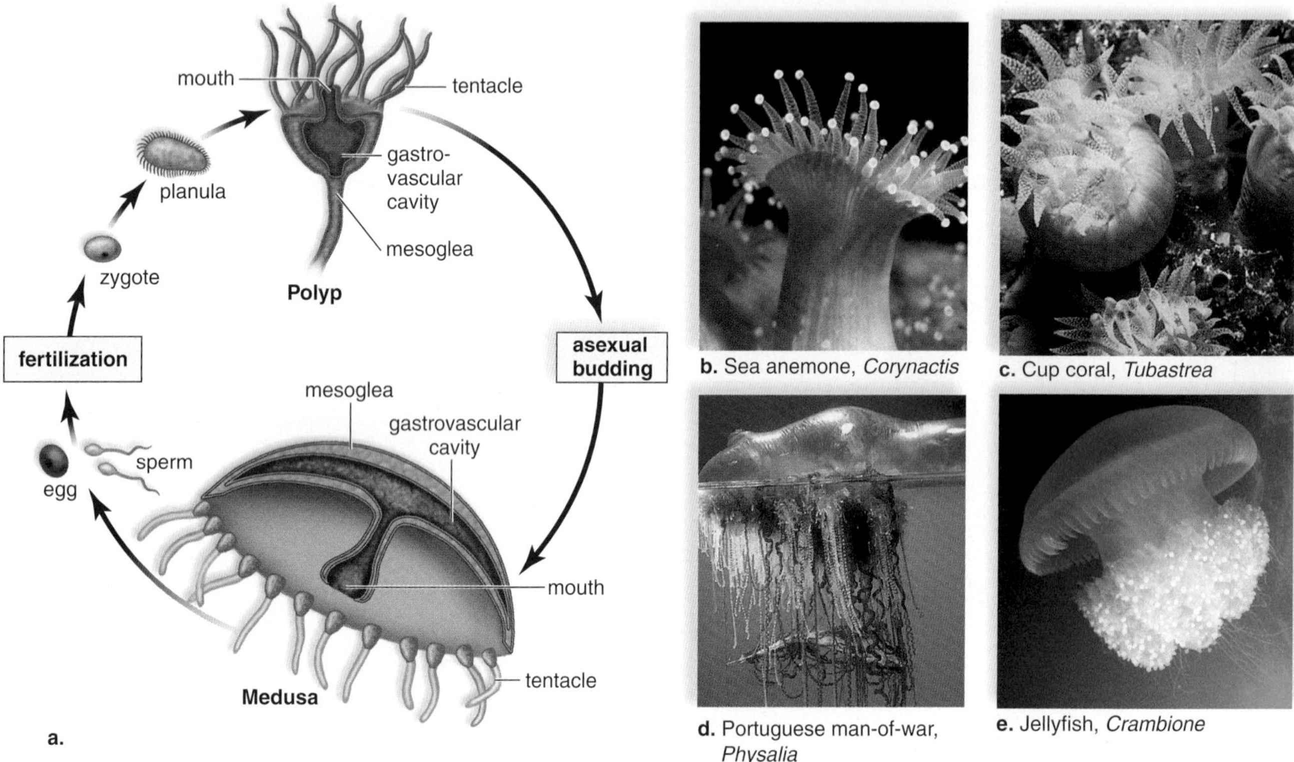

a.

b. Sea anemone, *Corynactis*

c. Cup coral, *Tubastrea*

d. Portuguese man-of-war, *Physalia*

e. Jellyfish, *Crambione*

Observation: Diversity of Cnidarians

Examine preserved, representative cnidarians, and with the help of Figure 28.4, complete Table 28.3.

Table 28.3 Cnidarian Diversity

Common Name of Specimen	Form (Polyp and/or Medusa)
1	
2	
3	
4	
5	

Anatomy of Cnidarians

Cnidarians have the following anatomical features.

1. **Tissue level of organization:** Animals can have a total of three developmental or germ layers (see Table 28.1); cnidarians have two. The germ layers have become an outer epidermis and an inner gastrodermis, cell layers separated by the jellylike mesoglea.
2. **Radial symmetry:** Any longitudinal cut gives two equal halves. Tentacles for capturing prey ring the mouth. Radial symmetry allows a relatively sessile animal to reach out in all directions to seek food.
3. **Sac body plan:** The mouth serves as both an incurrent and an excurrent opening.
4. **Gastrovascular cavity:** This plays an important role in digestion. Inner cells depend on the water within the cavity for gas exchange and excretion. The fluid also acts as a hydrostatic skeleton, offering resistance to contractile fibers.
5. **Nerve net:** Nerve cells interconnect to form a nerve net below the epidermis (Fig. 28.5).
6. **Stinging cells (cnidocytes):** These contain capsules **(nematocysts)** with a barbed or poisonous thread. When discharged, they help the animal capture prey or defend itself (Fig. 28.5).

Observation: Anatomy of Cnidarians

Hydra: *Living Specimen*

1. Examine a live specimen of *Hydra* with a binocular dissecting microscope or hand lens. Refer to Figure 28.5 for help in identifying the foot, tentacles, and mouth.
2. Touch one of the tentacles very gently with a dissecting needle.

   What happens? ____________________
3. Mount a living *Hydra* on a glass slide with a coverslip, and examine a tentacle. Note the stinging cells that appear as swellings on the tentacles.
4. Tap the coverslip, or add a drop of vinegar (5% acetic acid), if available, and note what happens to the cnidocytes (stinging cells). __________

   ____
5. Describe the structure and function of the cnidocytes and nematocysts.

   ____

   ____
6. With the aid of a hand lens, examine preserved specimens of *Hydra.* Note that some of these contain outgrowths or swellings along the trunk. *Hydra* reproduces both asexually and sexually. During asexual reproduction, buds form that develop directly into small hydras. During sexual reproduction, *Hydra* develops testes, which produce sperm, and ovaries, which produce eggs. The testes are generally located near the attachment of the tentacles; the ovaries appear farther down on the trunk, near the foot.

Figure 28.5 Anatomy of *Hydra.*
Hydra typifies the anatomy of a cnidarian as described on page 399.

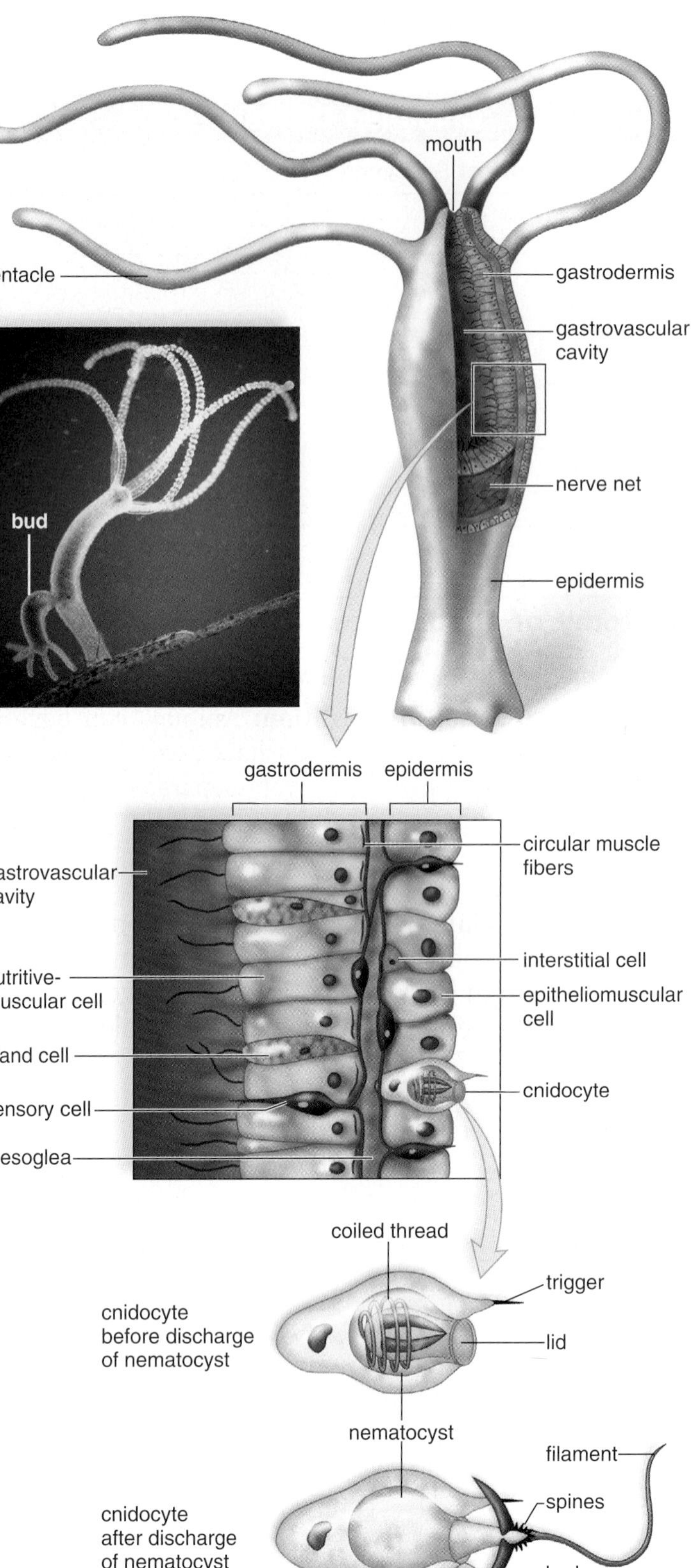

Hydra: *Prepared Slide*

1. Examine prepared slides of cross and longitudinal sections of *Hydra*. With the help of Figure 28.5, note the epidermis, mesoglea, gastrodermis, and gastrovascular cavity. Describe them. ______

______

2. Examine the epidermis and gastrodermis with high power.
   Do you find any cells? ______
   Describe them. ______

______

Obelia: *Prepared Slides*

Obelia's life cycle includes both a polyp phase and a medusa phase (Fig. 28.6).

1. Study a prepared slide showing a polyp colony of *Obelia*. The colony grows by budding; full-grown buds usually remain attached to the colony. Note the bell-shaped **feeding polyps** with a mouth and a ring of stinging cell-bearing tentacles. The other polyps, called **reproductive polyps,** develop small, bell-shaped medusae in their interior that, when mature, escape from the polyp and assume a free-swimming existence as sexually active individuals. Given that the polyp stage of *Obelia* is sessile, what function does the free-swimming medusa serve for the species? ______

______

2. Examine a prepared slide showing a medusa of *Obelia*. Note the tentacled bell and the central mouth. (Preparation of the slide unavoidably distorts the medusa—the umbrella is turned inside out.)

______

Figure 28.6 ***Obelia.***
An *Obelia* colony contains feeding polyps and reproductive polyps. The reproductive polyps asexually produce medusae, which carry out the sexual part of the life cycle.

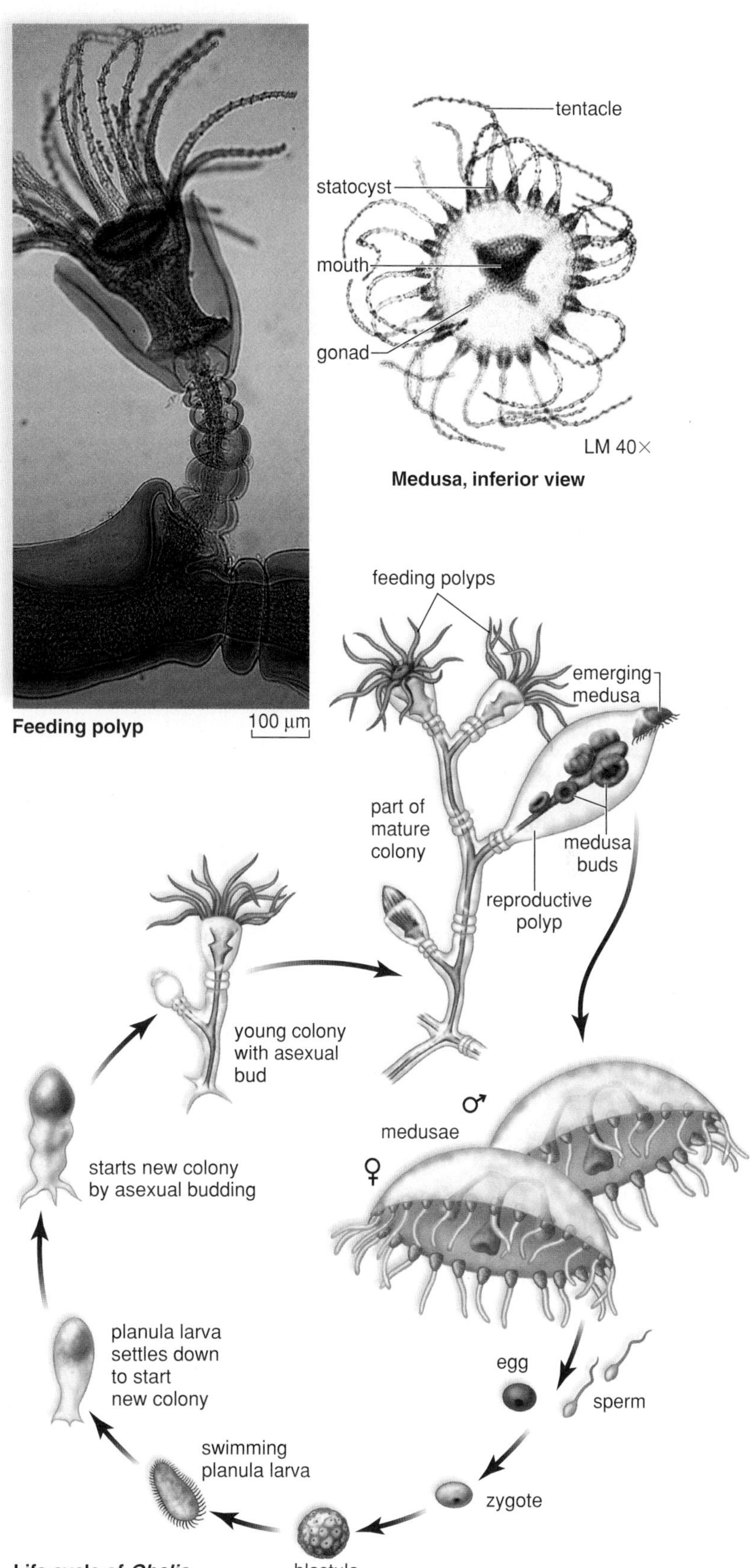

28.4 Phylum Platyhelminthes (Flatworms)

Flatworms (phylum Platyhelminthes) include three classes. Class Turbellaria contains planarians—black, gray, or brightly colored worms that are literally flat. Planarians live in fresh water or seawater, where they feed on protozoans, small crustaceans, snails, or other worms. Class Trematoda contains the flukes, which are internal parasites, and class Cestoda are the tapeworms, which live in digestive tracts. This Laboratory focuses on planarians and tapeworms.

Observation: Diversity of Flatworms

Examine the preserved flatworm specimens on display in the laboratory, and then complete Table 28.4.

Table 28.4 Flatworms

Sample	Common Name of Specimen	Description	Class
1			
2			
3			
4			
5			

Anatomy of Planarians

Planarians have the following anatomical features.

1. **Organ system level of organization:** Planarians have organs for digestion, excretion, reproduction, and nerve conduction.
2. **Acoelomates:** Planarians do not have a coelom.
3. **Bilateral symmetry:** Only one longitudinal cut produces two equal halves of the body. Bilateral symmetry is seen in active animals with a definite head and a posterior end.
4. **Sac body plan:** The mouth serves as both a mouth and an anus.
5. **Three germ layers:** The middle layer—the mesoderm—gives rise to parenchyma and muscles. Parenchyma completely fills the space between the epidermis and the intestine. There is no coelom.
6. **Nervous organization:** The brain and nerves are connected by longitudinal and lateral branches.
7. **Cephalization:** Planarians have a definite head with sense organs.

Observation: Anatomy of Planarians

Living Specimen

1. Examine a living specimen of a planarian (Fig. 28.7*a*) in water in a watch glass. Note the definite head region (cephalization). In Figure 28.7, locate the following structures.
   a. **Eyespots:** Areas sensitive to light that do not form images.
   b. **Auricle:** A lateral projection that contains nerve endings sensitive to touch and to chemicals.
   c. **Pharynx:** An extension from the midventral surface of the body when the animal is feeding. The mouth is at the free end of the pharynx.
   d. **Gastrovascular cavity:** Digestion is both extracellular and intracellular in a three-branched blind cavity.
2. Place a small piece of meat, such as hamburger or liver, in the watch glass, and carefully watch the planarian's reaction. Describe how a planarian feeds. ______________________________

   ____
3. Prepare a wet mount of a planarian, using a concave depression slide, and cover with a coverslip. Examine with a microscope, and note the planarian's mode of locomotion. Planarians have cilia on the ventral surface, and numerous gland cells secrete a mucous material that assists movement. Describe what you see.

   ____

   ____

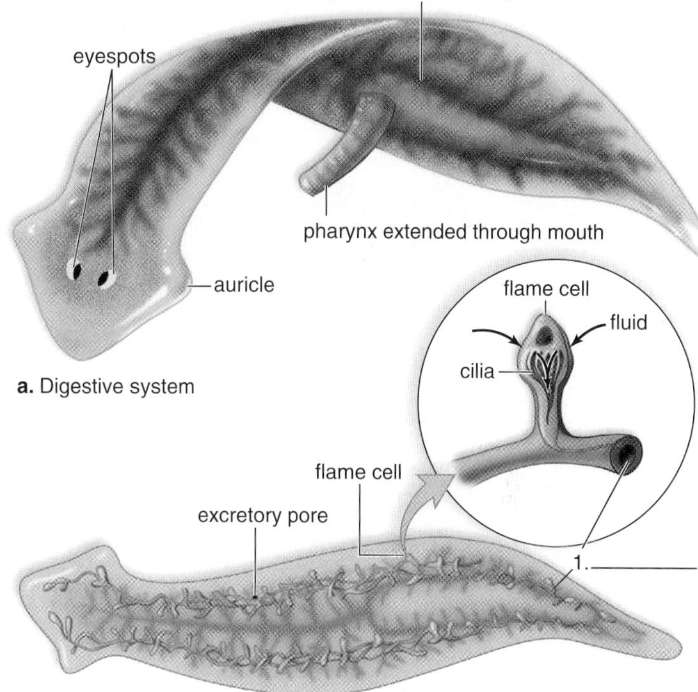

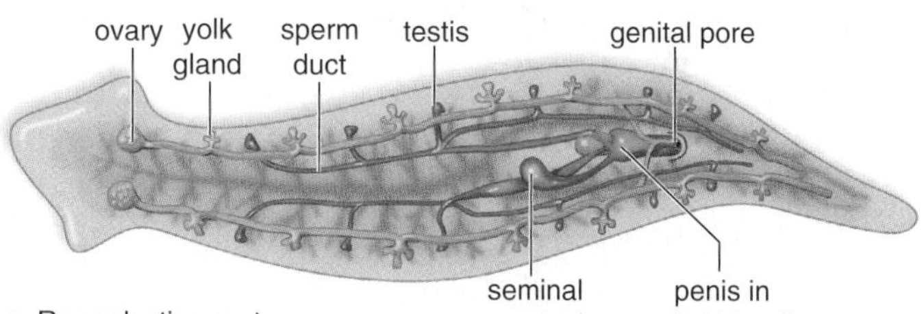

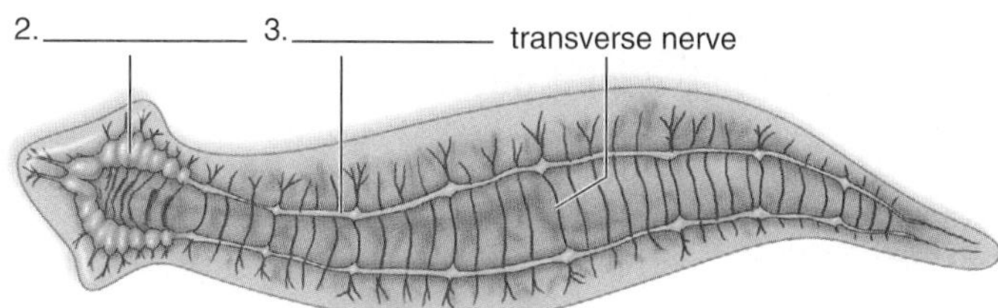

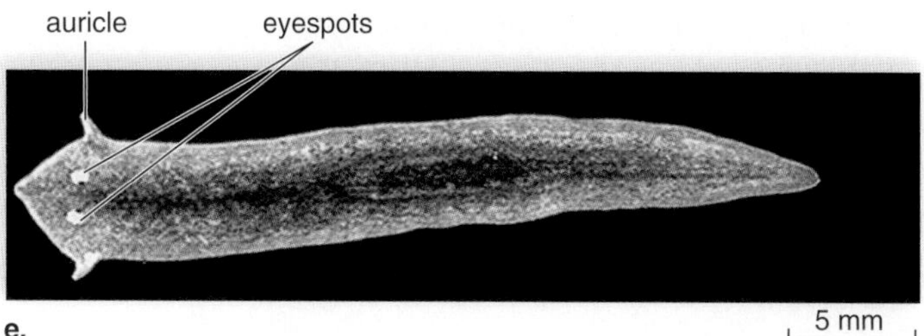

Figure 28.7 Planarian anatomy.
a. When a planarian extends the pharynx, food is sucked up into a gastrovascular cavity that branches throughout the body. **b.** The excretory system with flame cells is shown in detail. **c.** The reproductive system (shown in pink and blue) has both male and female organs. **d.** The nervous system has a ladderlike appearance. **e.** The photograph shows that a flatworm, *Dugesia,* is bilaterally symmetrical and has a head region with eyespots.

Figure 28.8 Planarian cross section.
Cross section through anterior end of a planarian at the region of the pharynx, as it would appear under the microscope.

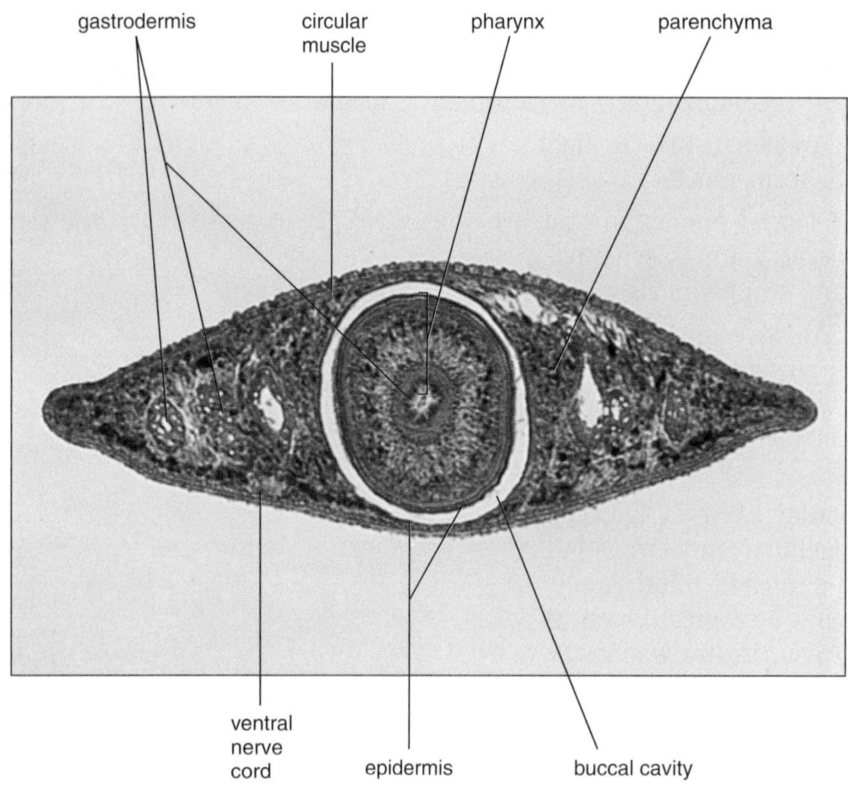

Prepared Slides

1. Examine a whole mount of a planarian that shows the branching gastrovascular cavity. Why is it sometimes called a tripartite organ? (Figure 28.7*a* may be helpful.) ______________________

____
2. Examine a cross section of a planarian under the microscope (Fig. 28.8). Do you see evidence of organs? ______________________
3. In Figure 28.7*b*, label an excretory canal, which ends in a **flame cell.** Flame cells collect water and wastes, which exit by way of excretory pores.
4. Note the reproductive system in Figure 28.7*c*. **Hermaphroditic animals** have both male and female sex organs.
5. Label the nerve cord and the brain in Figure 28.7*d*. Why are planarians said to have a ladderlike nervous system? ______________________

Conclusions

- Planarians, with three tissue layers, are more complex than cnidarians. Contrast a hydra with a planarian by stating in Table 28.5 any significant differences between them.
- Planarians have no respiratory or circulatory system. As with cnidarians, each individual ________ takes care of its own needs for these two life functions.

Table 28.5 Contrasts Between a Hydra and a Planarian

	Digestive System	Excretory System	Nervous Organization
Hydra			
Planarian			

Tapeworms

Tapeworms are parasitic flatworms in the class Cestoda. They live in the intestines of vertebrate animals, including humans (Fig. 28.9). The worms consist of a **scolex** (head), usually with suckers and hooks, and **proglottids** (segments of the body) (Fig. 28.10). Ripe proglottids detach and pass out with the host's feces, scattering fertilized eggs on the ground. If pigs or cattle happen to ingest these, larvae develop and eventually become encysted in muscle, which humans may then eat in poorly cooked or raw meat. Upon digestion, a bladder worm that escapes from the cyst develops into a new tapeworm that attaches to the intestinal wall.

1. How do humans get infected with the pig tapeworm? ______________________________

____

2. What is the function of a tapeworm's hooks (if present) and suckers? ______________________________

____

3. Proglottids mature into "bags of eggs." Given the life cycle of the tapeworm, why might a tapeworm produce so many eggs? ______________________________

____

Figure 28.9 Life cycle of the tapeworm *Taenia.*
The pig host is the means by which the worm is dispersed to the human host.

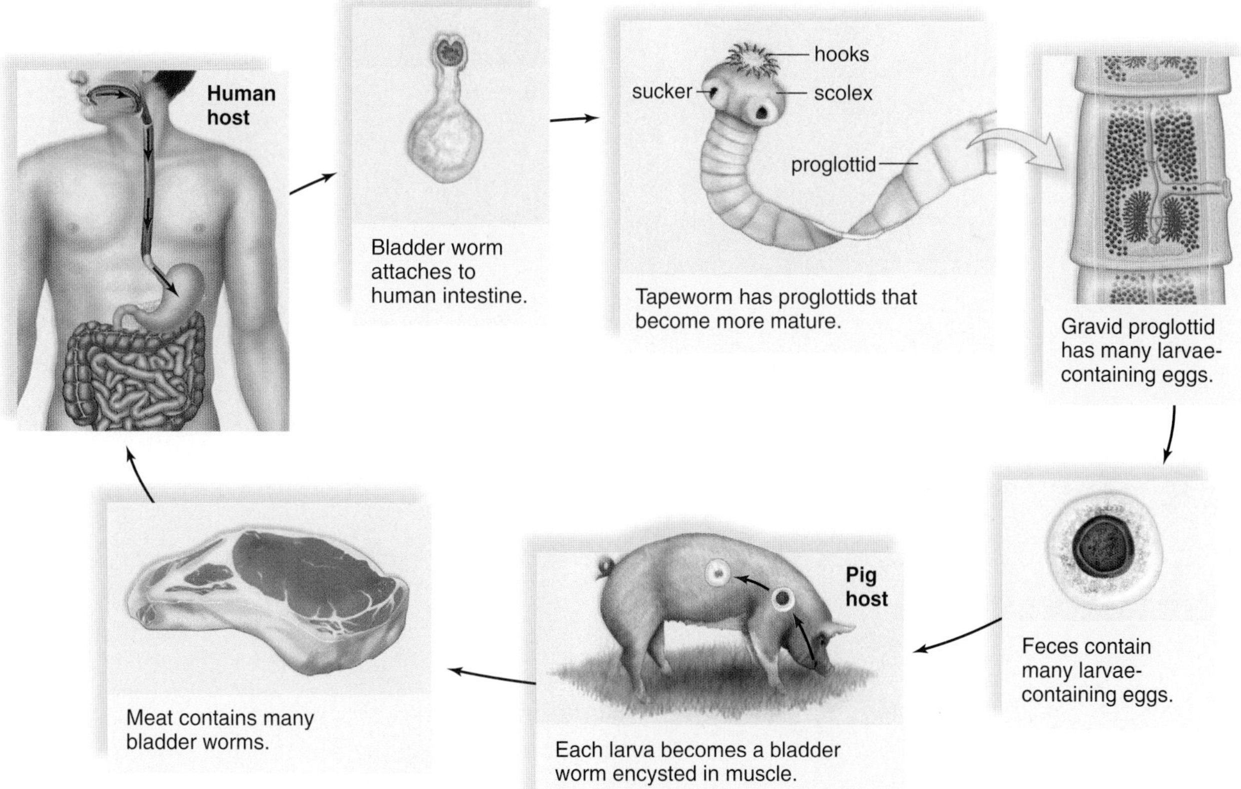

Human host
Bladder worm attaches to human intestine.
hooks
sucker
scolex
proglottid
Tapeworm has proglottids that become more mature.
Gravid proglottid has many larvae-containing eggs.
Feces contain many larvae-containing eggs.
Pig host
Each larva becomes a bladder worm encysted in muscle.
Meat contains many bladder worms.

Observation: Tapeworm Anatomy

1. Examine a preserved specimen and/or slide of *Taenia pisiformis,* a tapeworm.
2. With the help of Figure 28.10, identify the scolex, with hooks and suckers, and the proglottids.

Figure 28.10 Anatomy of *Taenia.*
The adult worm is modified for its parasitic way of life. It consists of a scolex and many proglottids, which become bags of eggs. (*a*: Magnification ×+5; *b*: Magnification ×7)

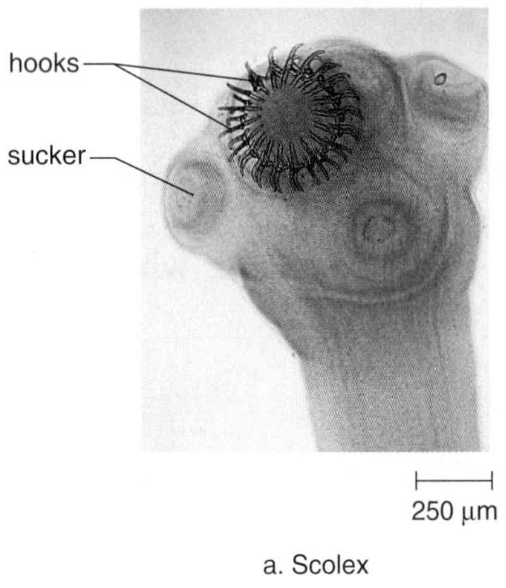

a. Scolex

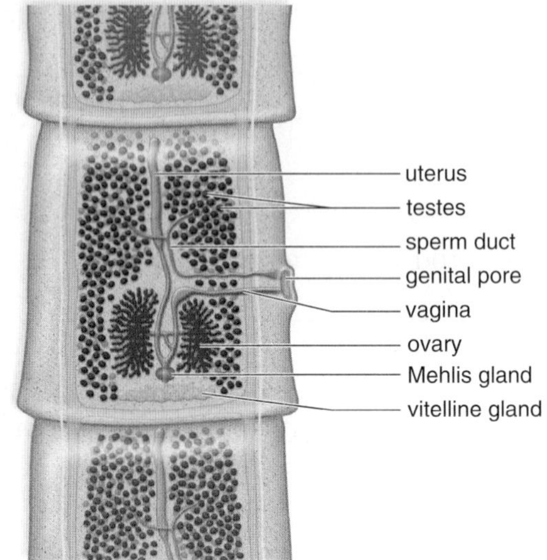

b. Proglottid

28.5 Phylum Nematoda (Roundworms)

Roundworms are found in all aquatic habitats and in damp soil. Some even survive in hot springs, deserts, and cider vinegar. They parasitize (take nourishment from) both plants and animals. They are significant crop pests and cause disease in humans. Both pinworms and hookworms are roundworms that cause intestinal difficulties; trichinosis and elephantiasis are also caused by roundworms. *Ascaris*, a large, primarily tropical intestinal parasite, is often studied as an example of this phylum (Fig. 28.11).

Figure 28.11 Roundworm anatomy.
a. Photograph of male *Ascaris*. **b.** Male reproductive system. **c.** Photograph of female *Ascaris*. **d.** Female reproductive system.

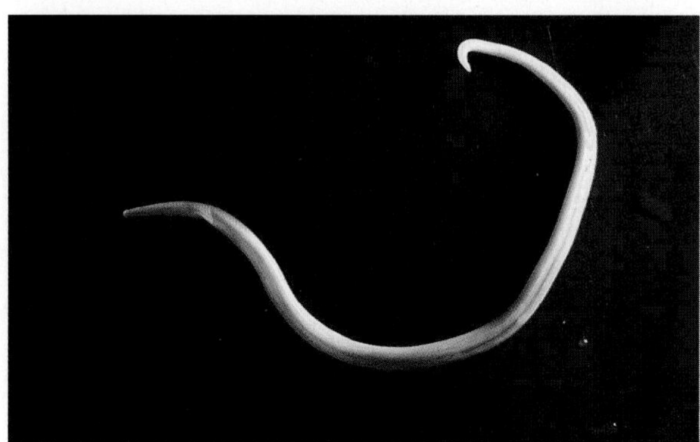

a. Male *Ascaris*

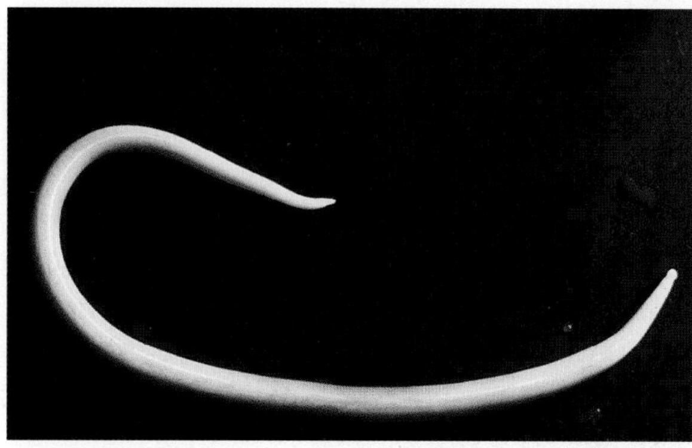

c. Female *Ascaris*

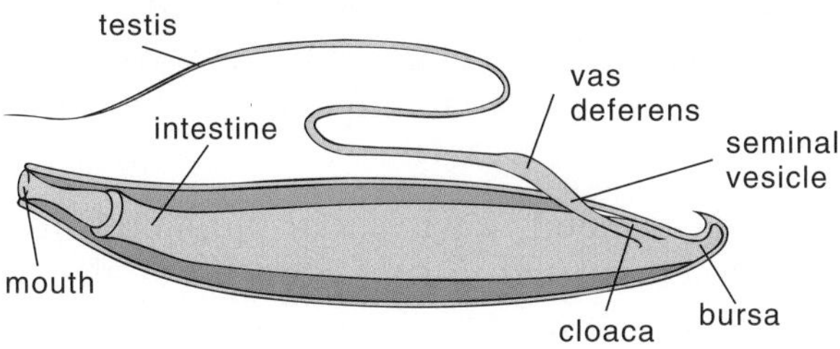

b. Male reproductive system

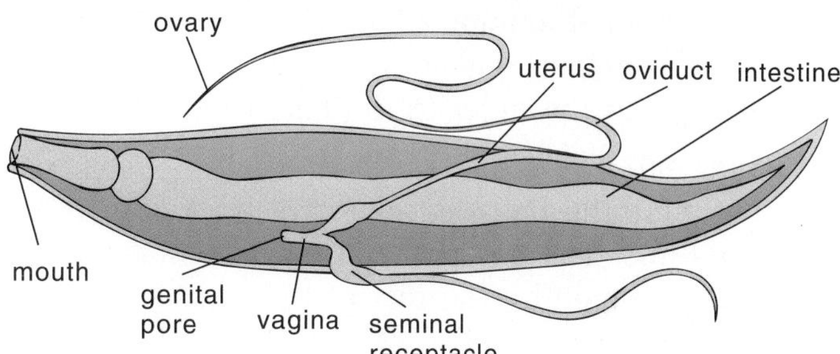

d. Female reproductive system

Observation: Diversity of Roundworms

Examine preserved, representative roundworms, and then complete Table 28.6.

Table 28.6 Roundworm Diversity

Common Name of Specimen	Description (Length, Thickness, Other)
1	
2	
3	
4	
5	

Anatomy of Roundworms

Like planarians, roundworms have an organ level of organization, three germ layers, and bilateral symmetry. In addition, roundworms have the following features.

1. **Tube-within-a-tube body plan:** The digestive tract has both a mouth and an anus.
2. **Pseudocoelom:** This body cavity, which allows space for the organs, is incompletely lined with mesoderm.
3. **Nervous organization:** Roundworms have a brain plus dorsal, ventral, and lateral nerves.
4. **Nonsegmented:** The body wall is smooth and not divided into segments.

Observation: Anatomy of Roundworms

Ascaris: *Dissection of Preserved Specimen*

1. Examine preserved specimens of *Ascaris,* both male (Fig. 28.11*a* and *b*) and female (Fig. 28.11*c* and *d*). Note the body shape and the smooth, tough cuticle that covers it. Find the mouth at the anterior end and the anus near the tip of the posterior end on the ventral surface. In roundworms, the sexes are separate. The male is smaller and has a curved posterior end. Be sure to examine specimens of each sex.
2. Place a specimen of *Ascaris* in a dissecting pan, pinning the anterior and posterior ends to the wax so that the dorsal surface is exposed.
3. Carefully slit open the worm longitudinally, cutting along the middorsal line.
4. Pin the body open to expose the internal organs.
5. Add a small amount of water to the pan to prevent drying out. Note the body cavity, or pseudocoelom, which is now evident. Note also the inner tube (consisting of the digestive tract) and the outer tube (made up of the muscular body wall). Does this explain the phrase *tube-within-a-tube body plan*? ______________________________
6. Examine the flattened digestive tract, which extends within the body cavity from the mouth to the anus. The anterior end forms a muscular pharynx, which is used to suck food materials into the mouth. The rest of the tract is a tube that passes to the anus.
7. Note the reproductive system. In the male, the reproductive structures are the **testis,** the **vas deferens,** and the **seminal vesicle** (Fig. 28.11*b*). These structures increase in size, from testes to vasa deferentia to seminal vesicles. In the female, the reproductive structures are the **ovary,** the **oviduct,** and the **uterus.** These structures increase in size, from ovaries to oviducts to uterus (Fig. 28.11*d*).

Vinegar Eels: Living Specimens

Vinegar eels are tiny, free-living nematodes that can live in unpasteurized vinegar.

1. Examine live vinegar eels, and observe their active, whiplike swimming movements. This thrashing motion may be a result of nematodes having longitudinal muscles only; they lack circular muscles.
2. Select a few larger vinegar eels for further study, and place them in a small drop of vinegar on a clean microscope slide. If the eels are too active for study, you can slow them by briefly warming them or by adding methyl cellulose.
3. Try to observe the tubular digestive tract, which begins with the mouth and ends with the anus. Also, you may be able to see some of the reproductive organs, particularly in a large female vinegar eel.

Figure 28.12 Larva of the roundworm *Trichinella* embedded in a muscle.
A larva coils in a spiral and is surrounded by a sheath derived from a muscle fiber.

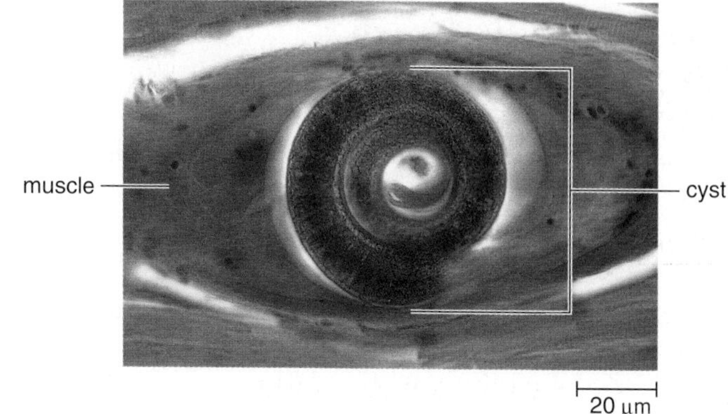

Trichinella

Trichinella is a parasitic roundworm that causes the disease **trichinosis.** When humans eat raw or undercooked pork infected with *Trichinella* cysts, the juvenile worms are released in the digestive tract and mature to reproduce sexually, producing other juvenile worms that form cysts in human muscle (Fig. 28.12). The juvenile worms have an extreme effect on muscle fibers. Muscle cell nuclei shrink; gene expression is halted and muscle proteins are inactive. The body responds by a massive production of eosinophils. In the end, patients often develop neurologic and cardiac abnormalities.

Observation: Trichinella

1. Examine preserved, infected muscle or a slide of infected muscle, and locate the *Trichinella* cysts, which contain the juvenile worms.
2. How can trichinosis be prevented? ______________________________

____

Conclusion

- Contrast roundworms (phylum Nematoda) with the other animals studied in the Laboratory by filling out the following table.

	Porifera	Cnidaria	Platyhelminthes	Nematoda
Example:				
Level of organization				
Germ layers				
Symmetry				
Body plan				
Nervous organization				
Coelom				
Segmentation				

Laboratory Review 28

1. Animals without an endoskeleton of bone or cartilage are called what?
2. Sponges are members of what phylum?
3. What does an acoelomate animal lack?
4. What type of gametes does a hermaphroditic animal produce?
5. What do the cnidocytes (stinging cells) of cnidarians contain?
6. A nerve net is characteristic of what group of animals?
7. Which of the animal phyla was the first to evolve three germ layers?
8. Which of the animal phyla studied today contains animals with radial symmetry?
9. What type of cell lines the interior cavity of a sponge?
10. On what basis are sponges classified?
11. The cnidarian life cycle often includes two phases. One phase is absent in *Hydra.* Which phase is present?
12. *Hydra* and *Obelia* use which two anatomical structures to obtain food?
13. What two characteristics of a planarian may be associated with cephalization?
14. What type of excretory system do planarians have?
15. What type of coelom does *Ascaris* have?
16. Upon examining a roundworm, how would you know it has a tube-within-a-tube body plan?
17. What type of meat must be cooked thoroughly to prevent trichinosis?
18. Which of the animal phyla studied today contain colonial organisms?
19. Which of the animal phyla studied today contain three germ layers?

Thought Questions

20. Explain the difference between radial and bilateral symmetry, and associate these with the lifestyle of one of the animals studied.

21. Relate the number of germ layers to the level of complexity.

22. Why are planarians considered more complex than cnidarians?

LABORATORY

29

Molluscs, Annelids, and Arthropods

Learning Outcomes

29.1 Phylum Mollusca (Molluscs)

- Describe the general characteristics of molluscs and the specific features of selected classes.
- Using preserved specimens, or images, identify and locate external and internal structures of a clam and a squid.
- Contrast the anatomy of the clam and the squid to show how each is adapted to its way of life.

29.2 Phylum Annelida (Annelids)

- Describe the general characteristics of annelids and the specific features of the three major classes.
- Contrast the features of these classes, indicating how the animals in each class are adapted to their way of life.
- Using preserved specimens, images, or models, identify and locate the anatomical structures of an earthworm.

29.3 Phylum Arthropoda (Arthropods)

- Describe the general characteristics of arthropods and the specific features of the groups studied.
- Using preserved specimens, images, or models, identify and locate external and internal structures of the crayfish and the grasshopper.
- Contrast the anatomy of the crayfish and the grasshopper, indicating how each is adapted to its way of life.

Introduction

Animals are often classified by their body type. The **acoelomates** have no body cavity, the **pseudocoelomates** have a body cavity incompletely lined with mesoderm, and the **coelomates** have a body cavity completely lined with mesoderm. A coelom offers many advantages:

- The digestive system and body wall can move independently.
- Internal organs can become more complex.
- Coelomic fluid can assist respiration, circulation, and excretion.
- The coelom also serves as a hydrostatic skeleton, allowing the muscles to work against a fluid-filled cavity, especially in those coelomates lacking rigid internal or external skeletal structures.

The coelomate phyla are divided into two groups. The molluscs, annelids, and arthropods are in one group, and the echinoderms and chordates are in the other group. The animals in the first group are the *protostomes* because the first (*protos*) embryonic opening becomes the mouth (*stoma*). The second group are the *deuterostomes* because the first opening becomes related to the anus, and the second (*deutero*) opening becomes the mouth. The deuterostomes are the topic of Laboratory 30.

29.1 Phylum Mollusca (Molluscs)

Most **molluscs** are marine, but there are also some freshwater and terrestrial molluscs (Fig. 29.1). All molluscs have a three-part body consisting of a ventral, muscular **foot** that is specialized for various means of locomotion; a **visceral mass** that includes the internal organs; and a **mantle,** a thin tissue that encloses the visceral mass and may secrete a shell. Molluscs also have a rasping tongue, or **radula,** which is modified to assist the specific feeding habits of the different forms of molluscs. Molluscs are classified as follows:

CLASSIFICATION: THE MOLLUSCS

PHYLUM MOLLUSCA (MOLLUSCS)
Body divided into a foot, visceral mass, and mantle; usually also a radula and shell; reduced coelom

Class Polyplacophora: chitons

Class Bivalvia: clams, scallops, oysters, and mussels

Class Cephalopoda: squid, nautiluses, and octopuses

Class Gastropoda: snails, slugs, and nudibranchs

Figure 29.1 Molluscan diversity.
a. Class Polyplacophora includes the chitons, which have a flattened foot and a shell that consists of eight articulating valves. **b.** Class Bivalvia includes scallops, which have a two-part shell. **c.** Class Cephalopoda includes nautiluses, which show marked cephalization. **d.** Class Gastropoda includes the snails, which usually have a spiral shell and an elongated body with cephalization.

a. Class Polyplacophora: chitons, *Tonicella* ______________________

b. Class Bivalvia: scallip, *Pecten*

____

c. Class Cephalopoda: nautilus, *Nautilus*

____

d. Class Gastropoda: snail, *Allogona*

____

Observation: Diversity of Molluscs

Most molluscs belong to one of four classes. Class Polyplacophora contains grazing marine herbivores, such as chitons, with a body flattened dorsoventrally covered by a shell consisting of eight plates (Fig 29.1*a*). Class Bivalvia contains marine and freshwater sessile filter feeders, such as clams, with a body enclosed by a shell consisting of two valves (Fig. 29.1*b*). These animals have a hatchet-shaped foot but no head or radula. Class Cephalopoda contains marine active predators, such as squids. The shell may be reduced or absent; tentacles are about the head, which is in line with an elongated visceral mass (Fig. 29.1*c*). The circulatory system is closed, and the well-developed nervous system is accompanied by cephalization. Class Gastropoda contains marine, freshwater, and often terrestrial herbivores, such as snails, in which the shell is coiled, if present, and the body symmetry is distorted by torsion (Fig. 29.1*d*). The head has tentacles and, like the cephalopods and polyplacophorans, the gastropods have a radula that assists their feeding.

1. Examine mollusc specimens, and complete the first two columns of Table 29.1.
2. Examine the foot, and complete the third column of Table 29.1. Some molluscs have a broad, flat foot, others a hatchet-shaped foot; in still others the foot has become tentacles and arms that assist in the capture of food.
3. Indicate in the last column if cephalization is present or not present.

Table 29.1 Molluscan Diversity

Common Name of Specimen	Class	Description of Foot	Cephalization (Yes or No)

Caution: This laboratory exercise includes several dissections. Dissections involve the use of sharp instruments and students are urged to exercise care when handling these instruments. Thorough washing of hands is advised after the dissections are complete.

Anatomy of Clam

Clams are bivalved because they have right and left shells secreted by the mantle. Clams have no head, and they burrow in sand by extending a "hatchet" foot between the valves. Clams are filter feeders and feed on debris that enters the mantle cavity. Clams have an open circulatory system; the blood leaves the heart and enters sinuses (cavities) by way of anterior and posterior aortas. There are many different types of clams. The one examined here is the freshwater clam *Venus*.

Observation: Anatomy of Clam

External Anatomy

1. Examine the external shell (Fig. 29.2) of a preserved clam (*Venus*). The shell is an **exoskeleton.**
2. Find the posterior and anterior ends. The more pointed end of the **valves** (the halves of the shell) is the posterior end.
3. Determine the clam's dorsal and ventral regions. The valves are hinged together dorsally. The raised **umbo** on the dorsal anterior region of the shell represents the oldest portion of the shell.
4. What is the function of a heavy shell? ______________________________

Figure 29.2 External view of clam shell.

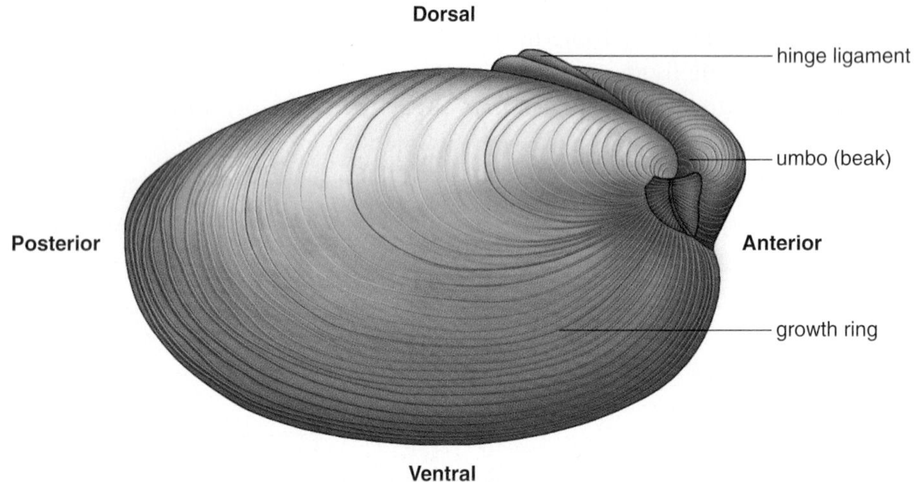

Internal Anatomy

1. Place the clam in the dissecting pan, with the **hinge ligament** and **umbo** (blunt dorsal protrusion) down. Carefully separate the **mantle** from the right valve by inserting a scalpel into the slight opening of the valves. What is a mantle? ______________________________

   ____
2. Insert the scalpel between the mantle and the valve you just loosened.
3. The **adductor muscles** hold the valves together. Cut the adductor muscles at the anterior and posterior ends by pressing the scalpel toward the dissecting pan. After these muscles are cut, the valve can be carefully lifted away. What is the advantage of powerful adductor muscles? ______

   ____
4. Examine the inside of the valve you removed. Note the concentric lines of growth on the outside, the hinge teeth that interlock with the other valve, the adductor muscle scars, and the mantle line. The inner layer of the shell is mother-of-pearl.
5. Examine the rest of the clam (Fig. 29.3) attached to the other valve. Notice the mantle, which lies over the visceral mass and foot, and also the adductor muscles.

Figure 29.3 Anatomy of a bivalve.
The mantle has been removed in the drawing **(a)** but is still visible in the photo of the dissected specimen **(b)**.

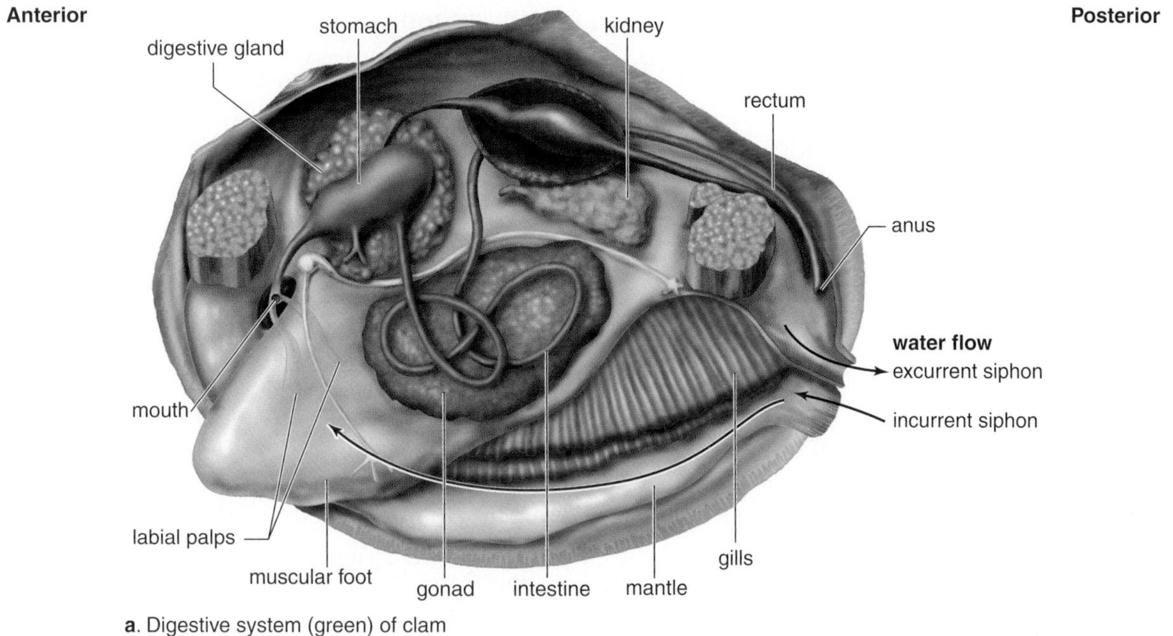

a. Digestive system (green) of clam

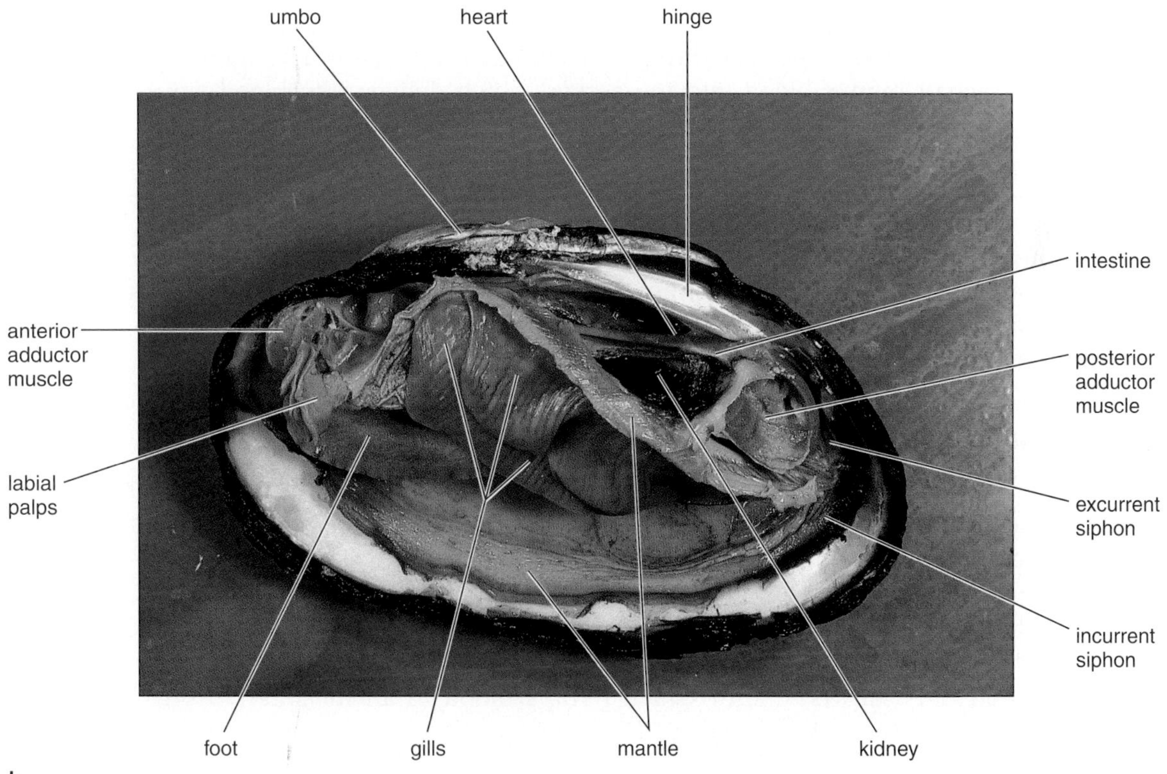

b.

6. Bring the two halves of the mantle together. Explain the term *mantle cavity.* ______

______

7. Identify the **incurrent** (more ventral) and **excurrent siphons** at the posterior end (Fig. 29.3). Explain how water enters and exits the mantle cavity. ______

______

8. Cut away the free-hanging portion of the mantle to expose the **gills.** Does the clam have a respiratory organ? ______

   What type of respiratory organ? ______
9. A mucous layer on the gills entraps food particles brought into the mantle cavity, and the cilia on the gills convey these food particles to the mouth. Why is the clam called a filter feeder?

______

______

10. The nervous system is composed of three pairs of ganglia (located anteriorly, posteriorly, and in the foot), which are all connected by nerves. The clam does not have a brain. A ganglion contains a limited number of neurons, whereas a brain is a large collection of neurons in a definite head region.
11. Identify the **foot,** a tough, muscular organ for locomotion, and the **visceral mass,** which lies above the foot and is soft and plump. The visceral mass contains the digestive and reproductive organs.
12. Identify the **labial palps** that channel food into the open mouth.
13. Identify the **anus,** which discharges into the excurrent siphon.
14. Find the **intestine** by its dark contents. Trace the intestine forward until it passes into a sac, the clam's only evidence of a coelom.
15. Locate the **pericardial sac (pericardium)** that contains the heart. The intestine passes through the heart. The heart pumps blood into the aortas, which deliver it to blood sinuses in the tissues.

    A clam has an **open circulatory system.** Explain. ______

______

16. Cut the visceral mass and the foot into exact left and right halves, and examine the cut surfaces. Identify the digestive glands, which are greenish-brown; the stomach, which is embedded in the digestive glands; and the intestine, which winds about in the visceral mass. Reproductive organs are also present.
17. How do the foot and adductor muscles work together to provide locomotion?

Anatomy of Squid

Squids are cephalopods because they have a well-defined head. The head contains a brain and bears sense organs. The squid moves quickly by jet propulsion of water, which enters the mantle cavity by way of a space that circles the head. When the cavity is closed off, water exits by means of a funnel. Then the squid moves rapidly in the opposite direction.

The squid seizes fish with its tentacles; the mouth has a pair of powerful, beaklike jaws and a **radula,** a beltlike organ containing rows of teeth. The squid has a closed circulatory system composed of vessels and three hearts, one of which pumps blood to all the internal organs, while the other two pump blood to the gills located in the mantle cavity.

Observation: Anatomy of Squid

1. Examine a preserved squid.
2. Refer to Figure 29.4 for help in identifying the mouth (defined by beaklike jaws and containing a radula, used to mechanically grind the food torn apart by the jaws), and the tentacles and arms, which encircle the mouth.
3. Locate the head with its sense organs, notably the large, well-developed eye.
4. Find the funnel, where water exits from the mantle cavity, causing the squid to move backward.
5. If the squid has been dissected, note the heart, gills, and blood vessels.

Figure 29.4 Anatomy of a squid.
The squid is an active predator and lacks the external shell of a clam. It captures fish with its tentacles and bites off pieces with its jaws. A strong contraction of the mantle forces water out the funnel, resulting in "jet propulsion."
a. Drawing. **b.** Dissected specimen.

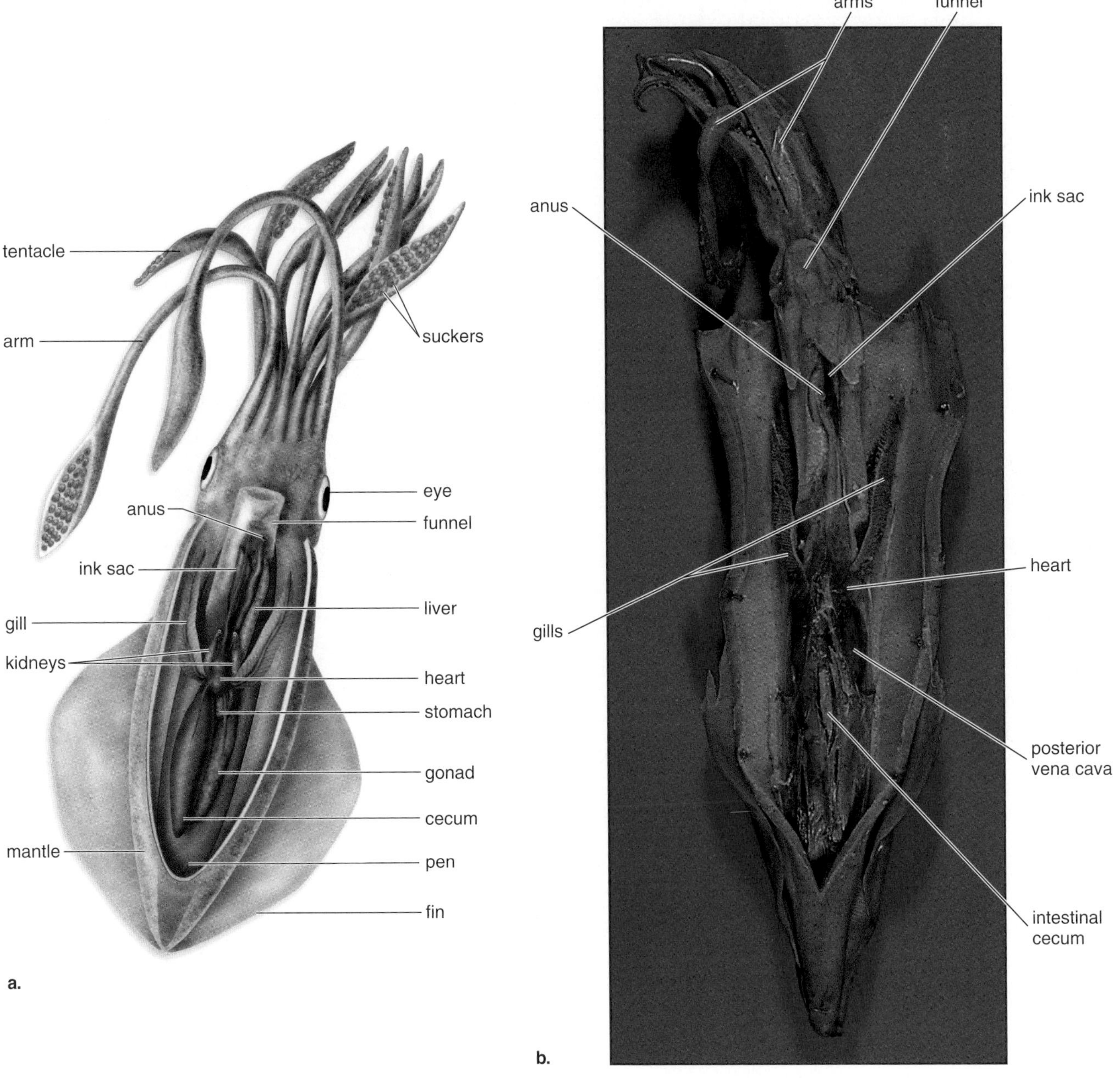

Clam Anatomy Compared with Squid Anatomy

1. Compare clam anatomy with squid anatomy by completing Table 29.2.
2. Explain how both clams and squids are adapted to their way of life.

____

____

____

Table 29.2 Comparison of Clam and Squid

	Clam	Squid
Feeding mode		
Skeleton		
Circulation		
Cephalization		
Locomotion		
Nervous system	Three separate ganglia	

29.2 Phylum Annelida (Annelids)

Annelids are the segmented worms, so called because the body is divided into a number of segments and has a ringed appearance (Fig. 29.5). The circular and longitudinal muscles work against the fluid-filled coelom to produce changes in width and length. Annelids are classified as follows:

CLASSIFICATION: THE ANNELIDS

PHYLUM ANNELIDA (ANNELIDS)
Segmented worms with a long, cylindrical, soft body; protostome coelomates with internal septa; specialized digestive tract; definite central nervous system with a brain and a ventral solid nerve cord; closed circulatory system

Class Polychaeta: clam worms and tube worms

Class Oligochaeta: earthworms

Class Hirudinea: leeches

Figure 29.5 Annelid diversity.
Aside from earthworms in the class Oligochaeta **(a)**, there are **(b)** clam worms and **(c)** fanworms in the class Polychaeta and **(d)** leeches in the class Hirudinea.

a. Class Oligochaeta: earthworms, *Lumbricus*, mating

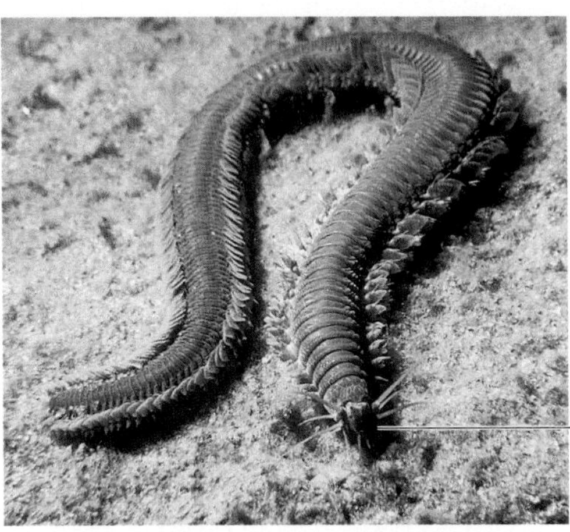

b. Class Polychaeta: clam worm, *Nereis*

c. Class Polychaeta: giant fanworm, *Eudistylia*

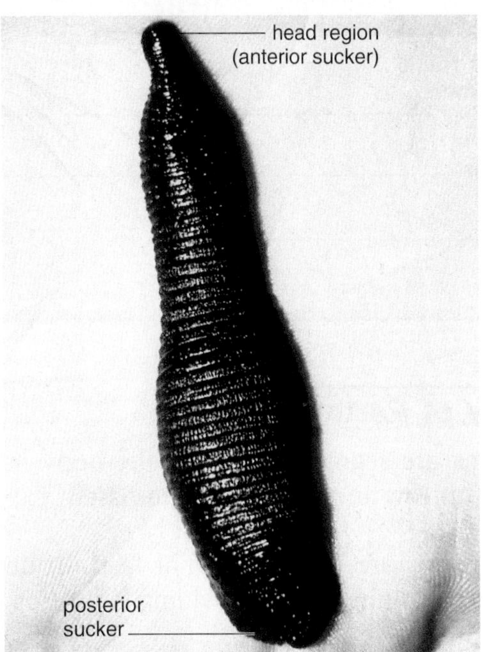

d. Class Hirudinea: leech, *Hirudo*

Observation: Diversity of Annelids

In phylum Annelida, the **polychaetes** (class Polychaeta) have many slender bristles called **setae;** the **earthworms** (class Oligochaeta) have fewer setae; and **leeches** (class Hirudinea) usually have no setae. Polychaetes, most of which are marine, are plentiful from the intertidal zone to the ocean depths. They are quite diverse, ranging from jawed forms that are carnivorous to fanworms that live in tubes and extend feathery filaments when filter feeding. Earthworms have a worldwide distribution in almost any soil, are often found in large numbers, and reach a length of as much as 3 meters. Leeches are known as bloodsuckers; the medicinal leech has been used in the practice of bloodletting. Annelids are classified according to the number of setae present.

1. Examine various specimens of annelids, and then complete the first two columns of Table 29.3.
2. Note that the clam worm, a polychaete, has a well-differentiated anterior region with specialized sense organs and paired, fleshy appendages, called **parapodia,** on most segments (see Fig. 29.5*b*). Complete the third column of Table 29.3.
3. Note in the last column if setae are present.

Table 29.3 Annelid Diversity

Common Name of Specimen	Class	Parapodia Present (Yes or No)	Setae Present (Yes or No)

Anatomy of Earthworm

Earthworms are segmented in that the body has a series of ringlike segments. Earthworms have no head and burrow in the soil by alternately expanding and contracting segments along the length of the body.

Earthworms are scavengers that feed on decaying organic matter in the soil. They have a well-developed coelomic cavity, providing room for a well-developed digestive tract.

Observation: Anatomy of Earthworm

External Anatomy

1. Examine a live or preserved specimen of an earthworm. Locate the small projection that sticks out over the mouth. Has cephalization occurred? ________ Explain. ______________________________

____

2. Count the total number of segments, beginning at the anterior end. The sperm duct openings are on segment 15 (somite XV) (Fig. 29.6). The enlarged section around a short length of the body is the **clitellum.** The clitellum secretes mucus that holds the worms together during mating. It also functions as a cocoon, in which fertilized eggs hatch and young worms develop. The anus is located on the worm's terminal segment.

3. Lightly pass your fingers over the earthworm's ventral and lateral sides. Do you feel the setae? ___ Earthworms insert these slender bristles into the soil. Setae, along with circular and longitudinal muscles, enable the worm to locomote. Explain the action. ______________________________

____

Figure 29.6 External anatomy of an earthworm.

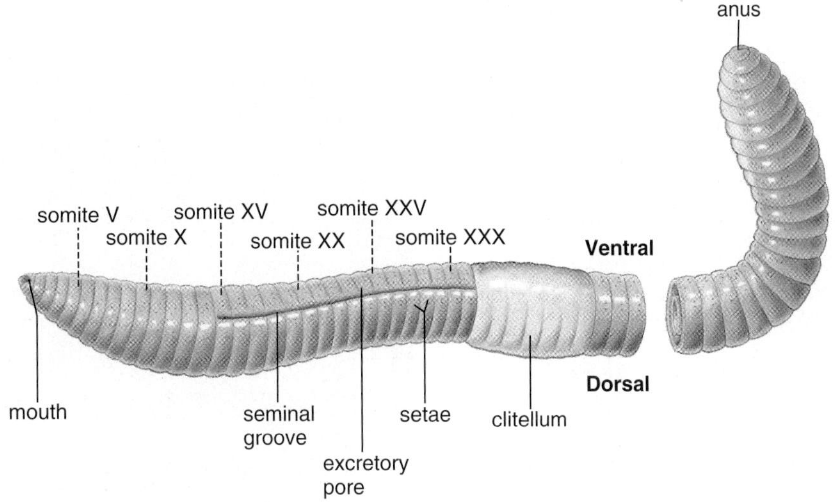

Internal Anatomy

1. Place a preserved earthworm on its ventral side in the dissecting pan. With a scalpel or razor blade, make a shallow incision slightly to the side of the blackish median dorsal blood vessel (Fig. 29.7*a*). Start your incision about ten segments after the clitellum, and proceed anteriorly to the mouth. If you see black ooze, you have accidentally cut the intestine.
2. Identify the thin partitions, the **septa,** between segments.
3. Lay out the body wall, and pin every tenth segment to the wax in your pan. Add water to prevent drying out.

Figure 29.7 Internal anatomy of an earthworm, dorsal view.
a. Drawing shows internal organs and a cross section. **b.** Dissected specimen.

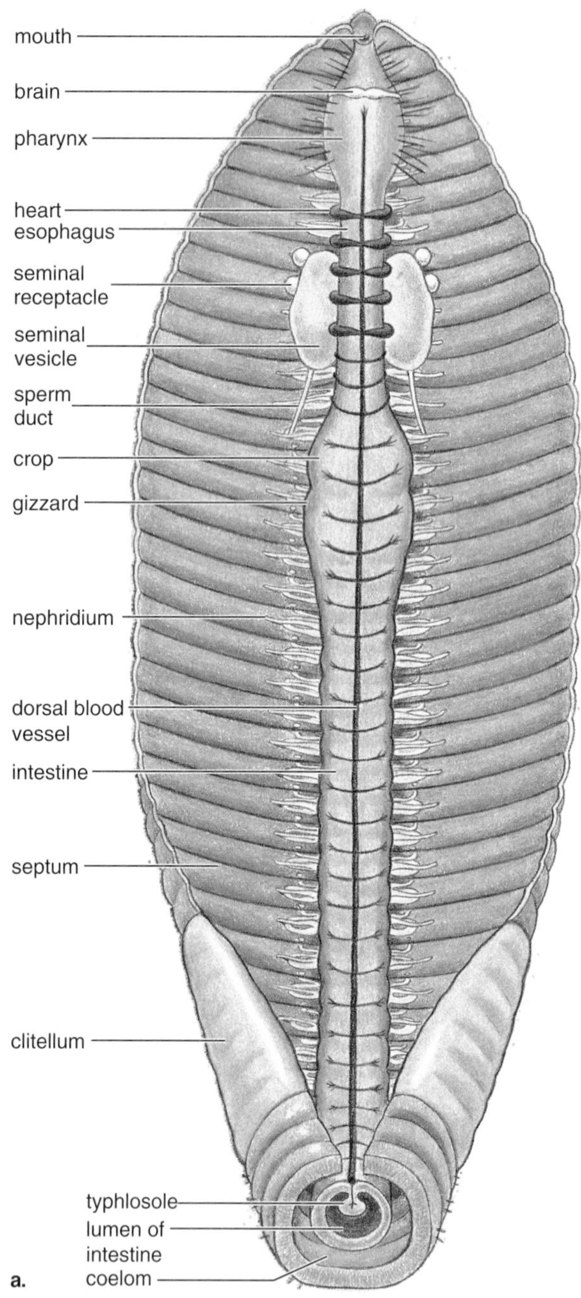

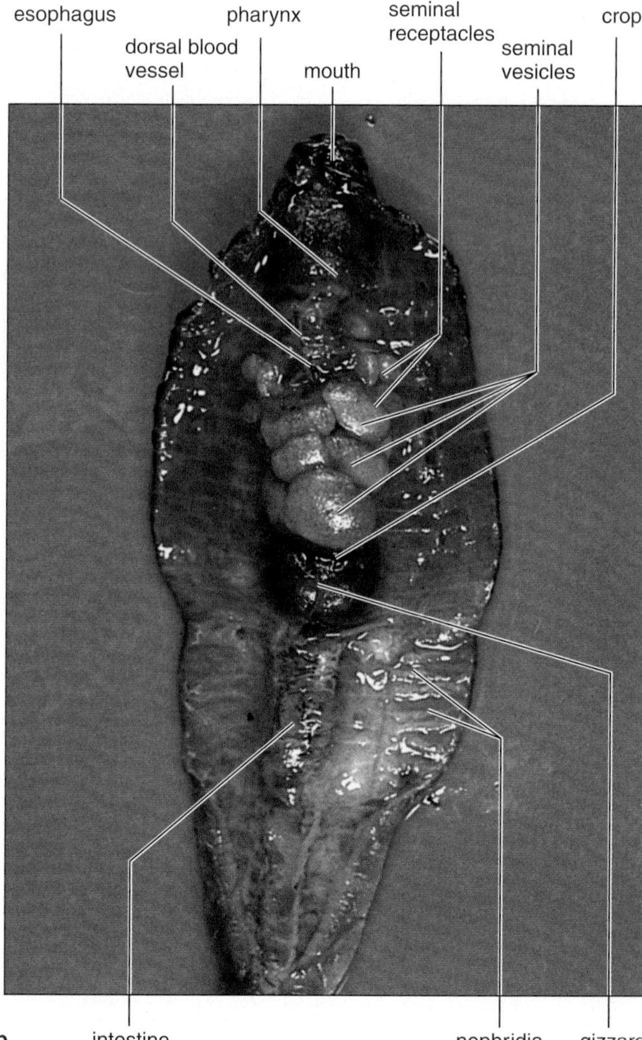

4. Identify the digestive tract, which begins at the mouth and extends through body segments 1, 2, and 3. It opens into the swollen, muscular, thick-walled **pharynx,** which extends from segment 3 to segment 6. The **esophagus** passes from the pharynx through segment 14 to the **crop.** Next is the **gizzard,** which lies in segments 17 through 28. The intestine extends from the gizzard to the anus. Does the digestive system show specialization of parts? ________

   Explain. ____

   ____

   An earthworm feeds on **detritus** (organic matter) in the soil.

5. Identify the earthworm's circulatory system. The blood is always contained within vessels and never runs free. The **dorsal blood vessel** is readily seen along the dorsal side of the digestive tract. A series of "hearts" encircles the esophagus between segments 7 and 11, connecting the dorsal blood vessel with the ventral blood vessel. Does the earthworm have an open or a closed circulatory system? ________ Explain. ____

   ____

6. Locate the earthworm's nervous system. The two-lobed brain is located on the dorsal surface of the pharynx in segment 3. Two nerves, one on each side of the pharynx, connect the brain to a ganglion that lies below the pharynx in segment 4. The **ventral solid nerve cord** then extends along the floor of the body cavity to the last segment.

7. Find the earthworm's excretory system, which consists of a pair of minute, coiled, white tubules, the **nephridia,** located in every segment except the first three and the last. Each nephridium opens to the outside by means of an excretory pore. Does the excretory system show that the earthworm is segmented? ________ Explain. ____

   ____

   ____

8. Identify the earthworm's reproductive system, including **seminal vesicles,** light-colored bodies in segments 9 through 12, which house maturing sperm that have been formed in two pairs of testes within them; **sperm ducts** that pass to openings in segment 15; and **seminal receptacles** (four small, white, special bodies that lie in segments 9 and 10), which store sperm received from another worm. **Ovaries** are located in segment 13 but are too small to be seen.

9. During mating, earthworms are arranged so that the sperm duct openings of one worm are just about, but not quite, opposite the seminal receptacle openings of the other worm. After being released, the sperm pass down a pair of seminal grooves on the ventral surface (see Fig. 29.6) and then cross over at the level of the seminal receptacles of the opposite worm. Once the worms separate, eggs and sperm are released into a cocoon secreted by the clitellum. Is the earthworm hermaphroditic? ________ Explain. ____

   ____

10. Does the earthworm have a respiratory system? ________ How does it exchange gases? ________

    ____

11. Why would you expect an earthworm to lack an exoskeleton? (*Hint:* Review question 10.) ____

    ____

12. Does the earthworm have an open or closed circulatory system? ________________________

Prepared Slide

1. Obtain a prepared slide of a cross section of an earthworm (Fig. 29.8). Examine the slide under the dissecting microscope and under the light microscope.
2. Identify the following structures:
   a. **Body wall:** A thick, outer circle of tissue, consisting of the **cuticle** and the **epidermis**
   b. **Coelom:** A relatively clear space with scattered fragments of tissue
   c. **Intestine:** An inner circle with a suspended fold
   d. **Typhlosole:** A fold that increases the intestine's surface area
   e. **Ventral nerve cord:** A white, threadlike structure
3. Does the typhlosole help in nutrient absorption? ________ Explain. ______________________

____

Figure 29.8 Cross section of an earthworm.
Cross-section slide as it would appear under the microscope.

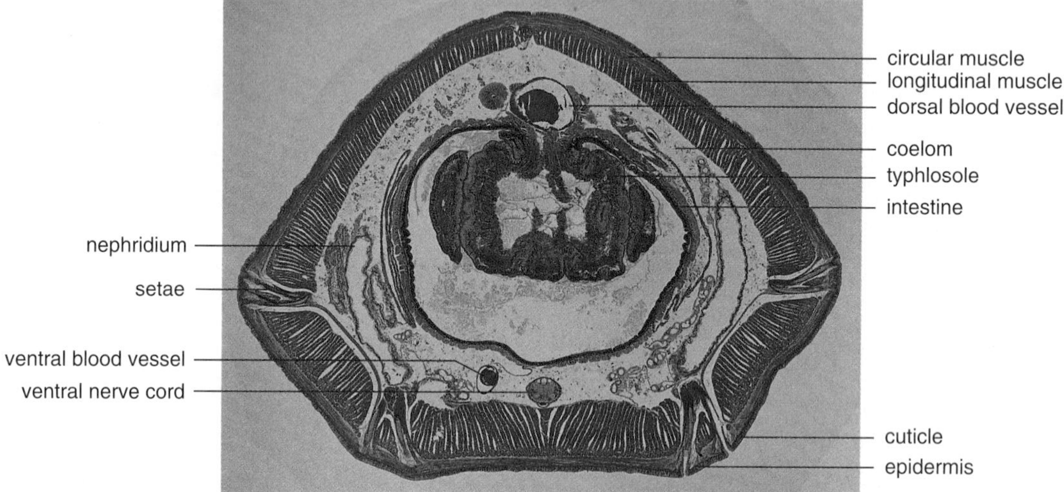

Conclusion

- Complete Table 29.4 to compare the anatomy of a clam to that of an earthworm.

Table 29.4 Comparison of Clam with Earthworm

	Clam	Earthworm
Nervous system		
Digestion		
Skeleton		
Excretory organ		
Circulation		
Respiratory organ		
Locomotion		
Reproduction		

29.3 Phylum Arthropoda (Arthropods)

Arthropods have paired, jointed appendages and a hard exoskeleton that contains chitin, a complex polysaccharide that is extremely durable. The chitinous exoskeleton consists of hardened plates separated by thin, membranous areas that allow movement of the body segments and appendages. Arthropods are segmented like the annelids, but specialization of segments has occurred. Explain. ____

____

Phylum Arthropoda includes the most common animals in the world—the insects—as well as centipedes and spiders, which are also terrestrial. The crustaceans, including lobsters and crabs, are aquatic. The arthropods are divided into three subphyla, as shown in Figure 29.9 and the following classification table.

Figure 29.9 Arthropod diversity.
a. Subphylum Uniramia contains insects, millipedes, and centipedes. **b.** Subphylum Chelicerata contains spiders, scorpions, and horseshoe crabs. **c.** Subphylum Crustacea contains crabs, shrimp, and barnacles, among others.

SUBPHYLUM UNIRAMIA

a. Class Insecta: honeybee, *Apis mellifera*

Class Diplopoda: millipede, *Ophyiulus pilosus*

Class Chilopoda: centipede, *Scolopendra* sp.

SUBPHYLUM CHELICERATA

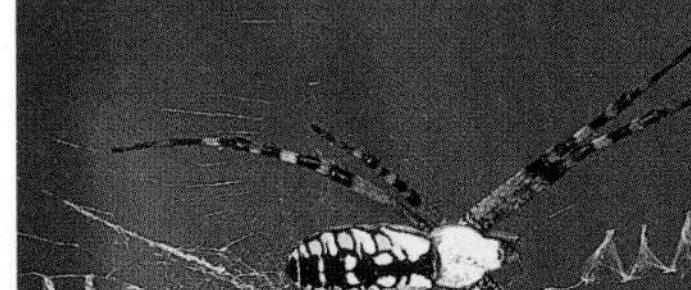

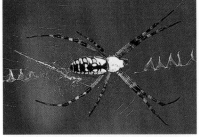

b. Class Arachnida: spider, *Argiope rafaria*

Class Arachnida: scorpion, *Hadrurus hirsutus*

Class Merostomata: horseshoe crab, *Limulus polyphemus*

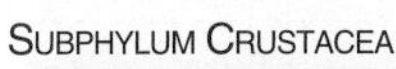

SUBPHYLUM CRUSTACEA

c. Class Malacostraca: crab, *Cancer productus*

Class Malacostraca: shrimp, *Stenopus* sp.

Class Maxillopoda: barnacles, *Lepas anatifera*

CLASSIFICATION: THE ARTHROPODS

PHYLUM ARTHROPODA
Chitinous exoskeleton with jointed appendages specialized in structure and function; well-developed central nervous system with brain and ventral solid nerve cord; reduced coelom; hemocoel

Subphylum Chelicerata
Chelicerae and pedipalp attached to head; no antennae, mandibles, or maxillae; four pairs of walking legs attached to a cephalothorax

Class Arachnida: spiders, scorpions

Class Merostomata: horseshoe crabs

Subphylum Uniramia
One pair of antennae, one pair of mandibles, and one or two pairs of maxillae attached to the head; uniramous appendages attached to the body

Superclass Myriapoda: millipedes, centipedes

Superclass Insecta: insects (bees, beetles, flies, grasshoppers)

Subphylum Crustacea
Compound eyes, antennae, antennules, mandibles, and maxillae attached to head; usually five pairs of walking legs attached to cephalothorax

Class Malacostraca: crabs, shrimp, lobsters, crayfish

Class Maxillopoda: barnacles

Observation: Diversity of Arthropods

1. Examine various specimens of arthropods, and complete Table 29.5.
2. In the last column, note the number and type of appendages attached to the thorax and/or abdomen.

Table 29.5 Arthropod Diversity

Common Name of Specimen	Subphylum	Appendages (Attached to Body)

Anatomy of Crayfish

Crayfish belong to a group of arthropods called crustaceans. Crayfish are adapted to an aquatic existence. They are known to be scavengers, but they also prey on other invertebrates. The mouth is surrounded by appendages modified for feeding, and there is a well-developed digestive tract. Dorsal, anterior, and posterior arteries carry hemolymph (blood plus lymph) to tissue spaces (hemocoel) and sinuses. In contrast to vertebrates, there is a ventral solid nerve cord.

Observation: Anatomy of Crayfish

External Anatomy

1. Obtain a preserved crayfish, and place it in a dissecting pan.
2. Identify the chitinous **exoskeleton.** With the help of Figure 29.10, identify the head, thorax, and abdomen. Together, the head and thorax are called the **cephalothorax;** the cephalothorax is covered by the **carapace.** Has specialization of segments occurred? ________ Explain. ________

   ____
3. Find the **antennae,** which project from the head. At the base of each antenna, locate a small, raised nipple containing an opening for the **green glands,** the organs of excretion. Crayfish excrete a liquid nitrogenous waste.

Figure 29.10 Anatomy of a crayfish.

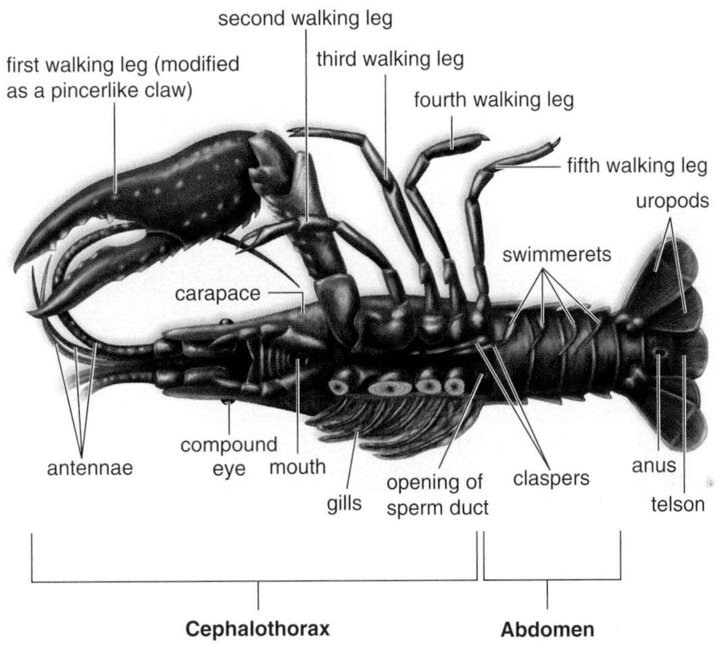

a.

b.

4. Locate the **compound eyes,** which are composed of many individual units for sight. Do crayfish demonstrate cephalization? ________ Explain. ____

____
5. Identify the six pairs of appendages around the mouth for handling food.
6. Find the five pairs of walking legs attached to the cephalothorax. The most anterior pair is modified as pincerlike claws.
7. Locate the five pairs of **swimmerets** on the abdomen. In the male, the anterior two pairs are stiffened and folded forward. They are claspers that aid in the transfer of sperm during mating.
8. In the female, identify the **seminal receptacles,** a swelling located between the bases of the third and fourth pairs of walking legs. Sperm from the male are deposited in the seminal receptacles. In the male, identify the opening of the sperm duct located at the base of the fifth walking leg. What sex is your specimen? ____
9. Examine the opposite sex also.
10. Find the last abdominal segment, which bears a pair of broad, fan-shaped **uropods** that, together with a terminal extension of the body, form a tail. Has specialization of appendages occurred? ________ Explain. ____

Internal Anatomy

1. Place the crayfish in the dissecting pan.
2. Cut away the lateral surface of the carapace with scissors to expose the **gills** (Fig. 29.10*b*). Observe that the gills occur in distinct, longitudinal rows. How many rows of gills are there in your specimen? ________ The outer row of gills is attached to the base of certain appendages. Which ones? ____
These outer gills are the **podobranchia** ("foot gills"). How many podobranchia do you find in your specimen? ____
3. Carefully separate the gills with a probe or dissecting needle, and locate the inner row(s) of gills. These inner gills are the **arthrobranchia** ("joint gills") and are attached to the chitinous membrane that joins the appendages to the thorax. How many rows of arthrobranchia do you find in the specimen? ____
4. Remove a gill with your scissors by cutting it free near its point of attachment, and place it in a watch glass filled with water. Observe the numerous gill filaments arranged along a central axis.
5. Carefully cut away the dorsal surface of the carapace with scissors and a scalpel. The epidermis that adheres to the exoskeleton secretes the exoskeleton. Remove any epidermis adhering to the internal organs.
6. Identify the diamond-shaped heart lying in the middorsal region. A crayfish has an open circulatory system. Carefully remove the heart.
7. Locate the **gonads** anterior to the heart in both the male and female. The gonads are tubular structures bilaterally arranged in front of the heart and continuing behind it as a single mass. In the male, the testes are highly coiled, white tubes.
8. Find the **mouth;** the short, tubular **esophagus;** and the two-part **stomach,** with the attached **digestive gland,** that precedes the intestine.
9. Identify the **green glands,** two excretory structures just anterior to the stomach, on the ventral segment wall.
10. Remove the thoracic contents previously identified.
11. Identify the **brain** in front of the esophagus. The brain is connected to the ventral nerve cord by a pair of nerves that pass around the esophagus.
12. Remove the animal's entire digestive tract, and float it in water. Observe the various parts, especially the connections of the digestive gland to the stomach.

13. Cut through the stomach, and notice in the anterior region of the stomach wall the heavy, toothlike projections, called the **gastric mill,** that grind up food. Do you see any grinding stones ingested by the crayfish? ____________________

If possible, identify what your specimen had been eating. ____________________

Anatomy of Grasshopper

The grasshopper is an **insect.** Insects are adapted to life on land. In insects with wings, such as the grasshopper, wings are attached to the thorax. Respiration is by a highly branched internal system of tubes, called tracheae.

Observation: Anatomy of Grasshopper

External Anatomy

1. Obtain a preserved grasshopper (*Romalea*), and study its external anatomy with the help of Figure 29.11. Identify the head, thorax, and abdomen.
2. The grasshopper's **thorax** consists of three fused segments: the large anterior **prothorax,** the middle **mesothorax,** and the hind **metathorax.** Identify the first pair of legs attached to the prothorax. Then find the second pair of legs and the outer pair of straight, leathery **forewings** attached to the mesothorax. Finally, locate the third pair of legs and the inner, membranous **hind wings** attached to the metathorax. Each leg consists of five segments. The hind leg is well developed and used for jumping. How many pairs of legs are there? ____________________

Figure 29.11 External anatomy of a female grasshopper, *Romalea.*
a. The legs and wings are attached to the thorax. **b.** The head has mouthparts of various types.

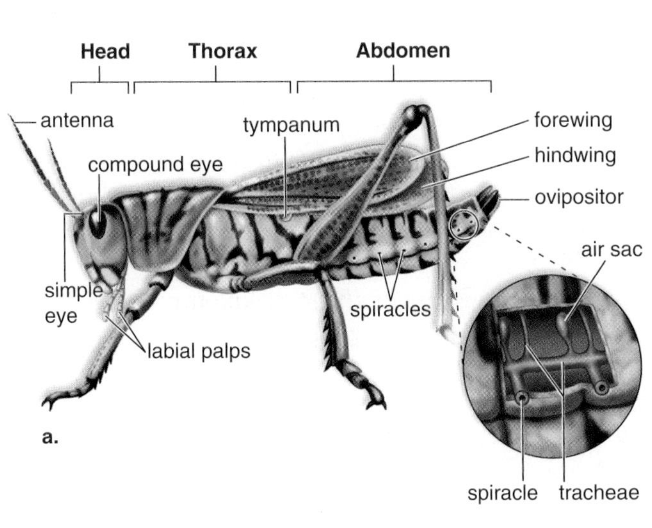

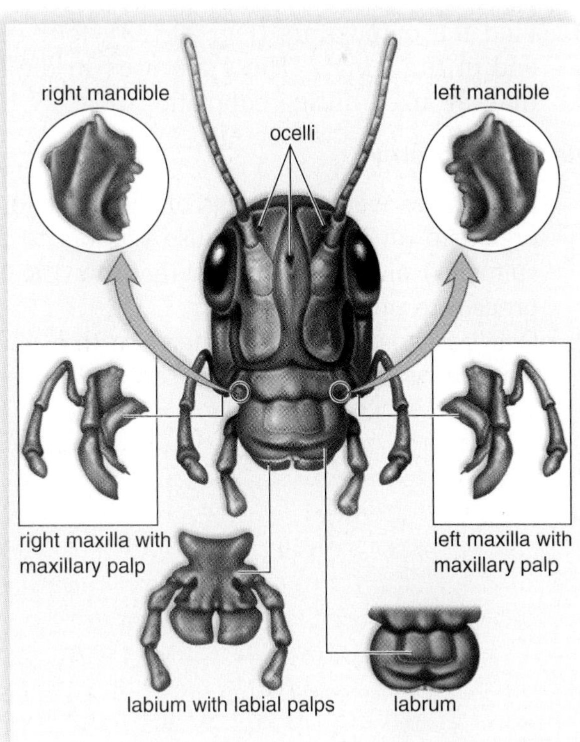

Figure 29.12 Grasshopper genitalia.
a. Females have an ovipositor, and **(b)** males have claspers.

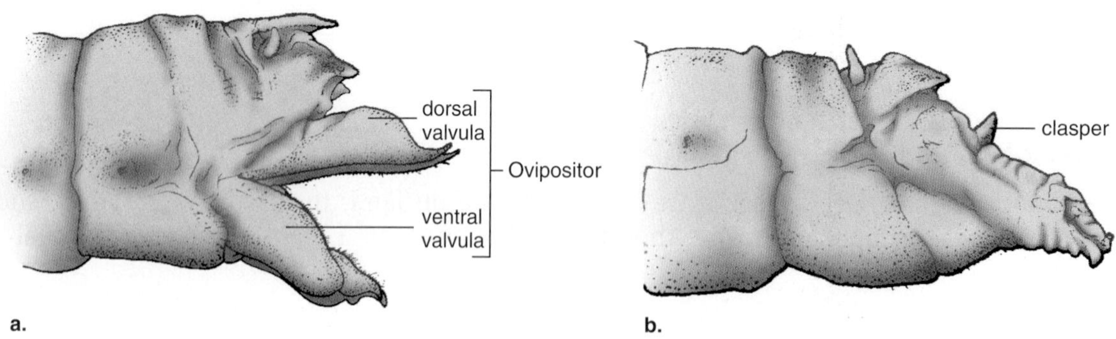

3. Is locomotion in the grasshopper adapted to land? ______________________
   Explain. ______________________
   ____
4. Use a hand lens or dissecting microscope to examine the grasshopper's special sense organs of the head. Identify the **antennae** (a pair of long, jointed feelers), the **compound eyes,** and the three, dotlike **simple eyes.**
5. Remove the **mouthparts** by grasping them with forceps and pulling them out. Arrange them in order on an index card, and compare them with Figure 29.11*b*. These mouthparts are used for chewing and are quite different from those of a piercing and sucking insect.
6. Identify the **tympana** (sing., **tympanum**), one on each side of the first abdominal segment (Fig. 29.11*a*). The grasshopper detects sound vibrations with these membranes.
7. Locate the **spiracles,** along the sides of the abdominal segments. These openings allow air to enter the tracheae, which constitute the respiratory system.
8. Find the **ovipositors** (Fig. 29.12*a*), four curved and pointed processes projecting from the hind end of the female. These are used to dig a hole in which eggs are laid. The male has **claspers** that are used during copulation (Fig. 29.12*b*).

Internal Anatomy

1. Detach the wings and legs of the grasshopper identified in number 2, page 429. Then turn the organism on its side, and use scissors to carefully cut through the exoskeleton (dorsal to the spiracles) along the full length (from the head to the posterior end) of the animal. Repeat this procedure on the other side.
2. Cut crosswise behind the head so that you can remove a strip of the exoskeleton. If necessary, reach in with a probe to loosen the muscle attachments and membranes.

Figure 29.13 Internal anatomy of a female grasshopper.
The digestive system of a grasshopper shows specialization of parts.

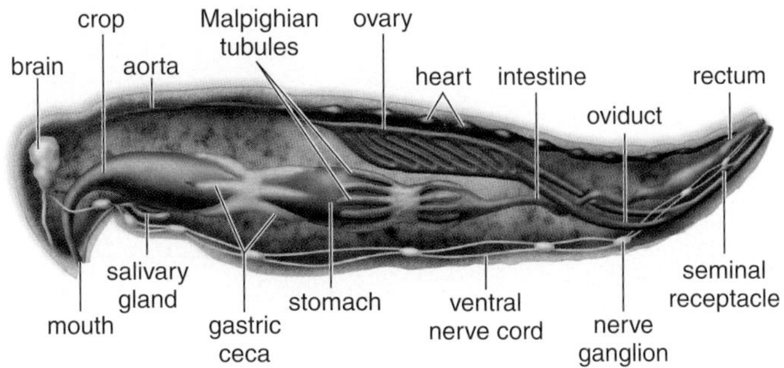

3. Pin the insect to the dissecting pan, dorsal side up. Cover the specimen with water to keep the tissues moist.
4. Identify the heart (Fig. 29.13) and aorta just beneath the portion of exoskeleton you removed. A grasshopper has an open circulatory system. Remove the heart and adjacent tissues.
5. Locate the **fat body,** a yellowish fatty tissue that covers the internal organs. Carefully remove it.
6. Find the **tracheae,** the respiratory system of insects. Using the dissecting microscope, look for glistening white tubules, which deliver air to the muscles.
7. Identify the reproductive organs that lie on either side of the digestive tract in the abdomen. If your specimen is a male, look for the testis, a coiled, elongated cord containing many tubules. If your specimen is a female, look for the ovary, essentially a collection of parallel, tapering tubules containing cigar-shaped eggs.
8. Locate the digestive tract and, in sequence, the **crop,** a large pouch for storing food (a grasshopper eats grasses); the **gastric ceca,** digestive glands attached to the stomach; the stomach and the intestine, which continues to the anus; and **Malpighian tubules,** excretory organs attached to the intestine. Insects secrete a solid nitrogenous waste. Is this an adaptation to life on land? ____________________

   Explain. ____________________
9. Work the digestive tract free, and move it to one side. Now identify the **salivary glands** that extend into the thoracic cavity.
10. Remove the internal organs. Now identify the ventral **nerve cord,** which is thickened at intervals by ganglia.
11. Remove one side of the exoskeleton covering the head. Identify the brain, which is anterior to the esophagus.

Conclusion

Compare the adaptations of a crayfish with those of a grasshopper by completing Table 29.6. Put a star beside each item that indicates an adaptation to life in the water (crayfish) and to life on land (grasshopper). How many did you identify? ________ Check with your instructor to see if you identified the maximum number of adaptations.

Table 29.6 Comparison of Crayfish and Grasshopper

	Crayfish	Grasshopper
Locomotion		
Respiration		
Nervous system		
Reproductive features		
Sense organs		

Insect Metamorphosis

Metamorphosis means a change, usually a drastic one, in form and shape. Some insects undergo what is called *complete metamorphosis,* in which case they have three stages of development: the larval stages, the pupa stage, and finally the adult stage. Metamorphosis occurs during the pupa stage, when the animal is enclosed within a hard covering. The animals that are best known for metamorphosis are the butterfly and the moth, whose larval stage is called a caterpillar and whose pupa stage is the cocoon; the adult is the butterfly or moth (Fig. 29.14*a*). Grasshoppers undergo *incomplete metamorphosis,* which is a gradual change in form rather than a drastic change. The immature stages of the grasshopper are called nymphs rather than larvae, and they are recognizable as grasshoppers even though they differ somewhat in shape and form (Fig. 29.14*b*).

If available, examine life cycle displays or plastomounts that illustrate complete and incomplete metamorphosis.

Observation: Insect Metamorphosis

Observe any insects available, and state in Table 29.7 whether they have complete metamorphosis or incomplete metamorphosis.

Table 29.7 Insect Metamorphosis

Common Name of Specimen	Complete or Incomplete Metamorphosis

Figure 29.14 Insect metamorphosis.
During **(a)** complete metamorphosis, a series of larvae leads to pupation. The adult hatches out of the pupa. During **(b)** incomplete metamorphosis, a series of nymphs leads to a full-grown grasshopper.

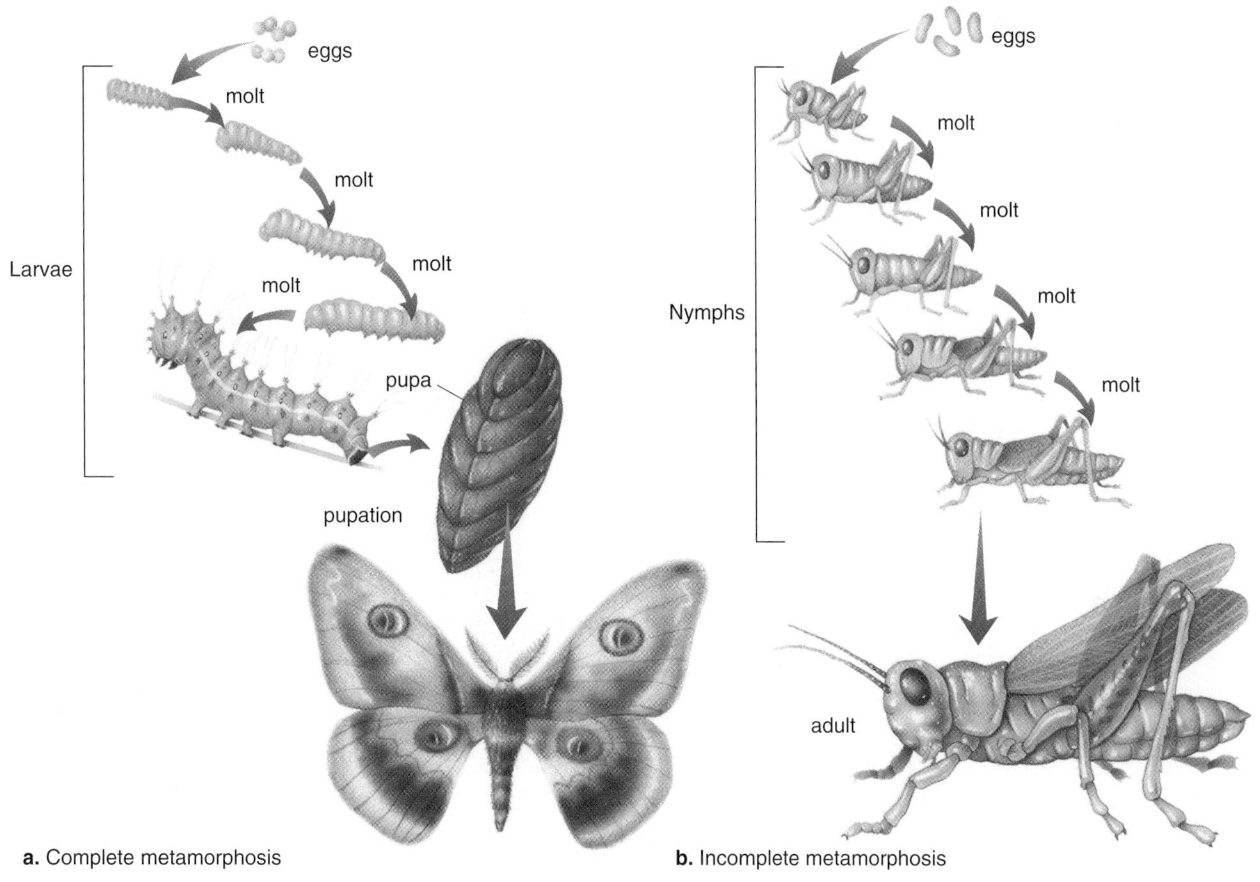

a. Complete metamorphosis

b. Incomplete metamorphosis

Conclusions

- With reference to Figure 29.14, what stage is missing when an insect does not have complete metamorphosis? ______________ What happens during this stage? ______________
- What form, the larvae or the adult, disperses offspring in flying insects? ______________ How is this a benefit? ______________

  ____
- In insects that undergo complete metamorphosis, the larvae and the adults utilize different food sources and habitats. Why might this be a benefit? ______________

  ____
- With reference to insects that undergo incomplete metamorphosis, which form, the nymphs or the adult, have better developed wings? ________ What is the benefit of wings to an insect?

  ____

  ____

Laboratory Review 29

_______________ 1. Jointed appendages and an exoskeleton are characteristic of what group of animals?

_______________ 2. Crayfish belong to what group of arthropods?

_______________ 3. A clam belongs to what group of molluscs?

_______________ 4. Molluscs, annelids, and arthropods are all what type of animal?

_______________ 5. A visceral mass, foot, and mantle are characteristic of what group of animals?

_______________ 6. All the animals studied today have what type of coelom?

_______________ 7. In a clam, which structure secretes the shell?

_______________ 8. The clam is a filter feeder, but the squid is a(n) _______________.

_______________ 9. The annelids are the first of the animal phyla studied to have what general characteristic?

_______________ 10. Which of the three classes of annelids has suckers as an adaptation to its way of life?

_______________ 11. What term indicates that earthworms have both male and female organs?

_______________ 12. The arthropods are the first of the animal phyla to have what general characteristic?

_______________ 13. What type of excretory organs are attached to the intestine of a grasshopper?

_______________ 14. Contrast the respiratory organ of a crayfish with that of a grasshopper.

_______________ 15. Identify the muscular organ used for locomotion in the clam.

_______________ 16. How do open and closed circulatory systems differ from one another?

Thought Questions

17. Compare respiratory organs in the crayfish and the grasshopper. How are these suitable to the habitat of each?

18. For each of the following characteristics, name an animal with the characteristic, and state the characteristic's advantages:
    a. Closed circulatory system
    b. Jointed appendages
    c. Exoskeleton
    d. Segmentation

LABORATORY

30

Echinoderms and Chordates

Learning Outcomes

30.1 Phylum Echinodermata (Echinoderms)

- Describe the general characteristics of echinoderms and the specific features of selected classes.
- Identify and locate external and internal structures of a sea star.
- Trace the path of water in the water vascular system of a sea star.

30.2 Phylum Chordata (Chordates)

- State the characteristics that all chordates have in common, and use the cross section of a lancelet to point out these characteristics.

30.3 Subphylum Vertebrata (Vertebrates)

- Name the vertebrate classes, give an example of an animal in each class, and state the characteristics of each.
- Trace the path of air, food, and urine in a frog, and explain the term *urogenital system.*
- Using a preserved specimen or image, identify and locate external and internal structures of a frog.

30.4 Comparative Vertebrate Anatomy

- Compare the organ systems (except musculoskeletal) of a frog with those of a perch, a pigeon, and a pig.
- Trace the path of blood through the heart of a fish, an amphibian, a bird, and a mammal.
- Relate aspects of an animal's respiratory system to the animal's environment and to features of the animal's circulatory system.

Introduction

As discussed in laboratories 28 and 29, animals may be grouped according to their type of coelom (body cavity). Some are acoelomates (having no body cavity), some are pseudocoelomates (body cavity incompletely lined with mesoderm), and some are coelomates (body cavity completely lined with mesoderm). The coelomates are divided into the protostomes, in which the first embryonic opening becomes the mouth, and the deuterostomes, in which the second (*deutero*) embryonic opening becomes the mouth (*stoma*).

Figure 30.1 Echinoderm larva compared with invertebrate chordate larva.
(a) Echinoderm larva. **(b)** invertebrate chordate larva.

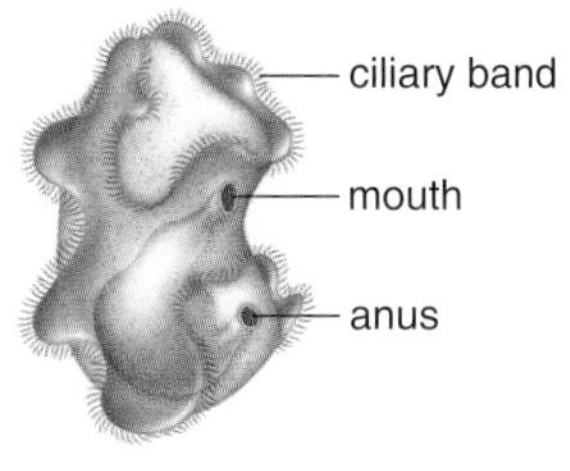

a. Echinoderm larva

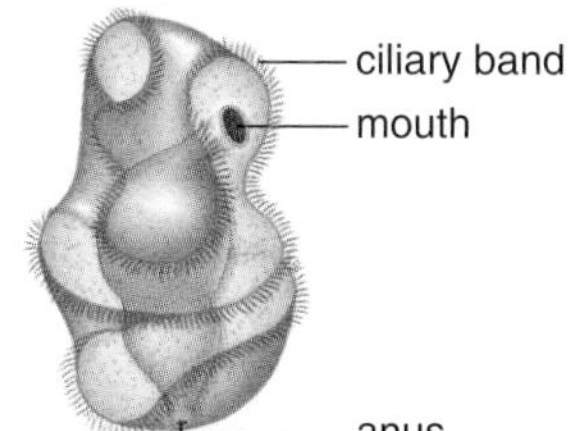

b. Invertebrate chordate larva

This laboratory focuses on the two phyla of deuterostomes: the **echinoderms** and the **chordates.** Based on embryological evidence, echinoderms are believed to be closely related to the invertebrate chordates (Fig. 30.1). Representatives of these groups will be evaluated in today's laboratory exercises. Most likely, these two groups share a common bilateral ancestor. The echinoderm larva is bilateral, while the adult echinoderm is radially symmetrical. The invertebrate chordates develop from a basic bilateral plan. Evidence suggests that they gave rise to the **vertebrates,** animals with an endoskeleton of bone and cartilage.

30.1 Phylum Echinodermata (Echinoderms)

Echinoderms are all marine, and they dwell on the seabed, either attached to it, like sea lilies, or creeping slowly over it. The name *echinoderm* means spiny-skinned, and most members of the group have defensive spines on the outside of their bodies. The spines arise from an **endoskeleton** composed of calcium carbonate plates. The endoskeleton supports the body wall and is covered by living tissue that may be soft (as in sea cucumbers) or hard (as in sea urchins). While the adult echinoderm is radially symmetrical, with generally five points of symmetry arranged around the axis of the mouth, the larvae are bilaterally symmetrical. The echinoderms' most unique feature is their **water vascular system.** In those echinoderms in which the arms make contact with the substratum, the **tube feet** associated with the water vascular system are used for locomotion. In other echinoderms, the tube feet are used for gas exchange and food gathering.

Echinoderms belong to one of five classes (Fig. 30.2):

CLASSIFICATION: THE ECHINODERMS

PHYLUM ECHINODERMATA (ECHINODERMS)
Radial symmetry; endoskeleton of spine-bearing plates; water vascular system with tube feet

Class Crinoidea: sea lilies, feather stars
Class Asteroidea: sea stars
Class Ophiuroidea: brittle stars
Class Echinoidea: sea urchins, sand dollars
Class Holothuroidea: sea cucumbers

Observation: Diversity of Echinoderms

Examine echinoderm specimens, and then complete Table 30.1. Try to locate the tube feet on each specimen. In Table 30.1, give a general description of the animal and the location of the tube feet.

Table 30.1 Echinoderm Diversity

Common Name of Specimen	Class	General Description	Location of Tube Feet

Figure 30.2 Echinoderm diversity.

a. Class Crinoidea: sea lily, *Comonthino* sp.

b. Class Asteroidea: sea star, *Pentaceraster cumingi*

c. Class Ophiuroidea: brittle stars, *Ophiopholis aculeata*

d. Class Echinoidea: sea urchin, *Stronglocentrotus pranciscanus*

e. Class Echinoidea: sand dollar, *Dendraster excentricus*

f. Class Holothuroidea: sea cucumber, *Pseudocolochirus* sp.

Anatomy of Sea Star

Sea stars (starfish) usually have five arms that radiate from a central disk, accounting for the class name, Asteroidea. The mouth is normally oriented downward, and when sea stars feed on clams, they use the suction of their tube feet to force the shells open a crack. Then they evert the cardiac portion of the stomach, which releases digestive juices into the mantle cavity. Partially digested tissues are taken up by the pyloric portion of the stomach; digestion continues in this portion of the stomach and in the digestive glands found in the arms.

Figure 30.3 Anatomy of a sea star.
a. Dissected animal. **b.** Isolated water vascular system. The water vascular system begins with a sieve plate and ends with tube feet.

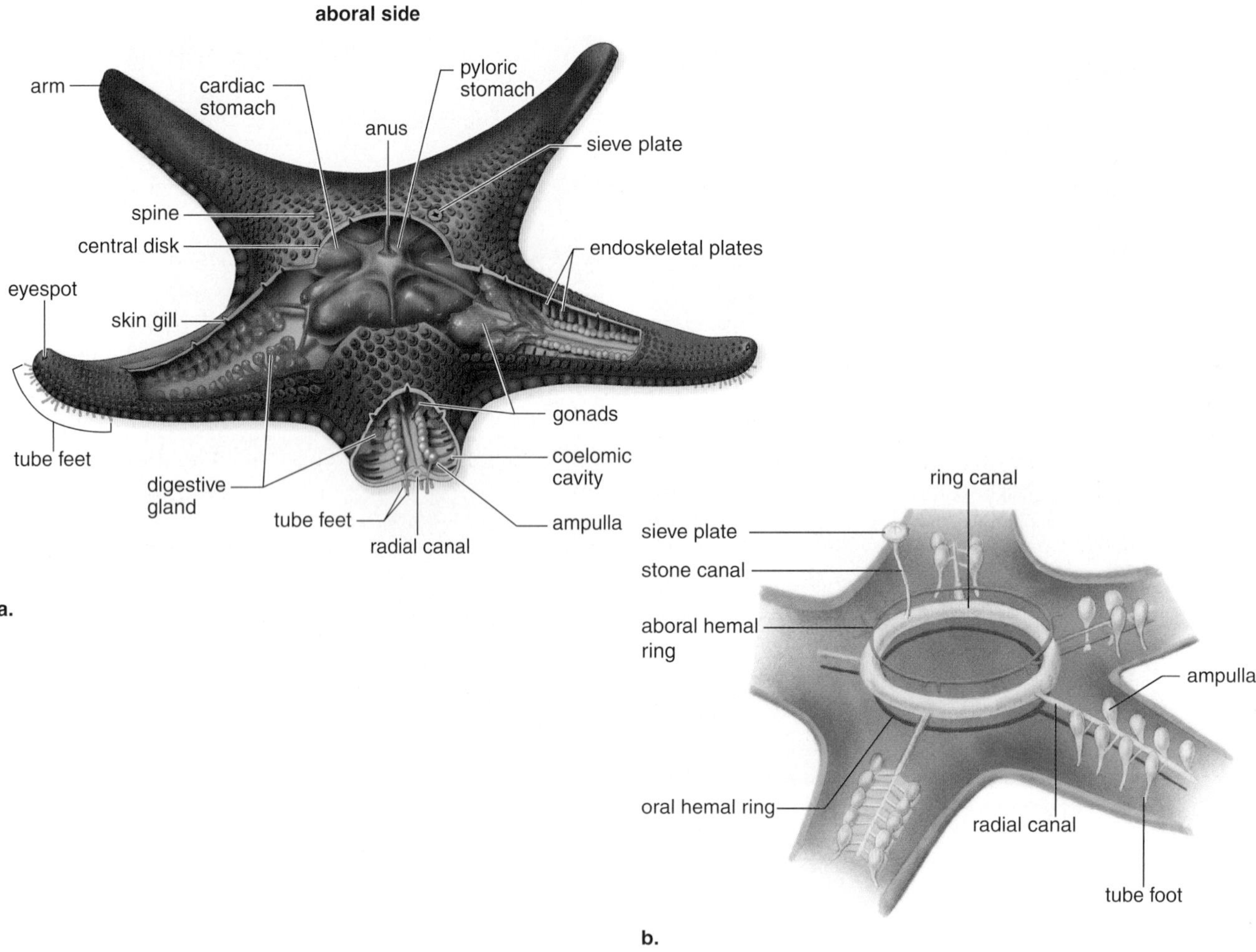

Observation: Anatomy of Sea Star

External Anatomy

1. Place a preserved sea star in a dissecting pan so that the oral side is uppermost.
2. With the help of Figure 30.3, identify the **central disk** and five arms. What type of symmetry does an adult sea star have? ______________
3. Find the mouth, located in the center and protected by spines. Why is this side of the sea star called the oral side? ____

4. Locate the **ambulacral groove** that runs along the middle of each arm and the **tube feet** (suctionlike disks) in rows on either side of the groove. Pluck away the tube feet from one area.

   How many rows of feet are there? ________ In the valley of each ambulacral groove, identify the **radial canal** that extends to the tip of each arm.
5. Turn the sea star over to its aboral side.
6. Locate the anal opening. Why is this side of the sea star called the aboral side? ______________

   ____
7. Lightly run your fingers over the surface spines. These extend from the calcium carbonate plates that lie buried in the body wall beneath the surface. The plates form an endoskeleton of the animal.
8. Identify the **sieve plate (madreporite),** a brownish, circular spot between two arms where water enters the water vascular system.

Internal Anatomy

1. Place the sea star so that the aboral side is uppermost. Refer to Figure 30.3 as you dissect the sea star according to the following instructions.
2. Cut the tip of one of the arms and, with scissors, carefully cut through the body wall along each side of this arm.
3. Carefully lift up the upper body wall. Separate any internal organs that may be adhering so that all internal organs are left intact.
4. Cut off the body wall near the central disk, but leave the sieve plate (madreporite) in place.
5. Remove the body wall of the central disk, being careful not to injure the internal organs.
6. Identify the digestive system. The mouth leads into a short **esophagus,** which is connected to the saclike **cardiac stomach.** When a sea star eats, the cardiac stomach sticks out through the sea star's mouth and starts digesting the contents of a clam or an oyster. Above the cardiac stomach is the **pyloric stomach,** which leads to a short intestine. Each arm contains one pair of **digestive glands.** To which stomach do the digestive glands attach? ______________
7. Remove the digestive glands of one arm.
8. Identify the **gonads** extending into the arm. What is the function of gonads? ______________

   ____

   It is not possible to distinguish male sea stars from females by this observation.
9. Remove both stomachs.
10. In the **water vascular system** (Fig. 30.3*b*), you have already located the sieve plate and tube feet. Now try to identify the following components:
    a. **Stone canal:** Takes water from the sieve plate to the ring canal.
    b. **Ring canal:** Surrounds the mouth and takes water to the radial canals.
    c. **Radial canals:** Send water into the ampulla. When the ampullae contract, water enters the tube feet. Each tube foot has an inner muscular sac called an ampulla. The ampulla contracts and forces water into the tube foot.

    What is the function of the water vascular system? ______________________

    ____
11. Cut off a portion of an arm, and examine the cut edge (Fig. 30.3). Identify the digestive glands, ambulacral groove, radial canal, ampullae, and tube feet.

30.2 Phylum Chordata (Chordates)

The **chordates** are organized into three subphyla; two of these contain invertebrates and the other contains the vertebrates. Despite their great diversity, all chordates at some time in their life history have four characteristics:

1. A **notochord,** a dorsal supporting rod extending the length of the body. The notochord is replaced during development by a vertebral column in the vertebrates.
2. A **dorsal tubular nerve cord.** In vertebrates, the nerve cord, more often called the **spinal cord,** is protected by the vertebrae.
3. **Pharyngeal pouches,** which become functioning **gills** in the invertebrate chordates, the fishes, and amphibian larvae. In terrestrial animals, the pouches are modified for various other functions.
4. A **post-anal tail**—as an embryo if not as an adult; a tail that extends beyond the anus.

CLASSIFICATION: THE CHORDATES

PHYLUM CHORDATA (CHORDATES)
A notochord, pharyngeal pouches, a dorsal tubular nerve cord, and a post-anal tail are all present at some time in the life history

Phylum Chordata: chordates
Subphylum Cephalochordata: lancelets
Subphylum Urochordata: tunicates
Subphylum Vertebrata

Invertebrate Chordates

The following are the two invertebrate chordate subphyla:

1. **Subphylum Urochordata.** This subphylum contains the tunicates, or sea squirts. These animals come in varying sizes and shapes, but all have incurrent and excurrent siphons. **Gill slits** are the only remaining chordate characteristic in adult tunicates (Fig. 30.4). Examine any examples of tunicates on display.

Figure 30.4 Subphylum Urochordata.
The gill slits of a tunicate are the only chordate characteristics remaining in the adult.

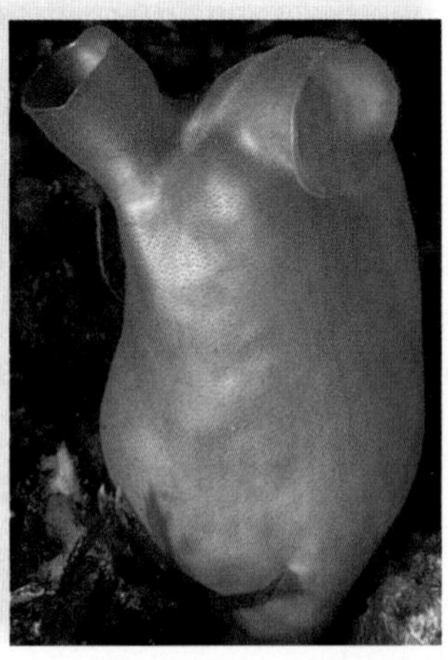

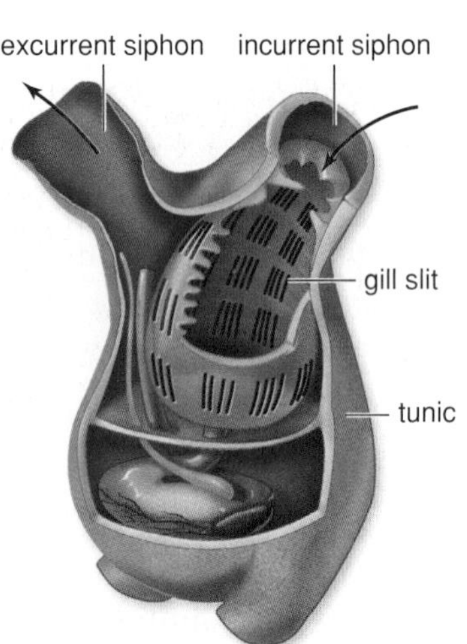

2. **Subphylum Cephalochordata.** Lancelets, which are also known as amphioxus (*Branchiostoma*), are small, fishlike animals that occur in shallow marine waters in most parts of the world (Fig. 30.5). They spend most of their time buried in the sandy bottom, with only the anterior end projecting.

Observation: Lancelet Anatomy

Preserved Specimen

1. Examine a preserved lancelet (Fig. 30.5).
2. Identify the **caudal fin** (enlarged tail) used in locomotion, the **dorsal fin,** and the short **ventral fin.**
3. Examine the lancelet's V-shaped muscles.
4. Find the tentacled **oral hood,** which is located anterior to the mouth and covers a vestibule. Water entering the mouth is channeled into the **pharynx,** where food particles are trapped before the water exits at the **atriopore.** Lancelets are filter feeders. Has cephalization occurred? ________

   Explain. ____

   ____

Figure 30.5 Anatomy of the lancelet, *Branchiostoma*.
Lancelets feed on microscopic particles filtered out of the constant stream of water that enters the mouth and exits through the gill slits into an atrium that opens at the atriopore.

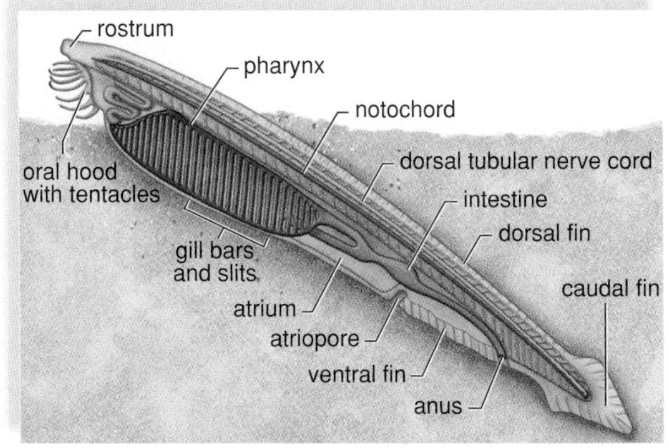

Prepared Slide

Examine a prepared cross section of a lancelet (Fig. 30.6), and note the three chordate characteristics: notochord, dorsal nerve cord, and gill slits.

Figure 30.6 Lancelet cross section.
Cross-section slide as it would appear under a microscope.

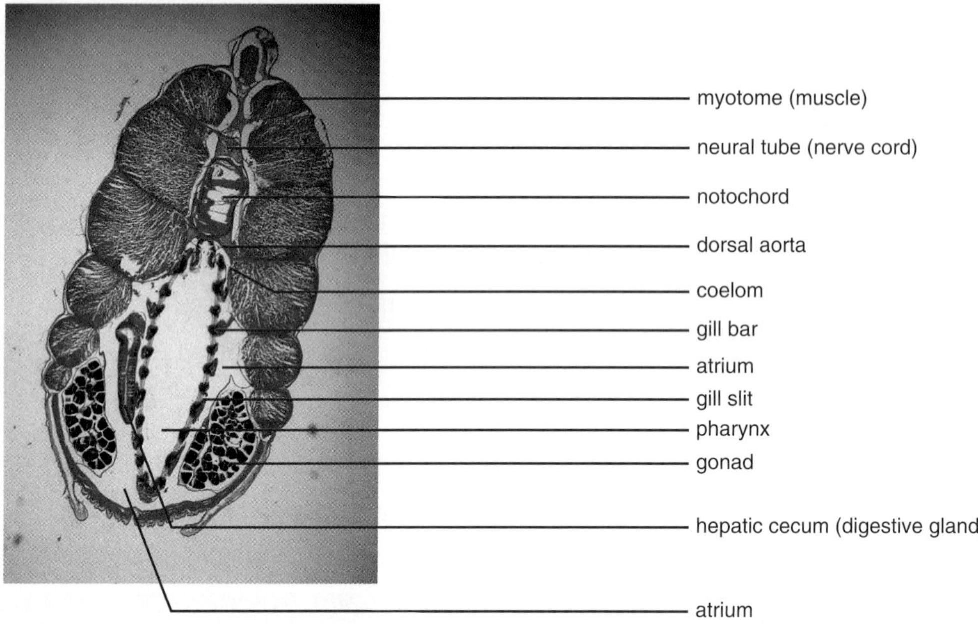

30.3 Subphylum Vertebrata (Vertebrates)

Vertebrates are segmented, and specialization of parts has occurred. Arthropods have an exoskeleton, while vertebrates have an endoskeleton, but they both have jointed appendages. In vertebrates, two pairs of appendages are characteristic. The vertebrate brain is more complex than that of arthropods and is enclosed by a skull. Among vertebrates, a high degree of cephalization is the rule. All organ systems are present and efficient.

Diversity of Vertebrates

There are both aquatic and terrestrial vertebrates (as shown in Figure 30.7 and the classification table on page 443. Fishes are aquatic; adult amphibians may be terrestrial, but most must return to an aquatic habitat to reproduce; and reptiles are terrestrial, although some reptiles, such as sea turtles, are secondarily adapted to life in the water. Mammals are adapted to a wide variety of habitats, including air (e.g., bats), sea (e.g., whales), and land (e.g., humans).

Figure 30.7 Vertebrate classes.
(a) Class Chondrichthyes contains the cartilaginous fishes. **(b)** Class Osteichthyes contains the bony fishes. **(c)** Class Amphibia contains the frogs and salamanders. **(d)** Class Reptilia contains the turtles, lizards, snakes, crocodiles, and alligators. **(e)** Class Aves contains the birds. **(f)** Class Mammalia contains the mammals.

a. Class Chondrichthyes: Blue shark

b. Class Osteichthyes: Blueback butterflyfish

c. Class Amphibia: Northern leopard frog

d. Class Reptilia: Pearl River redbelly turtle

e. Class Aves: Scissor-tailed flycatcher

f. Class Mammalia: Grey fox

CLASSIFICATION: THE VERTEBRATES

SUBPHYLUM VERTEBRATA
Notochord replaced by vertebrae that protect the nerve cord; skull that protects the brain; segmented with jointed appendages: Vertebrates

Superclass Agnatha
Marine and freshwater fishes; lack jaws and paired appendages; cartilaginous skeleton; notochord: lampreys and hagfishes

Superclass Gnathostomata
Hinged jaws; paired appendages: jawed fishes and all tetrapods

Class Chondrichthyes
Marine cartilaginous fishes; lack operculum and swim bladder; tail fin usually asymmetrical: sharks, skates, and rays

Class Osteichthyes
Marine and freshwater bony fishes; operculum; swim bladder or lungs; tail fin usually symmetrical: lungfishes, lobe-finned fishes, and ray-finned fishes (herring, salmon, sturgeon, eels, sea horse)

Class Amphibia
Tetrapod with nonamniotic egg; nonscaly skin; some show metamorphosis; three-chambered heart; ectothermic: urodeles (salamanders, newts) and anurans (frogs, toads)

Class Reptilia
Tetrapod with amniotic egg; scaly skin; ectothermic: squamata (snakes, lizards) and chelonians (turtles, tortoises)

Class Aves
Tetrapod with feathers; bipedal with wings; double circulation; endothermic: sparrows, penguins, and ostriches

Class Mammalia
Tetrapods with hair, mammary glands; double circulation; endothermic; teeth differentiated: monotremes (spiny anteater, duckbill platypus), marsupials (opossum, kangaroo), and placental mammals (whales, rodents, dogs, cats, elephants, horses, bats, humans)

Anatomy of Frog

Frogs are amphibians, a group of animals in which metamorphosis occurs. Metamorphosis includes a change in structure, as when an aquatic tadpole becomes a frog with lungs and limbs (Fig. 30.8). Amphibians were the first vertebrates to be adapted to living on land; however, they typically return to the water to reproduce. *Place a check in the margin below for every adaptation to a land environment.*

Observation: External Anatomy of Frog

1. Place a preserved frog (*Rana pipiens*) in a dissecting tray.
2. Identify the bulging eyes, which have nonmovable upper and lower lids but can be covered by a **nictitating membrane** that moistens the eye.
3. Locate the **tympanum** behind each eye (Fig. 30.8). What is the function of a tympanum? ______

   ____
4. Examine the external **nares** (sing., **naris,** or **nostril**) (Fig. 30.8). Insert a probe into an external naris, and observe that it protrudes from one of the paired, small openings, the internal nares, inside the mouth cavity. What is the function of the nares? ______________
5. Identify the paired limbs. The bones of the fore- and hind limbs are the same as in all tetrapods, in that the first bone articulates with a girdle and the limb ends in phalanges. The hind feet have five phalanges, and the forefeet have only four phalanges. Which pair of limbs is longest? ______

   How does a frog locomote on land? ______________________________

   ____

   What is a frog's means of locomotion in the water? ______________________

   ____

Figure 30.8 External frog anatomy.

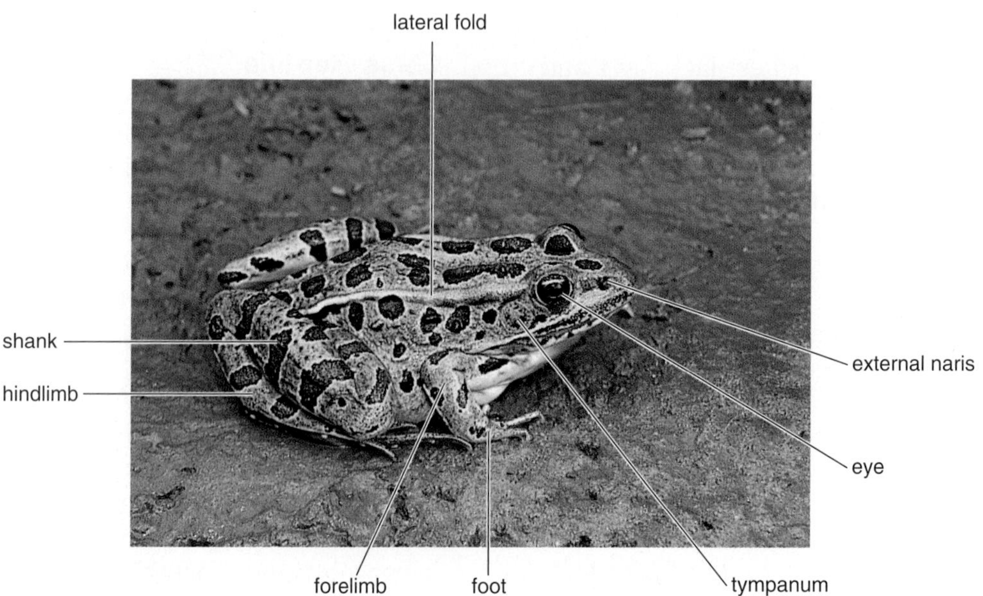

Observation: Internal Anatomy of Frog

Mouth

1. Open your frog's mouth very wide (Fig. 30.9), cutting the angles of the jaws if necessary.
2. Identify the tongue attached to the lower jaw's anterior end.
3. Find the **auditory (eustachian) tube** opening in the angle of the jaws. These tubes lead to the ears. Auditory tubes equalize air pressure in the ears.
4. Examine the **maxillary teeth** located along the rim of the upper jaw. Another set of teeth—**vomerine teeth**—is present just behind the midportion of the upper jaw.
5. Locate the **glottis,** a slit through which air passes into and out of the **trachea,** the short tube from glottis to lungs. What is the function of a glottis? ______________________________

   ____
6. Identify the **esophagus,** which lies dorsal and posterior to the glottis and leads to the stomach.

Figure 30.9 Mouth cavity of a frog.
a. Drawing. **b.** Dissected specimen.

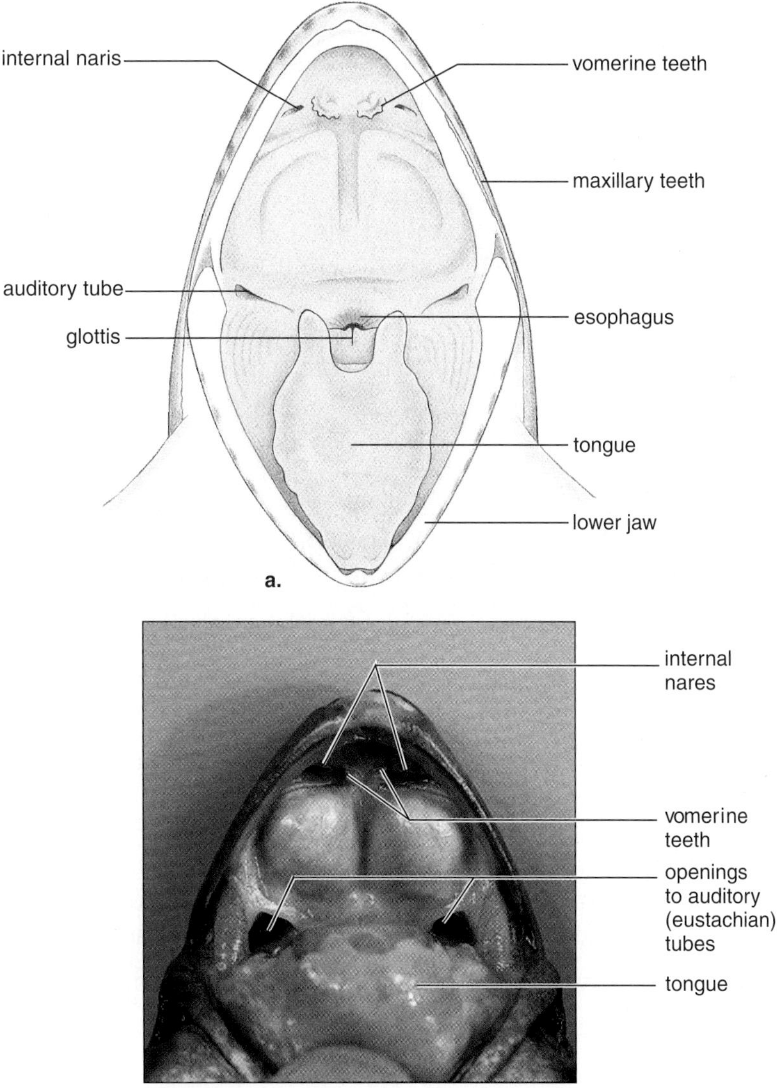

Opening the Frog

1. Place the frog ventral side up in the dissecting pan. Lift the skin with forceps, and use scissors to make a large, circular cut to remove the skin from the abdominal region as close to the limbs as possible. Cut only skin, not muscle.
2. Now, remove the muscles by cutting through them in the same circular fashion. At the same time, cut through any bones you encounter. A vein, called the abdominal vein, will be slightly attached to the internal side of the muscles.
3. Identify the **coelom,** or body cavity.
4. If your frog is female, the abdominal cavity is likely to be filled by a pair of large, transparent **ovaries,** each containing hundreds of black and white eggs. Gently lift the left ovary with forceps, and find its place of attachment. Cut through the attachment, and remove the ovary in one piece.

Respiratory System and Liver

1. Insert a probe into the glottis, and observe its passage into the trachea. Enlarge the glottis by making short cuts above and below it. When the glottis is spread open, you will see a fold on either side; these are the vocal cords used in croaking.
2. Identify the **lungs,** two small sacs on either side of the midline and partially hidden under the liver (Fig. 30.10). Trace the path of air from the external nares to the lungs. ________________

   ____
3. Locate the **liver,** the large, prominent, dark brown organ in the midventral portion of the trunk (Fig. 30.10). Between the right half and left half of the liver, find the **gallbladder.**

Circulatory System

1. Lift the liver gently. Identify the **heart,** covered by a membranous covering (the **pericardium**). With forceps, lift the covering, and gently slit it open. The heart consists of a single, thick-walled **ventricle** and two (right and left) anterior, thin-walled **atria.**
2. Locate the three large veins that join together beneath the heart to form the **sinus venosus.** (To lift the heart, you may have to snip the slender strand of tissue that connects the atria to the pericardium.) Blood from the sinus venosus enters the right atrium. The left atrium receives blood from the lungs.
3. Find the **conus arteriosus,** a single, wide arterial vessel leaving the ventricle and passing ventrally over the right atrium. Follow the conus arteriosus forward to where it divides into three branches on each side. The middle artery on each side is the **systemic artery,** which fuses behind the heart to become the **dorsal aorta.** The dorsal aorta transports blood through the body cavity and gives off many branches. The **posterior vena cava** begins between the two kidneys and returns blood to the sinus venosus. Which vessel lies above (dorsal to) the other? ________

   ____

Digestive Tract

1. Identify the **esophagus,** a very short connection between the mouth and the stomach. Lift the left liver lobe, and identify the stomach, which is whitish and J-shaped. The **stomach** connects with the esophagus anteriorly and with the small intestine posteriorly.
2. Find the **small intestine** and the **large intestine,** which enters the **cloaca** (see Figs. 30.11 and 30.12). The cloaca lies beneath the pubic bone and is a general receptacle for the intestine, the reproductive system, and the urinary system. It opens to the outside by way of the anus. Trace the path of food in the digestive tract from the mouth to the cloaca. ________________

   ____

Accessory Glands

1. You identified the liver and gallbladder previously. Now try to find the **pancreas,** a yellowish tissue near the stomach and intestine.
2. Locate the **spleen,** a small, pea-shaped body near the stomach.

Urogenital System

1. Identify the **kidneys,** which are long, narrow organs lying against the dorsal body wall (Fig. 30.11).
2. Locate the **testes** in a male frog (Fig. 30.11). Testes are yellow, oval organs attached to the anterior portions of the kidneys. Several small ducts, the **vasa efferentia,** carry sperm into kidney ducts that also carry urine from the kidneys. **Fat bodies,** which store fat, are attached to the testes.
3. Locate the ovaries in a female frog. The ovaries are attached to the dorsal body wall (Fig. 30.12). Fat bodies are also attached to the ovaries. Highly coiled **oviducts** lead to the cloaca. The ostium (opening) of the oviduct is dorsal to the liver.
4. Find the **mesonephric ducts**—thin, white tubes that carry urine from the kidney to the cloaca. In female frogs, you will have to remove the left ovary to see the mesonephric ducts.
5. Locate the **cloaca.** You will need to split through the bones of the pelvic girdle in the midventral line and carefully separate the bones and muscles to find the cloaca.
6. Identify the urinary bladder attached to the ventral wall of the cloaca. In frogs, urine backs up into the bladder from the cloaca.
7. Explain the term *urogenital system.* ____

   ____

8. The cloaca receives material from (1) ____,

   (2) ____________________________, and (3) ____________________________________.

Figure 30.10 Internal organs of a female frog, ventral view.

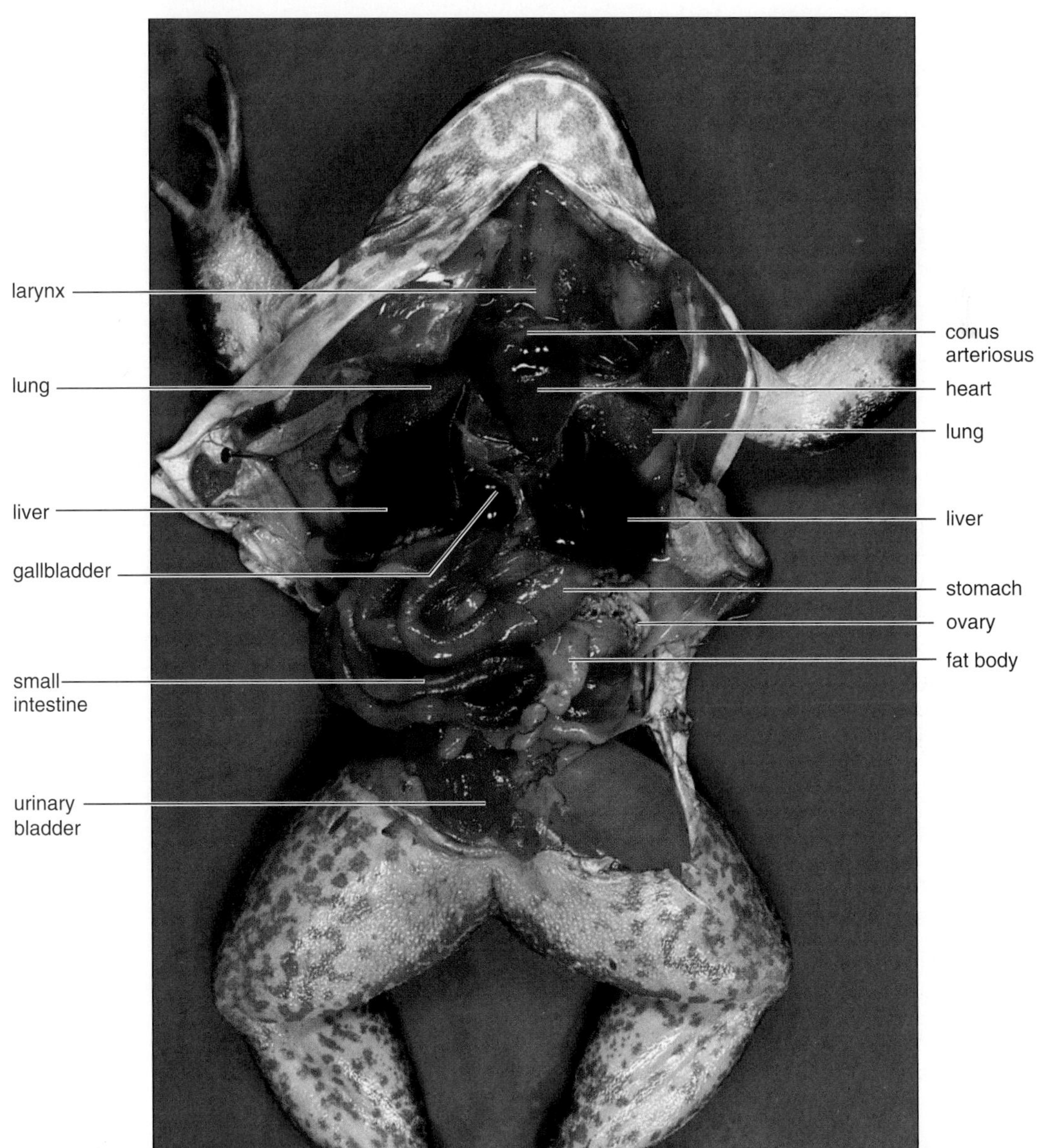

Figure 30.11 Urogenital system of a male frog.
a. Drawing. **b.** Dissected specimen.

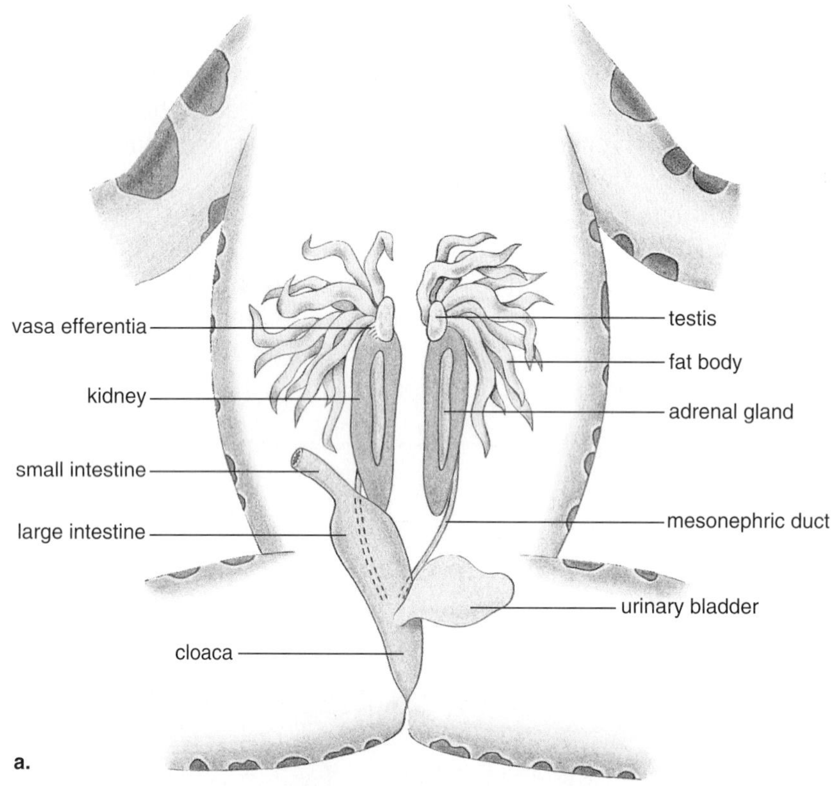

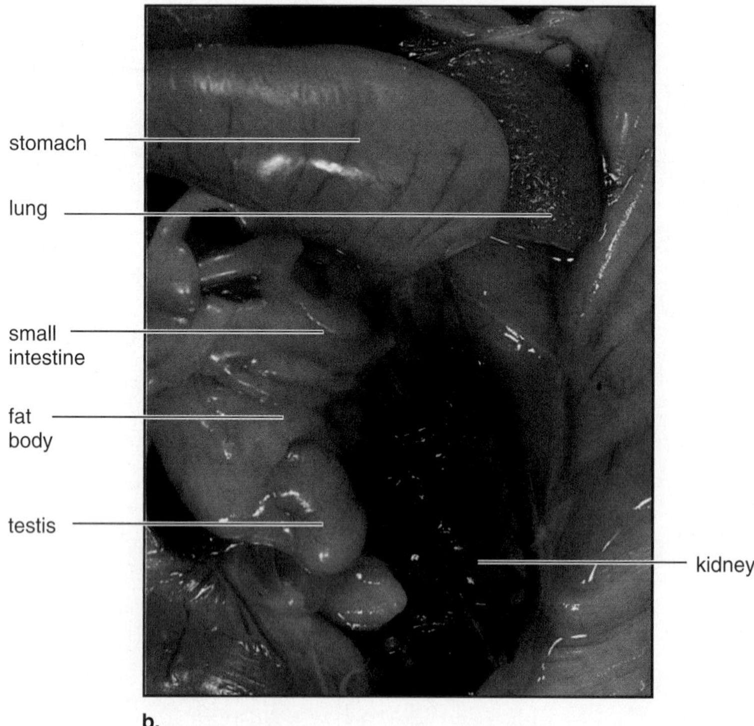

Figure 30.12 Urogenital system of a female frog. **a.** Drawing. **b.** Dissected specimen.

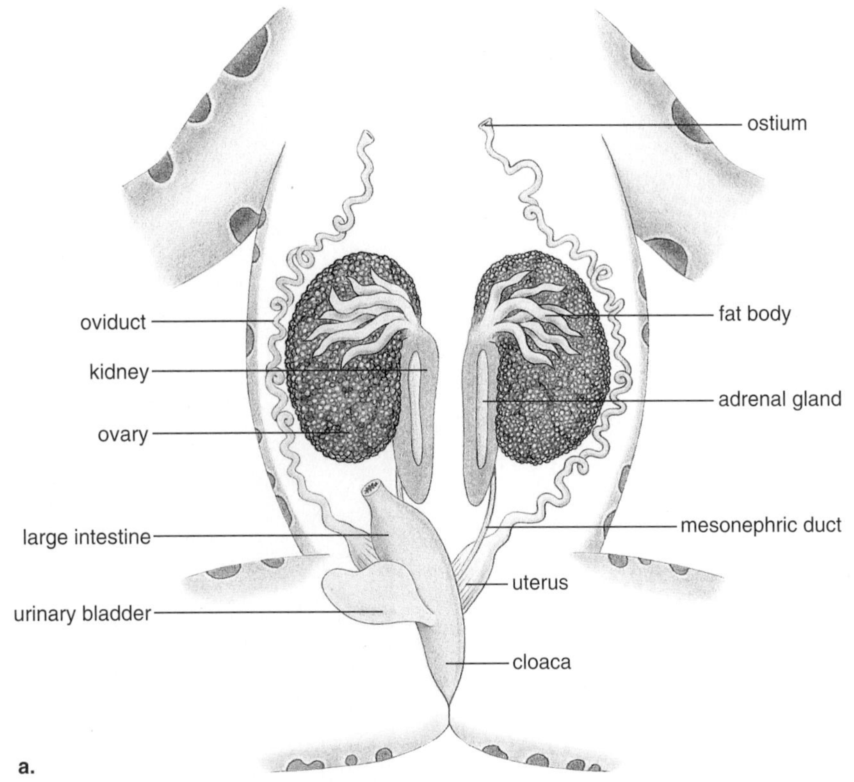

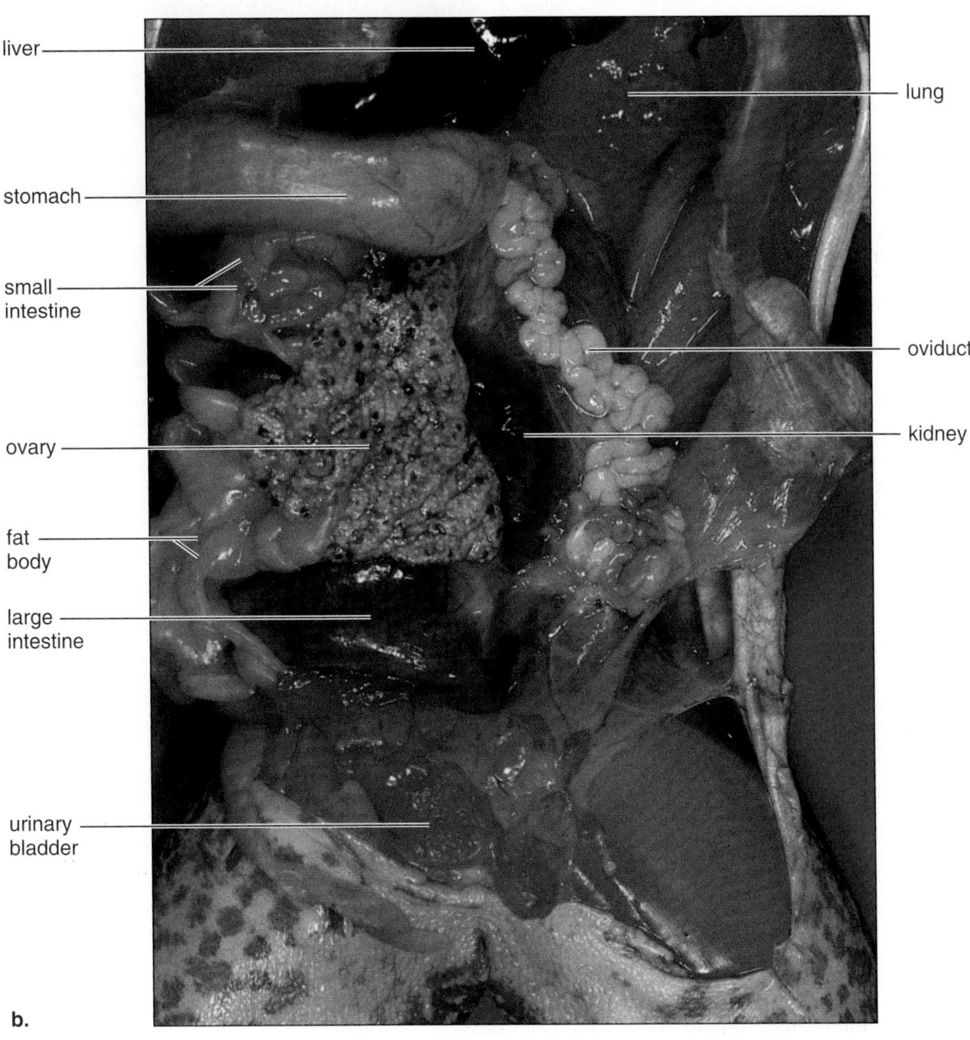

Nervous System

In the frog demonstration dissection, identify the **brain,** lying exposed within the skull. With the help of Figure 30.13, find the major parts of the brain.

Figure 30.13 Frog brain, dorsal view.

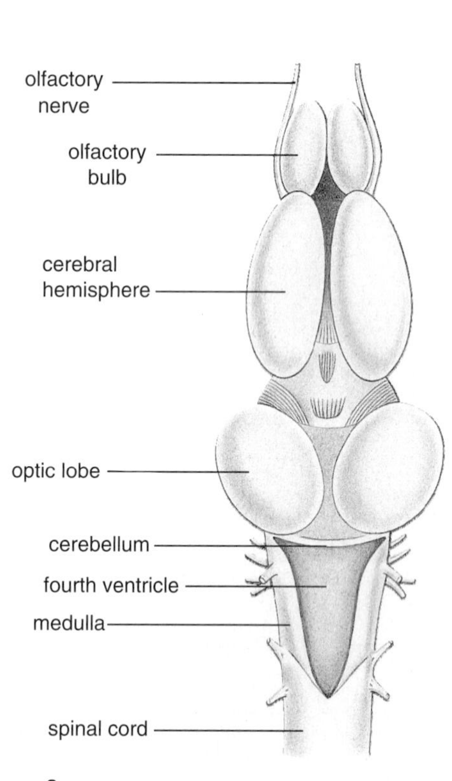

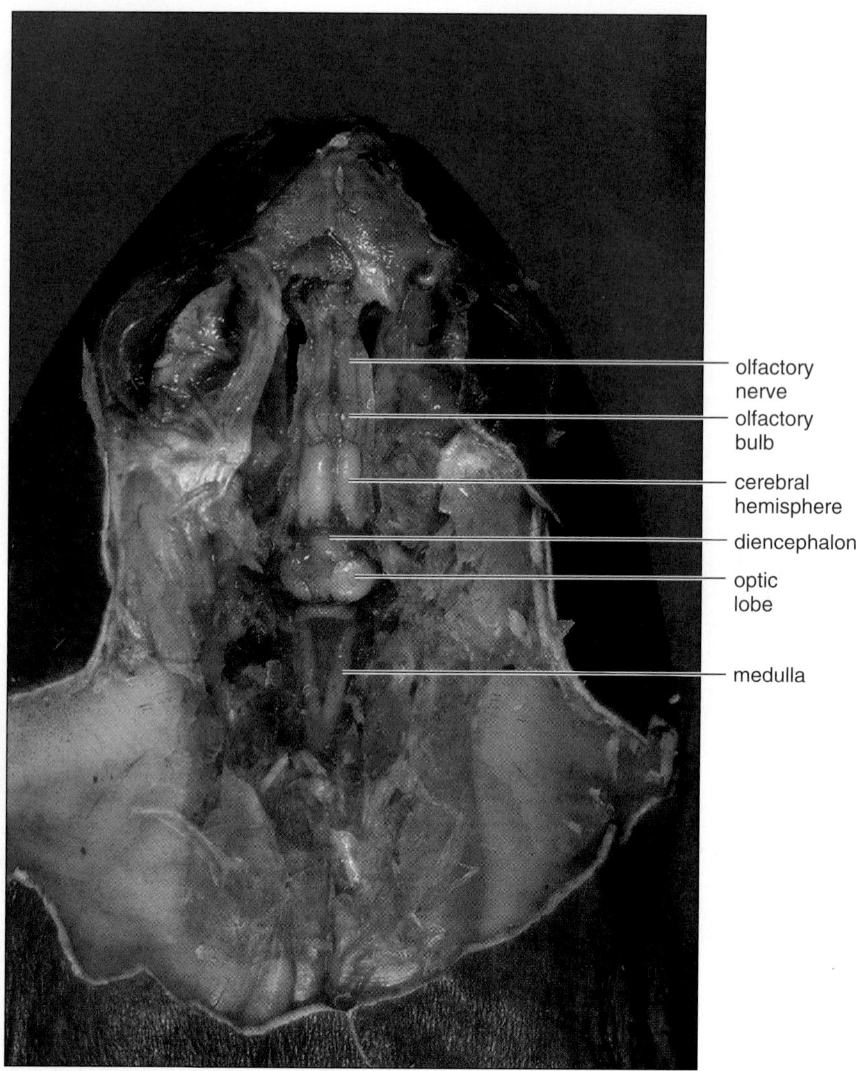

30.4 Comparative Vertebrate Anatomy

Examine the perch (fish), frog (amphibian), pigeon (bird), and pig (mammal) on display.

Observation: External Anatomy of Vertebrates

1. Compare the external features of the perch, frog, pigeon, and pig by answering the following questions and recording your observations in Table 30.2.
   - **a.** Is the skin smooth, scaly, hairy, or feathery?
   - **b.** Is there any external evidence of segmentation?
   - **c.** Are all forms bilaterally symmetrical?
   - **d.** Is the body differentiated into regions?
   - **e.** Is there a well-defined neck?
   - **f.** Is there a post-anal tail?
   - **g.** Are there nares?
   - **h.** Is there a cloaca, or are the urogenital and anal openings separate?
   - **i.** Are eyelids present? How many?

Table 30.2 Comparison of External Features

	Perch	Frog	Pigeon	Pig
a. Skin				
b. Segmentation				
c. Symmetry				
d. Regions				
e. Neck				
f. Post-anal tail				
g. Nares				
h. Cloaca				
i. Eyelids				
j. External ears				
k. Appendages				
l. Digits in forelimb				
m. Digits in hind limb				
n. Nails or claws				

j. Is there external evidence of an ear?
k. How many appendages are there? (Fins are considered appendages.)
l. How many digits are there in the forelimb?
m. How many digits are there in the hind limb?
n. Are nails or claws present?

2. The perch, pigeon, and pig have a nearly impenetrable covering. Why is this an advantage in each case? ______________________________

____

3. A frog uses its skin for breathing. Describe its skin in more detail.

____

Observation: Internal Anatomy of Vertebrates

1. Examine the internal organs of the perch, frog, pigeon, and pig.
2. If necessary, make a median longitudinal incision in the ventral body wall, from the jaws to the cloaca or anus. The body cavity is a coelom.
3. Which of these animals has a **diaphragm** dividing the body cavity into **thorax** and **abdomen?**

____

Circulatory Systems

1. Study heart models for a fish, an amphibian, a bird, and a mammal (Fig. 30.14).
2. Trace the path of the vessel that leaves the ventricle(s), and determine whether the animals have a **pulmonary system** (Fig. 30.14). The word *pulmonary* comes from the Latin *pulmonarius,* meaning "lungs." A pulmonary system contrasts with a **branchial system,** which involves gills.
3. Locate the **sinus venosus,** which enters the atrium in the perch and frog. In the frog, the sinus venosus is a triangular, membranous sac, located just behind the atria and ventricle, that enters the right atrium. The pig and pigeon do not have a sinus venosus.
4. Complete Table 30.3.
5. Do fish have a separate circulatory system to the gills? ______________________________
6. Would you expect blood pressure to be high or low after blood has moved through the gills? ______
7. What animals studied have a separate circulatory system for the respiratory organ? ____________
8. What is the advantage of having a separate circulatory system that returns blood to the heart? ____

____

9. Which of these animals have a four-chambered heart? ______________________________
10. What is the advantage of having separate ventricles? ______________________________

____

11. Contrast the body temperature of animals having a four-chambered heart to that of animals not having a four-chambered heart. ______________________________

____

____

Figure 30.14 Cardiovascular systems in vertebrates.
a. In a fish, the blood moves in a single loop. The heart has a single atrium and ventricle, which pumps the blood into the gill region, where gas exchange takes place. **b.** Amphibians have a double-loop system in which the heart pumps blood to both the gills and the body. **c.** In birds and mammals, the right side pumps blood to the lungs, and the left side pumps blood to the body.

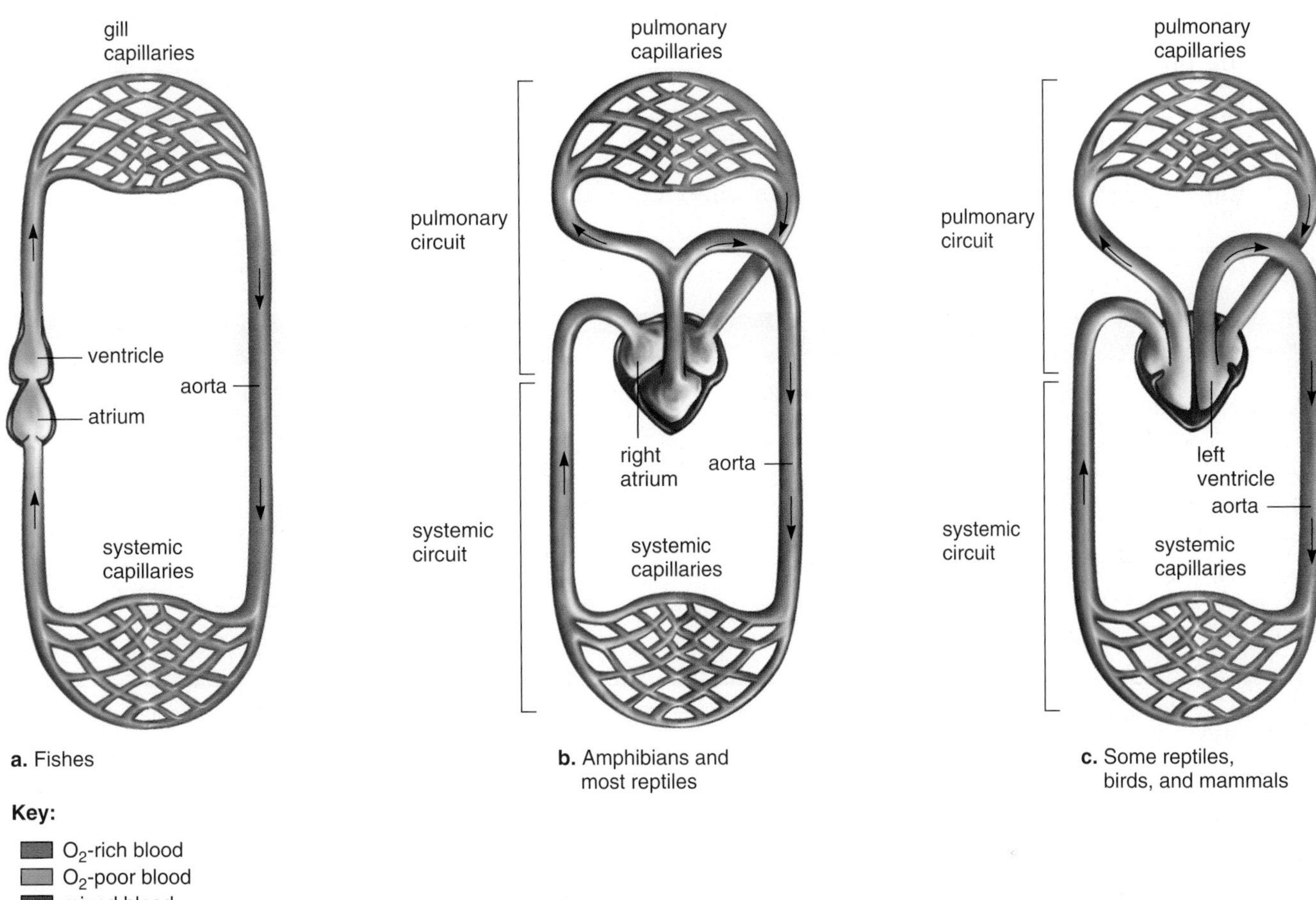

a. Fishes **b.** Amphibians and most reptiles **c.** Some reptiles, birds, and mammals

Table 30.3 Comparative Circulatory Systems

Animal	Number of Heart Chambers	Pulmonary System (Yes or No)
Perch		
Frog		
Pigeon		
Pig		

Respiratory Systems

1. Compare the respiratory systems of the frog, perch, pigeon, and pig (Figs. 30.10, 30.15, 30.16, and 30.17), and *complete Table 30.4 by checking the anatomical features that appear in each animal.*
2. On the basis of your examination, contrast the respiratory system of the perch with those of all the other animals. ____________________

   ____

   Can the differences be related to the environment of the perch compared with the environment of the other animals? __________ Explain. ____________________

   ____

Table 30.4 Respiratory Systems

	Glottis	Larynx	Trachea	Lungs	Rib Cage*	Diaphragm	Air Sacs
Perch							
Frog							
Pigeon							
Pig							

*A rib cage consists of ribs plus a sternum. Some ribs are connected to the sternum, which lies at the midline in the anterior portion of the rib cage.

3. What anatomical feature is present in the pig and pigeon but missing in the frog? __________

   ____

   Can this difference be related to the fact that frogs breathe by positive pressure, while birds and mammals breathe by negative pressure? __________ (A frog swallows air and then pushes the air into its lungs; in birds and mammals, the thorax expands first, and then the air is drawn in.) Explain. ____________________

   ____

   ____

4. What anatomical feature is present only in birds? ____________________ This feature allows the air to pass one way through the lungs of a bird and greatly increases the bird's ability to extract oxygen from the air.
5. What anatomical feature is present only in mammals? ____________________

   Of what benefit is this feature to mammals? ____________________

   ____

Digestive Systems

Figures 30.10, 30.15, 30.16, and 30.17 show the digestive systems of the animals you are studying.

1. Locate the stomach, small intestine, large intestine, rectum, and anus.
2. Is the position of the pancreas the same in all specimens? ______________________
3. Locate the liver.
4. Is a gallbladder present in all specimens? __________ Explain. ______________________

   ____

5. Is a spleen present in all specimens? __________ Does it have the same location in all cases? ____

   ____

Urogenital Systems

1. Identify the gonads and kidneys in the animals you are studying.
2. Try to trace their ducts to the urogenital sinus or cloaca, and check your previous external features observation (see Table 30.2) as to whether a cloaca is present. The urogenital system of the frog is described in Observation: Internal Anatomy of Frog on pages 447–449.

Figure 30.15 Perch anatomy.

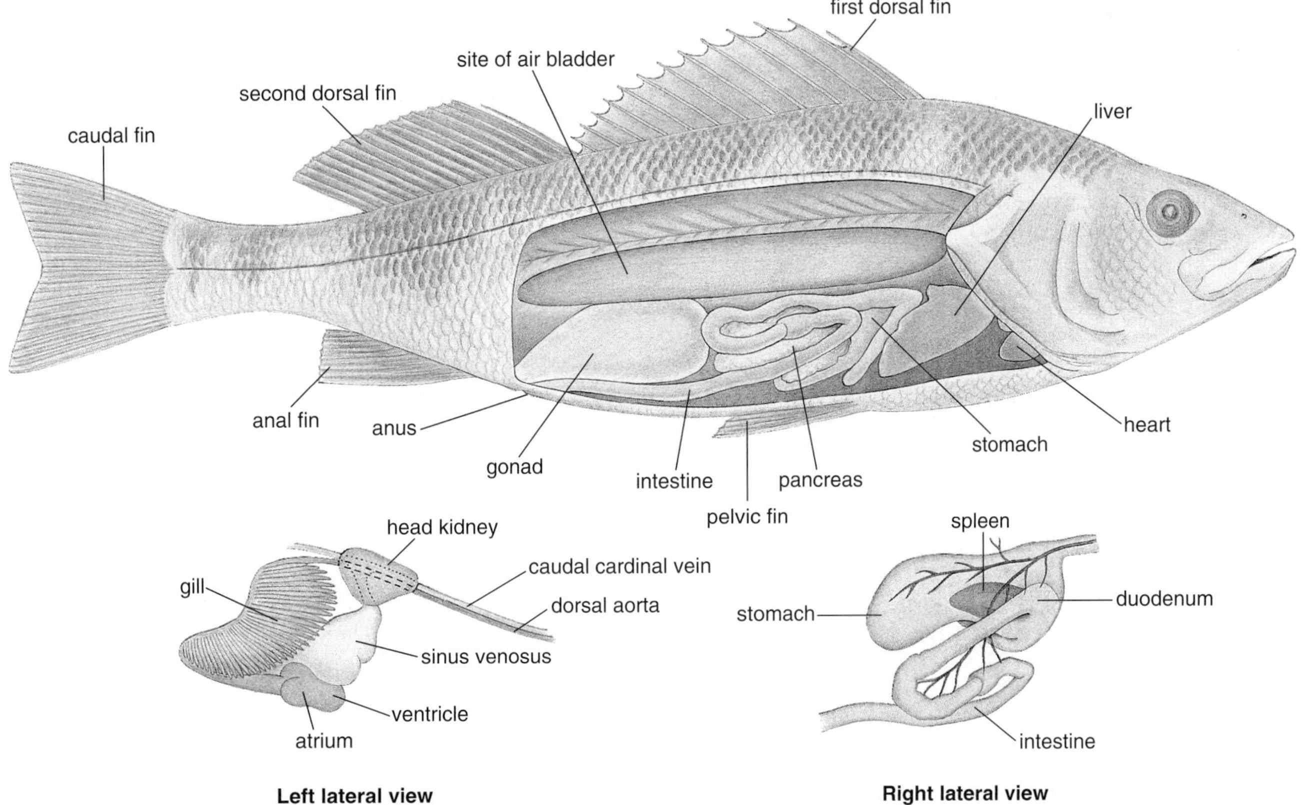

Figure 30.16 Pigeon anatomy.

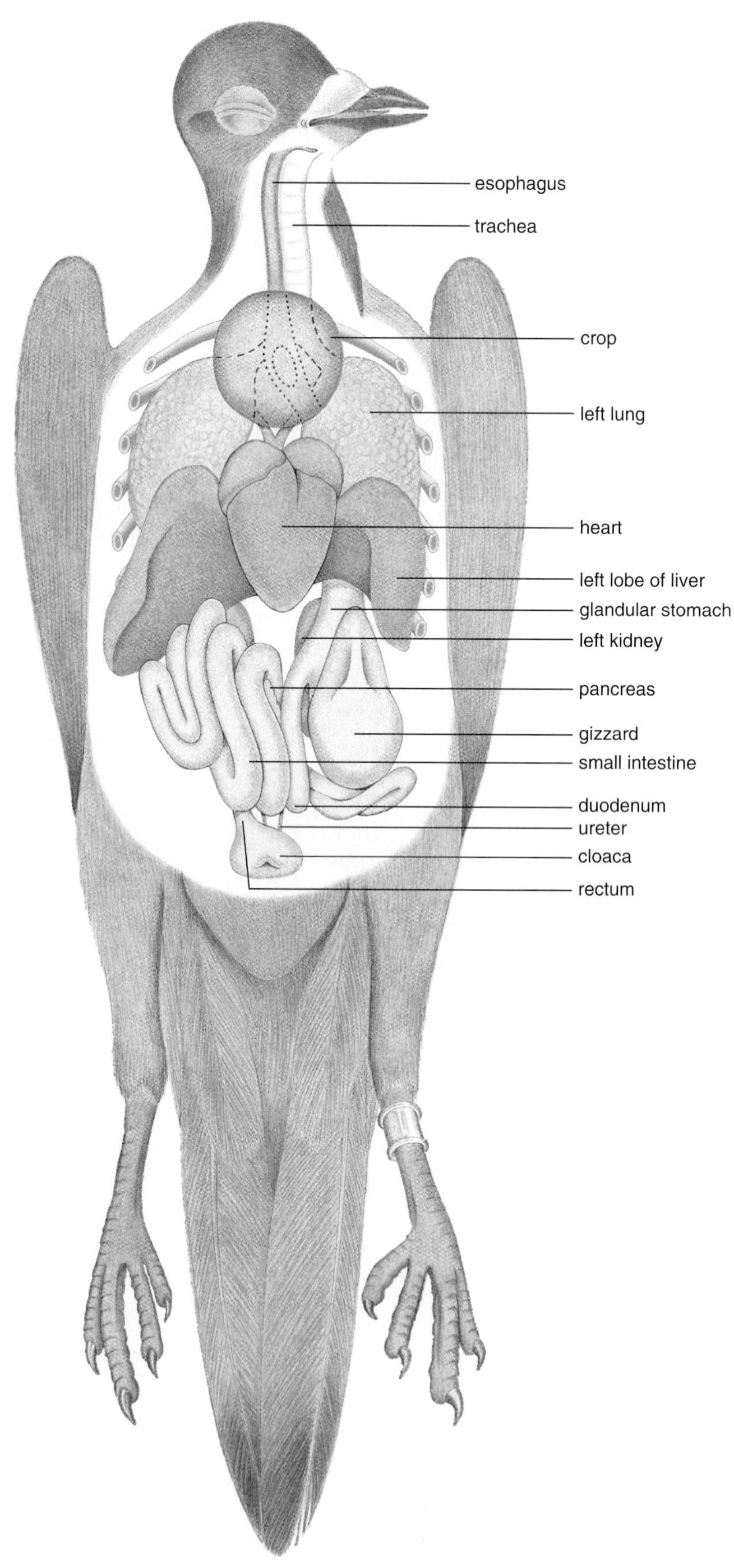

Figure 30.17 Fetal pig anatomy.

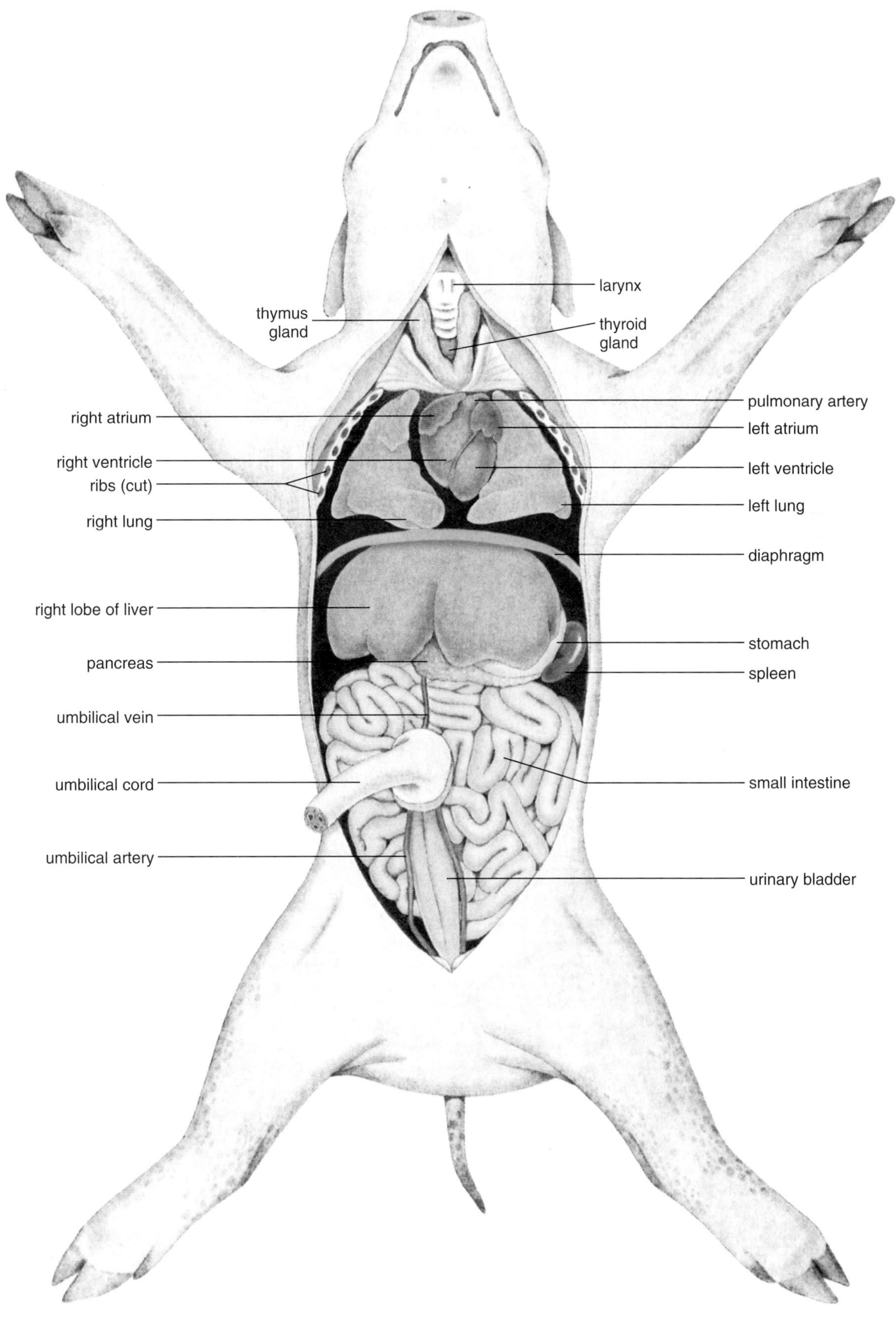

larynx
thymus gland
thyroid gland
pulmonary artery
right atrium
left atrium
right ventricle
left ventricle
ribs (cut)
left lung
right lung
diaphragm
right lobe of liver
stomach
pancreas
spleen
umbilical vein
umbilical cord
small intestine
umbilical artery
urinary bladder

Laboratory Review 30

_______________ 1. Which two phyla comprise the deuterostomes?

_______________ 2. Sea stars are members of what phylum?

_______________ 3. Fish breathe by gills, and reptiles breathe by what structure?

_______________ 4. In deuterostomes, which embryonic opening becomes the mouth?

_______________ 5. Echinoderms have what type of symmetry as adults?

_______________ 6. The tube feet in a sea star are powered by an influx of water through what external structure?

_______________ 7. The chordate invertebrates have what type of nerve cord?

_______________ 8. Why isn't the lancelet a vertebrate?

_______________ 9. In a frog, the glottis allows air to enter the _______.

_______________ 10. In a frog, the esophagus allows food to enter the _______.

_______________ 11. In a frog, the cloaca receives material from the intestine, the kidneys, and the _______.

_______________ 12. Which class of vertebrates is the first class to be fully adapted for reproduction on land?

_______________ 13. A frog's heart has one ventricle, but a pig's heart has how many ventricles?

_______________ 14. Birds have what anatomical feature not seen in any other class of vertebrates?

_______________ 15. What is the major difference in the respiratory system of a perch compared with that of a frog, a pigeon, and a pig?

_______________ 16. What features are associated with all chordates?

Thought Questions

17. Compare echinoderms with chordates on the basis of symmetry and segmentation.

18. What is the major difference between the heart of a frog and that of a pig?

19. What characteristics are associated with:
    a. Mammals but not birds?
    b. Fish but not amphibians?
    c. Birds but not fish?

LABORATORY

31

Sampling Ecosystems

Learning Outcomes

Introduction

- Define ecology, including a reference to the level of organization of an ecosystem.
- Identify the abiotic and biotic components of an ecosystem.

31.1 Terrestrial Ecosystems

- Define and give examples of producers in terrestrial ecosystems.
- Define and give examples of consumers in terrestrial ecosystems.
- Define and give examples of decomposers in terrestrial ecosystems.

31.2 Aquatic Ecosystems

- Define and give examples of producers in aquatic ecosystems.
- Define and give examples of consumers in aquatic ecosystems.
- Define and give examples of decomposers in aquatic ecosystems.

Introduction

Ecology is the study of interactions between organisms and their physical environment. An **ecosystem** consists of a **biotic** (living) community, together with associated **abiotic** (nonliving) components. The abiotic components of an ecosystem include soil, water, light, inorganic nutrients, and weather variables. The biotic components include producers and consumers.

Producers are autotrophic organisms with the ability to carry on photosynthesis and to make food for themselves (and indirectly for the other populations as well). In terrestrial ecosystems, the predominant producers are green plants, while in aquatic ecosystems, the dominant producers are various species of algae.

Consumers are heterotrophic organisms that eat available food. Three types of consumers can be identified, according to their food source:

1. **Herbivores** feed directly on green plants and are termed *primary consumers.* A caterpillar feeding on a leaf is a herbivore.
2. **Carnivores** feed on other animals and are therefore secondary or tertiary consumers. A blue heron feeding on a fish is a carnivore.
3. **Omnivores** feed on both plants and animals. A human who eats both leafy green vegetables and beef is an omnivore.

Decomposers are organisms of decay, such as bacteria and fungi, that break down **detritus** (nonliving organic matter) to inorganic matter, which can be used again by producers. In this way, the same chemical elements are constantly recycled in an ecosystem.

31.1 Terrestrial Ecosystems

Examining an ecosystem in a scientific manner requires concentrating on a representative portion of that ecosystem and recording as much information about it as possible. Representative areas or plots should be selected randomly. For example, random sampling of a terrestrial ecosystem often involves tossing a meterstick gently into the air in the general area to be sampled and then sampling the square meter where the stick lands. A terrestrial ecosystem sampling should include samples from the air above and from the various levels of plant materials growing on and in the ground beneath the selected square meter.

Study of Terrestrial Sampling Site

The objective of the next Experimental Procedure is to characterize the vertical profile of abiotic and biotic conditions in a terrestrial ecosystem and to test the influence of basic differences in ecosystem conditions (e.g., soil, temperature, pH, slope) on this profile. Although terrestrial ecosystems include deciduous forest, prairie, scrubland, and desert, a weedy field, if dominated by annual and perennial herbaceous plants up to a meter or so in height, is also a good site choice.

1. Gather all necessary equipment, such as metersticks, jars with labels for collecting specimens, nets, pH paper, thermometers, and other testing equipment, to take with you to the site. Do not forget data-recording materials. Number your collection jars so that you can use these numbers when recording data later (see Table 31.2).
2. When you arrive at the site, take several minutes to observe the general area. Describe what you observe, including the soil type and coloration.

   ____

   ____
3. Choose two sampling locations (each about 1 m square) at the site. The two locations should differ in significant features (e.g., northern or southern exposure, high or low slope position, time since last disturbance, native versus exotic species composition). Formulate hypotheses about differences between the two sampling locations in regard to the following variables:
   a. Temperature ____

      ____
   b. Plant biomass (quantity of living material) relations ______________________________

      ____
   c. Consumer biomass and diversity (variety of consumers and how many of each type) ______

      ____

Experimental Procedure: Terrestrial Ecosystems

Your instructor will organize class members into teams, and each team will be assigned specific tasks at each of the two chosen sampling locations.

Abiotic Components

Measure the following components, and record your data in Table 31.1:

1. Air temperature at 2 m, 1 m, 0.5 m, and 0.1 m and soil temperature at the soil surface and beneath ground level at −0.1 m, −0.2 m, and −0.3 m, in a hole dug with a soil corer.
2. Humidity at the same aboveground heights specified in step 1.
3. Wind velocity and wind direction at the same aboveground heights specified in step 1.
4. The pH of soil and any standing water at and below ground level, at the same depths specified in step 1.

Table 31.1 Abiotic Components—Terrestrial Ecosystem Analysis

Abiotic Factor	*Measurement*							
	Air				**Soil**			
	2 m	**1 m**	**0.5 m**	**0.1 m**	**Surface**	**− 0.1 m**	**− 0.2 m**	**− 0.3 m**
Temperature								
Humidity								
Wind: Velocity								
Direction								
pH								

Biotic Components—Plants

1. Collect all the dead litter from three 10 cm square areas into plastic bags.
2. Separately harvest live and standing dead plant material by the following height zones (in meters): 1–2, 0.5–1, 0.1–0.5, and 0–0.1 m.
3. Harvest root material by digging and washing soil samples from different depths (in meters): 0.01, 0.1–0.25, and 0–0.1 m.

Biotic Components—Animals

> **Caution:** Wear protective eyewear when working with ether, and keep away from open flame.

1. Cover a section of vegetation with an anesthesia box, and introduce ether. Remove the box, and search for insects, placing them in alcohol in your numbered collection jars. Alternatively, sweep the site with a net and empty the net into a jar filled with alcohol.
2. Collect litter samples for Berlesé funnel analysis.
3. Collect soil samples at the three depths (0.01 m, 0.1–0.25 m, 0–0.1 m) for Berlesé funnel analysis.
4. If instructed to do so, expose labelled sterile petri dishes containing bacterial medium to the air and to samples of soil and water to determine the presence of bacteria and fungi (decomposers).

Experimental Procedure: Terrestrial Ecosystem

Laboratory Work

1. Determine the wet biomass (weight) of plant material, or if a drying oven is available, determine the dry biomass. Dry biomass, although more time-consuming to measure, is preferable when comparing biomasses among sites.
2. Examine collected organisms with a binocular dissecting microscope or, if appropriate, a compound light microscope. Use identification keys provided by your instructor to identify the specimens and to classify them as producers, consumers, or decomposers. Complete Table 31.2.
3. Sort arthropods into herbivore, detritivore (a type of decomposer), and carnivore categories, and determine their wet (or dry, if a drying oven is available) biomass. Use identification keys to classify specimens.
4. Construct a graph showing the profiles of physical conditions, plant biomass, and biomasses of different animal groups.
5. Select any three producers and any three consumers from the organisms collected, and explain how each has adapted to its terrestrial environment.

   Producer 1 ____________________

   Producer 2 ____________________

   Producer 3 ____________________

   Consumer 1 ____________________

   Consumer 2 ____________________

   Consumer 3 ____________________

6. Return all living creatures, soil samples, and water samples to their respective collection sites, as explained by your instructor. If any organisms were preserved, ask your instructor what to do with them. Place any exposed petri dishes in a designated area for incubation until the next laboratory.

Table 31.2 Biotic Components—Terrestrial Ecosystem Analysis

			Abundance			
Constituents	**Location in Ecosystem**	**Rare**	**Occasional**	**Abundant**	**Classification (Phylum/Class)**	**Identifying Feature for Classification**
Producers						
Consumers						
Decomposers						

Conclusions

Compare the results of this Experimental Procedure with the hypotheses you formulated (see p. 460).

- Temperature ______________________________

  ____

- Plant biomass relations ______________________________

  ____

- Consumer biomass and diversity ______________________________

  ____

31.2 Aquatic Ecosystems

An aquatic sampling should include samples from the air above the water column, the column of water itself, and the soil beneath the water column.

Study of Aquatic Sampling Site

The objective of the next Experimental Procedure is to characterize the vertical profile of abiotic and biotic conditions in an aquatic ecosystem and to test the influence of a basic difference in ecosystem conditions on this profile. Aquatic ecosystems consist of freshwater ecosystems (e.g., lakes, ponds, rivers, and streams) and marine ecosystems (e.g., oceans). A good site for this study is a large pond, small lake, or reservoir (with a shallow margin having rooted aquatic plants and a deeper zone with water 2 to 3 or more meters deep).

1. Gather all necessary equipment, such as metersticks, collection jars with labels, nets, pH paper, thermometers, and other testing equipment, to take with you to the site. Do not forget data-recording materials. Number your collection jars so that you can use these numbers when recording data in Table 31.3.
2. When you arrive at the site, take several minutes to observe the general area. Describe what you observe.

   ____

   ____

3. Choose two sampling locations that differ in significant features (e.g., sheltered by trees versus unsheltered, near stream inflow versus far from stream inflow). Plan to sample conditions near shore (shallow-water zone) and in deeper water away from shore (or only one or the other, if logistics are limiting).

   Formulate hypotheses about differences between the two sampling locations in

   a. Temperature, humidity, and wind profiles above the surface ______________________________

   ____

   b. Temperature, oxygen, pH profiles, and visibility below the surface ______________________________

   ____

   c. Plant biomass relations ______________________________

   ____

   d. Consumer biomass and diversity ______________________________

   ____

Experimental Procedure: Aquatic Ecosystem

Your instructor will organize class members into teams, and each team will be assigned specific tasks at each of the two sampling locations.

Abiotic Components

Measure and record the following components in Table 31.3:

1. In the first column (under "Air"), record the temperature, humidity, and wind velocity at heights of 2 m, 1 m, 0.5 m, and 0.1 m above the surface.
2. In the second column (under "Water"), record the pH at the surface. Then record the temperature, oxygen content, and visibility at the surface and at depths below the surface of −0.1 m, −0.5 m, −1.0 m, and −2.0 m. (Use a measuring stick to determine the depth of visibility.)

Table 31.3 Abiotic Components—Aquatic Ecosystem Analysis

Abiotic Factor	*Measurement*								
	Air				Water				
	2 m	1 m	0.5 m	0.1 m	Surface	− 0.1 m	− 0.5 m	− 1 m	− 2 m
Temperature									
Humidity									
Wind velocity									
pH									
Oxygen content									
Visibility									

Biotic Components

Plankton contains phytoplankton and zooplankton. **Phytoplankton** consists of photosynthesizing organisms (e.g., algae), and **zooplankton** consists of minute consumers (e.g., protozoans) that live suspended in water. The **benthos** consists of organisms that live on the floor of a body of water.

1. In a shallow-water zone, pour a measured volume (e.g., 10 liters) of water collected at 0.01 m through plankton nets, and transfer plankton samples to collection bottles.
2. Collect plankton samples from various depths, using a Schindler trap and plankton filter nets suitable to phytoplankton and zooplankton. Or take water sampled at various depths with Niskin or Van Dorn bottles, and pour measured volumes through plankton nets.
3. Obtain samples of plant material.
4. Collect a bottom sediment sample with an Eckman dredge. Sieve, and preserve animals in alcohol bottles.
5. If instructed to do so, expose labelled sterile petri dishes containing bacterial medium to the air and to samples of soil and water to determine the presence of bacteria and fungi (decomposers).

Laboratory Work

1. Examine the collected organisms with a binocular dissecting microscope or, if appropriate, a compound light microscope. Examine water samples with a compound light microscope. Use identification keys provided by your instructor to identify the specimens and to classify them as producers, consumers, or decomposers. Complete Table 31.4.

2. Sort plankton and benthos into herbivore and carnivore categories, and determine their wet (or dry, if a drying oven is available) biomass.
3. Construct a graph showing the profiles of physical conditions, plant biomass, and biomasses of different animal groups.
4. Construct a biomass pyramid of producers, herbivores (including detritivores [some decomposers]), and carnivores.

Table 31.4 Biotic Components—Aquatic Ecosystem Analysis

		Abundance				
Constituents	**Location in Ecosystem**	**Rare**	**Occasional**	**Abundant**	**Classification (Phylum/Class)**	**Identifying Feature for Classification**
Producers						
Consumers						
Decomposers						

5. Select any three producers and any three consumers from the organisms collected, and explain how each has adapted to its aquatic environment.

   Producer 1 ______________________________

   Producer 2 ______________________________

   Producer 3 ______________________________

   Consumer 1 ______________________________

   Consumer 2 ______________________________

   Consumer 3 ______________________________

6. Return all living creatures, soil samples, and water samples to their respective collection sites, as explained by your instructor. If any organisms were preserved, ask your instructor what to do with them. Place any exposed petri dishes in a designated area for incubation until the next laboratory.

Conclusions

Compare the results of this Experimental Procedure with the hypotheses you formulated (see p. 463).

- Temperature, humidity, and wind profiles above the surface ______________________________

  ____

- Temperature, oxygen, pH profiles, and visibility below the surface ______________________

- Plant biomass relations ______________________

- Consumer biomass and diversity ______________________

Laboratory Review 31

______ 1. This laboratory studies organisms at what level of organization?

______ 2. Give an example of an abiotic component of an ecosystem.

______ 3. Give an example of a biotic component of an ecosystem.

______ 4. Which type of consumer feeds directly on green plants?

______ 5. Which type of consumer feeds on both plants and animals?

______ 6. If your assigned ecosystem contains plankton, it is most likely which type of ecosystem?

______ 7. Which type of ecosystem contains decomposers?

______ 8. In a forest ecosystem, what type of organisms are the predominant producers?

______ 9. What is the role of a producer?

______ 10. Give an example of a consumer in a terrestrial ecosystem.

______ 11. In an aquatic ecosystem, what type of organisms are the predominant producers?

______ 12. Give an example of a consumer in an aquatic ecosystem.

______ 13. What type of organisms are the predominant decomposers in an ecosystem?

______ 14. What does a decomposer take in for nutrients?

______ 15. How do phytoplankton differ from zooplankton?

Thought Questions

16. What is an ecosystem?

17. What is the role of decomposers in an ecosystem?

18. Explain how a significant change in temperature or pH might affect the biotic components of an ecosystem.

32

Effects of Pollution on Ecosystems

Learning Outcomes

32.1 Studying the Effects of Pollutants

- Based on your observations, generalize the effect of acid deposition on the growth of organisms.
- Based on your observations, generalize the effect of oxygen deprivation on organisms.
- Based on your observations, generalize the effect of thermal pollution on organisms.

32.2 Studying the Effects of Cultural Eutrophication

- Identify potential sources of pollution and predict the effect of cultural eutrophication on the food chains of a lake or pond in your vicinity.

Introduction

Pollutants are substances added to the environment that tend to disrupt normal biological processes. Pollutants often result from human activity, but they can be the result of natural processes. For example, acid rain occurs when oxides of nitrogen and sulfur mix with water in the atmosphere. While most of the atmospheric sulfur can be traced to human activities such as auto and industrial emissions, some sulfur is released naturally from steam vents and volcanic eruptions. When the sulfur compounds interact with water in the atmosphere, precipitation that has a low pH can form. Because most organisms are able to tolerate only narrow changes in environmental conditions, significant changes like this can significantly affect an organism's ability to survive. In this laboratory, we will examine the effect of two types of pollutants—acid deposition and thermal pollution—on the well-being of organisms and hypothesize how these pollutants might affect an ecological pyramid, such as the one shown in Figure 32.1.

Figure 32.1 Ecological pyramid.
The biomass at each succeeding trophic (feeding) level in an ecosystem decreases. At the right are examples of organisms at these trophic levels. They form a food chain in which owls eat shrews, which eat beetles, which eat green plants.

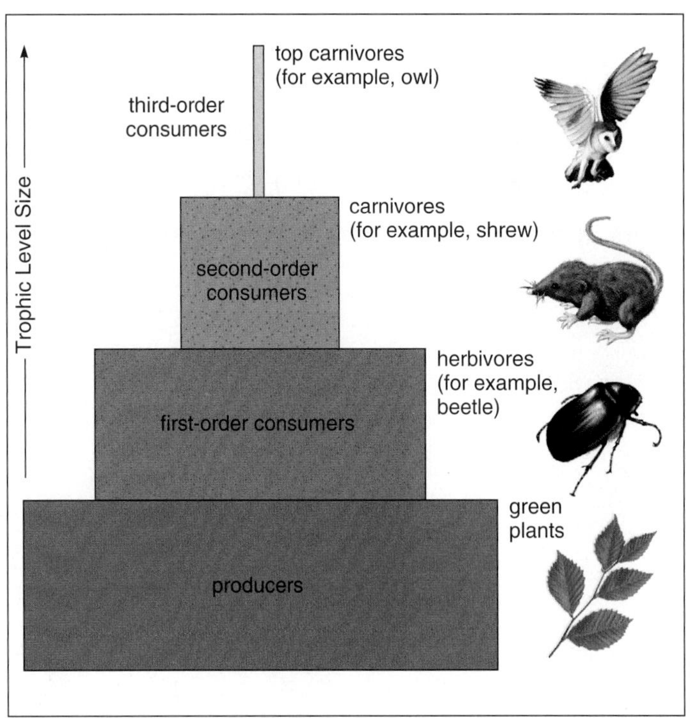

We will also explore how ecosystem interactions can change when excess nutrients are added to a system. Over time, bodies of water undergo a natural enrichment process called **eutrophication.** Eutrophication leads to overgrowth that can eventually cause a pond or lake to fill in and disappear.

In nature, eutrophication is typically the result of a long process. Sometimes, human activities add excess nutrients to bodies of water, often in relatively short periods of time, and this is called *cultural eutrophication*. As cultural eutrophication proceeds, it leads to a lack of oxygen and the death of most, if not all, of the organisms living in the water.

As our knowledge of human impact on the environment has grown, we have begun to recognize that even activities that might be viewed as environmentally friendly can have a down side. Recycling sewage sludge as fertilizer and reclaiming waste water have both been seen as examples of good environmental management. Now, scientists recognize that pharmaceuticals and personal care products pass through our systems into these reclaimed products, increasing their prevalence in the environment and having a significant impact other organisms.

Pollution is not limited to chemical changes in the environment. Temperature change, or thermal pollution, is a major issue in the world today. Not only are we concerned about the potential threat of global warming, but even localized thermal pollution can affect the ecosystem. Hot water discharges from power plants can alter the ecological communities of rivers, lakes, or coastal regions of the ocean shoreline.

32.1 Studying the Effects of Pollutants

In this laboratory exercise, we will study the effects of pollution by observing its effects on hay infusion cultures, on seed germination, and on an animal called *Gammarus*.

Study of Hay Infusion Cultures

A hay infusion culture (hay soaked in water) contains various microscopic organisms, such as those depicted in Figure 32.2. What do you predict will happen to organisms in a hay infusion culture when conditions are acidic, or when overenrichment occurs? Why? ______________________

____

____

Figure 32.2 Microorganisms in hay infusion cultures.

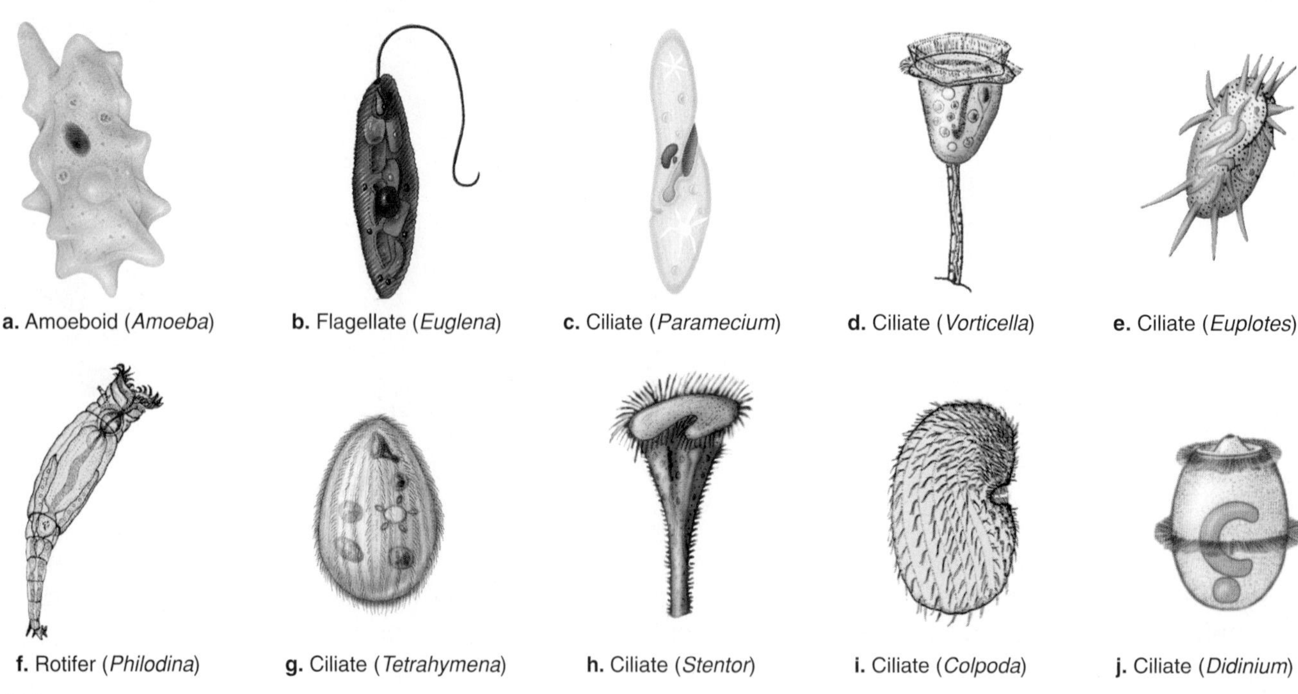

a. Amoeboid (*Amoeba*) **b.** Flagellate (*Euglena*) **c.** Ciliate (*Paramecium*) **d.** Ciliate (*Vorticella*) **e.** Ciliate (*Euplotes*) **f.** Rotifer (*Philodina*) **g.** Ciliate (*Tetrahymena*) **h.** Ciliate (*Stentor*) **i.** Ciliate (*Colpoda*) **j.** Ciliate (*Didinium*)

When nutrients produced by human activities enter a body of water, the algae overpopulate. Zooplankton are unable to significantly reduce the algal population, and bacteria use up all the available oxygen when decomposing them. What do you predict will happen next? Why? ___________

____

Experimental Procedure: Study of Hay Infusion Cultures

The following four hay infusion cultures have been provided:

1. **Control culture:** This hay infusion culture simulates the optimum conditions for normal growth.
2. **Enriched culture:** Same as control culture, but with more inorganic nutrients for the growth of algae.
3. **Oxygen-deprived culture:** Same as control culture, but with minimal oxygen to test the effect of lack of oxygen on organisms.
4. **Acidic culture:** Same as control culture, except that it is adjusted to pH 4 with sulfuric acid (H_2SO_4). This simulates the effect of acid deposition on organisms.

Examine each of these cultures by preparing a wet mount. Record in Table 32.1 the diversity of life that you observe and the relative quantity of organisms.

Table 32.1 Hay Infusion Cultures

Wet Mount	Type of Culture	Diversity of Life (List Organisms)	Relative Quantity of Organisms (High, Medium, or Low)
1	Control		
2	Enriched		
3	Oxygen-deprived		
4	Acidic		

Study of Seed Germination

Just like animals, plant seeds depend on proper conditions of temperature, oxygen, and moisture to germinate, grow, and reproduce. Many pollutants affect seeds' ability to germinate.

Experimental Procedure: Study of Seed Germination

Observe these petri dishes, and record your observations in Table 32.2.

1. **Control petri dish:** The seeds in this petri dish have been watered with a control solution having a neutral pH.
2. **Acidic petri dish:** The seeds in this petri dish have been watered with an acidic solution having a pH of 4 to simulate acid rain.

Table 32.2 Seed Germination

Petri Dish	Type of Solution	pH	Observations
1	Control		
2	Acidic		

Conclusions

- How might these observations of the hay infusion experiment relate to real ecosystems?
  ____
- What are the potential consequences of acid deposition on plant populations that reproduce by seeds? ____
- How does the addition of nutrients affect diversity and the relative quantity of organisms?
  ____

Study of *Gammarus*

We will study the effect of thermal pollution and acid rain on a small crustacean called *Gammarus,* which is found in ponds and streams (Fig. 32.3).

Experimental Procedure: Gammarus

Control Culture

1. Add 25 ml of spring water to a container.
2. Measure the pH of the spring water with a pH meter or litmus paper, and record it here: ________ . Add four *Gammarus* to the container.
3. Observe the behavior of *Gammarus* for 10 to 15 minutes, and then answer the following questions. Retain these *Gammarus* for subsequent exercises.
   - Where do the *Gammarus* spend their time in the container? ____
   - How do they spend their time? ____
   - What percentage of their time is spent moving? ____

- Do they use all their legs in swimming?

  ___

- Which legs are used in jumping and climbing?

  ___

- Do *Gammarus* avoid each other?

  ___

- What do *Gammarus* do when they "bump" into each other? ___

Figure 32.3 *Gammarus.*
Gammarus is a type of crustacean classified in a subphylum that also includes shrimp.

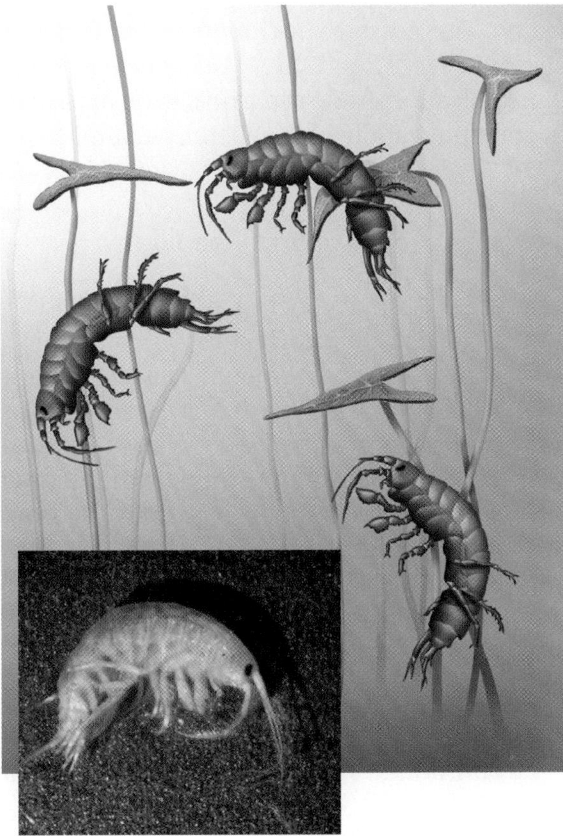

Thermal Pollution

1. To simulate thermal pollution, boil 100 ml of *water,* and then allow it to cool to 31°C.
2. Put 25 ml of this 31°C water in a beaker, and add two *Gammarus.*
3. Observe the behavior of *Gammarus,* and answer the following questions:
   - What is the effect of thermal pollution on the behavior of *Gammarus?*

     ___

     ___

   - Why might an increase in temperature affect the behavior of *Gammarus* in this way?

     ___

     ___

Acid Pollution

1. Put 25 ml of *acidic spring water* (adjusted to pH 4) in a beaker, and add two *Gammarus.*
2. Observe the animals' behavior, and answer the following questions:
   - What is the difference between the pH of the control culture and the pH of the acidic culture?

     ___

   - Compare the behavior of *Gammarus* in the control culture with its behavior in the acidic culture.

     ___

32.2 Studying the Effects of Cultural Eutrophication

This lab activity will simulate the cultural eutrophication of a fresh water lake or pond. *Chlorella,* the alga used in this study, is considered to be representative of phytoplankton in bodies of fresh water. The crustacean *Daphnia* is a zooplankton that feeds on *Chlorella*. First, you will observe how *Daphnia* feeds, and then you will determine the extent to which *Daphnia* could keep cultural eutrophication from occurring in a hypothetical example. Keep in mind that this case study is an oversimplification of a generally complex problem.

Observation: Daphnia *Feeding*

1. Place a small pool of petroleum jelly in the center of a small petri dish.
2. Use a dropper to take a *Daphnia* from the stock culture, place it on its back (covered by water) in the petroleum jelly, and observe it under the binocular dissecting microscope.
3. Note the clamlike carapace and the legs waving rapidly as the *Daphnia* filters the water.
4. Add a drop of *carmine solution,* and observe how the *Daphnia* filters the "food" from the water and passes it through the gut. The gut is more visible if you push the animal onto its side. In this position, you may also observe the heart beating in the region above the gut and just behind the head.
5. Allow the *Daphnia* to filter-feed for up to 30 minutes, and observe the progress of the carmine particles through the gut. Does the carmine travel completely through the gut in 30 minutes?

   ____

Experimental Procedure: Daphnia *Feeding on* Chlorella

This exercise requires the use of a spectrophotometer. Absorbance will be a measure of the algal population level; the greater the number of algal cells, the greater the absorbance. The higher the absorbance, the greater the amount of light absorbed and *not* passed through the solution.

1. Obtain two spectrophotometer tubes (cuvettes) and a Pasteur pipette.
2. Fill one of the cuvettes with distilled water, and use it to zero the spectrophotometer. Save this tube for number 6.
3. Use the Pasteur pipette to fill the second cuvette with *Chlorella.* Gently aspirate and expel the sample several times (without creating bubbles) to give a uniform dispersion of the algae.
4. Add ten hungry *Daphnia,* and following your instructor's directions, immediately measure the absorbance with the spectrophotometer. If a *Daphnia* swims through the beam of light, a strong deflection should occur; do not use any such higher readings—instead, use the lower reading for the absorbance. Record your reading in the first column of Table 32.3.
5. Remove the cuvette with the *Daphnia* to a safe place in a test-tube rack. Allow the *Daphnia* to feed for 30 minutes.
6. Rezero the spectrophotometer with the distilled water cuvette.
7. Measure the absorbance of the experimental cuvette again. Record your reading in the second column of Table 32.3, and explain your results in the third column.

Table 32.3 Spectrophotometer Data of *Daphnia* Feeding on *Chlorella*

Absorbance Before Feeding	Absorbance After Feeding	Explanation

Experimental Procedure: Case Study in Cultural Eutrophication

The following problem will test your understanding of the value of a single species—in this case, *Daphnia*. Please realize that this is an oversimplification of a generally complex problem.

1. Assume that developers want to build condominium units on the shores of Silver Lake. Homeowners in the area have asked the regional council to determine how many units can be built without altering the nature of the lake. As a member of the council, you have been given the following information:

   The present population of *Daphnia,* which is 10 animals/liter, presently filters 24% of the lake per day, meaning that it removes this percentage of the algal population per day. This is sufficient to keep the lake essentially clear. Predation—the eating of the algae—will allow the *Daphnia* population to increase to no more than 50 animals/liter. Therefore, 50 *Daphnia*/liter will be available for feeding on the increased number of algae that would result from building the condominiums.

   Using this information, complete Table 32.4.

Table 32.4 *Daphnia* Filtering

Number of *Daphnia*/Liter	Percent of Lake Filtered
10	24%
50	

2. The sewage system of the condominiums will add nutrients to the lake. Phosphorus output will be 1 kg per day for every ten condominiums. This will cause a 30% increase in the algal population. Using this information, complete Table 32.5.

Table 32.5 Cultural Eutrophication

Number of Condominiums	Phosphorus Added	Increase in Algal Population
10	1 kg	30%
20		
30		
40		
50		

Conclusion

- Assume that phosphorus is the only nutrient that will cause an increase in the algal population and that *Daphnia* is the only type of zooplankton available to feed on the algae. How many condominiums would you allow the developer to build? ____________________

  ____
- What other possible impacts could condominium construction have on the condition of the lake? ____________________

Laboratory Review 32

_______________ 1. Any effect on seeds would typify an effect on what trophic level in an ecosystem?

_______________ 2. Any effect on *Gammarus* would typify an effect on what type of population in an ecosystem?

_______________ 3. Stable organic chemicals are subject to what process as they move through food chains?

_______________ 4. What type of pollution results when water from rivers and ponds is used to cool industrial processes?

_______________ 5. In your experiment, did you add acid or base to adjust the hay infusion culture to pH 4?

_______________ 6. What condition does acid deposition cause that can be harmful to organisms?

_______________ 7. What trophic level is most subject to the effects of biological magnification?

_______________ 8. Cultural eutrophication begins with an excess of what type of substances?

_______________ 9. Overenrichment causes which types of populations to increase in size beyond the ordinary?

_______________ 10. In the case study, what two factors caused the producer population (*Chlorella*) to increase?

_______________ 11. What are free-floating algae called? Why?

_______________ 12. Why do excess nutrients alter some aquatic ecosystems?

Thought Questions

13. Does pollution affect all living things, including humans? How?

14. When pollutants enter the environment, they have far-ranging effects. Give an example from this laboratory.

15. Why might it be important to monitor predator/prey interactions when studying ecosystems?

APPENDIX

A

Preparing a Laboratory Report/ Laboratory Report Form

A laboratory report has the sections (and paragraphs) noted in the outline that follows. Use this description and a copy of the worksheet provided on pages 477 and 478 to help you write a report assigned by your instructor.

1. **Introduction:** Tell the reader what the experiment was about.
   a. **Background information:** Begin by giving an overview of the topic.
   Look at the introduction to the laboratory (and/or at the introduction to the section for which you are writing the report). *Do not copy the information,* but use it to indicate ideas about what information to include. For example, suppose you are doing a laboratory report on "Solar Energy" in Laboratory 8 (Photosynthesis). You might want to give a definition of photosynthesis and explain the composition of white light.
   b. **Purpose:** Next, state what the experiment was about.
   Think about the steps of the experiment, and then decide what its purpose was. Since you tested the effect of white and green light on photosynthesis, you might state that the purpose of this experiment was to determine the effect of white light versus green light on the photosynthetic rate.
   c. **Hypothesis:** Finally, state what results you expected from the experiment.
   Most likely, the introduction to the laboratory (and/or the particular section) will hint at the expected results. State this in the form of a hypothesis. For example, you might hypothesize that white light would be more effective for photosynthesis than green light.
2. **Procedure:** Tell the reader how you did the experiment.
   a. **Equipment used:** Help the reader envision the experimental setup.
   Look at any illustrations that accompany the experiment. For example, look at Figure 8.5, and read the Experimental Procedure, which describes how the experiment is conducted. Now, describe the equipment in your own words. You might state that a 150-watt lamp was the source of white light directed at *Elodea* placed in a test tube filled with a solution of sodium bicarbonate ($NaHCO_3$). A beaker of water placed between the lamp and test tube was a heat absorber.
   b. **Collection of data:** Tell the reader how the experiment was done.
   Think about what you did during the experiment. What did you observe, or what did you measure? You might state that the rate of photosynthesis was determined by the amount of oxygen released and was measured by how far water moved in a side arm of the test tube.

3. **Results:** Present the data in a clear manner.
   a. **Graph or table:** If at all possible, show your data in a graph or table.
      Look at any tables you filled in during the experiment, and reproduce those sections that pertain to data collection. (Do not include any columns that have to do with interpretation.)
      To continue with the Laboratory 8 example, you could reproduce Tables 8.2 and 8.3.
      You also might be able to draw a graph depicting results. For example, the results recorded in Table 6.2 might look like the graph at right.

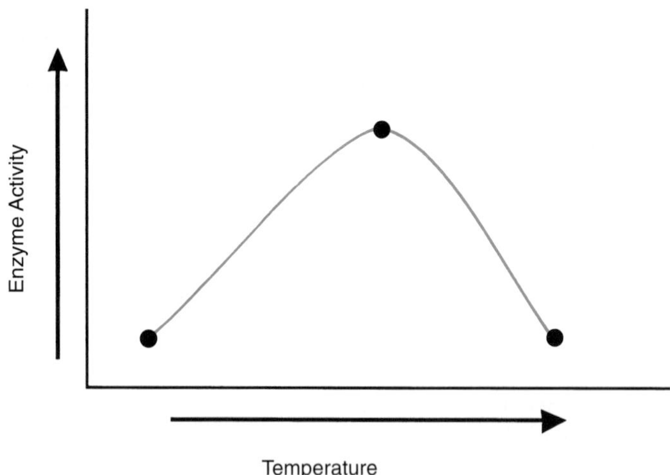

   b. **Description of data:**
      Examine your data, and decide what they tell you.
      For Table 8.3, you might state that the data in the table indicate that the rate of photosynthesis with white light was faster than with green light.
4. **Discussion:** Interpret your data and offer conclusions.
   a. **Support of hypothesis:** Tell whether the hypothesis was supported.
      Compare your hypothesis with your data. Do your data agree with the hypothesis?
      You might state that your results supported the hypothesis that white light gives a higher rate of photosynthesis than green light.
   b. **Explanation:** Explain why you think you got the results you did.
      Look at any questions you answered while in the laboratory, and use them to help you decide on an appropriate explanation. For example, the answers to the questions in Solar Energy section 8.2 might help you state that white light gives a higher rate of photosynthesis because it contains all the visual light rays. Green light gives a lower rate of photosynthesis because green plants do not absorb green light.
      (Be sure to report your data exactly. If your results fail to support the hypothesis, explain why you think this occurred.)
   c. **Conclusion:** Review the entire experiment in your mind, and then explain what you learned. You might state that you learned that the quality of light can affect the rate of photosynthesis.

POINTS TO REMEMBER

1. *Clarity.* Your report should be organized and presented so that the reader can easily understand your work.
2. *Good writing.* Pay particular attention to grammar and style. Your writing will be enhanced by careful proofreading and revision. Have someone else read and critique a draft of your report. Then consider his or her comments seriously in your revisions.
3. *Observations and interpretations.* Keep the distinction between observations and interpretations clear. These are frequently confused in student reports.
4. *Exercise good judgment.* As you prepare your report, you will constantly make difficult decisions about how best to present and interpret your data. Your attention to these decisions will be reflected in the quality of your report.

Date ______________________

Name ______________________

Laboratory Report for ____

____

1. **Introduction**
   a. Background information

   b. Purpose

   c. Hypothesis

2. **Procedure**
   a. Equipment used

   b. Collection of data

3. Results

a. Graph or table
(Place these on attached sheets.)

b. Description of data

4. Discussion

a. Support of hypothesis

b. Explanation

c. Conclusion

APPENDIX

Metric System

Unit and Abbreviation	Metric Equivalent	Approximate English-to-Metric Equivalents
Length		
nanometer (nm)	$= 10^{-9}$ m (10^{-3} μm)	
micrometer (μm)	$= 10^{-6}$ m (10^{-3} mm)	
millimeter (mm)	= 0.001 (10^{-3}) m	
centimeter (cm)	= 0.01 (10^{-2}) m	1 inch = 2.54 cm
		1 foot = 30.5 cm
meter (m)	= 100 (10^{2}) cm	1 foot = 0.30 m
	= 1,000 mm	1 yard = 0.91 m
kilometer (km)	= 1,000 (10^{3}) m	1 mi = 1.6 km
Weight (mass)		
nanogram (ng)	$= 10^{-9}$ g	
microgram (μg)	$= 10^{-6}$ g	
milligram (mg)	$= 10^{-3}$ g	
gram (g)	= 1,000 mg	1 ounce = 28.3 g
		1 pound = 454 g
kilogram (kg)	= 1,000 (10^{3}) g	= 0.45 kg
metric ton (t)	= 1,000 kg	1 ton = 0.91 t
Volume		
microliter (μl)	$= 10^{-6}$ l (10^{-3} ml)	
milliliter (ml)	$= 10^{-3}$ liter	1 tsp = 5 ml
	= 1 cm^3 (cc)	1 fl oz = 30 ml
	= 1,000 mm^3	
liter (l)	= 1,000 ml	1 pint = 0.47 liter
		1 quart = 0.95 liter
		1 gallon = 3.79 liter
kiloliter (kl)	= 1,000 liter	

Units of Temperature

°F — °C

212° — 100°
160° — 71°
134° — 57°
131°
105.8° — 41°
98.6° — 37°
56.66° — 13.7°
32° — 0°

To convert temperature scales:
°C = (F° − 32)/1.8
F° = 1.8(C°) + 32

Common Temperatures

°C	°F	
100	212	Water boils at standard temperature and pressure.
71	160	Flash pasteurization of milk
57	134	Highest recorded temperature in the United States, Death Valley, July 10, 1913
41	105.8	Average body temperature of a marathon runner in hot weather
37	98.6	Human body temperature
13.7	56.66	Human survival is still possible at this temperature.
0	32.0	Water freezes at standard temperature and pressure.

APPENDIX

Classification of Organisms

DOMAIN BACTERIA

Prokaryotic, unicellular organisms that lack a membrane-bounded nucleus and reproduce asexually. Metabolically diverse, being heterotrophic by absorption; autotrophic by chemosynthesis or by photosynthesis. Motile forms move by flagella consisting of a single filament.

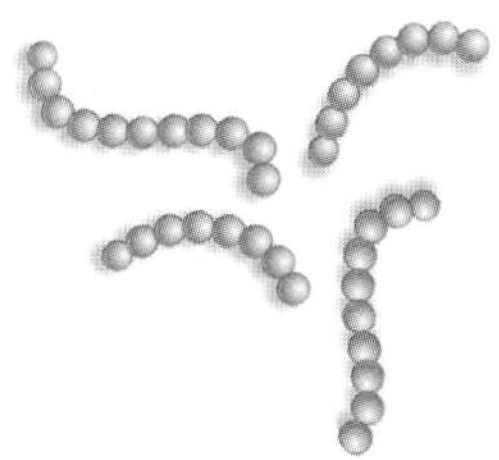

DOMAIN ARCHAEA

Prokaryotic, unicellular organisms that lack a membrane-bounded nucleus and reproduce asexually. Many are autotrophic by chemosynthesis; some are heterotrophic by absorption. Most live in extreme or anaerobic environments. Archaea are distinguishable from bacteria by their unique rRNA base sequence and their distinctive plasma membrane and cell wall chemistry.

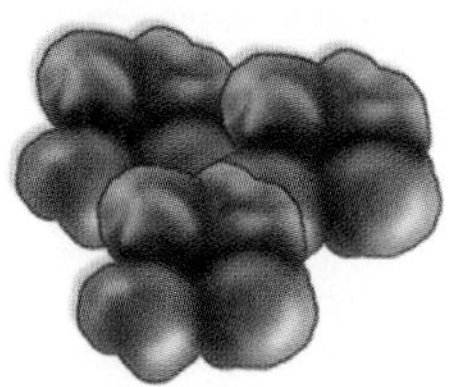

DOMAIN EUKARYA

Eukaryotic, unicellular to multicellular organisms that have a membrane-bounded nucleus containing several chromosomes. Sexual reproduction is common. Phenotypes and nutrition are diverse; each kingdom has specializations that distinguish it from the other kingdoms. Flagella, if present, have a 9 + 2 organization.

KINGDOM PROTISTA

Eukaryotic, unicellular organisms and their immediate multicellular descendants. Asexual reproduction is common, but sexual reproduction as a part of various life cycles does occur. Phenotypically and nutritionally diverse, being either photosynthetic or heterotrophic by various means. Locomotion, if present, utilizes flagella, cilia, or pseudopods.

PHOTOAUTOTROPHS*

Phylum Chlorophyta: green algae
Phylum Rhodophyta: red algae
Phylum Phaeophyta: brown algae
Phylum Bacillariophyta: diatoms, golden-brown algae
Phylum Pyrrophyta: dinoflagellates
Phylum Euglenophyta: euglenoids

HETEROTROPHS BY INGESTION OR PARASITIC*

Phylum Zoomastigophora: zooflagellates
Phylum Rhizopoda: amoeboids
Phylum Foraminifera: foraminiferans
Phylum Actinopoda: radiolarians
Phylum Ciliophora: ciliates
Phylum Apicomplexa: sporozoans
Phylum Myxomycota: plasmodial slime molds
Phylum Acrasiomycota: cellular slime molds

HETEROTROPHS BY ABSORPTION OR PARASITIC*

Phylum Oomycota: water molds

KINGDOM FUNGI

Multicellular eukaryotes that form nonmotile spores during both asexual and sexual reproduction as a part of the haploid life cycle. The only multicellular forms of life to be heterotrophic by absorption. They lack flagella in all life cycle stages.

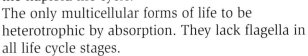

**Not a classification category, but added for clarity*

Phylum Zygomycota: zygospore fungi
Phylum Ascomycota: sac fungi
Phylum Basidiomycota: club fungi

KINGDOM PLANTAE

Multicellular, primarily terrestrial, eukaryotes with well-developed tissues. Plants have an alternation of generations life cycle and are usually photosynthetic. Like green algae, they contain chlorophylls *a* and *b*, carotenoids; store starch in chloroplasts; and have a cell wall that contains cellulose.

NONVASCULAR PLANTS*

Phylum Anthocerophyta: hornworts
Phylum Hepatophyta: liverworts
Phylum Bryophyta: mosses

SEEDLESS VASCULAR PLANTS*

Phylum Lycophyta: club mosses
Phylum Pterophyta: ferns
Phylum Psilotophyta: whisk ferns
Phylum Sphenophyta: horsetails

GYMNOSPERMS*

Phylum Cycadophyta: cycads
Phylum Ginkgophyta: maidenhair tree
Phylum Coniferophyta: conifers (pines, firs, spruces)
Phylum Gnetophyta: gnetophytes

ANGIOSPERMS*

Phylum Anthophyta: flowering plants
- Class Monocotyledones: monocots
- Class Eudicotyledones: eudicots

KINGDOM ANIMALIA

Multicellular organisms with well-developed tissues that have the diploid life cycle. Animals tend to be mobile and are heterotrophic by ingestion, generally in a digestive cavity. Complexity varies; the more complex forms have well-developed organ systems. More than a million species have been described.

INVERTEBRATES*

Phylum Porifera: sponges
Phylum Cnidaria: hydras, jellyfishes
- Class Anthozoa: sea anemones, corals
- Class Hydrozoa: *Hydra, Obelia*
- Class Scyphozoa: *Aurelia*

Phylum Ctenophora: comb jellies, sea walnuts
Phylum Platyhelminthes: flatworms
- Class Turbellaria: planarians
- Class Trematoda: flukes
- Class Cestoda: tapeworms

Phylum Nemertea: ribbon worms
Phylum Nematoda: roundworms *(Ascaris)*
Phylum Rotifera: rotifers
Phylum Mollusca: molluscs
- Class Polyplacophora: chitons
- Class Bivalvia: clams, scallops, oysters, mussels
- Class Cephalopoda: squids, nautiluses, octopuses
- Class Gastropoda: snails, slugs, nudibranchs

Phylum Annelida: annelids
- Class Polychaeta: clam worms, tube worms
- Class Oligochaeta: earthworms
- Class Hirudinea: leeches

Phylum Arthropoda: arthropods
- Subphylum Trilobitomorpha: trilobites**
- Subphylum Crustacea: crustaceans
- Subphylum Uniramia: millipedes, centipedes, insects
- Subphylum Chelicerata: spiders, scorpions, horseshoe crabs

Phylum Echinodermata: echinoderms
- Class Crinoidea: sea lilies, feather stars
- Class Asteroidea: sea stars
- Class Ophiuroidea: brittle stars
- Class Echinoidea: sea urchins, sand dollars
- Class Holothuroidea: sea cucumbers

Phylum Chordata: chordates
- Subphylum Cephalochordata: lancelets
- Subphylum Urochordata: tunicates

VERTEBRATES*

Subphylum Vertebrata: vertebrates
- Superclass Agnatha: jawless fishes
- Superclass Gnathostomata: jawed fishes, tetrapods
  - Class Chondrichthyes: cartilaginous fishes (sharks, rays)
  - Class Osteichthyes: bony fishes (lobe-finned fishes, ray-finned fishes)
  - Class Amphibia: amphibians (frogs, toads)
  - Class Reptilia: reptiles (turtles, snakes, lizards, crocodiles)
  - Class Aves: birds (owls, woodpeckers, kingfishers, hawks)
  - Class Mammalia: mammals
    - Monotremes: duckbill platypus, spiny anteater
    - Marsupials: opossums, kangaroos, koalas
    - Placental mammals: shrews, whales, rats, cats, humans
      - Order Primates: primates (prosimians, monkeys, apes, hominids)
      - Family Hominidae: hominids
        - Genus *Homo:* humans

***extinct*

Name ______________________ Section ____________ Date ____________

Practical Examination Answer Sheet

1. ______________________
2. ______________________
3. ______________________
4. ______________________
5. ______________________
6. ______________________
7. ______________________
8. ______________________
9. ______________________
10. ______________________
11. ______________________
12. ______________________
13. ______________________
14. ______________________
15. ______________________
16. ______________________
17. ______________________
18. ______________________
19. ______________________
20. ______________________
21. ______________________
22. ______________________
23. ______________________
24. ______________________
25. ______________________
26. ______________________
27. ______________________
28. ______________________
29. ______________________
30. ______________________
31. ______________________
32. ______________________
33. ______________________
34. ______________________
35. ______________________
36. ______________________
37. ______________________
38. ______________________
39. ______________________
40. ______________________
41. ______________________
42. ______________________
43. ______________________
44. ______________________
45. ______________________
46. ______________________
47. ______________________
48. ______________________
49. ______________________
50. ______________________

Credits

Text and Line Art

Chapter 4

Figure 4.10: Originally in P. Anderson and S. Mader, *General Biology*, Copyright © 1973 Kendall/Hunt Publishing Company, Dubuque, Iowa.

Chapter 21

Table 21.9: Data from W.T. Keeton, et al., *Laboratory Guide for Biological Science*, 1968, p. 189.

Chapter 32

Figure 32.3: Drawing by Kristine A. Kohn. From Harriet Stabbs, "Acid Precipitation Awareness Curriculum Materials in the Life Sciences." *The American Biology Teacher* (1983), 45(4) 221. With permission from the National Association of Biology Teachers.

Photo Credits

Laboratory 1

Figure 1.1: © James Robinson/Animals Animals/Earth Scenes.

Laboratory 2

Figure 2.5a: © Michael Ross/Photo Researchers, Inc.; **2.5b:** © CNRI/SPL/ Photo Researchers, Inc.; **2.5c:** © Steve Gschmeissner/Photo Researchers, Inc.; **2.6, 2.7:** Courtesy Leica, Inc. Deerfield, IL; **2.8a:** © Biodisc/Visuals Unlimited; 2.8b: © Kevin and Betty Collins/Visuals Unlimited; **2.9a:** © Biophoto Associates/Photo Researchers, Inc.; **2.9b:** © Dr. Gopal Murti/Photo Researchers, Inc.; **2.12a:** © Tom Adams/Peter Arnold, Inc.

Laboratory 3

Figure 3.4a: © Jeremy Burgess/SPL/Photo Researchers, Inc.; **3.7:** © Ed Reschke

Laboratory 4

Figure 4.1: © Ralph Slepecky/Visuals Unlimited; **4.2:** © Ed Reschke; **4.5a, b:** Courtesy Ray F. Evert, University of Wisconsin; **4.9a, b, c:** © David M. Phillips/ Visuals Unlimited; **4.10a:** © Dwight Kuhn; **4.10b:** © Alfred Owczarzak/Biological Photo Service

Laboratory 5

Figure 5.2a: © Andrew Syred/Photo Researchers, Inc.; **5.3(Animal cells-early prophase, prophase, prometaphase, anaphase, telophase):** © Ed Reschke; **(Animal cells-metaphase):** © Michael Abbey/Photo Researchers, Inc.; **5.3(Plant cells-early prophase, prometaphase):** © Ed Reschke; **(Plant cells-prophase, metaphase, anaphase):** © Robert Calentine/Visuals Unlimited; **(Plant cells-telophase):** © Jack M. Bostrack/Visuals Unlimited; **5.4(both):** © R.G. Kessel and C.Y. Shih, "Scanning Electron Microscopy in Biology: A Students' Atlas on Biological Organization," 1974 Springer-Verlag, New York; **5.5:** © B.A. Palevitz & E.H. Newcomb/BPS/Tom Stack & Associates; **5.6(all):** © Ed Reschke; **5.10b(top):** © Secchi-Leaque/CNR/SPL/ Photo Researchers, Inc.; **5.10b(bottom):** © Ed Reschke; **5.11:** © Ed Reschke/Peter Arnold, Inc.

Laboratory 7

Figure 7.1: Courtesy Dr. Keith Porter.

Laboratory 9

Figure 9.3b: © Andrew Syred/SPL/Photo Researchers, Inc.; **9.5a:** © Ed Reschke; **9.5b:** Courtesy Ray F. Evert/University of Wisconsin Madison; **9.6(left):** © Carolina Biological Supply/PhotoTake; **9.6(right):** © Kingsley Stern; **9.7b:** © Carolina Biological Supply/PhotoTake; **9.10b:** © Ed Reschke; **9.11:** © J. Robert Waaland/Biological Photo Service.

Laboratory 10

Figure 10.1(Pollen): Courtesy Graham Kent: **10.1(Embryo sac):** © Ed Reschke; **10.2: (Proembryo, globular, heart-shaped):** Courtesy Dr. Chun-Ming Liu; **10.2(Torpedo):** © Biology Media/Photo Researchers, Inc.; **10.2(Embryo):** © Jack Bostrack/Visuals Unlimited; **10.4b:** © Ed Reschke; **10.5b(Coleoptile):** © James Mauseth; **10.5b(First leaf):** © Barry L. Runk/Grant Heilman, Inc.

Laboratory 11

Figure 11.1(Simple squamous, pseudostratified, cuboidal, columnar, cardiac, skeletal, nervous, adipose, bone, cartilage): © Ed Reschke; **11.1(Smooth muscle, dense fibrous):** © The McGraw Hill Companies, Inc./Dennis Strete, photographer; **11.1(Blood):** © National Cancer Institute/ Photo Researchers, Inc.; **11.2, 11.3, 11.4, 11.5, 11.6, 11.8, 11.9, 11.10, 11.11a-d, 11.13, 11.15a, 11.16:** © Ed Reschke; **11.7:** © The McGraw Hill Companies, Inc./Dennis Strete, photographer; 11.11e: © R. Kessel/Visuals Unlimited; **11.14:** © The McGraw Hill Companies, Inc./Dennis Strete, photographer.

Laboratory 12

Figure 12.2b: © Ed Reschke/Peter Arnold, Inc.

Laboratory 13

Figures 13.3b, 13.6: © Ken Taylor/Wildlife Images

Laboratory 14

Figure 14.9: © CNR/SPL/Photo Researchers, Inc.

Laboratory 15

Figure 15.1a, b, d, e: © Ed Reschke; **15.1c:** © Biophoto Associates/Photo Researchers, Inc.; **15.6:** Courtesy Omron Healthcare.

Laboratory 16

Figures 16.4, 16.7: © The McGraw Hill Companies, Inc./Carlyn Iverson, photographer; **16.9, 16.10:** © Ed Reschke; **16.11:** © Ken Taylor/Wildlife Images.

Laboratory 17

Figure 17.2a: © Dwight Kuhn.

Laboratory 18

Figure 18.1(all): Courtesy Dr. J. Timothy Cannon; **18.4:** © Manfred Kage/Peter Arnold, Inc.; **18.5a:** Copyright © 2005 The Regents of the University of California. All Rights Reserved. Used by permission; **18.5b:** © P.H. Gerbier/SPL/Photo Researchers, Inc.

Laboratory 19

Figure 19.2(Hyaline, compact bone): © Ed Reschke; **19.2(Osteocyte):** © Biophoto Associates/Photo Researchers, Inc.; **19.7:** © Biology Media/Photo Researchers, Inc.

Laboratory 20

Figures 20.1a, c–h: © Carolina Biological Supply/PhotoTake; **20.1b:** © Dr. Robert Calentine/Visuals Unlimited; **20.1i:** © Ed Reschke/Peter Arnold, Inc.; **20.2a:** © Martin Rotker/Phototake; **20.2b–h:** © Carolina Biological Supply/PhotoTake; **20.2i:** © Alfred Owczarzak/BPS; **20.6a, 20.7a, 20.8a, 20.9a:** © Carolina Biological Supply/PhotoTake; **20.11:** © Petit format/Photo Researchers, Inc.

Laboratory 21

Figures 21.5(all): © Carolina Biological Supply/PhotoTake; **21.6a:** © Superstock; **21.6b:** © Michael Grecco/Stock Boston; **21.6c–f, h:** © The McGraw-Hill Companies, Inc./Bob Coyle, photographer; **21.6g:** Royalty-Free/Corbis; **21.7:** © Carolina Biological Supply/PhotoTake

Laboratory 23:

Figures 23.2c: © James King-Holmes/SPL/Photo Researchers, Inc.; **23.2d:** © CNRI/SPL/Photo Researchers, Inc.; **23.10(both):** © Bill Longcore/Photo Researchers, Inc.

Laboratory 24

Figures 24.4(both): © 2001 Time Inc. reprinted by permission

Laboratory 25

Figures 25.2a: © Henry Aldrich/Visuals Unlimited; **25.3a:** © Dr. Richard Kessel & Dr. Gene Shih/Visuals Unlimited; **25.3b:** © Gary Gaugler/Visuals Unlimited; **25.3c:** © SciMAT/Photo Researchers, Inc.; **25.4c:** © Leon Lebeau; **25.5a:** © M.I. Walker/Photo Researchers, Inc.; **25.5b:** © R. Knauft/Photo Researchers, Inc.; **25.6b:** © M.I. Walker/Photo Researchers, Inc.; **25.7(left):** © John D. Cunningham/Visuals Unlimited; **25.7(right):** © Carolina Biological Supply/Visuals Unlimited; **25.8:** © Carolina Biological Supply/PhotoTake; **25.9c:** Image acquired by Prof. Michael Duszenko, University of Tubingen and Eye of Science, Reutlingen (Germany); **25.11(left):** © Carolina Biological Supply/Visuals Unlimited; **25.11(right):** © V. Duran/Visuals Unlimited; **25.12(Zygospore):** © David M. Phillips/Visuals Unlimited; **25.12(Sporangium):** © Carolina Biological Supply/PhotoTake; **25.13a:** © James W. Richardson/Visuals Unlimited; **25.13b:** © David M. Phillips/Visuals Unlimited; **25.14:** © SciMAT/Photo Researchers, Inc.; **25.15a:** © Dr. P. Marazzi/SPL/Photo Researchers, Inc.; **25.15b:** © John Hadfield/SPL/Photo Researchers, Inc.; **25.15c:** © Evertt S. Beneke/Visuals Unlimited

Laboratory 26

Figure 26.3: © Heather Angel/Biofotos; **26.4, 26.5:** © J.R. Waaland/Biological Photo Service; **26.6b:** © Ed Reschke; **26.7a:** © Ed Reschke/Peter Arnold, Inc.; **26.7b:** © J.M. Conrarder/Nat'l Audubon Society/Photo Researchers, Inc.; **26.7c:** © R. Calentine/Visuals Unlimited; **26.8a:** © Carolina Biological Supply/PhotoTake; **26.10:** © Matt Meadows/Peter Arnold, Inc.; **26.11:** © The McGraw-Hill Companies/Carlyn Iverson, photographer; **26.13(Archegonia):** © R. Knauft/Biology Media/Photo Researchers, Inc.; **26.13(Antheridia):** © Carolina Biological Supply/PhotoTake

Laboratory 27

Figure 27.2a: © D. Cavagnaro/Visuals Unlimited; **27.2b:** © Kingsley Stern; **27.2c:** © D. Giannechini/Photo Researchers, Inc.; **27.3:** © PhotoTake; **27.6a:** © J.R. Waaland/Biological Photo Service; **27.6b:** © Ed Reschke; **27.7:** © Carolina Biological Supply/PhotoTake; **27.10:** © Ed Reschke/Peter Arnold, Inc.

Laboratory 28

Figure 28.2a: © Doug Perrine/SeaPics.com; **28.2b:** © Andrew J. Martinez/Photo Researchers, Inc.; **28.2c:** © OSF/Bernard/Animals Animals/Earth Scenes; **28.4b:** © Azure Computer & Photo Services/Animals Animals/Earth Scenes; **28.4c:** © Ron & Valerie Taylor/Bruce Coleman; **28.4d:** Runk/Schoenberger/Grant Heilman Photography; **28.4e:** © Under Watercolours; **28.5:** © Carolina Biological Supply/Visuals Unlimited; **28.6(left):** © Runk/Schoenberger/Grant Heilman Photography; **28.6(right):** © Biodisc/Visuals Unlimited; **28.7:** © Tom E. Adams/Peter Arnold, Inc.; **28.8:** © Carolina Biological Supply/PhotoTake; **28.10a:** © R. Calentine/Visuals Unlimited; **28.11a, c:** © Larry Jenson/Visuals Unlimited; **28.12:** © James Solliday/Biological Photo Service, Inc.

Laboratory 29

Figure 29.1a: © Fred Bavendam/Peter Arnold, Inc.; **29.1b:** © Biophoto Associates; **29.1c:** © Milton Rand/Tom Stack & Associates; **29.1d:** © Douglas Faulkner/Photo Researchers, Inc.; **29.3b, 29.4b:** © Ken Taylor/Wildlife Images; **29.5a:** © Roger K. Burnard/BPS; **29.5b:** © R. DeGoursey/Visuals Unlimited; **29.5c:** © Brian Parker/Tom Stack & Associates; **29.5d:** © C.P. Hickman/Visuals Unlimited; **29.7b:** © Ken Taylor/Wildlife Images; **29.8:** © John Cunningham/Visuals Unlimited; **29.9a(Honeybee):** © R. Williamson/Visuals Unlimited; **29.9a(Millipede):** © Bill Beatty/Visuals Unlimited; **29.9a(Centipede):** © Adrian Wenner/Visuals Unlimited; **29.9b(Spider):** © W.J. Weber/Visuals Unlimited; **29.9b(Scorpion):** © David M. Dennis/Tom Stack & Associates; **29.9b(Horseshoe crab):** © E.R. Degginger/Photo Researchers, Inc.; **29.9c(Crab):** © Tom McHugh/Photo Researchers, Inc.; **29.9c(Shrimp):** © Alex Kerstitch/Visuals Unlimited; **29.9c(Barnacles):** © Kjell Sandved/Visuals Unlimited; **29.10b:** © Ken Taylor/Wildlife Images

Laboratory 30

Figure 30.2a: © Daniel Gotshall/Visuals Unlimited; **30.2b:** © Hal Beral/Visuals Unlimited; **30.2c:** © Robert Dunne/Photo Researchers, Inc.; **30.2d:** © Neil McDan/Photo Researchers, Inc.; **30.2e:** © Robert Clay/Visuals Unlimited; **30.2f:** © Alex Kerstitch/Visuals Unlimited; **30.4:** © Rick Harbo; **30.5:** © Heather Angel/Biofotos; **30.7:** © Stan Sims/Visuals Unlimited; **30.8(Shark):** © Hal Beral/Visuals Unlimited; **30.8(Butterflyfish):** © Patrice/Visuals Unlimited; **30.8(Frog):** © Rod Planck/Photo Researchers, Inc.; **30.8(Turtle):** © Suzanne and Joseph Collins/Photo Researchers, Inc.; **30.8(Flycatcher):** © Robert and Linda Mitchell Photography; **30.8(Fox):** © Craig Lorenz/Photo Researchers, Inc.; **30.9:** © Rod Planck/Photo Researchers, Inc.; **30.10b:** © Carolina Biological Supply/PhotoTake; **30.11, 30.12b, 30.13b, 30.14b:** © Ken Taylor/Wildlife Images.

Laboratory 32

Figure 32.3: © NOAA/Visuals Unlimited.

Index

Page numbers followed by an *f* or *t* indicate figures and tables.

B

C

F

G

H

I

J

K

L

M

N

O

P

Q

R

S

T

U

V

W

X

Y

Z